AF658389

STRATEGIES FOR BREEDING CLIMATE CHANGE READY CROPS

THE AUTHOR

Prof. Pradip C, Deka was the founder Vice-Chancellor of Sir Padampat Singhania University, Udaipur, Rajasthan, India. Earlier he served as the Vice-Chancellor of Tezpur University, Tezpur, Assam, India. He served as the Dean, Faculty of Agriculture, and was the founder Professor & Head, Department of Agricultural Biotechnology, Assam Agricultural University, Jorhat, Assam, India. Prof. Deka served as a Visiting Professor at the Institute of Plant Genomics, ETH, Basel, Switzerland, and at Seoul National University, Seoul, South Korea.

After graduating from Assam Agricultural University, Prof. Deka obtained his Master's and Doctorate degrees in 'Genetics and Plant Breeding' from Banaras Hindu University, Varanasi, India. He did post-doctorate research in Molecular Genetics at the University of Florida, Gainesville, USA, and at the University of California, Berkeley, USA.

Prof. Deka published more than 200 research papers in national and international journals, and authored eight books on Molecular Genetics and Biotechnology. His research interest is in Plant Molecular Genetics and Plant Tissue Culture.

STRATEGIES FOR BREEDING CLIMATE CHANGE READY CROPS

Pradip C. Deka

2025

Associated Publishing Company
A Division of
Astral International Pvt. Ltd.
New Delhi – 110 002

ISBN: 978-93-5919-394-6

Published by : **Associated Publishing Company**
A Division of
Astral International Pvt. Ltd.
– ISO 9001:2015 Certified Company –
4736/23, Ansari Road, Darya Ganj
New Delhi-110 002
Ph. 011-43549197, 23278134
E-mail: info@astralint.com
Website: www.astralint.com

Printed at : **Replika Press Pvt Ltd**

Acknowledgment

I extend my gratitude to Dr. Pallavi Dwivedi, and Mr. Lalitesh Sharma for their help in the preparation of some of the figures in the book. I would like to thank my wife Nanda, son Gautam, daughter Namrata (Juhi), and son-in-law Abhishek for their support and encouragement during the preparation of the manuscript of this book. I am also thankful to my parents and my brothers (Mrigen and Dileep) and sisters (Binu, Anu, and Anee) for their altruistic love, affection, support, and encouragement during my studies and career development as a teacher, researcher, and administrator, which helped me to do what I love to do best.

I feel immense pleasure in thanking Astral International Pvt. Ltd., New Delhi, India for bringing out the book in its present form.

Pradip C Deka

Preface

The changes in climatic events such as temperature and rainfall significantly affect the yield of crops. As influenced by climatic variables, crop productivity depends on adaptation abilities and crop type, climate scenario, and CO2 fertilization effect. The temperature rise is likely to reduce the yield of many crops by reducing their duration. If the average temperature rises continuously over time, the aggregate production of all the crops including wheat, rice, and maize is expected to decrease in both the temperate and tropical regions. Projected climate variations are not limited to only increases in drought conditions. Climate change has also resulted in increased and sporadic incidences of floods. Floods have resulted in the washout of topsoil and nutrients from the soil, resulting in low productivity. This in turn has affected crop yield. Following are the approaches adopted by plant breeders to overcome the adverse effects of climate change on crop plants.

Re-sequencing (RS) of a large number of germplasm accessions not only provides information on the origin, domestication, and population structure, but also identifies lines with deleterious mutations in the genomes that can be eliminated to minimize the genetic load in the crop species. Next-generation sequencing (NGS) technologies together with precise phenotyping have been used for the identification of marker-trait associations in several crops.

Advances in DNA sequencing technology and the development of high-throughput systems for multi-parallel interrogation of thousands of single nucleotide polymorphisms (SNPs), provide a suite of technological platforms for the analysis of thousands of SNPs, within a short time. Deployment of these technologies should help in the analysis of plant genetic resources, in order to identify patterns of genetic diversity, map quantitative traits, and mine novel alleles from the vast number of genetic resources maintained in genebanks around the world. Wild ancestors and traditional landraces are reservoirs of valuable traits, including diverse forms of resistance to both biotic and abiotic stresses, which remain crucial for the adaptation of modern cultivars to current and future climates. Crop wild relatives (CWRs) have been used for decades in crop improvement to enhance plant performance.

The domestication of wild species led to a wide variety of crops adapted to a range of climatic and edaphic conditions, which allowed the expansion of cultivation to larger areas and over longer periods. Subsequent crop breeding led to higher yields and facilitated population growth. Until recent times, both domestication and breeding occurred empirically with little understanding of the underlying biological mechanisms. Genetic variation is created by mutation and breeding operates by stacking favourable mutations in a single plant through recombination. Technical advances will eventually allow controlled manipulation of mutation and recombination, and thus de novo domestication of wild species by targeted modification of their genomes. Such tools will also aid in accelerating the improvement of traditional, semi-domesticated "orphan" crops that perform poorly in modern agricultural systems. The synergistic combination of classical archaeobotany and genetics with high-throughput genomics reveals that a variety of different processes may have operated in the domestication of crops. This knowledge, combined with increasingly powerful gene-editing toolkits, sets the stage for the continual domestication of crop wild relatives and other lesser-known plant species by defining an ideal plant type ("ideotype") that is more resilient, nutritious, and productive in a specific environment. Wild plants that are naturally resistant to biotic (insects and diseases) and tolerant of abiotic (drought and heat) stresses can be selected and manipulated by introducing mutations that mimic the domestication events that led to improved yield and agronomic performance of the major crops of today. The recent progress in the understanding of crop domestication and technical breakthroughs in gene-editing technology could be combined to produce better crops for the future. During the process of domestication, different crops have been selected on the basis of analogous traits such as favorable plant architecture, simultaneous flowering, ease of harvesting, and larger fruit size for higher yield, etc. With the increase in our understanding of the genetic basis for these domestication traits, it has become easier to identify several so-called domestication genes. By targeting these genes with CRISPR, the domestication process has been shown to be accelerated in several crop plants.

The acquisition of large-scale phenotypic data has become one of the major bottlenecks hindering crop breeding and functional genomics studies. Recent technological advances provide great potential solutions to relieve this bottleneck and explore advanced methods for large-scale phenotyping data acquisition and processing. The major progress on high-throughput phenotyping in controlled environments and field conditions as well as its use for post-harvest yield and quality assessment in the past decades have generated prospects to overcome this bottleneck. High-throughput image-based plant phenotyping facilitates the computation of phenotypes by analyzing a large number of plants in short time intervals with precision, nullifying the need for time-consuming physical human labour. The process is generally non-destructive, allowing the same traits to be quantified repeatedly at multiple times and scales during a plant's life cycle. It is an interdisciplinary approach involving computer science, biology, remote sensing, statistics, and genomics. The multi-omics research combining high-throughput phenotyping with genetic studies has also contributed to this direction.

Genome editing is defined as a collection of advanced molecular biological techniques that facilitates precise, efficient, and targeted modifications at genomic loci. Genomic editing using zinc-finger nucleases (ZNFs) and transcription activator-like effector nucleases (TALENs) has been around for two decades. Recently, a new technology called the clustered regularly interspersed short repeats (CRISPR/Cas) system has been introduced, which provides simplicity and ease of targeted gene editing. Many gene knockout mutants and some gene replacement and insertion mutants have been produced using genome-editing technologies in a wide variety of plants, and many of these mutants have shown to be useful for crop improvement. The risk involved in altering genomes through the use of genome-editing technology is significantly lower than those associated with GM crops because most edits alter only a few nucleotides, producing changes that are not unlike those found throughout naturally occurring populations. Very recently, a modified version of the CRISPR/Cas genome editing system called "Prime Editing" was developed, which could overcome all the limitations posed by the CRISPR/Cas system. Thus, the introduction of genome editing into modern plant breeding programs should facilitate rapid and precise crop improvement. Several recent studies have demonstrated the potential of CRISPR/Cas to generate a broad range of allelic diversity at specific loci, thereby helping to create large genetic diversity. The creation of novel genetic diversity shall help in resolving diverse plant breeding objectives. Recently CRISPR-like activities have been reported in eukaryotes. This has created new hope for crop breeders.

Speed breeding is a suite of techniques that involves the manipulation of environmental conditions under which crop genotypes are grown, aiming to accelerate flowering and seed set, to advance to the next breeding generation as quickly as possible. The method saves breeding time and resources through rapid generation advancement. Various selection methods can be integrated into speed breeding, such as the single seed descent (SSD), single pod descent (SPD), single plant selection (SPS), clonal selection, and marker-assisted selection (MAS) to shorten the breeding cycle and for efficient resource use. Speed breeding results in 3 to 9 generations per year compared to 1 to 2 generations per year achieved with conventional selection approaches. Accordingly, speed breeding offers opportunities to rapidly develop homozygous and stable genotypes, and to facilitate rapid generation advancement, resulting in accelerated development and release of new cultivars. Speed breeding technology also fits well with MAS and high-throughput phenotyping methodologies for multiple trait selection.

A sensor is a gadget that perceives and responds to certain inputs which could be illumination, locomotion, pressure, heat, or moisture, and transforms it into a representation or signals that can be read by humans for further reading and processing. There are various types of sensors used in breeding that assist various smart breeding operations. Sensors are very efficient in precision breeding because they transmit data that helps breeders not only to monitor but also to improve their products and keep abreast of changes in the field and ecosystem. Using smart sensors in breeding, it is now possible to record the various components related to the growth and management of the crops, keep an eye on their effectiveness remotely, and make swift decisions to meet the breeding objectives.

The Internet of Things (IoT) is part of the technological advancements that are changing the mode of human operations worldwide. The Internet of Things network comprises systems and a network of web-connected intelligent devices that employ encoded networks like sensors, processors, and interactive hardware to receive, send, and store data. The network in the IoT uses the stored data. Connected devices on the IoT network interact with each other and use the shared data they acquire to solve problems. The interactions within the network are done without human supervision or interference, and most of the job is carried out by the devices. Even though humans may communicate with the devices to set them up, writing scripts for protocols and commands provides access to information. Breeders can use sensors and other IoT devices to gather data and carry out analyses for accurate estimates. With IoT capabilities, breeders can optimize functions and improve crop plants at a lower cost.

Artificial intelligence (AI) has emerged as a revolutionary field, providing a great opportunity to shape modern crop breeding. Advances in crop phenomics, enviromics, together with the other "omics" approaches are paving the way for elucidating the detailed complex biological mechanisms that motivate crop functions in response to environmental trepidations. These "omics" approaches have provided plant breeders with precise tools to evaluate the important agronomic traits for larger-sized germplasm at a reduced time interval in the early growth stages. However, the big data and the complex relationships within impede the understanding of the complex mechanisms behind genes driving the agronomic trait formations. AI brings huge computational power and many new tools and strategies for future breeding. It is important to understand how applications of AI technology can be utilized for current breeding practices, and assist in solving the problem in high-throughput phenotyping, and gene functional analysis. Advances in AI technologies have brought new opportunities for future breeding, and made envirotyping data widely utilized in breeding programmes. In the current breeding methods, linking genotype to phenotype remains a massive challenge and impedes the optimal application of high-throughput field phenotyping, genomics, and enviromics. Therefore, the application of AI will be the preferred tool to increase the accuracy in high-throughput crop phenotyping, genotyping, and envirotyping data. Integration of AI with "omics" tools shall assist rapid gene identification and accelerate the process of crop-improvement programmes.

Plant breeders are usually faced with the problem of predicting the performance of new individuals with untested gene combinations. Therefore, it is important to follow an integrated breeding approach by combining molecular tools, molecular mapping, and MAS. It is also required to develop tools for modelling and simulation analysis by utilizing all pre-existing and newly generated data. Several software tools have been developed that integrate breeding simulations and phenotype prediction models using genomic information. Reliable phenotype prediction models for the simulation were constructed from actual genotype and phenotype data. Such a simulation-based genome-assisted approach to breeding will help optimize plant breeding in all important agricultural crops. Software tools have also been developed for designing target sites or evaluating the outcome of genome/gene editing systems.

An attempt has been made to provide up-to-date information on all the topics described above with illustrations, wherever necessary. Comparative assessment of various techniques, their relevance, and prospects are also discussed.

The approach made in this book should primarily be suitable for the students, researchers, and teachers of Plant Breeding, Genetics, Molecular Biology, Molecular Taxonomy, and Biotechnology. Being an interdisciplinary subject, the book should also be useful for related disciplines such as Agronomy, Plant Pathology, Entomology, Soil Science, Computer Science, and Electronics.

Pradip C Deka

Contents

Chapter-1

Climate Change and Crop Breeding

Abstract

Climate change is one of the most defining concerns of today's world and is in the process of altering the earth's ecosystems. Although climate change has been a constant process on Earth, in recent times, (approximately the last 100 years), the pace of this variation has increased manifolds. Due to anthropogenic activities, the average temperature has risen by 0.9° C since the nineteenth century, mainly due to greenhouse gas (GHG) emissions in the atmosphere. As per estimates, this rise is expected to be 1.5° C by 2050 and 2^0 C by 2100. As deforestation is occurring, GHG emission is increasing, and pollution of soil, water bodies, and air has also increased. The unprecedented hike in temperature has resulted in increased events of droughts, floods, irregular patterns of precipitation, heat waves, and other extreme happenings throughout the globe.

Climate change is a global threat to the food and nutritional security of the world. As greenhouse gas emissions in the atmosphere are increasing, the temperature is also rising due to the greenhouse effect. The concentration of CO_2, which accounts for a major proportion of greenhouse gases, is increasing at an alarming rate, and has led to higher growth and plant productivity due to increased photosynthesis, but increased temperature offsets this effect as it leads to increased crop respiration rate and evapotranspiration, higher pest infestation, a shift in weed flora, and reduced crop duration. Climate change also affects the microbial population and their enzymatic activities in soil. This chapter deals with the issue of climate change, its possible causes, its projection in the near future, its impact on agriculture as an influence on physiological and metabolic activities of plants, and its potential and reported *implications for growth and plant productivity, pest infestation, and mitigation strategies and their economic impact.*

Keywords: *Climate change, food security, nutrition security, mitigation, economic impact.*

Introduction

Climate change is one of the biggest challenges to the world in present times. It is defined as significant changes in the average values of meteorological elements, such as precipitation and temperature, for which averages have been computed over a long period (World Meteorological Organization, 1992). The past few decades indicate that significant changes in climate at a global level were the result of enhanced human activities that altered the composition of the global atmosphere (IPCC Climate Change 207). The concentration of greenhouse gases such as methane (CH_4), carbon dioxide (CO_2), and nitrous oxide (N_2O) has increased by 150%, 40%, and 20%, respectively since 1750 (IPCC Climate Change 2014). Carbon dioxide emissions, which account for the maximum proportion of greenhouse gases (Sathaye et al. 2006), rose to 36.14 billion metric tons in 2014 from 22.15 billion metric tons in 1990 (Abeydeera et al. 2019). The average global temperature has increased at an average rate of 0.15–0.20^0 C per decade since 1975, and is expected to increase by 1.4–5.8 0°C by 2021 (Arora et al. 2005). Greenhouse gas (GHG) emissions, particularly CO_2 from the combustion of fossil fuels and non-CO_2 GHGs such as nitrous oxide, methane, and CFCs add to global warming. The CO_2 concentration in the atmosphere has increased to 412 ppm in 2022 from 280 ppm in 1750 (Figure 1.1) (NOAA 2020). CO_2 constitutes a major proportion of greenhouse gases in the atmosphere: 65% from fossil fuels and industrial processes and 11% from forestry and other land use, followed by methane (16%), nitrous oxide (6%), and fluorinated gases (2%) (IPCC Climate Change 2014).

Before 1750, CO_2 emissions from fossil fuels were negligible, but it increased rapidly with industrialization. The world has emitted around 1.5 trillion metric tons of CO_2 since 1751. However, there are regional variations in the emission. Europe is the largest contributor of CO_2 having around 514 billion metric tons of CO_2 emissions, followed by Asia and the North American continent, which have recorded cumulative CO_2 emissions of 457 billion metric tons each. The USA is the largest contributor to CO_2 emissions (399 billion metric tons), and has contributed 25% of total historical emissions since 1751, followed by China (200 billion metric tons). The European Union (EU-28), a union of 28 countries that sets collaborative targets, has contributed 22% of historical emissions of CO_2. Africa contributes only 3% of global cumulative CO_2 emissions due to low per-capita emissions. However, countries like Brazil and India, whose historical emissions are less, significantly add to the total emissions in the current context (CDIAC 2020). With an increased level of CO_2 in the atmosphere, the fertilization of crops is increased along with decreased energy requirements due to warming. These are certain positive impacts of climate change, whereas water resources are negatively impacted due to climate change. In the 20th century, the impact of climate change was mainly positive. Most countries benefitted until 1980, after which the trend remained the same for the developed world, while the Third-World countries were negatively impacted. In the 21st century, climate change will become a severe problem, and both rich and developing countries will face negative externalities (TOI 2013).

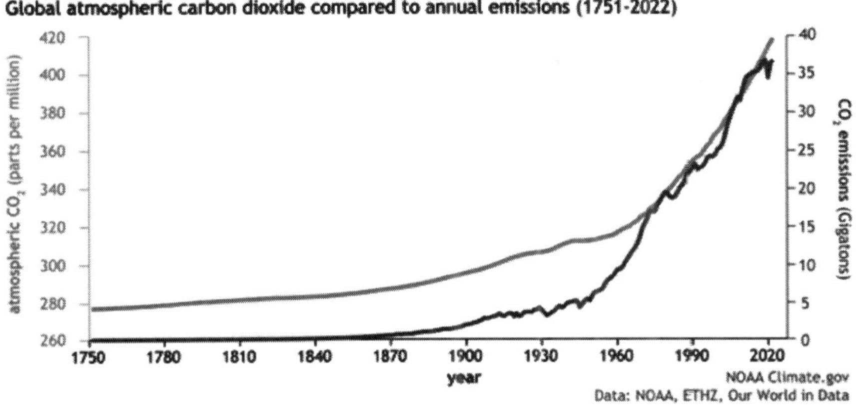

Figure 1.1. The amount of carbon dioxide in the atmosphere (blue line) has increased along with human emissions (gray line) since the start of the Industrial Revolution in 1750. Emissions rose slowly to about 5 billion tons per year in the mid-20th century before skyrocketing to more than 35 billion tons per year by the end of the century. (Reproduced from Annual Report of NOAA Global Monitoring Lab 2023).

The rise in greenhouse gases has implications for the rising temperature of the atmosphere. These infrared active gases, mainly carbon dioxide (CO_2), ozone (O3), and water vapor (H_2O) absorb the thermal radiations emitted by the atmosphere and the surface of the earth, which in turn warms the earth. This phenomenon is known as the greenhouse effect. The average global temperature anomaly is shown in Figure 1.2. This illustrates a significant increase in the global temperature compared to the average temperature of the base period (1901–2000). The global average temperature has increased in the range of 1–1.2^0 C since 1850. Still, since the temperature changes in landmasses are much more prominent, the global land temperature has increased around twice that of the oceans. The temperature of land across the world has risen by 1.32 ± 0.04^0 C compared to the 1951–1980 average, while the increase was 0.59 ± 0.06^0 C for the ocean surface temperature (excluding areas of sea ice). Moreover, since the Northern Hemisphere constitutes the larger portion of landmasses, it has shown a higher average temperature than the Southern Hemisphere. The temperature of the Northern Hemisphere and the Southern Hemisphere has risen 1.31^0 C and 0.91^0 C, respectively, with a global average of 1.11^0 C since 1850. The extreme rise in temperature has been observed in the polar regions, and has detrimental effects, like glacial melting (Ghosh et al. 2018).

As the global temperature is rising, there is a need to reduce greenhouse-gas emissions to limit the temperature increase of 2^0 C relative to pre-industrialization. The developed countries have around a 60–80% contribution to the global temperature rise, sea-ice reduction, and upper-ocean warming, compared to 20–40% for developing countries, since 2005 (Wel et al. 2012). The average global temperature is expected to rise by 2^0 C by 2100 and 4.2^0 C by 2400, as predicted by probabilistic computations of the IPCC's range of climate sensitivity. However, surpassing 2^0 C by 2100 at the present radiative-force level does not seem likely to happen. But the risk is increasing, mainly due to the stabilizing of radiative forces above 400 ppm of CO_2. Moreover, it is exceptionally unlikely that the temperature will rise by 2^0 C if anthropogenic emissions were ceased tomorrow (Hora and Meinshausen 2006).

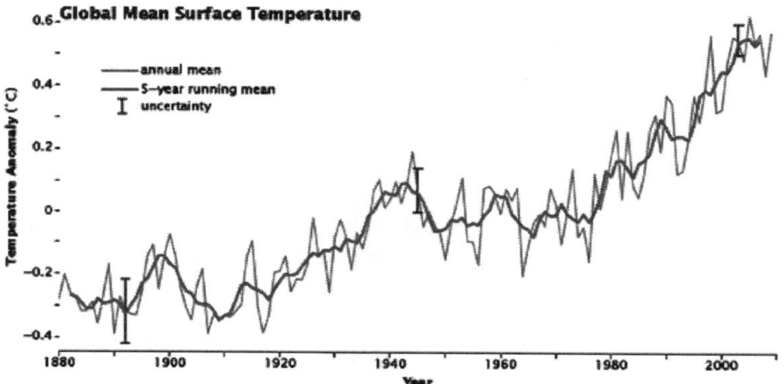

Figure 1.2. Global mean surface temperature. (Source: NASA: Goddard Institute for Space Studies)

The global precipitation anomalies over the base period (1901–2000) are shown in Figure 1.3. This shows that the precipitation change is showing a positive trend, but these vary according to the region. There was an absolute change of 1.98 cm in precipitation across the world from 1901 to 2015 (Our World Data 2020). However, temperature and precipitation extremes are more likely to be witnessed in the near future due to global warming. Extreme precipitation phenomena, either heavy rainfall or drought, are dependent on a region's geography. The increased average river flows due to prolonged heavy rainfall are more likely to be observed in South and East Asia, while the drought in southern Africa and South America will be less severe. There is a probability of more warm extremes, lesser cold extremes, and more strengthened precipitation extremes in the future period in the northeastern United States. Higher emissions will intensify these changes (Ning et al. 2015). The increased intensity and frequency of precipitation also impacts soil erosion, and will have more adverse consequences in northeast China if greenhouse-gas emissions increase (Zhang et al. 2010).

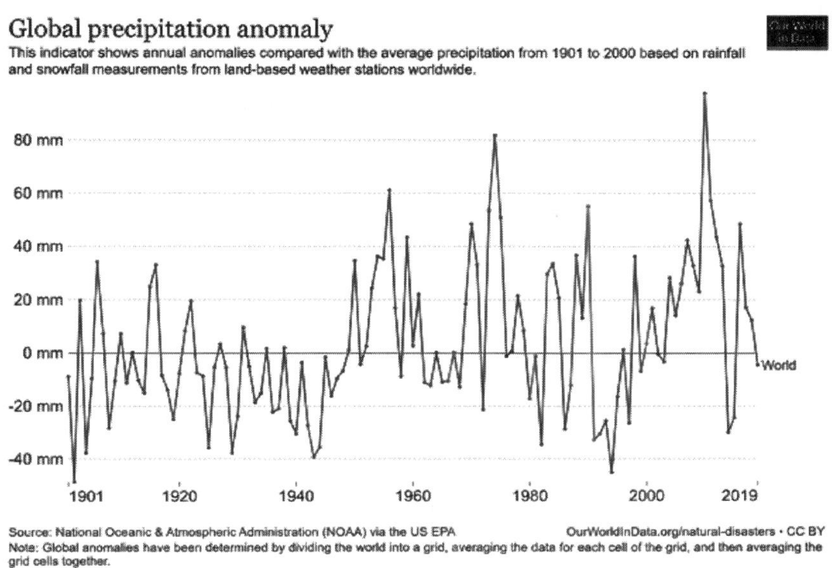

Figure 1.3. Global precipitation anomaly. (Source: National Oceanic and Atmospheric Administration)

Precipitation anomalies have detrimental effects on agriculture, mainly in developing nations. Apart from affecting crop yields, it significantly influences cropland areas. There is evidence suggesting that the approximately 9% rate of cropland expansion in the developing world over the last two decades is due to dry anomalies as farmers expand the area to compensate for yield losses (Zaveri et al. 2020). Global warming will pose a severe threat to the world's food security, but if it is limited to 1.5^{0} C, 76% of developing countries' vulnerability will be reduced compared with the same regions at 2^{0} C (Betts et al. 2018).

Ensuring food for the world's population in the face of climate change is not an easy task, owing to its huge impact on agriculture production (World Population Review 2020). There must be an annual increase in the world's agricultural production by 60% from 2005/2007 to 2050, comprising a rise of 77% in developing and 24% in developed countries, to fulfil the food and nutritional requirements of the population by 2050 (Alexandratos and Bruinsma 2012). Climate change is known to have an adverse effect on agricultural production, and is projected to reduce the global cereal production of maize and wheat by 3.8% and 5.5%, respectively (Lobell et al. 2011). Because of climatic factors, plants have to face several abiotic stresses such as salinity, drought, heat stress, cold stress, etc. (Malhi et al. 2020). Shortage of water availability, soil fertility loss, and pest infestations in crops are the significant undesirable impacts of climate change (Baul and McDonald 2015).

Climate change is resulting in a very high rate of land degradation causing enhanced desertification and nutrient-deficient soils. The menace of land degradation is increasing by the day and has been characterized as a major global threat. According to the Global Assessment of Land Degradation and Improvement (GLADA) a quarter of land area around the globe can now be marked as degraded. Land degradation is supposed to influence the lives of 1.5 billion people and 15 billion tons of fertile soil is lost every year due to anthropogenic activities and climate change. Land degradation is resulting in mass migrations and as per a report published by the United Nations Environment Programme in 2017, 500 million hectares of farmland have been abandoned due to drought and desertification resulting in major social and environmental constraints. Extreme drought conditions, frequently occurring due to climate change, exacerbate the productivity of crops by causing nutrient immobilization and salt accumulation in soils making them dry, unhealthy, saline, and finally infertile. Such barren lands become non-arable over time and are eventually abandoned by farmers leading to economic losses and social issues.

Projected climate variations are not limited to only increases in drought conditions. Climate change has also resulted in increased and sporadic incidences of food in the last few years. Reports by European Academies' Science Advisory Council (EASAC) suggest that extreme events including foods have increased by 50% in the last 10 years and are now occurring at a rate of four times higher than in comparison to 20 years back. Foods have resulted in the washout of topsoil and nutrients from the soil, resulting in low productivity for several years to come, unless and until corrective and proactive remediation strategies are not worked upon. Rise in sea levels or prevalence of heavy rainfall could result in the decline of agricultural lands in coastal regions. This has also resulted in the salinity of soils in coastal regions leading to stresses in crops such as decreased respiration, photosynthesis, and transpiration, ultimately jeopardizing food availability and security in such regions.

Why Carbon Dioxide Matters

Carbon dioxide is Earth's most important greenhouse gas: a gas that absorbs and radiates heat. Unlike oxygen or nitrogen (which make up most of our atmosphere), greenhouse gases absorb heat radiating from the Earth's surface and re-release it in all directions—including back toward Earth's surface. Without carbon dioxide, Earth's natural greenhouse effect would be too weak to keep the average global surface temperature above freezing. By adding more carbon dioxide to the atmosphere, people are supercharging the natural greenhouse effect, causing global temperature to rise. According to observations by the NOAA Global Monitoring Lab, in 2021 carbon dioxide alone was responsible for about two-thirds of the total heating influence of all human-produced greenhouse gases.

Another reason carbon dioxide is important in the Earth system is that it dissolves into the ocean like the fizz in a can of soda. It reacts with water molecules, producing carbonic acid and lowering the ocean's pH (raising its acidity). Since the start of the Industrial Revolution, the pH of the ocean's surface waters has dropped from 8.21 to 8.10. This drop in pH is called ocean acidification.

Causes of Climate Change

Temperature changes are caused by natural phenomena and anthropogenic activities on Earth, which ultimately initiate the concentration of GHGs (Stern and Kaufmann 2014). Anthropogenic activities lead to the emission of greenhouse gas such as CO2, methane, and nitrous oxide, as well as other substances that lead to ozone depletion in the atmosphere (Montzka et al. 2011). The increased CO2 concentration in the atmosphere can affect microbial activities in the soil, along with implications on water content. Increased atmospheric CO2 (463–780 ppm) can stimulate nitrous oxide and methane emissions from upland soil and wetlands, respectively. This nullifies the 16.6% mitigation effect of climate change as predicted by increasing terrestrial carbon sink (Groenigen et al. 2011).

The agriculture sector contributes 15% of total emissions, primarily methane and nitrous oxide. The global emission of non-agricultural greenhouse gases is predicted to rise until 2055 if the dietary preferences and consumption of food energy are kept constant at 1995 levels. However, with changing preferences toward high-value foods such as milk and meat, emissions are predicted to rise at an even higher rate. The emission can be reduced either with technological mitigation or by reduced meat consumption, or both (Popp et al. 2010). The livestock sector is the main contributor to greenhouse-gas emissions, and according to the IPCC, it generates around 8–10.8% of emissions. However, it can contribute up to 18% of GHG emissions based on lifecycle analysis (O'Mara 2011). The main sources of greenhouse-gas emissions by the livestock sectors include enteric fermentation, N2O emissions, liming, fossil fuels, organic farming, and fertilizer production (Lesschen et al. 2-11). The use of nitrogenous chemical fertilizers also leads to greenhouse-gas emissions (Kahri et al. 2010). With better management of crop production, N fertilizer use can be lowered by 38%. Better crop management also leads to the consumption of 11% less input energy with 33% increased yields, leading to a reduction in greenhouse-gas emissions by 20% (Soltani et al. 2013).

The increasing demand for food due to the ever-growing population has resulted in intensive agricultural practices including unprecedented use of agrochemicals, livestock generation (for meat and other source of income), exploitation of water resources etc. This has further aggravated the situation by the release of GHG (due to agricultural activities) and resulting in pollution of natural resources. Forests act as a sink to the increasing amount of CO2 but the unchecked rate of deforestation (mainly for development and agriculture) has imbalanced the natural process of the carbon cycle. This has increased the amount of carbon footprint and induced uneven patterns of climate resulting in a variety of adverse effects which have a major impact on agricultural production.

Climate Change and Agriculture

Agriculture is the most vulnerable sector to climate change, owing to its huge size and sensitivity to weather parameters, thereby causing huge economic impacts (Mendelsohn 2013). The changes in climatic events such as temperature and rainfall significantly affect the yield of crops. The effect of rising temperatures, precipitation variation, and CO2 fertilization varies according to the crop, location, and magnitude of change in the parameters. The temperature increase is found to reduce the yield, while the precipitation increase is likely to offset or reduce the impact of increasing temperature (Adams et al. 1998). As influenced by climatic variables, crop productivity depends on adaptation abilities and crop type, climate scenario, and CO2 fertilization effect (Karim et al. 2018). The net revenue of farmers is found to decrease significantly with a decrease in precipitation or an increase in temperature. Statistical evidence shows that temperature affects coffee yield in Veracruz, Mexico. It was also found that coffee production may not remain economically viable for the producers in the coming years, as there is an indication of a 34% reduction in current production (Gay et al. 2006). The effect of climate change on crop yields varies according to the area and irrigation application. Crop yields can be increased by expanding irrigated areas, which can have a detrimental effect on the environment (Kang et al. 2009). The rise in temperature is likely to reduce the yield of many crops by reducing their duration (Mahato 2014). The aggregate production of wheat, rice, and maize is expected to decrease if both the temperate and tropical regions experience warming of 2^0 C (Chalinor et al. 2014).

In general, climate change has more impact on tropical regions, as tropical crops remain closer to their high-temperature optima, and thereby experience high-temperature stress during elevated levels of temperature. Moreover, insect pests and diseases are more prevalent in humid and warmer regions (Rosenzweig and Liverman 1992). Other parameters such as humidity and wind speed, along with temperature and rainfall, also impact crop yields, and in the absence of these parameters, there has been a chance of over-prediction of the cost of climate change. Moreover, it was found that climate change is likely to reduce the yields of wheat, corn, and rice in China by 18.26 ± 12.13, 45.10 ± 11.55, and 36.25 ± 10.75% until 2100 (Zhang et al. 2017). Extreme weather events have become more frequent since the 1900s in the Netherlands, and have significantly affected the wheat yield in the Dutch region. The week in which an extreme weather event occurred determined the extent of yield reduction in wheat (Pongel and Reinhard 2016). There has been a projection of higher droughts in the near future due to climate change in most of the regions of the world, and an increase in drought-affected area from 15.4 to 44.0% is projected by 2100. Africa is cited as the most vulnerable area. The yield of major crops in drought areas is expected to be reduced by more than 50% by 2050 and by almost 90% by 2100 (Li et al. 2009).

As per Food and Agriculture Organization (FAO) data published in 2016, if the current situation of GHG emissions and climate change continues then by the year 2100 there will be a decline in the production of major cereal crops (20–45% in maize yields, 5–50% in wheat and 20–30% in rice). Hence if the trends continue, in the very near future crop losses may increase at an unprecedented rate which will substantially contribute to reduced production, spiked food prices, and it will become difficult to cope with the rising needs of a growing population.

The loss of crop yields can increase food prices, and can have an absurd effect on agriculture welfare globally, with a 0.3% annual loss of future GDP globally by 2100 (Stevanovic et al. 2016). However, Bosello and Zhang (2005) reported that climate change has limited influence on the world food supply, but developing countries will face severe negative consequences. In India, the temperature is predicted to rise between 2.33^0 C and 4.78^0 C along with a doubling of CO_2 concentration and longevity of heat waves, which could have a detrimental effect on the agriculture sector (Kumar and Gautam 2014). The yield losses in three cereal grains (rice, maize, and wheat) are projected to worsen by 10 to 25% with a 1^0 C increase in mean surface temperature globally (Duetsch et al. 2018). In sub-Saharan Africa, the average crop yield is projected to be reduced by 6–24% due to climate change (Waha et al. 2013). The total fish demand in Solomon Island is also projected to exceed the production in 2050, which will have serious implications for food security, as per-capita consumption will be reduced (Dey et al. 2016).

Plant–water relations are highly susceptible to changes in temperature and precipitation, and physiological changes are more likely to be impacted by extreme changes in these parameters than by changes in mean climate (Reyer et al. 2012). Plants' response to climate change varies according to the plant species and developmental stage of the plant. There are species-specific thresholds for different plants, and their responses, such as elongation of roots, disturbance in growth angle of roots, and reduction in yield, vary among different species of plants (Gray and Brady 2016). With increased CO2 content in the atmosphere, reduced transpiration was found in plants, leading to an increase in air temperature of 0.42 ± 0.02 K. This indirect physiological effect of raised CO2 and a direct radiative effect can increase the warming of land surfaces by 3.33 ± 0.03 K (Cao et al. 2010). With a rising CO_2 level in the atmosphere, the harvestable produce of crops is expected to increase. The developmental changes in plants depend upon the type of crops. Under climate change conditions, the C3 crops are expected to produce more, and, the water requirements of both C3 and C4 crops are expected to be lowered. But these beneficial impacts of raised CO2 are likely to be offset by elevated temperatures and altered precipitations (DaMatta et al. 2010).

A positive impact of climate change on agriculture production is also observed in some areas. But these regional changes, whether increases or decreases, would not result in massive changes, and these changes will be more prominent in some low latitudes only. However, if the temperature is increased beyond that equivalent to doubled CO_2, this can cause substantial economic losses (Aydinalp and Cresser 2008). The detrimental impact of climate change will be enormous in developing countries' tropical regions, but it will largely depend on the region's climate scenario. The drier region of Sri Lanka (north and east) will experience huge losses in agriculture compared to the cooler central highland region, the output of which is expected to remain the same or even increase with rising temperatures (Seo et al. 2005). The pace of climate change determines its impact, thereby determining the cost of adjustment, so environmental policies must be dynamic and implemented with adaptation and flexibility (Ziberman et al. 2004)

A sensitivity analysis using CERES (crop estimation through resources and environmental synthesis) has also shown that wheat and rice yields in northwest India have the potential to increase by 28% and 15%, respectively, at double the levels of CO_2. However, the increased thermal stress due to elevated levels of temperatures associated with high CO_2 nearly cancels out the positive impact. There will be an increased yield in both rice and wheat by 21% and 4%, respectively, if the irrigation scheduling followed at present is practiced even under the combined effect of enhanced CO_2 and thermal stress. But in case of acute water shortage along with thermal stress, the yield of rice and wheat is projected to decline even under the positive influence of raised CO_2 in the future (Lal et al. 1998). The increased CO_2 concentration has the potential to offset the losses in crop yields due to rising temperatures and reduced soil moisture (Long et al. 2006). The raised CO_2 concentration considerably reduces the global yield losses by mainly decreasing agricultural consumptive water use (4–17%). The differences in the yield of crops in different regions are mostly due to different growing environments (Derying et al. 2016). The concentration of nutrients (N, Fe, Zn, and S) mainly found in proteins is reduced in non-leguminous C3 crops with an elevated level of CO_2 (Udding et al. 2018). At the increased level of CO_2, enhanced vegetative and reproductive growth and enhanced seed yield are observed in rice crops at an ambient air temperature of 29^0 C. However, with increased temperature, the seed set was decreased (Madan et al. 2012).

At an elevated level of CO_2, the zinc and iron content of C3 grain crops and legumes decreases, which has detrimental effects on human health. C3 plants and legumes' protein concentration is also observed to be lowered, while C4 plants are unaffected by a raised CO_2 level (Mayers et al. 2014). The changes in climate also impact the microbial population present in the soil, and their enzymatic activities. When assessed from a temperature gradient tunnel with a $4–5^0$ C higher temperature, the microbial population was found to be significantly higher than in field conditions. The population of nitrogen-fixing and P-solubilizers bacteria and fungus, as well as enzymatic activities, were significantly higher under a wide range of temperatures, but the highest parameters were found on or near the optimum temperature (Kaur et al. 2014). In contrast, the growth of endophytic fungus and plant growth-promoting bacteria has a positive, negative, or neutral impact, depending on the temperature range (Compant et al. 2010),

Projected changes in climate are most likely to affect the development and survival of pathogens (Elad 2018). The change in climate or weather pattern of an area is predicted to increase a crop's susceptibility to various pests, diseases, and weeds. Increased yields are projected in countries of high and mid-latitudes, while yields are projected to decrease at lower latitudes (Rosenzweig et al. 2001). However, there are projections of a 10–25% increase in losses due to insect pest infestation with an increased temperature of one degree (Shrestha 2019). Climate change has the potential to increase the pest population and its migration, which can have an adverse impact on agricultural yields and even viability, as the pest population depends mainly on abiotic factors such as humidity and temperature. In Brazil, the infestation of coffee nematodes and leaf miners is expected to increase due to an increase in the number of generations in a month compared to the climatic conditions of 1961–1990 (Ghini et al. 2008). Pest infestation thereby has led to huge pesticide costs for pest management. There is statistical evidence that increased rainfall and temperatures increased the costs of pesticides for crops such as corn, potatoes, and soybean, in contrast to a reduction in wheat in the USA (Chen and McCarl 2001). The ratio of arable land affected by the European corn borer and the Colorado potato beetle is expected to increase by 43 and 48%, respectively, for the second generations in HadCM3-high 2050 scenario, and the unoccupied areas of high altitudes are also found vulnerable to these pests in the scenario of increasing temperature in Central Europe (Kocmankova et al. 2016). In the current global-warming scenario, an expansion of the suitable areas for wheat aphids (*Schizaphis graminum*) has been predicted to upper latitudes in the northern hemisphere by 2030 (Aljaryian and Kumar 2016). The frequency of the insect outbreak of 30 pest species is also expected to rise. It is likely to affect new areas with the increasing temperature in Sweden and affect its forestry sector (Hof and Svahin 2015). The future projection of the potato tuber moth (*Phthorimaea operculella*) when done through GIS modelling, reported an estimated increase in the pest's damage potential in tropical and subtropical warmer regions, where the pest already prevails. It is also predicted to expand in temperate and mountainous regions, with a slightly increased damage potential (Kroschel et al. 2013).

The life cycle of pathogens such as *Puccinia striiformis* f. sp. trictici is projected to be limited by increasing temperature, while an increase in the concentration of atmospheric CO_2 is estimated to provide advantageous conditions for *Fusarium pseudograminearum* (Luck et al. 2011). Climate change impacts the geographical distribution and growth rate of populations, along with increasing the number of generations. Climate change can extend the development season of the pests and change the synchronization of crops and pests. It can also increase the risk of migrant pest invasion. The efficiency of plant protection measures such as host plant resistance, natural enemies, transgenic plants, synthetic chemicals, or biopesticides is also likely to be reduced due to climate change (Reddy 2013). There could be unpredictable interactions between cropping systems, weather, and pests due to climate change and globalization (Lamichhane et al. 2015). Climate change is likely to impact insects' growth and their metabolic rates, particularly in temperate regions (Deutsch et al. 2018). The region of suitability for pest infestation is also increasing because of climate change. The three common insect species of Africa, *Tuta absoluta*, *Ceratitis cosyra*, and *Bactrocera invadens*, have increasing habitat suitability across the entire continent, particularly in regions close to its most suitable habitat (Biber-Freudenberger et al. 2016). Moreover, the increased CO2 level and the rising temperature are increasing the threat of late blight of potato, blast, and sheath blight of rice, which could pose a serious threat to the world's food security (Gautam et al. 2013).

Weed infestation of crops is also affected by climate change. C3 weeds respond more strongly to an increase in CO_2 concentration, with increased leaf area and biomass. C3 weeds are a major problem in C4 plants, while C4 weeds in C3 plants become less competitive (Kores et al. 2016). Weeds compete with crops for water and nutrients, as they have higher nutrient requirements than crop plants (Malhi et al. 20290). Climate change also influences the dynamics of crop–weed competition. Apart from weed growth, climate change also significantly influences herbicide efficacy, as it affects the herbicidal mode of action (Varanasi et al. 2016). Climate change is projected to have a favourable influence on the weeds of wheat crops, which are very vital to world food security (Bajwa et al. 2020). In the wake of climate change, new geographical horizons are being opened up for weeds, and their management can only be possible if new management practices are being planned while considering climate change.

Mitigation and the Role of Plant Breeders

All these climate-related disturbances would not only have drastic impacts on food distribution patterns but also on its quality and access. It will also raise humanitarian concerns as food security is deeply entwined with public health and will create a vicious cycle of hunger, diseases, and crime. FAO reports that the rise in world hunger level is constantly increasing since 2014 and an increment in the number of undernourished people has been observed from 804 million in 2016 to 824 million in 2017. These reports established the fact that achieving the target of sustainable development goals (SDG) by 2030, to eradicate hunger is still difficult to achieve. Therefore, implementing and adapting to actions/policies to mitigate the effects caused by climate variability are urgently required. This can be achieved by adopting a climate-smart agriculture system which is in line with FAO's vision of Sustainable Food and Agriculture goals. This includes developing resilient varieties of crops that can withstand abrupt stresses of temperature and precipitation, implementing biotic and sustainable agricultural methods, for example, application of bioinoculants/biofertilizers using tolerant plant growth promoting rhizobacteria (PGPR) and their metabolites and other methods of organic farming in order to promote yield of crops in extreme environments in a holistic and a sustainable way.

The use of biofertilizers and biopesticides will result in decreased reliance on chemical fertilizers and pesticides. This will result in reduced emissions of GHG and a decrease in levels of soil, water, and air pollution. Changes in crop cultivars, sowing time, and cultivation techniques (such as mixed cropping systems) to provide greater durability to crops in extreme events should also be considered. Reclamation of marginal and degraded lands using resilient crops, by afforestation, or using beneficial soil microbes will not only result in increased agriculture productivity but also help in combating climate change.

The marginal or drought-impacted lands are characterized by very low soil organic matter and soil microbial biomass. As per United Nations Convention to Combat Desertification (UNCCD) "restoring the soils of degraded ecosystems has the potential to store up to 3 billion tonnes of carbon annually". Phyto or rhizoremediation of such soils increases the organic matter which results in enhanced CO_2 fixation in the soil. Soil and oceans are the biggest sinks of CO_2, thus by increasing the soil organic matter and subsequent enhancement in yield or planting of perennial crops or through planned agro-forestry can result in decreasing the levels of GHGs, helping in combating the climate change and simultaneously resulting in the reclamation of marginal or degraded lands.

Furthermore, extensive data collection and field trials to assess the impact of prevailing climatic conditions in different agroecological zones can help in designing future strategies through computational simulation. Remote sensing and satellite imaging can also help in future predictions for vulnerable agroecosystems and suggest corrective measures by involving a multi-disciplinary approach. This can also help in working out responses, preparedness, and planning of managing the agroecosystems for extreme events such as water scarcity, heat waves, floods, and so on. Connecting and sensitizing farmers to sustainable technologies and activities is of utmost importance as they are the ones who can play a major role in the implementation of ecological goals.

Climate-smart agriculture is also an important requirement for enhancing the yields and quality of production. Climate change and agriculture are strongly co-related. Apparently, the fast pace of climate change will have a far-reaching impact on agroecosystems and their productivity. Hence it is high time for the plant breeders to prepare an action plan to meet the upcoming challenges to combat the impact of climate change and ensure food security not only for humans but other living beings as well.

Strategies adopted by the breeders for developing climate change-ready crops include:

- Resequencing of the plant germplasm,
- De novo domestication of wild relatives,
- High-throughput phenotyping,
- Genome editing, and Speed breeding

Technological advancements in various fields such as: Sensors, The Internet of Things (IoT), Artificial Intelligence (AI) and Software Tools have contributed immensely towards achieving the goals of various crop breeding programmes. Role of these aspects on crop breeding are discussed in the following chapters of this book.

Economic Impact of Climate Change

As per the annual report of Weather, Climate, and Catastrophe Insight, natural disasters alone have caused economic losses to the tune of USD 225 billion across the world in 2018, and since 2016 the losses due to natural calamities have crossed USD 200 billion per year. About 95% of these losses are attributed to weather-related incidences, of which cyclones, floods, and droughts are the key players and are directly related to climate change. Altogether, the impact of climate change is very comprehensive but its far-reaching effects are now clearly visible on the agricultural sector, on which relies the food production and economy of the world. It is also worth noting that the world population is expected to reach 9.7 billion by 2050 which would magnify the pressure on agricultural lands to meet the growing food demands already affected by the impact of climate change. As climate change and agriculture have inextricable links, abrupt changes in climatic conditions at such a rapid pace have threatened food security at global scale. World Food Programme (WFP) report of 2018 revealed that the increase in crop yield per hectare is significantly slower as compared to rates of rising population.

Climate change has initially had certain positive impacts, but the environment's unavoidable warming is a negative externality. A rise in temperature beyond 3^0 C has net negative results, and more than 7^0 C can cause total welfare loss. The world's social cost of carbon emission is expected to be USD 29/tC (tonnes of Carbon) in 2015, and to raise at 2% per year (Tol et al. 2012). The net economic gains in the fishery sector of Solomon Island would be considerable if mitigation strategies for climate change are adopted. Climate change also will severely affect agricultural markets, causing a reduction of 0.26% in global GDP (Cosfinot et al. 2016). There would be a projected annual loss of 0.2–1% in household welfare if the climate predicted for the 2080s occurred today (Ciscar et al. 2011) Both market and non-market damages increase in quadratic progression and are expected to cost 1.2% of GDP with a 1^0 C increase in mean global temperature (Hsiang et al. 2017). If future mitigation strategies follow the adaptation of strategies used in the past, global income is projected to show a 23% decrease by 2100 and a wider gap in income inequality (Burke et al. 2015). Global economic growth is projected to be reduced by 0.28% per year (Carlton and Hsiang 2016).

Conclusions

An increasing population has put a lot of pressure on agriculture to ensure the food and nutritional security of the world, which is further worsening with climate change. Even though there are uncertainties regarding the future climate scenario and its possible impacts, various studies report that climate change will decrease agricultural productivity in the coming years. The key factors of climate, namely temperature, precipitation, and greenhouse gases, significantly hampered pest infestation, soil fertility, irrigation resources, physiology, and plants' metabolic activities.

A number of mitigation and adaptation strategies have been developed to offset the deleterious impact of climate change on agricultural sustainability. These technologies include water-smart practices (laser land levelling, rainwater harvesting, micro-irrigation, crop diversification, raised-bed planting, direct-seeded rice), nutrient-smart practices (precision nutrient application, leaf colour charts, crop residue management), weather-smart activities (stress-tolerant varieties, ICT-based agrometeorological services), carbon-smart activities (zero tillage, legumes, crop residue management) and knowledge-smart activities (agricultural extensions to enhance capacity-building). These technologies significantly reduce the effects of climate change on crops, and make them more suited to the climate by minimizing the unfavourable impacts. Climate change is predicted to cause huge economic losses at both the micro and macro levels that can be mitigated through these interventions. But these interventions must be organized at the regional or local level to improve their efficacy.

Mitigation and adaptation strategies are expected to increase farmers' income without compromising agricultural production sustainability. The future of climate change and its associated impacts is highly unpredictable, which makes planning for mitigation and adaptation a bit complex. This necessitates the formulation of climate-resilient technologies involving an interdisciplinary approach according to the region. Suitable varieties need to be developed that could adapt to climatic variations, along with planned agronomic management and crop pest control. Farmers need to be educated regarding various climate-smart technologies, and be provided training to simplify their use at the field level.

References

Abeydeera, L.H.U.W., Mesthrige, J.W., & Samarasinghalage, T.I. (2019). Global research on carbon emissions: a scientometric review. *Sustainability, 11*: 3972.

Adams, R.M., Hurd, B.H., Lenhart, S., & Leary, N. (1998). Effects of global climate change on agriculture: an interpretative review. *Clim. Res. 11*: 19–30.

Aggarwal, P.K. (2008). Global climate change and Indian agriculture: Impacts, adaptation and mitigation. Indian J. Agric. Sci. 78: 911–919.

Alexandratos, N., & Bruinsma, J. (2012). World Agriculture: Towards 2030/2050; *ESA Working Paper* No. 12-03; FAO: Rome, Italy.

Aljaryian, R., & Kumar, L. (2016). Changing global risk of invading greenbug *Schizaphis graminum* under climate change. *Crop Prot. 88*: 137–148.

Altieri, M.A., & Nicholls, C.I. (2017). The adaptation and mitigation potential of traditional agriculture in a changing climate. *Clim. Chang. 140*: 33–45.

Anderson, A. 2012. Climate change education for mitigation and adaptation. *J. Educ. Sustain. Dev. 6*: 191–206.

Arbuckle, J.G., Morton, L.W., & Hobbs, J. (2015). Understanding farmer perspectives on climate change adaptation and mitigation: The roles of trust in sources of climate information, climate change beliefs, and perceived risk. *Environ. Behav. 47*: 205–234.

Arora, M., Goel, N.K., & Singh, P. (2005). Evaluation of temperature trends over India. *Hydrol. Sci. J. 50*: 81–93.

Arora, N.K. (2019). Impact of climate change on agriculture production and its sustainable solutions. *Environmental Sustainability, 2*: 95–96.

Aryal, J.P., Mehrotra, M.B., Jat, M.L., & Sidhu, H.S. (2015). Impacts of laser land leveling in rice-wheat systems of the north-western indo-gangetic plains of India. *Food Sec. 23*: 453-460.

Aryal, J.P., Sapkota, T.B., Jat, M.L., & Bishnoi, J.K. (2015). On-farm economic and environmental impact of zero-tillage wheat: A case of north-west *India. Exptl. Agric. 51*: 1–16.

Atlin, G.N., Cairns, J.E., & Das, B. (2017). Rapid breeding and varietal replacement are critical to adaptation of cropping systems in the developing world to climate change. *Glob. Food Sec. 12*: 31–37.

Aydinalp, C., & Cresser, M. (2008). The effects of global climate change on agriculture. *Am. Eur. J. Agric. Environ. Sci. 3*: 672–676.

Bajwa, A.A., Farooq, M., Al-Sadi, A.M., Nawaz, A., Jabran, K., & Siddique, K.H.M. 2020. Impact of climate change on biology and management of wheat pests. *Crop Prot.137*: 105304.

Baul, T.K., & McDonald, M. (2015). Integration of Indigenous knowledge in addressing climate change. *Indian J. Tradit. Knowl. 1*: 20–27.

Betts, R.A., Alfieri, L., Bradshaw, C., Caeser, J., ... Feyen, L. (2018). Changes in climate extremes, fresh water availability and vulnerability to food insecurity projected at 1.5 ?C and 2 ?Cglobal warming with a higher-resolution global climate model. *Phil. Trans. R. Soc. A, 376*: 20160452.

Bhatia, A., Pathal, H., Jain, N., Singh, PK., & Tomer, R. (2012). Greenhouse gas mitigation in rice-wheat system with leaf color chart-based urea application. *Environ. Monit. Assess. 184*: 3095–3107.

Bhullar, M.S., Singh, S., Kumar, S., & Gill, G. (2018). Agronomic and economic impacts of direct seeded rice in Punjab. *Agric. Res. J. 55*: 236–242.

Biber-Freudenberger, L., Ziemacki, J., Tonnang, H.E.Z., & Borgemeister, C. (2016). Future risks of pest species under changing climatic conditions. *PLoS ONE. 11*: e0153237.

Birkenholtz, T. (2017). Assessing India's drip-irrigation boom: efficiency, climate change and groundwater policy. *Water Int. 42*: 663–677.

Birthal, P.S., Nigam, A.N., Narayanan, A.V., & Kareem, K.A. (2012). Potential economic benefits from adoption of improved drought tolerant groundnut in India. *Agric. Econ. Res. Rev. 25*: 1–14.

Bokhari, S.A.A., Rasul, G., Ruane, A.C., Hoogenboom, G., & Ahmad, A. (2017). The past and future changes in climate of the rice-wheat cropping zone in Punjab, Pakistan. *Pak. J. Meteorol.13*: 9–23.

Bongiovanni, R., & Lowenberg-Deboer, J. (2004). Precision agriculture and sustainability. *Precis. Agric. 5*: 359–387.

Bosello, F., & Zhang, J. (2005). Assessing climate change impacts: agriculture; FEEM Working Paper No. 94.05; CMCC Research Paper No. 2; Elsevier: New York, NY, USA. https://ssrn.com/abstract=771245

Brown, B., Llewellyn, R., & Nuberg, I. (2018). Global learnings to inform the local adaptation of conservation agriculture in Eastern and Southern Africa. *Glob. Food Sec. 17*: 213–220.

Burke, M. Hsiang, S.M., & Miguel, E. (2015). Global non-linear effect of temperature on economic production. *Nature. 527*: 235–239.

Cai, X., Wang, D., & Laurent, R. (2009). Impact of climate change on crop yield: a case study of rainfed corn in Central Illinois. *J. Appl. Meteorol. Climatol. 48*: 1868–1881.

Cao, L., Bala, G., Caldeira, K., Nemani, R., & Ban-Weiss, G. (2010). Importance of carbon dioxide physiological forcing to future climate change. *Proc. Natl. Acad. Sci. USA. 107*: 9513–9518.

Carleton, TA., & Hsiang, SM. (2016). Social and economic impacts of climate. *Science. 353*: 9837.

CDIAC. (2020). Carbon Dioxide Information Analysis Center. www.cdiac.ess-dive.lbl.gov

Challinor, A.J., Watson, J., Lobell, D.B., Howden, S.M., Smith, D.R., & Chhetri, N. (2014). A meta-analysis of crop yield under climate change and adaptation. *Nat. Clim. Chang. 4:* 287–291.

Chen, C.C., & McCarl, B.A. (2001). An investigation of the relationship between pesticide usage and climate change. *Clim. Chang. 50*: 475–487.

Chen, H., & Sun, J. 2018. Projected changes in climate extremes in China in a 1.5C warmer world. Int. J. Climatol. 38: 3607–3617.

Chhogyell, N., Pradhan, N., Ghimiray, M., & Bajgai, Y. (2016). Evaluation of short duration rice (*Oryza sativa*) varieties as a strategy to cope with climate change. *Proc. Bhutan Ecol. Soc. 1*: 91–103.

Ciscar, J.C., Iglesias, A., Feyen, L., Szabo, L., ... Regemorter, D.V. (2011). Physical and economic consequences of climate change in Europe. *Proc. Natl. Acad. Sci. USA. 108*: 2678–2683.

Compant, S., Heijden, M.G.A.V.D., & Sessitsch, A. (2010). Climate change effects on beneficial plant-microorganism interactions. *FEMS Microbiol. Ecol. 73*: 197–214.

Costinot, A., Donaldson, D., & Smith, C. (2016). Evolving comparative advantage and the impact of climate change in agricultural markets: evidence from 1.7 million fields around the world. *J. Pol. Econ. 124*: 20–25.

DaMatta, F.M., Grandis, A., Arenque, B.C., & Buckeridge, M.S. (2010). Impacts of climate changes on crop physiology and food quality. *Food Res. Int. 43*: 1814–1823.

Deryng, D., Elliott, J., Folberth, C., Muller, C., ... Pugh, T.A.M. (2016). Regional disparities in the beneficial effects of rising CO2 concentrations on crop water productivity. *Nat. Clim. Chang. 6*: 786–790.

Deutsch, C.A., Tewksbury, J.J., Tigchelaar, M., ... Battisti, D.S. (2018). Increase in crop losses to insect pests in a warming climate. *Science. 361*: 916–919.

Dey, M.M., Gosh, K., Valmonte-Santos, R., Rosegrant, M.W., & Chen, O.L. (2016). Economic impact of climate change and climate change adaptation strategies for fisheries sector in Solomon Islands: Implication for food security. *Mar. Policy. 67*: 171–178.

Elad, Y., & Pertot, I. (2014). Climate change impact on plant pathogens and plant diseases. *J. Crop Improv. 28*: 99–139.

Elliott, J., Deryng, D., Muller, C., ... Frieler, K.(2014). Constraints and potentials of future irrigation water availability on agricultural production under climate change. *Proc. Natl. Acad. Sci. USA. 111*: 3239–3244.

Erenstein, O., Sayre, K., Wall, P., Hellin, J., & Dixon, J. (2012). Conservation agriculture in Maize and Wheat based systems in the (sub)tropics: Lessons from adaptation initiatives in South Asia, Mexico, and Southern Africa. *J. Sustain. Agric. 36*: 180–206.

Gautam, H.R., Bhardwaj, M.L., & Kumar, R. (2013). Climate change and its impact on plant diseases. *Curr. Sci. 105*: 1685–1691.

Gay, C., Estrada, F., Conde, C., Eakin, H., & Villers, L. (2006). Potential impacts of climate change on agriculture: a case of study of coffee production in Veracruz, Mexico. *Clim. Chang. 79*: 259–288.

Ghini, R., Hamada, E., Junior, M.J.P., Marengo, J.A., & Goncalves, R.R.V. (2008). Risk analysis of climate change on coffee nematodes and leaf miner in Brazil. *Pesqui. Agropecu. Braileira. 43*: 187–194.

Gray, S.B., & Brady, S.M. (2016). Plant developmental responses to climate change. *Dev. Biol. 419*: 64–77.

Gregorio, G.B., Islam, M.R., Vergara, G.V., & Thirumeni, S. (2013). Recent advances in rice science to design salinity and other abiotic stress tolerant rice varieties. *SABRAO J. Breed. Genet. 45*: 31–41.

Groenigen, K.J.V., Osenberg, C.W., & Hungate, B.A. (2011). Increased soil emissions of potent greenhouse gases under increased atmospheric CO2. *Nature. 475*: 214–216.

Hare, B., & Meinshausen, M. (2006). How much warming are we committed to and how much can be avoided? *Clim. Chang. 75*: 111–149.

Hof, A.R., & Svahlin, A. (2015). The potential effect of climate change on insect pest species' geographical distribution in the Swedish boreal forest. *Scand. J. For. Res. 31*: 1–15.

Hsiang, S., Kopp, R., Jina, A., ... Rising, J. (2017). Estimating economic damage from climate change in the United States. *Science. 356*: 1362–1369.

IPCC. Climate change (2007): Impacts, adaptation and vulnerability. In: *Working Group II Contribution to the Fourth Assessment Report of the Intergovernmental Panel on Climate Change; Cambridge University Press: Cambridge, UK.*

IPCC. Climate Change (2014): Synthesis Report. Pachauri, R.K., Meyer, L.A., (Eds.); Contribution of Working Groups I, II and III to the Fifth Assessment Report of the Intergovernmental Panel on Climate Change, *IPCC: Geneva, Switzerland*, p.151.

Kahrl, F., Li, Y., Su, Y., Tennigkeit, T. Wilkes, A., & Xu, J. (2010). Greenhouse gas emissions from nitrogen fertilizer use in China. *Environ. Sci. Policy. 13*: 688–694.

Kang, Y., Khan, S., & Ma, X. (2009). Climate change impacts on crop yield, crop water productivity and food security: a review. *Prog. Nat. Sci. 19*: 1665–167.

Karimi, V., Karimi, E., & Keshavarz, M. (2018). Climate change and agriculture: Impacts and adaptive responses in Iran. *J. Integr. Agric. 17*: 1–15.

Kaur, J., Gosal, S.K., & Kaur, P. (2014). Effects of climate change on plant associated microbial communities and enzyme activities. *Afr. J. Microbiol. Res. 8*: 3087–3093.

Kaur, N., & Kaur, P. (2016). Projected climate change under different scenarios in central region of Punjab, India. *J. Agrometeorol. 18*: 88–92.

Khatri-Chhetri, A., Aryal, J.P., Sapkota, T.B., & Khurana, R. (2016). Economic benefits of climate-smart agricultural practices to smallholder farmers in the Indo-Gangetic Plains of India. *Curr. Sci. 110*: 1251–1256.

Kocmankova, E., Trnka, M., Eitzinger, J., ... Formayer, H. (2010). Estimating the impact of climate change on the occurrence of selected pests in Central European region. *Clim. Res. 44*: 95–105.

Korres, N.E., Norsworthy, J.K., Tehranchian, P., ... Gitsopoulos, T.K. (2016). Cultivars to face climate change on crops and weeds: a review. *Agron. Sustain Dev. 36*: 12–22.

Kroschel, J., Sporleder, M., Tonnang, H.E.Z., ... Juarez, H. (2013). Predicting climate-change ca used changes in global temperature on potato tuber moth

Phthorimaea operculella (Zeller) distribution and abundance using phenology modeling and GIS mapping. *Agric. For. Meteorol. 170*: 228–241.

Kukal, M.S., & Irmak, S. (2018). Climate-driven crop yield and yield variability and climate change impacts on the U.S. great plains agricultural production. *Sci. Rep. 8*: 1–18.

Kumar, R., & Gautam, H.R. (2014). Climate change and its impact on agricultural productivity in India. *J. Climatol. Weather Forecast. 2*: 1–3.

Lal, M., Singh, K.K., Rathore, L.S., Srinivasan, G., & Saseendram, S.A. (1998). Vulnerability of rice and wheat yields in NW India to future changes in climate. *Agric. For. Meteorol. 89*: 101–114.

Lal, R. (2004). Soil carbon sequestration impacts on global climate change and food security. *Science. 304*: 1623–1627.

Lal, R., Delgado, J.A., Groffman, P.M., ... Millar, N. (2011). Management to mitigate and adapt to climate change. *J. Soil Water Conserv. 66*: 276–285.

Lamichhane, J.R., Barzman, M., Booij, K., ... Boonekamp, P. (2015). Robust cropping systems to tackle pests under climate change. a review. *Agron. Sustain. Dev. 35*: 443–459.

Lee, J., Gryze, S.D., & Six, J. (2011). Effect of climate change on field crop production in California's central valley. *Clim. Chang. 109*: S335–S353.

Lesschen, J.P., Berg, M.V.D., Westhoek, H.J., & Oenema, O. (2011). Greenhouse gas emission profiles of European livestock sectors. *Anim. Feed Sci. Technol. 166–167*: 16–28.

Li, Y., Ye, W., Wang, M., & Yan, X. (2009). Climate change and drought: a risk assessment of crop-yield impacts. *Clim. Res. 39*, 31–46.

Lipper, L., Thornton, P., Campbell, B.M., ... Baedeker, T. (2014). Climate-smart agriculture for food security. *Nat. Clim. Chang. 4*: 1068–1072.

Liu, L., & Basso, B. (2020). Impacts of climate variability and adaptation strategies on crop yields and soil organic carbon in US Midwest. *PLoS ONE. 15*: e0225433.

Lobell, D.B., Schlenker, W., & Costa-Roberts, J. (2011). Climate trends and global crop production since 1980. *Science. 333*: 616–620.

Long, S.P., Ainsworth, E.A., Leakey, A.D.B., Nosberger, J., & Ort, D.R. (2006). Food for thought: Lower-than-expected crop yield stimulation with rising CO2 concentrations. *Science. 312*: 1918–1921.

Lu, Y., Chadwick, D., Norse, D., Powlson, D., & Shi, W. (2015). Sustainable intensification of China's agriculture: The key role of nutrient management and climate change mitigation and adaptation. *Agric. Ecosyst. Environ. 209*: 1–4.

Luck, J., Spackman, M., Freeman, A., ... Trebicki, P. (2011). Climate change and diseases of food crops. *Plant Pathol. 60*: 113–121.

Lybbert, T.J., & Sumner, D.A. (2012). Agricultural technologies for climate change in developing countries: Policy options for innovation and technology diffusion. *Food Policy. 37*: 114–123.

Madan, P., Jagadish, S.V.K., Craufurd, P.Q., & Fitzgerald, M. (2012). Effect of elevated CO2 and high temperature on seed-set and grain quality of rice. *J. Exp. Bot. 63*: 3843–3852.

Mahato, A. (2014). Climate change and its impact on agriculture. *Int. J. Sci. Res. Publ. 4*: 1–6.

Malhi, G.S., Rana, M.C., Rana, S.S., & Kaushik, P. (2020). Effect of individual or combined application of herbicide imazethapyr on nutrient uptake by black gram (*Vigna mungo* L.). *J. Exp. Biol. Agric. Sci. 8*: 441–446.

Malhi, G.S., Kaur, M., & Kaushik, P. (2021). Impact of climate change on agriculture and its mitigation strategies: a review. *Sustainability. 13*: 1318. https://doi.org/10.3390/su13031318

Mendelsohn, R. (2009). The impact of climate change on agriculture in developing countries. *J. Nat. Res. Policy Res. 1*: 5–19.

Mishra, A.K., Khanal, A.R., & Pede, V. (2017). Economic and resource conservation perspectives of direct seeded rice planting methods: evidence from India. In: Proceedings of the Agricultural and Applied Economics Association. p. 568-572.

Moher, D., Liberati, A., Tetzlaff, J., & Altman, D.G. (2010). The prisma group. Preferred reporting items for systematic reviews and meta-analyses: The PRISMA statement. *Int. J. Surg. 8*: 336–341.

Molua, E.L., & Lambi, C.M. (2007). The Economic impact of climate change on agriculture in Cameroon: policy research working paper. The World Bank: Washington, DC, USA. p. 1–31.

Montzka, S.A., Dlugokencky, E.J., & Butler, J.H. (2011). Non-CO2 greenhouse gases and climate change. *Nature. 476*: 43–50.

Myers, S.S., Zanobetti, A., Kloog, I., ... Huybers, P. (2014). Increasing CO2 threatens human nutrition. *Nature. 510*: 139–142.

Narayanamoorthy, A., Bhattarai, M., & Joshi, P. (2018). An assessment of the economic impact of drip irrigation in vegetable production in India. *Agric. Econ. Res. Rev. 31*: 105–112.

Narayanamoorthy, A., & Devika, N. (2017). Economic and resource impacts of drip method of irrigation on okra cultivation: An analysis of field survey data. *J. Land Rural Stud. 6*: 15–33.

NASA (2020). Earth Observatory. Goddard Space Flight Centre United States. www.earthobservatory.nasa.gov

NCDC. (2020) National Climatic Data Center-National Center for Environmental Information (NOAA). www.ncdc.noaa.gov

Ning, L., Riddle, E.E., & Bradley, R.S. (2015). Projected changes in climate extremes over the Northeastern United States. *J. Climate. 28*: 3289–3310.

NOAA. (2020) Earth System Research Laboratory (NOAA). www.esrl.noaa.gov

O'Mara, F.P. (2011). The significance of livestock as a contributor to global greenhouse gas emissions today and in the near future. *Anim. Feed Sci. Technol. 166–167*: 7–15.

Ogada, M.J., Rao, E.J.O., Radeny, M., Recha, J.W., & Solomon, D. (2020). Climate-smart agriculture, household income and asset accumulation among smallholder farmers in Nyando basin of Kenya. *World Dev. Perspect.*

Pal, B.D., Kapoor, S., Saroj, S., ... Jat, M.L. (2020). Impact of laser land levelling on food production and farmers' income: evidence from drought prone semi-arid tropics in India: *IFPRI discussion paper 01960; IFPRI: Washington, DC, USA.*

Pampolino, M.F., Manguiat, I.J., Ramanathan, S., ... Gines, H.C. (2007). Environmental impact and economic benefits of site-specific nutrient management (SSNM) in irrigated rice systems. *Agric. Syst. 93*: 1–24.

Parthasarathi, T., Vanitha, K., Lakshamanakumar, P., & Kalalyarasi, D. (2012). Aerobic rice-mitigating water stress for the future climate change. *Int. J. Agr. Plant Prod. 3*: 241–254.

Pisante, M., Stagnari, F., Acutis, M., ... Bindi, M. (2014). Conservation agriculture and climate change. In: Conservation Agriculture; Farooq, M., Siddique, K., Eds.; Springer: Cham, Switzerland.

Popp, A., Lotze-Campen, H., & Bodirsky, B. (2010). Food consumption, diet shifts and associated non-CO2 greenhouse gases from agricultural production. *Glob. Environ. Chang. 20*: 451–462.

Powell, J.P., & Reinhard, S. (2016). Measuring the effects of extreme weather events on yields. *Weather Clim. Extrem. 12*: 69–79.

Powlson, D.S., Stirling, C.M., Jat, M.L., ... Gerard, B.G. (2014). Limited potential of no-till agriculture for climate change mitigation. *Nat. Clim. Chang. 4*: 678–683.

Rajbhandari, R., Shrestha, A.B., Kulkarni, A., Patwardhan, S.K., & Bajracharya, S.R. (2015). Projected changes in climate over the Indus river basin using a high resolution regional climate model (PRECIS). *Clim. Dyn. 44*: 339–357.

Ray, D.K., West, P.C., Clark, M., ... Gerber, J.S. (2019). Climate change has likely already affected global food production. *PLoS ONE. 14*: e0217148.

Reddy, P.P. (2013). Impact of climate change on insect pests, pathogens and nematodes. *Pest Manag. Hortic. Ecosyst. 19*: 225–233.

Reyer, C.P.O., Leuzinger, S., Rammig, A., ... Wolf, A. (2012). A plant's perspective of extremes: terrestrial plant responses to changing climatic variability. *Glob. Chang. Biol. 19*: 75–89.

Richie, H., & Roser, M. (2017). Our World in Data. CO2 and Greenhouse Emissions. https://ourworldindata.org/co2-and-other-greenhouse-gas-emissions

Rosenzweig, C., Iglesius, A., Yang, X.B., Epstein, P.R., & Chivian, E. (2001). Climate change and extreme weather events—Implications for food production, plant diseases, and pests. *Glob. Chang. Human Health. 2*: 90–104.

Rosenzweig, C., & Liverman, D. (1992). Predicted effects of climate change on agriculture: A comparison of temperate and tropical regions. In: Global Climate Change: Implications, Challenges, and Mitigation Measures; Majumdar, S.K., (Eds.). *The Pennsylvania Academy of Sciences: Grove City, PA, USA.* p. 342–361.

Sandhu, S.S., Kaur, P., Gill, K.K., & Vashisth, B.B. (2019). The effect of recent climate shifts on optimal sowing windows for wheat in Punjab, India. *J. Water Clim. Chang. 11*: 1177–1190.

Sathaye, J., Shukla, P.R., & Ravindranath, N.H. (2006). Climate change, sustainable development and India: global and national concerns. *Curr. Sci. 90*: 314–325.

Schlenker, W., & Roberts, M.J. (2009). Nonlinear temperature effects indicate severe damages to U.S. crop yields under climate change. *Proc. Natl. Acad. Sci. USA. 106*: 15594–15598.

Semenza, J.C., Ploubidis, G.B., & George, L.A. (2011). Climate change and climate variability: Personal motivation for adaptation and mitigation. *Environ. Health. 10*: 1–12.

Seo, S.N., Mendelsohn, R., & Munasinghe, M. (2005). Climate change and agriculture in Sri Lanka: a Ricardian valuation. *Environ. Dev. Econ. 10*: 581–596.

Shakoor, U., Saboor, A., Ali, I., & Mohsin, A.Q. (2011). Impact of climate change on agriculture: empirical evidence from arid region. *Pak. J. Agri. Sci. 48*: 327–333.

Shrestha, S. (2019). Effects of climate change in agricultural insect pest. *Acta Scient. Agric. 3*: 74–80.

Singh, V.K., Shukla, A.K., Singh, M.P., ... Majumdar, K. (2015). Effect of site-specific nutrient management on yield, profit and apparent nutrient balance under pre-dominant cropping systems of Upper Gangetic Plains. *Indian J. Agric. Sci. 85*: 335–343.

Singh, Y., Singh, B., Ladha, J.K., ... Bains, J.S. (2007). On-farm evaluation of leaf color chart for need-based nitrogen management in irrigated transplanted rice in northwestern India. *Nutr. Cycl. Agroecosyst. 78*: 167–176.

Smit, B., & Skinner, M.W. (2002). Adaptation options in agriculture to climate change: a typology. *Mitig. Adapt. Strateg. Glob. Chang. 7*: 85–114.

Smith, P., & Olesen, J.E. (2010). Synergies between the mitigation of, and adaptation to, climate change in agriculture. *J. Agric. Sci. 148*: 543–552.

Soltani, A., Rajabi, M.H., Zeinali, E., & Soltani, E. (2013). Energy inputs and greenhouse gases emissions in wheat production in Gorgan. *Iran. Energy. 50*: 54–61.

Stern, D.I., & Kaufmann, R.K. (2014). Anthropogenic and natural causes of climate change. *Clim. Chang. 122*: 257–269.

Stevanovic, M., Popp, A., Campen, H.L., & Dietrich, J.P. (2016). The impact of high-end climate change on agricultural welfare. *Sci. Adv. 2*: e1501452.

Taneja, G., Pal, B.D., Joshi, P.K., Aggarwal, P.K., & Tyagi, N.K. (2019). Farmers' preferences for climate-smart agriculture-an assessment in the Indo-Gangetic plain. In: Climate Smart Agriculture in South Asia. (Pal, B., Kishore, A., Joshi, P., Tyagi, N., (Eds.). *Springer: Singapore.*

Tao, F., Yokozawa, M., Xu, Y., Hayashi, Y., & Zhang, Z. (2006). Climate changes and trends in phenology and yields of field crops in China, 1981–2000. *Agric. For. Meteorol. 138*: 82–92.

Tao, Y., Chen, Q., Peng, S., Wang, W., & Nie, L. (2016). Lower global warming potential and higher yield of wet direct-seeded rice in Central China. *Agron. Sustain. Dev. 36*: 1–9.

Tol, R.S.J. (2012). On the uncertainty about the total economic impact of climate change. *Environ. Res. Econ. 53*: 97–116.

Tol, R.S.J. (2013). The economic impact of climate change in the 20th and 21st centuries. *Clim. Chang. 117*: 795–808.

Tripathi, R.S., Raju, R., & Thimmappa, K. (2013). Impacts of zero tillage on economics of wheat production in Haryana. *Agric. Econ. Res. Rev. 26*: 101–108.

Uddling, J., Broberg, M.C., Feng, Z., & Pleijel, H. (2018). Crop quality under rising atmospheric CO2 . *Curr. Opin. Plant Biol. 45*: 262–267.

Varanasi, A., Prasad, P.V.V., & Jugulam, M. (2016). Impact of climate change factors on weeds and herbicide efficacy. *Adv. Agr. 135*: 108–138.

Ventakeswarlu, B., & Shanker, A.K. (2009). Climate change and agriculture: adaptation and mitigation strategies. *Indian J. Agril. 54*: 226–230.

Verchot, L.V., Noordwijk, M.V., Kandji, S., ... Tomich, T. (2007). Climate change: linking adaptation and mitigation through agroforestry. *Mitig. Adapt. Strat. Glob. Chang. 12*: 901–918.

Vuuren, D.P.V., Issac, M., Kundzewicz, Z.W., ... Arnell, N. (2010). The use of scenarios as the basis for combined assessment of climate change mitigation and mitigation. *Glob. Environ. Chang. 21*: 575–591.

Wagan, S.A., Memon, Q.U.A., Wagan, T.A., Memon, I.H., & Wagan, A. (2015). Economic analysis of laser land leveling technology water use efficiency and crop productivity of wheat crop in Sindh, Pakistan. *J. Environ. Earth Sci. 5*: 21–25.

Waha, K., Muller, C., Bondeau, A., ... Dietrich, J.P. (2013). Adaptation to climate change through the choice of cropping system and sowing date in sub-Saharan Africa. *Glob. Environ. Chang. 23*: 130–143.

Wei, T., Yang, S., Moore, J.C., & Shi, P. (2012). Developed and developing world responsibilities for historical climate change and CO2 mitigation. *Proc. Natl. Acad. Sci. USA. 109*: 12911–12915.

World Meteorological Organization (WMO). (1992). International Meteorological Vocabulary, 2nd ed, *WMO: Geneva, Switzerland.*

World Population Review. Walnut, United States. (2022). www.world populationreview.com

Zaveri, E., Russ, J., & Damania, R. (2020). Rainfall anomalies are a significant driver of cropland expansion. *Proc. Natl. Acad. Sci. USA. 117*: 10225–10233.

Zhang, P., Zhang, J., & Chen, M. (2017). Economic impacts of climate change on agriculture: the importance of additional climatic variables other than temperature and precipitation. *J. Environ. Econ. Manag. 83*: 8–31.

Zhang, Y.G., Nearing, M.A., Zhang, X.C., Xie, Y., & Wei, H. (2010). Projected rainfall erosivity changes under climate change from multi-model and multi-scenario projections in Northeast China. *J. Hydrol. 384*: 97–106.

Zhao, C., Liu, B., Piao, S., ... Wang, X. (2017). Temperature increase reduces global yields of major crops in four independent estimates. *Proc. Natl. Acad. Sci. USA. 114*: 9326–9331.

Zilberman, D., Liu, X., Holst, D.R., & Sunding, D. (2004). The economics of climate change in agriculture. *Mitig. Adapt. Strateg. Glob. Chang. 9*: 365–382.

Zou, X., Li, Y., Cremades, R., Gao, Q., Wan, Y., & Qin, X. (2013). Cost-effectiveness analysis of water-saving irrigation technologies based on climate change response: a case study of China. Agric. *Water Manag. 129*: 9–20.

Zougmore, R., Jalloh, A., & Tioro, A. (2014). Climate-smart soil water and nutrient management options in semiarid West Africa: A review of evidence and analysis of stone bunds and zai techniques. *Agric. Food Sec. 3*: 1–8.

Chapter-2

Resequencing of Plant Germplasm

Abstract

Global climate change has threatened world crop production and food security. Climate change exacerbates the stress in plants due to their sessile nature, which constrains worldwide agricultural production and threatens global food security. Decoding the adaptive genetic basis of wild relatives provides an invaluable genomic resource for climate-smart crop breeding. It is an efficient strategy to integrate adaptive genes into crop cultivars to enhance the productivity of crops. Compared with the cultivated plants, their wild relatives harbour higher genetic diversity and show stronger environmental adaptability. However, the genetic basis of adaptability, as well as how the environments have shaped genomic diversity, has been less understood. Mining the genes associated with environmental adaptation from the wild relatives of crops is critical for understanding the adaptation mechanism and climate-smart cultivar development for feeding the growing global population. With the rapid development of modern molecular biology and high-throughput sequencing technologies, whole-genome resequencing and genome-wide association studies (GWAS) have become common methods to study population genetic diversity and locate phenotype-related quantitative trait loci (QTLs) or genes. To exploit the full potential of germplasms available in the genebanks worldwide, it has become essential to resequence them to identify novel genes having the potential to meet the challenges posed due to climate change.

Keywords: *Resequencing, germplasm, wild relatives, genomic diversity, genebank*

Introduction

Crop genebanks are an essential part of the worldwide effort to conserve, characterize, and use biological diversity to improve agricultural productivity, sustainability, and crop nutritional value (McCouch et al., 2013; McCouch, McNally, Wan, and Hamilton, 2012). Approximately 7.4 million accessions are stored in 1700 seed banks globally, offering breeders a wealth of natural variation. Genebank passport data coupled with climatic data could be used as surrogates for abiotic stresses to identify genotypes that harbour important haplotypes which can then be integrated into breeding programs. To enhance the genetic gains through the incorporation of new sources of genetic variation, novel, and superior genes need to be identified from crop wild relatives (CWR) and landraces available in genebanks (Bohra et al. 2022). There are many examples where NGS-based approaches are being successfully used to identify DNA polymorphisms associated with traits of interest. For instance, the 3000 rice genome (Lv et al. 2020) and the 3000 chickpea genome sequencing Initiative (Thudi et al. 2016) offer opportunities to identify novel variations for a large number of genes through genotype-phenotypic associations.

Re-sequencing (RS) of a large number of germplasm accessions not only provides information on the origin, domestication, and population structure, but also identifies lines with deleterious mutations in the genomes that can be eliminated to minimize the genetic load in the crop species (Xu et al. 2011, Huang et al. 2015). NGS technologies together with precise phenotyping have been used for the identification of marker-trait associations in several crops. For example, rice (Lv et al. 2020), soybean (Yang et al. 2022), pigeonpea (Varshney et al. 2017), tomato (Tripodi et al. 2022), sesame (Segla et al. 2023), and rapeseed (Wu et. al. 2019). Insights gained from these studies include new information on the genetic architecture of agriculturally important traits and the identification of valuable and sometimes novel alleles/haplotypes for morphological, agronomic, developmental, and quality-related traits for enhancing genetic gains. In the context of developing climate change-ready crops, it is essential to mine and incorporate superior alleles adapted to harsh and extreme environmental conditions by resequencing and phenotyping many germplasm accessions for the crop species.

More than 70 years after the first ex-situ genebanks were established, major efforts in this field are still concerned with issues related to further completion of individual collections and securing their storage. Attempts regarding the valorization of ex-situ collections for plant breeders have been hampered by the limited availability of phenotypic and genotypic information. With the advent of molecular marker technologies, first efforts were made to fingerprint genebank accessions, albeit on a very small scale and mostly based on inadequate DNA marker systems. Advances in DNA sequencing technology and the development of high-throughput systems for multi-parallel interrogation of thousands of single nucleotide polymorphisms (SNPs) now provide a suite of technological platforms facilitating the analysis of several hundred Gigabases per day using state-of-the-art sequencing technology or, at the same time, of thousands of SNPs. Deployment of these technologies should help in the analysis of plant genetic resources, identify patterns of genetic diversity, map quantitative traits, and mine novel alleles from the vast number of genetic resources maintained in genebanks around the world (Kilian and Garner 2012).

During the past decades, remarkable advancements have been made in the development of next-generation sequencing technologies, making possible a deepened knowledge of the various plant genomes. The decline in costs and the solutions offered by diverse NGS platforms paved the way for the exploration of the diversity enclosed in the germplasm collection. The research community is benefiting from the knowledge to understand the role of genes underlying major traits of agricultural interest and for genomic-assisted breeding.

Resequencing of Wild and Cultivated Plant Germplasm

Wild ancestors and traditional landraces are reservoirs of valuable traits, including diverse forms of resistance to both biotic and abiotic stresses, which remain crucial for the adaptation of modern cultivars to current and future climates. Crop wild relatives (CWRs) have been used for decades in crop improvement to enhance plant performance.

The incorporation of genes from CWRs into the cultivated gene pool is not straightforward. Breeders are reluctant to use CWRs in commercial breeding programs owing to the challenges presented by crossability barriers, linkage drag, poor agronomic performance, and a wide range of phenotyping challenges. Advances in plant molecular biology have introduced new opportunities for overcoming several of these problems. For instance, genes can now be 'edited' in situ such that long-lost ancestral alleles known to confer desirable traits or phenotypes can be reintroduced into modern, elite cultivars without disrupting the constellation of genes in the genetic background that confer valuable characteristics essential to productivity in modern agricultural systems.

Advances in sequencing technologies and analysis tools have facilitated the generation of high-quality reference genomes in many crops. These reference genome sequences have, in turn, catalyzed numerous downstream applications such as the development of genome-wide maps of genetic variants, the identification of gene–trait associations, and the cloning and characterization of genes of interest. During recent years, the cost of sequencing has decreased and throughput and accuracy have increased and the idea that a single reference genome was sufficient as a template to describe the full landscape of genetic diversity within a species was soon debunked. The growing amount of high-quality genome sequencing data for diverse accessions demonstrated that reference genome-based methods were insufficient to detect critical structural variations (SVs), including presence–absence variation (PAV) and copy-number variation (CNV), that distinguish one accession from another. One implication of importance to breeders is that genomic segments introgressed from wild or exotic donors, may actually carry genes that are not present in either the reference genome or the elite recurrent parent, or, conversely, exotic donors may fail to carry the genes that were targeted for introgression based on a reference genome.

Whole genome re-sequencing (WGRS) involves sequencing the entire genome of a plant and comparing the sequence to that of a known reference genome. Re-sequencing of the plant genome will identify genetic variations such as SNPs and InDels, and is often applied for the identification of functional genes and markers of important traits to facilitate molecular breeding and to improve agricultural production and conservation. Recent reports on resequencing so important plant germplasms are presented below.

RS of Soybean Germplasm

Several reports have been published on the domestication and improvement of soybean at the genome-wide level. These include genes and genetic networks related to agronomic traits and functions of soybean (Carter et al. 2004, Doebley et al. 2006, Hylen et al. 2006). However, our knowledge of the soybean genome and its functional genes is still limited compared to rice and maize (Hufford et al. 2012, Jiao et al. 2012, Qi et al. 2013), which is attributed to the diversity of soybean varieties and their complex genetic backgrounds. Therefore, a large number of soybean varieties need to be further explored at the genomic level, particularly in relation to molecular traits associated with edible quality, ideal plant architecture, and the underlying genetic network of high-yielding varieties. Understanding soybean (*Glycine max*) domestication and improvement at a genetic level is important to inform future efforts to further improve a crop that provides the world's main source of oilseed.

Zhou et al. (2015) detected 230 selective sweeps and 162 selected copy number variants by analysis of 302 resequenced wild, landrace, and improved soybean accessions at >11× depth. A genome-wide association study using these new sequences reveals associations between 10 selected regions and 9 domestication or improvement traits, and identifies 13 previously uncharacterized loci for agronomic traits including oil content, plant height, and pubescence form. Combined with previous quantitative trait loci (QTL) information, they reported that, of the 230 selected regions, 96 correlate with reported oil QTLs, and 21 contain fatty acid biosynthesis genes. Moreover, they observed that some traits and loci are associated with geographical regions, which indicates that soybean populations are structured geographically. These results are expected to provide resources for genomics-enabled improvements in soybean breeding.

To investigate the genetics of the Brazilian soybean germplasm, Vitor et al. (2016) selected soybean cultivars based on the year of commercialization, geographical region, and maturity group and resequenced their genomes. They resequenced the genomes of 28 Brazilian soybean cultivars with an average genome coverage of 14.8X. A total of 5,835,185 single nucleotide polymorphisms (SNPs) and 1,329,844 InDels were identified across the 20 soybean chromosomes, with 541,762

SNPs, 98,922 InDels and 1,093 CNVs that were exclusive to the 28 Brazilian cultivars. In addition, 668 allelic variations of 327 genes were shared among all the Brazilian cultivars, including genes related to DNA-dependent transcription-elongation, photosynthesis, ATP synthesis-coupled electron transport, cellular respiration, and precursors of metabolite generation and energy. A very homogeneous structure was also observed for the Brazilian soybean germplasm. They observed 41 regions putatively influenced by positive selection. They also detected 3,880 regions with copy-number variations (CNVs) that could help to explain the divergence among the accessions evaluated. The large number of allelic and structural variations identified in this study can be used in marker-assisted selection programs to detect unique SNPs for cultivar fingerprinting.

Yang et al. (2022) resequenced the genomes of 250 soybean landraces and cultivars from China, America, and Europe, and investigated their population structure, genetic diversity and architecture, and the selective sweep regions of these accessions. Five novel agronomically important genes were identified, and the effects of functional mutations in respective genes were examined. The candidate genes GSTT1, GL3, and GSTL3 associated with the isoflavone content, CKX3 associated with yield traits, and CYP85A2 associated with both architecture and yield traits were found. The phenotype–gene network analysis revealed that hub nodes play a crucial role in complex phenotypic associations (Yang et al. 2022). Results thus obtained describe novel agronomic trait-associated genes and a complex genetic network, providing a valuable resource for future soybean molecular breeding.

RS of Pigeonpea Germplasm

Pigeonpea (*Cajanus cajan*), a tropical grain legume with low input requirements, is expected to continue to have an important role in supplying food and nutritional security in developing countries in Asia, Africa, and the tropical Americas. Varshney et al. (2017) carried out whole genome resequencing of 292 Cajanus accessions encompassing breeding lines, landraces and wild species, and characterized genome-wide variation. Based on a scan for selective sweeps, they found several genomic regions that were likely targets of domestication and breeding. Using genome-wide association analysis, they identified associations between several candidate genes and agronomically important traits. Candidate genes for these traits in pigeonpea have sequence similarity to genes functionally characterized in other plants for flowering time control, seed development and pod dehiscence (Varshney et al. 2017). These findings are expected to accelerate genetic gains for key traits to improve yield and sustainability in pigeonpea.

RS of Chickpea Germplasm

Chickpea (Cicer arietinum L.) is the second most important grain legume cultivated by resource poor farmers in South Asia and Sub-Saharan Africa. In order to harness the untapped genetic potential available for chickpea improvement, Thudi et al. (2016) re-sequenced 35 chickpea genotypes representing parental lines of 16 mapping populations segregating for abiotic (drought, heat, salinity), biotic stresses (Fusarium wilt, Ascochyta blight, Botrytis grey mould, *Helicoverpa armigera*) and nutritionally important (protein content) traits using whole genome re-sequencing approach. They generated a total of 192.19 Gb data on 35 genotypes of chickpea, comprising 973.13 million reads, with an average sequencing depth of ~10 X for each line. On average 92.18 % reads from each genotype were aligned to the chickpea reference genome with 82.17 % coverage. A total of 2,058,566 unique single nucleotide polymorphisms (SNPs) and 292,588 Indels were detected while comparing with the reference chickpea genome. The highest number of SNPs were identified on the Ca4 pseudomolecule. In addition, copy number variations (CNVs) such as gene deletions and duplications were identified across the chickpea parental genotypes, which were minimum in PI 489777 (1 gene deletion) and maximum in JG 74 (1,497). A total of 164,856 line-specific variations (144,888 SNPs and 19,968 Indels) with the highest percentage were identified in coding regions in ICC 1496 (21 %) followed by ICCV 97105 (12 %). Of 539 miscellaneous variations, 339, 138, and 62 were inter-chromosomal variations (CTX), intra-chromosomal variations (ITX), and inversions (INV) respectively (Thudi et al. 2016).

Genome-wide SNPs, Indels, CNVs, PAVs, and miscellaneous variations identified in different mapping populations are a valuable resource in genetic research and helpful in locating genes/genomic segments responsible for economically important traits in cheakpea. Further, the genome-wide variations identified by Thudi et al. (2016) can be used for developing high-density SNP arrays for genetics and breeding applications.

RS of Rapeseed Germplasm

Rapeseed (*Brassica napus*), an important oilseed crop, has adapted to diverse climate zones and latitudes by forming three main ecotype groups, namely winter, semi-winter, and spring types. However, genetic variations underlying the divergence of these ecotypes are largely unknown. Wu et al. (2019) reported the global pattern of genetic polymorphisms in rapeseed determined by resequencing a worldwide collection of 991 germplasm accessions. A total of 5.56 and 5.53 million single nucleotide polymorphisms (SNPs) as well as 1.86 and 1.92 million InDels were identified by mapping reads to the reference genomes of "Darmor-bzh" and "Tapidor," respectively. They generated a map of allelic drift paths that shows splits and mixtures of the main populations, and revealed an asymmetric evolution of the two sub genomes of *B. napus* by calculating the genetic diversity and linkage disequilibrium parameters. Selective-sweep analysis revealed genetic changes in genes orthologous to those regulating various aspects of plant development and response to stresses. A genome-wide association study identified SNPs in the promoter regions of FLOWERING LOCUS T and FLOWERING LOCUS C orthologs that corresponded to the different rapeseed ecotype groups. This study provided important insights into the genomic footprints of rapeseed evolution and flowering-time divergence among three ecotype groups, which is expected to facilitate screening of molecular markers for accelerating rapeseed breeding (Wu et al. 2019).

RS of Maize Germplasm

Grain drying rate (FDR) after physiological maturity in maize (*Zea mays* L.) is an essential determinant of grain moisture at harvest. Investigating the genetic structure of FDR is of great significance to breeding maize hybrids with low grain moisture at harvest that are suitable for mechanical harvesting. In this study, 80 elite maize inbred lines from Jilin Province (China) were selected as an association population, and their FDR after physiological maturity was determined by the oven-drying method in 2014 and 2015. At the same time, whole-genome resequencing was performed for the association population, and 1,490,007 high-quality single nucleotide polymorphisms (SNPs) were obtained for a subsequent genome-wide association study. The results showed high genetic variation for FDR in different maize inbred lines, with a broad-sense heritability of 0.76, indicating that the trait is mainly controlled by genetic factors. Thirteen independent genomic positions, which contain 19 SNP markers significantly associated with FDR in maize ($P < 0.000001$), were detected by a genome-wide association study in 2014 and 2015. Seven genomic positions were located within the quantitative trait loci previously mapped for FDR, whereas one position was detected in both years by association analysis. Six candidate genes located in chromosome bins 4.05, 5.04 and 7.02 were identified within the 13 genomic positions. These genes were predicted to encode vegetative cell wall gp1-like glycoprotein, phosphopyruvate hydratase, mitochondrial inner membrane protease subunit 1-like protein, Cucumber mosaic virus 1a interacting protein 1 and two hypothetical proteins, which may be closely associated with FDR. The results lay a foundation for molecular marker-assisted selection of maize materials with rapid grain drying and for the cloning of relevant functional genes.

RS of Cotton Germplasm

Upland cotton (*Gossypium hirsutum*) is the world's largest source of natural fibre and dominates the global textile industry. Hybrid cotton varieties exhibit strong heterosis that confers high fibre yields, yet the genome-wide effects of artificial selection that have influenced Upland cotton during its breeding history are poorly understood. Ma et al. (2019) resequenced Upland cotton genomes and constructed a variation map of an intact breeding pedigree comprising seven elite and 19 backbone parents. Compared to wild accessions, the 26 pedigree accessions underwent strong artificial selection during domestication that has resulted in reduced genetic diversity but stronger linkage disequilibrium and higher extents of selective sweeps. In contrast to the backbone parents, the elite parents have acquired significantly improved agronomic traits, with an especially pronounced increase in the lint percentage. Notably, identify by descent (IBD) tracking revealed that the elite parents inherited abundant beneficial trait segments and loci from the backbone parents. The combined analyses led to the identification of a core genomic segment that was inherited in the elite lines from the parents Zhong 7263 and Ejing 1 and that was strongly associated with lint percentage. Additionally,

SNP correlation analysis of this core segment showed that a non-synonymous SNP (A-to-G) site in a gene encoding the cell wall-associated receptor-like kinase 3 (GhWAKL3) protein was highly correlated with increased lint percentage (Ma et al. 2019). These findings is expected to substantially increase the valuable genomics resources available for future genetic and functional genomics studies of cotton and reveal insights that will facilitate yield increases in the molecular breeding of cotton.

RS of Rice Germplasm

Rice accounts for 20% of the world's dietary energy supply (FAO, 2004) and because of this, there is immense global interest in using genetic resources for crop improvement. In fact, there are over 773,948 rice germplasm accessions housed in some 179 world genebanks (FAO, 2010) including the USDA–ARS NPGS.

Through genomic technology, it is now possible to further characterize rice germplasm according to species, subspecies, and subpopulation. Genotyping with simple sequence repeat (SSR) markers confirmed *O. glaberrima* was a distinct species from *O. sativa*, although they share the AA genome (Semon et al. 2005). Also, the two major subspecies of *O. sativa*, Indica, and Japonica, which have historically been differentiated based on morphology and crossability (Kato et al. 1930), can now be further subdivided into the indica (IND) and aus (AUS) subpopulations within Indica and into the aromatic (ARO), temperate japonica (TEJ), and tropical japonica (TRJ) subpopulations within Japonica (Garris et al. 2003: Garris et al. 2005; McCouch et al., 2002). Although not currently available in GRIN-Global, having access to accurate species, subspecies, and subpopulation classification, all of which have been presumed based on the country of collection or pedigree, offers a new means of increasing the relevance of the rice NSGC for cultivar development since most breeding programs focus on germplasm within certain subpopulations and ecological adaptation zones (McClung et al. 2020).

Genotyping of diversity panels (Agrama et al. 2009, Li et al. 2010, Bryant et al. 2011, Zhao et al. 2011, Eizenga et al. 2014, McCouch et al. 2016; Wang et al. 2016; Wang et al. 2018a, Wang et al. 2018b) and mapping in various biparental populations led to the development of TSMs important to rice cultivar development. Some of the markers relevant to breeding programmes include five rice blast (*Magnaporthe oryzae* B. Couch) resistance genes (Fjellstrom et al. 2004, 2006, Wang et al. 2010), pubescence (Li et al. 2010, Eizenga et al. 2019), apparent amylose content (Ayres et al., 1997, Chen et al. 2008), starch pasting properties (Chen et al. 2010), starch gelatinization temperature (Bao et al. 2006b; Huggins et al., 2019), the fragrance (frg) gene associated with 2-acetyl-1-pyrroline (2-AP), the characteristic compound found in aromatic rice (Fitzgerald et al. 2008, Kovach et al. 2009), red pericarp colour as determined by genes involved in proanthocyanidin synthesis (Sweeney et al. 2006, Furukawa et al., 2007, Wu et al. 2019), and an induced mutation for herbicide resistance (Kadaru et al. 2008). Although molecular markers are available for these economically important traits, they have not been used to systematically characterize the rice NSGC prior to this study.

Characterizing the rice NSGC through genotyping and phenotyping will result in a more searchable and efficient resource for breeding and will create new opportunities to mine the collection for novel alleles and germplasm needed to broaden breeding gene pools. To demonstrate this, approximately 2,000 rice NSGC *O. sativa* accessions with limited phenotypic data and no genotypic data were selected to (a) assess species, subspecies, and subpopulation structure using a small panel of markers; (b) determine the efficacy of using selected molecular markers to reveal desired alleles for traits of economic importance in cultivar development; (c) improve the efficiency and integrity of collection management by identifying redundant or mislabelled accessions and developing a rigorous protocol for addressing germplasm mixtures; and (d) determine the utility of these molecular markers to distinguish among commercialized U.S. cultivars derived from a narrow TRJ gene pool.

Obtaining genetic variation information from indica rice hybrid parents and identification of loci associated with heterosis are important for hybrid rice breeding. Lv et al. (2020) resequenced 1,143 indica accessions mostly selected from the parents of superior hybrid rice cultivars of China, to identify genetic variations, and perform kinship analysis. They found different hybrid rice crossing patterns between 3- and 2-line superior hybrid lines. By calculating frequencies of parental variation differences (FPVDs), a more direct approach for studying rice heterosis, they could identify loci that are linked to heterosis, which include 98 in superior 3-line hybrids and 36 in superior 2-line hybrids. They found two accessions harbouring a deletion in OsNramp5, a previously reported gene functioning in cadmium absorption, which can be used to mitigate rice grain cadmium levels through hybrid breeding. Resource of indica rice genetic variation is expected to be valuable to geneticists and breeders involved in rice breeding programmes.

RS of Watermelon Germplasm

Fruit characteristics of sweet watermelon are largely the result of human selection. Guo et al. (2019) reported an improved watermelon reference genome and whole-genome resequencing of 414 accessions representing all extant species in the *Citrullus* genus. Population genomic analyses reveal the evolutionary history of *Citrullus*, suggesting independent evolutions in Citrullus amarus and the lineage containing *Citrullus lanatus* and *Citrullus mucosospermus*. Our findings indicate that different loci affecting watermelon fruit size have been under selection during speciation, domestication, and improvement. A non-bitter allele, arising in the progenitor of sweet watermelon, is largely fixed in *C. lanatus*. Selection for flesh sweetness started in the progenitor of *C. lanatus* and continues through modern breeding on loci controlling raffinose catabolism and sugar transport. Fruit flesh coloration and sugar accumulation might have co-evolved through shared genetic components including a sugar transporter gene. This study provides valuable genomic resources and sheds light on watermelon speciation and breeding history (Guo et al. 2019).

RS of Flex Germplasm

Seed size and weight are key traits determining crop yield, which often undergo strongly artificial selection during crop domestication. Although seed sizes differ significantly between oil flax and fiber flax, the genetic basis of morphological differences and artificial selection characteristics in seed size remains largely unclear. Guo et al (2020) re-sequenced 200 flax, cultivated accessions to generate a genome variation map based on chromosome assembly reference genomes. They provided evidence that the oil flax group is the ancestor of cultivated flax, and the oil-fiber dual purpose group (OF) is the evolutionary intermediate transition state between oil and fiber flax. Genome-wide association studies (GWAS) were combined with LD Heatmap to identify candidate regions related to seed size and weight, thereafter candidate genes were screened based on detailed functional annotations and estimated the nucleotide polymorphism effects. Using this strategy, they obtained 13 candidate genes related to seed size and weight. Selective sweeps analysis indicated the human-involved selection of small seeds during the oil-to-fiber flax transition. They showed the existence of elite alleles for seed size and weight in flax germplasm and provided molecular insights into approaches for further improvement.

RS of *Rosa* Germplasm

Rosa rugosa is a shrub that originated in China and has economic and ecological value. During the development of *R. rugosa*, the genetic background was chaotic, and the genetic structure among different wild populations was unclear, as well as wild and cultivated accessions. Zang et al (2023) report whole-genome resequencing of wild and cultivated *R. rugosa* accessions. A total of 19,041,284 SNPs were identified in 188 *R. rugosa* accessions and 3 *R. chinensis* accessions by resequencing. Population genetic analysis revealed that cultivated and wild groups were separated very early. All *R. rugosa* accessions were divided into 8 categories based on genetic structure: (i) Weihai, Yantai, and Liaoning category, (ii) Jilin category, and (iii) Hammonasset category (above three are wild); (iv) traditional varieties, (v) hybrids between *R. rugosa* and *R. chinensis*, (vi) Zizhi Rose, (vii) Kushui Rose, (viii) hybrids between *R. rugosa* and *R. multiflora*. They found that the heterozygosity and genetic diversity of wild accessions were generally lower than those of cultivated individuals. The genes that were selected during cultivation were identified, and it was found that these genes were mainly related to environmental adaptation and growth.

Based on the above findings, Zang et al (2023) concluded that the Jilin population was the oldest population and later migrated to Liaoning and then migrated to Yantai and Weihai by sea regression in the Bohai Basin. The Hammonasset naturalized population probably originated from the Jilin population and then experienced separate differentiation. The long-term asexual reproduction pattern of *R. rugosa* decreased genetic diversity in the wild population. During *R. rugosa* cultivation, the ancestors of the Jilin population were involved in breeding traditional varieties, after which almost no wild individuals were engaged in breeding. However, in recent decades, cross breeding of *R. rugosa* started the utilization of wild germplasms. In comparison, some other species play important roles in variety formation. Few genes related to economic traits were selected, suggesting no directional domestication in the *R. rugosa* cultivation process.

RS of Melon Germplasm

The importance of Indian germplasm as the origin and primary center of diversity of cultivated melon is widely accepted. Genetic diversity of several collections from India has been studied previously, although an integrated analysis of these collections from a global diversity perspective has not been possible. Gonzalo et al. (2019) analyzed the genetic diversity structure based on Genotype and Sequence data of a sample of Indian collections of melon together with a selection of worldwide cultivars. They generated a set of 6158 informative Single Nucleotide Polymorphism (SNP) in 175 melon accessions. Gonzalo et al. (2019) classified Melon germplasm into six major groups, in concordance with horticultural groups. Indian group was in the center of the diversity plot, with the highest genetic diversity. No strict genetic differentiation between wild and cultivated accessions was appreciated in this group. Genomic regions likely involved in the process of diversification were also found. Interestingly, some SNPs differentiating inodorus and cantalupensis groups are linked to Quantitative Trait Loci involved in ripening behaviour (a major characteristic that differentiates those groups). Linkage disequilibrium was found to be low (17 kb), with a more rapid decay in euchromatic (8 kb) than heterochromatic (30 kb) regions, demonstrating that recombination events do occur within heterochromatin, although at a lower frequency than in euchromatin. Concomitantly, haplotype blocks were relatively small (59 kb). Some of those haplotype blocks were found fixed in different melon groups, being, therefore, candidate regions that are involved in the diversification of melon cultivars.

Based on the above findings, Gonzalo et al. (2019) confirmed the hypothesis that India is the primary center of diversity of melon, Occidental and Far-East cultivars have been developed by divergent selection. Indian germplasm is genetically distinct from African germplasm, supporting independent domestication events. The current set of traditional Indian accessions may be considered as a population rather than a standard collection of fixed landraces with high intercrossing between cultivated and wild melons.

RS of Spinach Germplasm

Spinach is a nutritious leafy vegetable belonging to the family Chenopodiaceae. Cai et al. (2021) reported a chromosome-scale reference genome assembly of spinach, which has a total size of 894.3 Mb and an N50 contig size of 23.8 Mb, with 98.3% anchored and ordered on the six chromosomes. Reconstruction of ancestral Chenopodiaceae karyotype indicates substantial genome rearrangements in spinach after its divergence from ancestral Chenopodiaceae, coinciding with high repeat content in the spinach genome. Resequencing the genomes of 305 cultivated and wild spinach accessions provides insights into spinach genetic diversity and population differentiation. Genome-wide association studies of 20 agronomical traits identify numerous significantly associated regions and candidate genes for these traits. Domestication sweeps in the spinach genome were identified, some of which were associated with important traits (e.g., leaf phenotype, bolting, and flowering), demonstrating the role of artificial selection in shaping spinach phenotypic evolution. These findings provided not only insights into spinach evolution and domestication but also valuable resources for facilitating spinach breeding (Cai et al. 2021).

RS of Barley Germplasm

Barley (*Hordeum vulgare*) was domesticated ~10,000 years ago in the Fertile Crescent (Pourkheirandish et al., 2015). Domesticated barley diversified during its distribution around the world and its adaptation to new environments (Bothmer et al., 2003), developing varieties with differing levels of nutrients and minerals. Human preferences also affected barley traits such as seed dormancy. The major genetic variants in cultivated barley are preserved as accessions in crop gene banks or maintained as active germplasm in breeding programs (Sato et al., 2014), allowing us to access the grain quality profiles that are preferred for a variety of uses, including malting for production of alcoholic beverages and direct consumption as grain. Recent progress in genome analysis and genotyping tools has enabled plant breeders to measure variation in grain quality–related traits and the associated genetic factors (Sato, 2020).

Grain quality is a major focus in barley research. Robustness is important; physical, chemical, or biological damage degrades grain quality. If a seed is degraded to the point where it cannot germinate, it cannot be used as malting barley. Different quality profiles are required for each category of use; for example, a high beta-glucan content is required for food barley (Garcia-Gimenez et al., 2019) but a low beta-glucan level is better suited to malting barley.

Some mineral elements in barley grain are essential for human health; however, others, such as the heavy metal cadmium (Cd), are toxic to humans and animals. Due to its extensive industrial use (for example, in batteries), many farmlands are contaminated with Cd. Cd contamination of food causes cancer, kidney failure, osteoporosis, and osteomalacia in humans (Bertin and Averbeck, 2006; Nawrot et al., 2006). In addition, a mass Cd poisoning in Japan in the 20th century led to many people developing itai-itai disease, which causes severe pain due to softening of the bones and kidney failure (Horiguchi et al., 1994). Cd levels in the plow layer of agricultural soils increase 0.03–0.05% per year in Sweden (Olsson et al., 2005). Feed from barley grain and straw also contributes substantial amounts of Cd into local agricultural systems. Therefore, it is important to reduce the transfer of Cd from the soil into barley grains. Strategies to reduce Cd include soil amendments, water management, and phytoremediation (Ma et al., 2021), but the most effective method to prevent the contamination of food by this metal is to develop cultivars with low Cd accumulation rates (Lei et al., 2020).

According to Knüpffer (2009), there are ~485,000 accessions of Hordeum in over 200 different gene banks around the world. Since the barley accessions are too numerous to analyze in their entirety, researchers developed the international Barley Core Collection (BCC) consisting of ~1500 accessions representing the greatest genetic diversity possible while still remaining manageable for phenotyping. In addition to the BCC, other sets of diverse cultivated core collections are available in several gene banks. For example, the United States Department of Agriculture established the National Small Grains Collection Barley Core, consisting of 2298 accessions passing the quality control for single-nucleotide polymorphism (SNP) calls (Muñoz-Amatriaín et al., 2014). At the Leibniz Institute of Plant Genetics and Crop Plant Research (IPK) in Germany, Pasam et al. (2014) established a diverse panel of 1485 spring landraces. Since the core collections contain too many accessions for a detailed analysis, smaller sets of accessions are also used, such as the 274 landrace accessions from diverse collection regions maintained at Okayama University, Japan (Pourkheirandish et al., 2015; Nakamura et al., 2016; Sato et al., 2020).

Evaluating genetic variation in barley (*Hordeum vulgare*) germplasm, combined with genome-wide genotyping, is vital for identifying genes controlling important grain-quality traits. For example, in addition to traditional grain quality properties such as starch and protein contents, grain safety parameters such as heavy metal content, are important in the use of barley for human food and animal feed. Several genes affecting grain quality have been identified by map-based cloning strategies and functionally analyzed by genetic transformation experiments. Moreover, germplasm evaluation yielded information that enabled the introgression of a key gene controlling grain cadmium accumulation into an elite barley cultivar, reducing the content of this heavy metal in grain. Genotyping of molecular markers and resequencing of germplasm accessions may provide information about how grain quality–related loci evolved and how the current allelic diversity was established. Sato et al. (2021) described germplasm resources for barley grain quality–related traits and the methods used to analyze the functions of genes controlling these traits, illustrating cadmium accumulation as an example. They also provided directions for the efficient identification of grain quality–related genes.

RS of Sesame Germplasm

Sesame is a promising oilseed crop that produces specific lignans of clinical importance. Hence, a molecular description of the regulatory mechanisms of lignan biosynthesis is essential for crop improvement. Seqla et al. (2023) resequenced 410 sesame accessions and identify 5.38 and 1.16 million SNPs (single nucleotide polymorphisms) and InDels, respectively. Population genomic analyses reveal that sesame has evolved a geographic pattern categorized into northern (NC), middle (MC), and southern (SC) groups, with potential origin in the southern region and subsequent introduction to the other regions. Selective sweeps analysis uncovers 120 and 75 significant selected genomic regions in MC and NC groups, respectively. By screening these genomic regions, they unveiled 184 common genes positively selected in these subpopulations for exploitation in sesame improvement. Genome-wide association study identifies 17 and 72 SNP loci for sesamin and sesamolin variation, respectively, and 11 candidate causative genes. The major pleiotropic SNPC/A locus for lignans variation is located in the exon of the gene SiNST1. Further analyses revealed that this locus was positively selected in higher lignan content sesame accessions, and the "C" allele is favorable for a higher accumulation of lignans. Overexpression of SiNST1C in sesame hairy roots significantly up-regulated the expression of SiMYB58, SiMYB209, SiMYB134, SiMYB276, and most of the monolignol biosynthetic genes. Consequently, the lignans content was significantly increased, and the lignin content was slightly increased (Seqla et al. (2023). The results thus obtained provide insights into lignans and lignin regulation in sesame and will facilitate molecular breeding of elite varieties and marker-traits association studies.

RS of Buckwheat Germplasm

Background Tartary buckwheat (*Fagopyrum tataricum*) is a nutritionally balanced and flavonoid-rich crop plant that has been in cultivation for 4000 years and is now grown globally. Despite its nutraceutical and agricultural value, the characterization of its genetics and its domestication history is limited. Zhang et al. (2021) reported a comprehensive database of Tartary buckwheat genomic variation based on whole-genome resequencing of 510 germplasms. They suggested that two independent domestication events occurred in southwestern and northern China, resulting in diverse characteristics of modern Tartary buckwheat varieties. Genome-wide association studies for important agricultural traits identify several candidate genes, including FtUFGT3 and FtAP2YT1 that significantly correlate with flavonoid accumulation and grain weight, respectively. They described the domestication history of Tartary buckwheat and provided a detailed resource of genomic variation to allow for genomic-assisted breeding in the improvement of elite cultivars.

RS of *Cannabis* Germplasm

Cannabis sativa has long been an important source of fiber extracted from hemp and both medicinal and recreational drugs based on cannabinoid compounds. Here, we investigated its poorly known domestication history using whole-genome resequencing of 110 accessions from worldwide origins. Ren et al. (2021) reported that *C. sativa* was first domesticated in early Neolithic times in East Asia and that all current hemp and drug cultivars diverged from an ancestral gene pool currently represented by feral plants and landraces in China. They identified candidate genes associated with traits differentiating hemp and drug cultivars, including branching patterns and cellulose/lignin biosynthesis. They also found evidence for the loss of function of genes involved in the synthesis of the two major biochemically competing cannabinoids during selection for increased fiber production or psychoactive properties. The results provided a unique global view of the domestication of *C. sativa* and offered valuable genomic resources for ongoing functional and molecular breeding research (Ren et al. 2021).

RS of Eggplant Germplasm

Whole-genome resequencing provides information of great relevance for crop genetics, evolution, and breeding. Gramazio et al (2019) reported the first whole-genome resequencing study using seven eggplant (*Solanum melongena*) and one wild relative (*Solanum incanum*) accessions. These eight accessions were selected for displaying a high phenotypic and genetic diversity and for being the founder parents of an eggplant multiparent advanced generation intercrosses population. By resequencing at an average depth of 19.8× and comparing to the high-quality reference genome "67/3", over 10 million high-reliable polymorphisms were discovered, of which over 9 million (84.5%) were single nucleotide polymorphisms and more than 700,000 (6.5%) InDels. However, while for the *S. melongena* accessions, the variants identified ranged from 0.8 to 1.3 million, over 9 million were detected for the wild *S. incanum*. This confirms the narrow genetic diversity of the domesticated eggplant and suggests that introgression breeding using wild relatives can efficiently contribute to broadening the genetic basis of this crop. Differences were observed among accessions for the enrichment of genes regulating important biological processes. By analyzing the distribution of the variants, they identified the potential footprints of old introgressions from wild relatives that can help to unravel the unclear domestication and breeding history. The comprehensive annotation of these eight genomes represents a landmark in eggplant genomics and allows the development of tools for eggplant genetics and breeding (Gramazio et al. 2019).

RS of Cassava Germplasm

Heterozygous genomes are widespread in outcrossing and clonally propagated crops. However, the variation in heterozygosity underlying key agronomic traits and crop domestication remains largely unknown. Cassava is a highly heterozygous genome, and a staple crop in Africa and other tropical regions.

Hu et al. (2021) reported a genomic variation map from 388 resequenced genomes of cassava cultivars and wild accessions. They identify 52 loci for 23 agronomic traits through a genome-wide association study. Eighteen allelic variations in heterozygosity for nine candidate genes are significantly associated with seven key agronomic traits. They further detect 81 selective sweeps with decreasing heterozygosity and nucleotide diversity, harboring 548 genes, which are enriched in multiple biological processes including growth, development, hormone metabolisms and responses, and immune-related processes. Artificial selection for decreased heterozygosity has contributed to the domestication of the large starchy storage root of cassava. Selection for homozygous GG allele in MeTIR1 during domestication contributes to increased starch content. Selection of homozygous AA allele in MeAHL17 is associated with increased storage root weight and cassava bacterial blight (CBB) susceptibility. Hu et al. (2021) verified the positive roles of MeTIR1 in increasing starch content and MeAHL17 in resistance to CBB by transient overexpression and silencing analysis. The allelic combinations in MeTIR1 and MeAHL17 may result in high starch content and resistance to CBB. These findings provide insights into allelic variation in heterozygosity associated with key agronomic traits and cassava domestication. It also offers valuable resources for the improvement of cassava and other highly heterozygous crops.

RS of Tomato Germplasm

The first reference genome was released by the International Tomato Genome Consortium in 2012. The full sequence of S. lycopersicum cultivar Heinz 1706 and the draft of S. pimpinellifolium were presented, combining both first and second-next-generation sequencing technologies. The cultivated tomato genome was assembled in 91 scaffolds and covered 760 MB of the estimated 900 MB for a total of 34,727 predicted protein-coding genes. Comparative analysis with the wild ancestor and other members of the family Solanaceae, identified two triplication events that reshaped the functionality of genes underlying key fruit traits. Since then, efforts have been pushed forward genome resequencing projects. The first attempt was performed by Causse et al. (2013) on 4 S. lycopersicum and 4 S. l. cerasiforme genotypes. The Illumina GAIIx platform generated 970 million paired-end reads mostly of 101 bp for an average sequence depth of coverage of 11.2×. A total of over 4 million unique SNPs, ⍰128 thousand InDels and 1,686 putative copy-number variations were detected, thus providing novel genomic variation to use for QTL and gene mapping. Thereafter, the 100 Tomato Genome Sequencing Consortium resequenced the genome of 84 cultivated and wild tomato accessions and produced de novo assemblies for the reference Heinz 1706 and for three wild parents of crossing populations (S. habrochaites LYC4, S. pennellii LA0716, and S. arcanum LA2157) by combining Illumina HiSeq 2000 and 454 FLX technology, thus opening new perspectives for genetic diversity, phylogenetics, and hybridization breeding (Aflitos et al. 2014). A mean genome coverage of 36-fold per accession was found, with several SNPs ranging from 0.5 to 4 million in the cultivated accessions up to 10–11 million in wild species, indicating the low level of nucleotide diversity enclosed in domesticated tomatoes due to genetic erosion. Furthermore, a higher number of non-synonymous SNPs was found in cultivated tomatoes when compared to wild relatives due to breeding activities and selective pressure that led to differential allele fixation. After a few months, additional resequencing data for 360 tomato accessions were released (Lin et al. 2014). Overall, 2.6 trillion base pairs of sequence, with a median depth of 5.7× and a final set of about 12 million SNPs and 1.3 million small indels were used to reconstruct the evolutionary history of tomato breeding. The authors corroborated previous findings that support S. pimpinellifolium as the ancestor of the cultivated tomato and S. l. cerasiforme as an intermediate species (Blanca et al. 2015), thus defining a two steps process: the domestication phase where the cerasiforme types evolved from the wild relative, and the improvement phase from which the big fruit sizes were selected in the cultivated types. At the genome level, the evolution of fruit mass was evidenced by the presence of 186 domestication sweeps and 133 improvement sweeps that covered 8.3 % and 7.0 % of the genome respectively for a total of 10,412 genes. Within these were located the major QTLs involved in the enlargement of tomato fruits. Furthermore, the overlap found between both domestication and improvement sweeps suggested loci that had undergone an additional round of selection toward the improvement of fruit size. The broad information provided from these studies has been expanded by Tieman et al. (2017), which resequenced additional 231 accessions, defining through genome-wide association the genetic components underlying the improvement of fruit flavour. The authors demonstrated how breeding programs have been unbalanced toward the selection for alleles that favored fruit weight and acid content, while loci correlated with compounds contributing to consumer preferences and aroma intensity have been lost in modern varieties. Additional investigations that used the sequencing data previously obtained, were performed to identify novel QTLs for plant architecture, leaves, and fruit size (Mata-Nicolas et al. 2020). The last population study at a deep genome scale was done by Razifard et al. (2020) who performed a comprehensive analysis combining de novo genome sequences of 166 accessions of S.lycopersicum, S. l. cerasiforme and S pimpinellifolium retrieved from different regions of South and Mesoamerica, with previously published genomes. The whole dataset generated over 23 million SNPs and ⍰4.3 million indels used to reconstruct the evolutionary history of the cultivated tomato. The authors provided new insight with respect to the two-step domestication process previously reported using both SNP array (Blanca et al. 2015) and whole genome resequencing (Lin et al. 2014), which saw the S. l. cerasiforme as the intermediated species that led the transition from wild to cultivated tomato. In this study, a new link between the semi-wild ancestor and the modern tomato was found. The origin of S. l. cerasiforme has been reported to be an antecedent to domestication, diverging from the wild S. pimpinellifolium thousands of years before humans were present in Latin America. The authors also found two ancestral groups for S. l. cerasiforme based on their provenance in the South and Central America regions. Therefore, this species may have originated in Ecuador as a wild and then spread in other surrounding regions finally toward Central America. During this transition, native medium-size fruits were selected, however, some traits were lost once the partially domesticated forms were spread northward. By reshaping the role of the semi-domesticated berry tomato, this study opened new questions on the movement of this nightshade species in the origin centre highlighting the importance to explore the gene flow during the different steps of domestication and identify lost alleles, hence opening new frontiers for plant breeding (Tripodi 2022).

Genome sequencing data of 725 tomato accessions, including 561 from already published studies have been used to construct the first tomato pan-genome by second-generation sequencing technologies (Gao et al. 2019). The tomato pan-genome included both cultivated tomatoes and close relatives and consisted of

40,369 protein-coding genes for a total size of 1179 Mb, hence capturing additional 4,873 genes and 30 % more sequences compared to the reference. Comparative analysis of the presence/absence variation that occurred at determined loci of some genomes compared to others (Tan et al. 2012), confirmed genetic erosion during the domestication and subsequent improvement phases, thus suggesting a higher percentage of genes lost during the former process. Indeed, wild relatives showed more genes when compared to the semi-domesticated tomato, likewise in the latter when compared to the cultivated form. Furthermore, within cultivated tomatoes, modern cultivars retained a higher gene content compared to heirlooms, mostly due to introgression breeding for resistance to pathogens. A 4 kilobases substitution was found in the promoter region of TomLoxC, a gene involved in apocarotenoid production and influencing fruit flavor. The non-reference allele was found in 90 % of the wild relative *S. pimpinellifolium* and only in 2 % of the cultivated tomato, highlighting a strong negative selection during the domestication and improvement. The construction of this first pan-genome represents the next step to precise mining the natural variation to be exploited for breeding purposes.

The first results of the applications of long-read sequencing in tomato have been presented from 2020 onward. PacBio, ONT, HiCi and Bionano have been used singularly or in combination to generate high-quality reference genomes for individual cultivated lines or wild relatives, including S. pimpinellifolium acc. LA2093 and LA1589 (Wang et al. 2020), S. pimpinellifolium acc. LA1670 and S. lycopersicum var. cerasiforme acc. LA1673 (Takei et al. 2021), S. lycopersicum cv. Moneyberg (van Rengs et al. 2022), S. lycopersicum cv. (Heinz 2022), and S. lycopersicoides acc. LA2951 (Powel et al. 2022). Genomes generated by long-read sequencing guaranteed a sequencing depth 5 times higher than those of the short-read, beyond 100x. By linking PacBio to ONT was possible to improve the assembly contiguity at the whole chromosome level avoiding the occurrence of any gap and ensuring, furthermore, scaffold chromosome lengths ten time higher than those generated by second-generation technologies (Takei et al. 2021). The new references allowed to detect with high precision the existing structural variants (SVs) between wild and cultivated species, determining their role in the evolution of key traits during domestication and breeding, thus facilitating their use as potential targets for the improvement of important horticultural traits (Wang et al. 2020). The breakthrough of third-generation sequencing in tomato genomics has been achieved for pan-genome investigations, providing a broader view of structural and genetic changes underlying the diversity. Alonge et al. (2020), applied nanopore sequencing to detect structural variations in 100 tomato accessions and released de novo assemblies for 14 reference genomes combining both short- and long-read sequencing. ONT produced a total of ⍰8 Terabases of long-read data with a minimum of 40× genome coverage. In total, over 238 thousand SVs were detected, mostly insertions and deletions. In agreement with previous studies (Aflitos et al. 2014, Gao et al. 2019), the wild species had more structural variants compared to the cultivated one, as a result of the loss of diversity during the evolution of tomato. Within the cultivated tomato, different clusters were detected reflecting a level of admixture with wild and semi-domesticated accessions, thus suggesting the occurrence of many SVs still undiscovered. Indeed, authors detected several SV associated with resistances introgressed from wild relatives. The functional analysis allowed to link the SV to three major domestication and breeding traits: the smoky flavour which negatively influences consumer preferences, sb1 responsible of the partial suppressor of branching and fw3.2 a major QTL controlling fruit mass. The two latter were found associated with hidden duplications that altered gene expression, thus reshaping the phenotypic diversity. The study opened new insights into the possibility of integrating the study of pan-genomes with long-read sequencing to reveal, with high reliability, the presence and function of structural variants in complex genomes (Tripodi 2022).

The release of the graphical pan-genome is the latest milestone achieved in tomato genomics. The concept of graph pan-genomes has been first applied in bacteria, aiming to represent the genome of a species rather than a single accession (Tettelin et al. 2005). Thanks to the quality of long-read sequences and progresses in bioinformatics, this approach is increasingly applied in crop species (Della Coletta et al. 2021). Zhou et al. (2022), developed a large pan-genome consisting of 832 genotypes. Authors integrated 32 de novo assemblies obtained through PacBio and HiCi technologies and 806 previously sequenced accessions, of which 100 using ONT (Alonge et al. 2020) and 706 using short-read (Causse et al. 2013, Aflitos et al. 2014, Bolger et al 2014, Lin et al. 2014, Tieman et al. 2017, Gao et al. 2019). This approach allowed for generating high-quality assemblies representing as much as possible of the whole diversity occurring in tomato. The graph pan-genome spanned 1007.56 Mbp for a total of 51,155 genes, 28.3 % of which new respect the reference genome. In total, over 19 million variants were detected including ~18 million SNPs, ~1.5 million indels, and about 200 thousand SVs (Zhou et al. 2022). The authors then tested the potentiality of the graph pan-genome for the identification of variants associated with metabolic traits. Compared to single reference genomes, graph pan-genome offered a higher mapping resolution in GWA studies, enabling the detection of the missing heritability and providing comprehensive coverage of all genetic variants. Such advancement paves the way for new opportunities for genomic-assisted breeding and for the selection of target traits. Although some constraints for comparative analysis, genome mapping, and visualization still occur, it is expected that graphical pan-genomes will be obtained for several other species in further years thanks to the development of novel dedicated analytic tools (Edwards and Batley 2022).

Although in tomato high-quality genomes at deep sequencing scale have been obtained, the complete genomic information on the entire existing variability is still lacking. Among Solanaceous, such efforts have been done only in pepper (Tripodi et al. 2021). Unlocking the diversity enclosed in over 10,000 existing varieties will provide novel opportunities for harnessing the potential stored in germplasm resources, facilitating their use in genetic improvement programs, and accelerating gene discovery and transfer (Tripodi 2022).

RS of Tea Germplasm

Tea plant is an important economic crop, which is used to produce the world's oldest and most widely consumed tea beverages. Xia et al (2020) reported a high-quality reference genome assembly of the tea plant (*Camellia sinensis* var. sinensis) consisting of 15 pseudo-chromosomes. LTR retrotransposons (LTR-RTs) account for 70.38% of the genome, and LTR-RTs play critical roles in genome size expansion and the transcriptional diversification of tea plant genes through preferential insertion in promoter regions and introns. Genes, particularly those coding for terpene biosynthesis proteins, associated with tea aroma and stress resistance were significantly amplified through recent tandem duplications and exist as gene clusters in tea plant genome. Phylogenetic analysis of the sequences of 81 tea plant accessions with diverse origins revealed three well-differentiated tea plant populations, supporting the proposition for the southwest origin of the Chinese cultivated tea plant and its later spread to western Asia through the introduction (Xia et al. (2020). Domestication and modern breeding left significant signatures on hundreds of genes in the tea plant genome, particularly those associated with tea quality and stress resistance. The genomic sequences of the reported reference and resequenced tea plant accessions provide valuable resources for future functional genomics study and molecular breeding of improved cultivars of tea plants.

RS of Wheat Germplasm

Hao et al (2020) observed that controlled pedigrees and the multi-decade timescale of national crop plant breeding programs offer a unique experimental context for examining how selection affects plant genomes. More than 3000 wheat cultivars have been registered, released, and documented since 1949 in China. A set of 145 elite cultivars selected from historical points of wheat breeding in China were re-sequenced (Hao et al 2020). A total of 43.75 Tb of sequence data were generated with an average read depth of 17.943 for each cultivar, and more than 60.92 million SNPs and 2.54 million InDels were captured, based on the Chinese Spring RefSeq genome v1.0. Seventy years of breeder-driven selection led to dramatic changes in grain yield and related phenotypes, with distinct genomic regions and phenotypes targeted by different breeders across the decades. There are very clear instances illustrating how introduced Italian and other foreign germplasm was integrated into Chinese wheat programs and reshaped the genomic landscape of local modern cultivars. Importantly, the resequencing data also highlighted significant asymmetric breeding selection among the three sub-genomes: this was evident in both the collinear blocks for homeologous chromosomes and among sets of three homeologous genes. Accumulation of more newly assembled genes in newer cultivars implied the potential value of these genes in breeding. Conserved and extended sharing of linkage disequilibrium (LD) blocks were highlighted among pedigree-related cultivars, in which fewer haplotype differences were detected. Fixation or replacement of haplotypes from founder genotypes after generations of breeding was related to their breeding value. Based on the haplotype frequency changes in LD blocks of pedigree-related cultivars, a strategy was proposed (Hao et al 2020) for evaluating the breeding value of any given line on the basis of the accumulation (pyramiding) of beneficial haplotypes. Collectively, their study demonstrated the influence of "founder genotypes" on the output of breeding efforts over many decades and also suggests that founder genotype perspectives are in fact more dynamic when applied in the context of modern genomics-informed breeding.

Bread wheat (*Triticum aestivum* L.) is one of the most important crops worldwide, providing approximately 20% of the calories consumed globally. However, the genetic diversity within wheat cultivars has been drastically reduced during the process of domestication and breeding. Crop progenitors and their relatives represent a large gene repository for the improvement of desirable traits. Thus, genomic characterization of wheat progenitors and identification of genes underlying phenotypic variations will provide important resources for wheat breeding. Triticum urartu is the progenitor of the A subgenome in tetraploid and hexaploid wheat. Wang et al. (2022) proposed that uncovering the landscape of genetic variations in T. urartu will help to understand the evolutionary and polyploid characteristics of wheat. Accordingly, they investigated the population genomics of T. urartu by genome-wide sequencing of 59 representative accessions collected around the world. A total of 42.2 million high-quality single-nucleotide polymorphisms and 3 million insertions and deletions were obtained by mapping reads to the reference genome. The ancient T. urartu population experienced a significant reduction in effective population size (Ne) from ~3 000 000 to ~140 000 and subsequently split into eastern Mediterranean coastal and Mesopotamian-Transcaucasian populations during the Younger Dryas period. A map of allelic drift paths displayed splits and mixtures between different geographic groups, and a strong genetic drift towards hexaploid wheat was also observed, indicating that the direct donor of the A subgenome originated from northwestern Syria. Genetic changes were revealed between the eastern Mediterranean coastal and Mesopotamian-Transcaucasian populations in genes orthologous to those regulating plant development and stress responses. A genome-wide association study identified two single-nucleotide polymorphisms in the exonic regions of the SEMI-DWARF 37 ortholog that corresponded to the different T. urartu ecotype groups. Their study provided novel insights into the origin and genetic legacy of the A subgenome in polyploid wheat and contributes a gene repertoire for genomics-enabled improvements in wheat breeding (Wang et al. 2022).

Conclusions

Crop wild relatives (CWRs) have provided breeders with several 'game-changing' traits or genes that have boosted crop resilience and global agricultural production. Advances in breeding and genomics have accelerated the identification of valuable CWRs for use in crop improvement. The enhanced genetic diversity of breeding pools carrying optimum combinations of favourable alleles for targeted crop-growing regions is crucial to sustaining genetic gain. In parallel, growing sequence information on wild genomes in combination with precise gene-editing tools provide a fast-track route to transform CWRs into ideal future crops. Data-informed germplasm collection and management strategies together with adequate policy support will be equally important to improve access to CWRs and their sustainable use to meet food and nutrition security targets (Bohra et al. 2022).

References

Aflitos, S., Schijlen, E., de Jong, H., de Ridder, D., Smit, S., ... Finkers, R. (2014). Exploring genetic variation in the tomato (Solanum section Lycopersicon) clade by whole-genome sequencing. Plant J. 80:136–48. https://doi.org/10.1111/tpj.12616

Alonge, M., Wang, X., Benoit, M., Soyk, S., Pereira, L., ... Zhang, L. (2020). Major impacts of widespread structural variation on gene expression and crop improvement in tomato. Cell. 182:145–61.

Bohra, A., Kilian, B., Shoba Sivasankar, S., ... Caccamo, M. (2022). Reap the crop wild relatives for breeding future crops. Trends Biotech., 40: 412-431. https://doi.org/10.1016/j.tibtech.2021.08.009

Bolger, A., King, P., Hudson, J., ... Hill, K. (2014). The genome of the stress-tolerant wild tomato species Solanum pennellii. Nat. Genet. 46:1034–8. https://doi.org/10.1038/ng.3046.

Cai, X., Sun, X., Xu, C., Sun, H., ... Xiaoli Wang, X. (2021). Reference genome and resequencing of 305 accessions provide insights into spinach evolution, domestication and genetic basis of agronomic traits. Nature Communications. doi: 10.1038/s41467-021-27432-z

Cao, Z.Z., Lin, X.Y., Yang, Y.J., Guan, M.Y., & Chen, M.X. (2019). Gene identification and transcriptome analysis of low cadmium accumulation rice mutant (lcd1) in response to cadmium stress using MutMap and RNA-seq. BMC Plant Biol. 19: 250.

Causse, M., Desplat, N., Pascual, L., ... Brown, P. (2013). Whole genome resequencing in tomato reveals variation associated with introgression and breeding events. BMC Genomics. 14:791. https://doi.org/10.1186/1471-2164-14-791.

Chen, E., Huang, X., Tian, Z., Wing, R. A., & Han, B. (2019). The genomics of Oryza species provides insights into rice domestication and heterosis. Annu. Rev. Plant Biol. 70: 639–665.

Chen, Y.W., & Nelson, R.L. (2004). Evaluation and classification of leaflet shape and size in wild soybean. Crop Sci. 44: 671–677.

Cheng, L., Kim, K.W., & Park, Y.J. (2019). Evidence for selection events during domestication by extensive mitochondrial genome analysis between japonica and indica in cultivated rice. Sci. Rep. 9: 10846.

Chung, W.H., Prague, P., Handreson, L., ... Morgon, H. (2014). Population structure and domestication revealed by high-depth resequencing of Korean cultivated and wild soybean genomes. DNA Res. 21: 153–167.

Clough, S.J., Ku, S., Hang, C., ... King, J. (2004). Features of a 103-kb gene-rich region in soybean include an inverted perfect repeat cluster of CHS genes comprising the I locus. Genome. 47: 819–831.

Cook, D.E., Brown, B., Buttler, H., ... Watson, B. (2012). Copy number variation of multiple genes at Rhg1 mediates nematode resistance in soybean. Science 338: 1206–1209.

Dai, L., Wu, L., Dong, Q., ... Li, W. (2017). Genome-wide association study of field grain drying rate after physiological maturity based on a resequencing approach in elite maize germplasm. Euphytica. 213: 182. https://doi.org/10.1007/s10681-017-1970-9

Della Coletta, R., Qiu, Y., Ou, S., ... Mu, O. (2021). How the pan-genome is changing crop genomics and improvement. Genome Biol. 22:3. https://doi.org/ 10.1186/ s13059-020-02224-8.

Doebley, J.F., Gaut, B.S., & Smith, B.D. (2006). The molecular genetics of crop domestication. Cell. 127: 1309–1321.

Du, H. Su, D., Li., ... Wang, O. (2017). Sequencing and de novo assembly of a near complete indica rice genome. Nat. Commun. 8: 15324.

Edwards, D., & Batley J. 2022. Graph pangenomes find missing heritability. Nat. Genet. 54: 919–20. https://doi.org/10.1038/s41588-022-01099-8.

Funatsuki, H., & Ohnishi, S. (2009). Recent advances in physiological and genetic studies on chilling tolerance in soybean. Jpn. Agric. Res. Q. 43: 95–101.

Gao, H., Ki, R., Sin, P., ... Vasist, N. (2013). Ehd4 encodes a novel and Oryza-genus-specific regulator of photoperiodic flowering in rice. PLoS Genet. 9: e1003281.

Gao, L., Gonda, I., Sun, H., ... Gun, K. (2019). The tomato pan-genome uncovers new genes and a rare allele regulating fruit flavor. Nat. Genet. 51:1044–51.

Gomi, K., Kaul, E., Srivastava, G., ... Kim, F. (2010). Role of hydroperoxide lyase in white-backed planthopper (Sogatella furcifera Horváth)-induced resistance to bacterial blight in rice, Oryza sativa L. Plant J. 61: 46–57.

Gonzalo, M.J., Díaz, A., Dhillon, N.P.S., ... Mirza, D. (2019). Re-evaluation of the role of Indian germplasm as center of melon diversification based on genotyping-by-sequencing analysis. BMC Genomics. 20: 448. https://doi.org/10.1186/ s12864-019-5784-0

Gramazio, P., Yan, H., Hasing, T., Vilanova, S., ... Su, S. (2019). Whole-genome resequencing of seven eggplant (Solanum melongena) and one wild relative (S. incanum) accessions provides new insights and breeding tools for eggplant Enhancement. Front. Plant Sci. 10: https://doi.org/10.3389/fpls.2019.01220

Grant, D., Nelson, R.T., Cannon, S.B., & Shoemaker, R.C., (2010). SoyBase, the USDA-ARS soybean genetics and genomics database. Nucleic Acids Res. 38: D843–D846.

Guo, D., Jiang, H., Yan, W., ... Yang, L. (2020). Resequencing 200 flax cultivated accessions identifies candidate genes related to seed size and weight and reveals signatures of artificial selection. Front. Plant Sci.10. https://doi.org/10.3389/fpls.2019.01682

Guo, J., Hi, J., Ko, S., ... Hu, H. (2010). A single origin and moderate bottleneck during domestication of soybean (Glycine max): implications from microsatellites and nucleotide sequences. Ann. Bot. 106: 505–514.

Guo, S., Zhao, S., Sun, H., ... Zheng, R. (2019). Resequencing of 414 cultivated and wild watermelon accessions identifies selection for fruit quality traits. Nat Genet 51: 1616–1623. https://doi.org/10.1038/s41588-019-0518-4

Hao C., Jiao C., Hou J., Li T., Liu H., Wang Y., Zheng J., ... Liu H. (2020). Resequencing of 145 landmark cultivars reveals asymmetric sub-genome selection and strong founder genotype effects on wheat breeding in China. Mol. Plant. 13: 1733–1751.

Hu, W., Ji, C., Liang, Z., ... Ku, E. (2021). Resequencing of 388 cassava accessions identifies valuable loci and selection for variation in heterozygosity. Genome Biol 22: 316. https://doi.org/10.1186/s13059-021-02524-7

Huang, X., Wang, K., Lu, l., ... Li, J. (2010). Genome-wide association studies of 14 agronomic traits in rice landraces. Nat. Genet. 42: 961–967.

Huang, X., Singh, P., Ku., L., ... Wang, P. (2011). Genome-wide association study of flowering time and grain yield traits in a worldwide collection of rice germplasm. Nat. Genet. 44: 32–39.

Huang, X., Zi, H., Wang, K., ... Zheng, L. (2012). A map of rice genome variation reveals the origin of cultivated rice. Nature. 490: 497–501.

Huang, X., Li., Su., O., ... Mo, H. (2015). Genomic analysis of hybrid rice varieties reveals numerous superior alleles that contribute to heterosis. Nat. Commun. 6: 6258.

Jeong, N., Ku, G., Mi, O., ... Lu, J. (2012). Ln is a key regulator of leaflet shape and number of seeds per pod in soybean. Plant Cell. 24: 4807–4818.

Kawahara, Y., Bastide, M.D.L., Hamilton, J.P., & Kanamori, H. (2013). Improvement of the Oryza sativa Nipponbare reference genome using next generation sequence and optical map data. Rice. 6: 1–10.

Kilian, B., & Graner, A. (2012). NGS technologies for analyzing germplasm diversity in gene banks. Briefings Functional Genomes. 11: 38-50. doi:10.1093/bfgp/elr046

Kim, M.Y., Ku, K., Hu, K., ... Lu, H. (2010). Whole-genome sequencing and intensive analysis of the undomesticated soybean (Glycine soja Sieb. and Zucc.) genome. Proc. Natl. Acad. Sci. USA. 107: 22032–22037.

Komatsu, K., Maekawa, M., Shimamoto, K., & Kyozuka, J. (2001). The LAX1 and FRIZZY PANICLE 2 genes determine the inflorescence architecture of rice by controlling rachis-branch and spikelet development. Dev. Biol. 231: 364–373.

Lam, H.M., Fe, J., Mi, R., ... Ku, V. (2010). Resequencing of 31 wild and cultivated soybean genomes identifies patterns of genetic diversity and selection. Nat. Genet. 42: 1053–1059.

Lee, Y. S., Hua, K., Sun, L., ... Mi, H. (2010). OsCOL4 is a constitutive flowering repressor upstream of Ehd1 and downstream of OsphyB. Plant J. 63: 18–30.

Li, H., Peng, S., Nui, I., ... Lin, J. (2009). The sequence alignment/map format and SAMtools. Bioinformatics. 25: 2078–2079.

Li, H., & Durbin, R. (2009). Fast and accurate short read alignment with Burrows–Wheeler transform. Bioinformatics. 25: 1754–1760.

Li, Y.H., Mu, J., Peng, C., ... King, P. (2008). Genetic structure and diversity of cultivated soybean (Glycine max (L.) Merr.) landraces in China. Theo. Appl. Genet. 117: 857–871.

Li, Y.H., Sing, K., Hu, R., ... Wang, K. (2013). Molecular footprints of domestication and improvement in soybean revealed by whole genome re-sequencing. BMC Genomics. 14: 579.

Lin, T., Zhu, G., Zhang, J., Xu, X., Yu, Q., ... Zheng, Z. (2014). Genomic analyses provide insights into the history of tomato breeding. Nat Genet. 46:1220–6. https://doi.org/10.1038/ng.3117.

Liu, B., Xu, B., Ku, S., ... Zing, W. (2007). QTL mapping of domestication-related traits in soybean (Glycine max). Ann. Bot. 100: 1027–1038.

Liu, B.H., Sun, K., Bu, K., ... Wang, K. (2010). The soybean stem growth habit gene Dt1 is an ortholog of Arabidopsis TERMINAL FLOWER1. Plant Physiol. 153: 198–210.

Liu, S., Ku, M., Xu, D., ... Lu, K. (2019). Characterization and evaluation of OsLCT1 and OsNramp5 mutants generated through CRISPR/Cas9-mediated mutagenesis for breeding low Cd rice. Rice Sci. 26: 88–97.

Liu, T., Liu, H., Zhang, H., & Xing, Y. (2013). Validation and characterization of Ghd7.1, a major quantitative trait locus with pleiotropic effects on spikelets per panicle, plant height, and heading date in rice (Oryza sativa L.). J. Integr. Plant Biol. 55: 917–927.

Liu, X., Bai, X., Wang, X., & Chu, C. (2007). OsWRKY71, a rice transcription factor, is involved in rice defense response. J. Plant Physiol. 164: 969–979.

Lv, Q., Li, W., Sun, Z., Ouyang, N., ... Jing, X. (2020). Resequencing of 1,143 indica rice accessions reveals important genetic variations and different heterosis patterns Nature Communication. https://doi.org/10.1038/s41467-020-18608-0

Ma, X., Wang, Z., Li, W., ... Zhang, Y. (2019). Resequencing core accessions of a pedigree identifies derivation of genomic segments and key agronomic trait loci during cotton improvement. Plant Biotechnology J. 17: 762–775.

Mata-Nicolás, E., Montero-Pau, J., Gimeno-Paez, E., ... Garcia-Carpintero, V. (2020). Exploiting the diversity of tomato: the development of a phenotypically and genetically detailed germplasm collection. Hortic Res. 7, 299-304.

McClung, A.M., Edwards, J.D., Jia, M.H., ... Huggins, T.D. (2020). Enhancing the searchability, breeding utility, and efficient management of germplasm accessions in the USDA–ARS rice collection. Crop Science. https://doi.org/10.1002/csc2.20256.

McKenna, A., Tyson, J., DeCosta, J., ... Watanabe, T. (2010). The genome analysis toolkit: a MapReduce framework for analyzing next-generation DNA sequencing data. Genome Res. 20: 1297–1303.

McNally, K. L., Brown, K., Watson, B., ... Lal, P. (2009). Genomewide SNP variation reveals relationships among landraces and modern varieties of rice. Proc. Natl Acad. Sci. USA.106: 12273–12278.

Paradis, E., Claude, J., & Strimmer, K. (2004). APE: analyses of phylogenetics and evolution in R language. Bioinformatics. 20: 289–290.

Peng, H., Xi, M., Bu, K., ... Baker, P. (2020). MBKbase for rice: an integrated omics knowledgebase for molecular breeding in rice. Nucleic Acids Res. 48: D1085–D1092.

Powell, A.F., Feder, A., Li, J., Schmidt, M.H.W., Courtney, L., ... Alseekh, S. (2022). A Solanum lycopersicoides reference genome facilitates insights into tomato specialized metabolism and immunity. Plant J. 20: 564-572.

Purcell, S., Black, W., Dyson, P., ... Kutty. I. (2007). PLINK: a tool set for whole-genome association and population-based linkage analyses. Am. J. Hum. Genet. 81: 559–575.

Qi, J., Li, A., Nu, O., ... Ze, K. (2013). A genomic variation map provides insights into the genetic basis of cucumber domestication and diversity. Nat. Genet. 45: 1510–1515.

Quinlan, A.R., & Hall, I.M. (2010). BEDTools: a flexible suite of utilities for comparing genomic features. Bioinformatics. 26: 841–842.

Ray, D.K., Mueller, N.D., West, P.C., & Foley, J.A. (2013). Yield trends are insufficient to double global crop production by 2050. PLoS One 8: e66428.

Razifard, H., Ramos, A., Della Valle, A.L., Bodary, C., Goetz, E., ... Manser, E.J. (2020). Genomic evidence for complex domestication history of the cultivated tomato in Latin America. Mol. Biol. Evol. 37:1118–1132.

Ren, G., Zhang, X., Li, Y., Ridout, K., Serrano-Serrano, M.L., ... Yang, Y. (2021). Large-scale whole-genome resequencing unravels the domestication history of Cannabis sativa. Science Advances 7. Doi: 10.1126/sciadv.abg2286

Sato, K., Takeda, K., & Ma J.F. (2021). Germplasm evaluation for crop improvement: analysis of grain quality and cadmium accumulation in barley. J. Cereal Sci. 101: 103297

Sayama, T., Gaur, F., Gill, C., ... Sharma, B. (2012). The Sg-1 glycosyltransferase locus regulates structural diversity of triterpenoid saponins of soybean. Plant Cell. 24: 2123–2138.

Schmutz, J., Hog, P., Gates, K., ... Hog, P. (2010). Genome sequence of the palaeopolyploid soybean. Nature. 463: 178–183.

Segla, S., Dossou, K., Song, S., Liu, A., & Li, D. (2023). Resequencing of 410 Sesame Accessions Identifies SINST1 as the Major Underlying Gene for Lignans Variation. Int. J. Mol. Sci. 24: 1055; https://doi.org/10.3390/ijms24021055

Srivastava, S.K., Wolinski, P., & Pereira, A. (2014). A strategy for genome-wide identification of gene based polymorphisms in rice reveals non-synonymous variation and functional genotypic markers. PLoS One 9: e105335.

Strickler, S.R., Toyada, P., Wong, S., ... Munz, J. (2015). Comparative genomics and phylogenetic discordance of cultivated tomato and close wild relatives. Peer J. 3: e793.

Su, X., Wang, B., Geng, X., ... Ju, P (2021). A high-continuity and annotated tomato reference genome. BMC Genomics. 22:898. https://doi.org/10.1186/ s12864-021-08212-x.

Sun, C., Hu, J., Wong, D., ... Nei, M. (2017). RPAN: rice pan-genome browser for ~3000 rice genomes. Nucleic Acids Res. 45: 597–6012.

Takei, H., Shirasawa, K., Kuwabara, K., Toyoda, A., Matsuzawa, Y., ... Iioka, S. (2021). De novo genome assembly of two tomato ancestors, Solanum pimpinellifolium and Solanum lycopersicum var. cerasiforme, by long-read sequencing. DNA Res. 28: dsaa029.

Tan, S., Zhong, Y., Hou, H., ... Mu, Z. (2012). Variation of presence/absence genes among Arabidopsis populations. BMC Evol. Biol. 12: 86. https://doi.org/10.1186/ 1471-2148-12-86.

Tang, L., Zi, F., Von, R., ... Wang, D. (2017). Knockout of OsNramp5 using the CRISPR/ Cas9 system produces low Cd-accumulating indica rice without compromising yield. Sci. Rep. 7: 14438.

Tanksley, S.D., & McCouch, S.R. (1997). Seed banks and molecular maps: unlocking genetic potential from the wild. Science. 277: 1063–1066.

Tettelin, H., Masignani, V., Cieslewicz, M.J., Donati, C., Medini, D., ... Ward, N.L. (2005). Genome analysis of multiple pathogenic isolates of Streptococcus agalactiae: implications for the microbial "pan-genome". Proc. Natl. Acad. Sci. USA.102:13950–13955.

Thudi, M., Khan, A.W., Kumar, V. .... Tang, N. (2016). Whole genome re-sequencing reveals genome-wide variations among parental lines of 16 mapping populations in chickpea (Cicer arietinum L.). BMC Plant Biol. 16 (Suppl 1): 10. https://doi.org/10.1186/s12870-015-0690-3

Tieman, D., Zhu, G., Resende Jr, M.F., Lin, T., Nguyen, C., ... Bies, D. (2017). A chemical genetic roadmap to improved tomato flavor. Science. 355: 391–394.

Tilman, D., Balzer, C., Hill, J., & Befort, B.L. (2011). Global food demand and the sustainable intensification of agriculture. Proc. Natl. Acad. Sci. USA. 108: 20260–20264.

Toda, K., Akasaka, M., Dubouzet, E.G., Kawasaki, S., & Takahashi, R. (2005). Structure of flavonoid 3⍰-hydroxylase gene for pubescence color in soybean. Crop Sci. 45: 2212–2217.

Todd, J.J., & Vodkin, L.O. (1996). Duplications that suppress and deletions that restore expression from a chalcone synthase multigene family. Plant Cell. 8: 687–699.

Tomato Genome Consortium. (2012). The tomato genome sequence provides insights into fleshy fruit evolution. Nature. 485: 635-64.

Tripodi, P. (2022). Next generation sequencing technologies to explore the diversity of germplasm resources: achievements and trends in tomato. Computational Structural Biotech. J. 20: 6250-6258.

Tripodi, P., Rabanus-Wallace, M.T., Barchi, L., Kale, S., Esposito, S., ... Acquadro, A. (2021). Global range expansion history of pepper (Capsicum spp.) revealed by over 10,000 genebank accessions. Proc. Natl. Acad. Sci. USA.118: 34.

Tuteja, J.H., Clough, S.J., Chan, W.C., & Vodkin, L.O. (2004). Tissue-specific gene silencing mediated by a naturally occurring chalcone synthase gene cluster in Glycine max. Plant Cell. 16: 819–835.

van Rengs, W.M., Schmidt, M.H.W., Effgen, S., Le, D.B., Wang, Y., ... Zaidan, M.W.A.M. (2022). A chromosome scale tomato genome built from complementary PacBio and Nanopore sequences alone reveals extensive linkage drag during breeding. Plant J. 110: 572–88.

Varshney, R.K., Saxena, R.K., Hari, D., ... Hazra, S. (2017). Whole-genome resequencing of 292 pigeonpea accessions identifies genomic regions associated with domestication and agronomic traits. Nature Genetics. 1: 8. doi:10.1038/ng.3872

Vitor, J., Santos, M., Valliyodan, B., ... Joshi T. (2016). Evaluation of genetic variation among Brazilian soybean cultivars through genome resequencing. BMC Genomics. 17:110-118.

Wang, W., Xi, X., Bu, L., ... Zu, M. (2018). Genomic variation in 3,010 diverse accessions of Asian cultivated rice. Nature. 557: 43–49.

Wang, X., Gao, L., Jiao, C., ... Ku, K. (2020). Genome of Solanum pimpinellifolium provides insights into structural variants during tomato breeding. Nat. Commun. 11: 5817. https://doi.org/10.1038/s41467-020-19682-0.

Wang, X., Hu, Y., He, W., Yu, K., Zhang, C., ... Yiwen Li, Y. (2022). Whole-genome resequencing of the wheat A subgenome progenitor Triticum urartu provides insights into its demographic history and geographic adaptation. Plant Communications. 3: 100354.

Watanabe, S., Suzuki, M., Velva, Z., ... Suzuki, M. (2011). A map-based cloning strategy employing a residual heterozygous line reveals that the GIGANTEA gene is involved in soybean maturity and flowering. Genetics. 188: 395–407.

Wu, D., Liang, Z., Yan, T., Xu, Y., Xuan, L., Tang, J., ... Zhou, G. (2019). Whole-genome resequencing of a worldwide collection of rapeseed accessions reveals the genetic basis of ecotype divergence. Mol. Plant. 12: 30–43.

Xia, E., Tong W., Hou Y., An Y., ... Chen L. (2020). The reference genome of tea plant and resequencing of 81 diverse accessions provide insights into its genome evolution and adaptation. Molecular Plant. 13: 1013-1026.

Xia, Z., Moi, V., Kung, E., ... Nu, J. (2012). Positional cloning and characterization reveal the molecular basis for soybean maturity locus E1 that regulates photoperiodic flowering. Proc. Natl. Acad. Sci. USA. 109: E2155–E2164.

Xia, Z., Zhai, H., Lü, S., Wu, H., & Zhang, Y. (2013). Recent achievements in gene cloning and functional genomics in soybean. Scientific World J. 13: 1–7.

Xie, W., Cui, N., Wang, M., ... Wang, P. (2015). Breeding signatures of rice improvement revealed by a genomic variation map from a large germplasm collection. Proc. Natl Acad. Sci. USA. 112: E5411–E5419.

Xu, X., Su, B., Bui, W., ... Nai, K. (2011). Resequencing 50 accessions of cultivated and wild rice yields markers for identifying agronomically important genes. Nat. Biotechnol. 30: 105–111.

Yamada, T., Takagi, K., & Ishimoto, M. (2012). Recent advances in soybean transformation and their application to molecular breeding and genomic analysis. Breed. Sci. 61: 480–494.

Yang, C., Yan, J., Jiang, S., Li, X., ... Min, H. (2022). Resequencing 250 soybean accessions: new insights into genes associated with agronomic traits and genetic networks. Genomics, Proteomics Bioinformatics. 20: 29-41.

Yang, C.H., Zhang, Y., & Huang, C.F. (2019). Reduction in cadmium accumulation in japonica rice grains by CRISPR/Cas9-mediated editing of OsNRAMP5. J. Integr. Agric. 18: 688–697.

Zabala, G., & Vodkin, L. (2007). A rearrangement resulting in small tandem repeats in the F3?5?H gene of white flower genotypes is associated with the soybean W1 locus. Crop Sci. 47: 113–124.

Zang, F., Ma, Y., Wu, Q., ... Hu, V. (2023). Resequencing of Rosa rugosa accessions revealed the history of population dynamics, breed origin, and domestication pathways. BMC Plant Biol. 23: 235. https://doi.org/10.1186/s12870-023-04244-5

Zhang, K., He, M., Fan, Y., ... Zhao, H. (2021). Resequencing of global tartary buckwheat accessions reveals multiple domestication events and key loci associated with agronomic traits. Genome Biology. 22: 23. Doi:10.1186/s13059-020-02217-7

Zhou, Y., Zhang, Z., Bao, Z., ... Kung, P. (2022). Graph pangenome captures missing heritability and empowers tomato breeding. Nature. 606: 527–34. https://doi.org/ 10.1038/s41586-022-04808-9.

Zhou, Z., Jiang, Y., Wang, Z., Gou, Z., Lyu, J., ... Li, W. (2015). Resequencing 302 wild and cultivated accessions identifies genes related to domestication and improvement in soybean. Nature Biotechnology. 33: 408 – 414.

Chapter-3

De Novo Domestication of Wild Relatives

Abstract

The domestication of wild species led to a wide variety of crops adapted to a range of climatic and edaphic conditions, which allowed expansion of cultivation to larger areas and over longer periods. Subsequent crop breeding led to higher yields and facilitated population growth. However, both domestication and breeding occurred empirically with little understanding of the underlying biological mechanisms. It is now known that genetic variation is created by mutation and that breeding operates by stacking favorable mutations in a single plant through recombination. Technical advances will eventually allow controlled manipulation of mutation and recombination, and thus de novo domestication of wild species by targeted modification of their genomes. Such tools will also aid in accelerating the improvement of traditional, semi-domesticated "orphan" crops that perform poorly in modern agricultural systems. The combination of these approaches will help broaden the narrow genetic basis of crops on which humankind currently relies. Recent breakthroughs have shown that targeted control of gene expression is an even faster avenue to produce desirable phenotypes, thus bypassing the need for mutation and recombination. However, the deliberate effort to create new crops or improve existing ones requires a thorough understanding of the genetic basis of domestication. The synergistic combination of classical archaeobotanical and genetics with high-throughput genomics is revealing that a variety of different processes may have operated in the domestication of crops. Further, in combination with increasingly powerful gene-editing toolkits, sets have been staged for the continual domestication of crop wild relatives and other lesser-known plant species by defining an ideal plant type ("ideotype") that is more resilient, nutritious, and productive in a given environment. Wild plants that are naturally resistant to biotic (insects and diseases) and tolerant of abiotic (drought and heat) stresses can be selected and manipulated by introducing mutations that mimic the domestication events that led to improved yield and agronomic performance of the major crops of today. Proof-of-concept for the potential of this de novo domestication approach was provided using gene editing to create agronomically important traits in wild relatives of the tomato, the orphan crop Physalis, and in a polyploid wild relative of rice. Recent progress in the understanding of crop domestication and technical breakthroughs in gene-editing technology could be combined to produce better crops for the future.

Keywords: *Crop domestication, De novo domestication, wild relatives, domesticated genes*

Introduction

Most of the crops and vegetables currently under cultivation were domesticated from their wild progenitors during the past 12,000 years. The process of domestication was based on choosing agricultural products according to the needs of our ancestors. Such simple choices ultimately led to the pyramiding of valuable mutations and recombinants in key genes that rendered cultivated crops easier to breed, culture, and store seeds from. Intriguingly, many different crop species share common domesticated traits such as seed shattering. In rice, the seed number was greatly increased by domestication, whereas the weight of a single kernel was not (Si et al., 2016). By contrast, in tomatoes and eggplants, the number of fruits per plant did not increase greatly upon domestication, but the size and weight of the individual fruit did (Zhu et al., 2018). These species represent only a few examples of the 150 or so plants commonly cultivated today.

The domestication of wild species led to a wide variety of crops adapted to a range of climatic and edaphic conditions, which allowed the expansion of cultivation to larger areas and over longer periods. Subsequent crop breeding led to higher yields and facilitated population growth (Evans, 1998). Until recent times, both domestication and breeding occurred empirically with little understanding of the underlying biological mechanisms. Genetic variation is created by mutation and breeding operates by stacking favourable mutations in a single plant through recombination. Technical advances will eventually allow controlled manipulation of mutation and recombination (Taagen et al., 2020; Nasti and Voytas, 2021), and thus de novo domestication of wild species by targeted modification of their genomes (Gasparini et al., 2021). Such tools will also aid in accelerating the improvement of traditional, semi-domesticated "orphan" crops that perform poorly in modern agricultural systems (Tadele, 2019). The combination of these approaches will help broaden the narrow genetic basis of crops on which humankind currently relies (Milla and Osborne, 2021). Furthermore, recent breakthroughs have shown that targeted control of gene expression is an even faster avenue to produce desirable phenotypes (Pan et al., 2021a), thus bypassing the need for mutation and recombination. However, the deliberate effort to create new crops or improve existing ones requires a thorough understanding of the genetic basis of domestication (Kantar et al., 2017). The synergistic combination of classical archaeobotany and genetics (Denham et al., 2020) with high-throughput genomics (Purugganan and Jackson, 2021) reveals that a variety of different processes may have operated in the domestication of crops. This knowledge, combined with increasingly powerful gene-editing toolkits, sets the stage for the continual domestication of crop wild relatives and other lesser-known plant species by defining an ideal plant type ("ideotype") that is more resilient, nutritious, and productive in a specific environment (Zsögön et al., 2017). Wild plants that are naturally resistant to biotic (insects and diseases) and tolerant of abiotic (drought and heat) stresses can be selected and manipulated by introducing mutations that mimic the domestication events that led to improved yield and agronomic performance of the major crops of today. Proof-of-concept for the potential of this de novo domestication approach was provided using gene editing to create agronomically important traits in wild relatives of the tomato (*Solanum lycopersicum*) (Li et al., 2018; Zsögön et al., 2018), the orphan crop Physalis (Lemmon et al., 2018) and, more recently, in a polyploid wild relative of rice (*Oryza* species) (Yu et al., 2021). The recent progress in the understanding of crop domestication and technical breakthroughs in gene-editing technology could be combined to produce better crops for the future.

The History of Crop Domestication

Human-guided domestication began approximately 12,000 years ago in the Middle East and Fertile Crescent, and subsequently occurred in different parts of the world including China, Mesoamerica and the Andes, Near Oceania, sub-Saharan Africa, and eastern North America (Purugganan and Fuller, 2009, Meyer and Purugganan, 2013). Fernie and Gutierrez-Marcos (2019) divide the history of crop improvement into four generations. The first generation of breeding began with phenotype-based selection by local independent farmers, which slowly led to the dramatic changes observed in modern crops (Doebley et al., 2006, Wallace et al., 2019). With the world population massively increasing by the turn of the 20th century, until hybrid breeding was developed in the 1980s and mating designs and statistical analyses were incorporated (Fisher, 1919, Wright, 1921), the use of fertilizer and pesticide became widespread (Carvalho, 2017), and the first "green revolution" occurred at the end of the 1950s, resulting in dwarf plants with consequently higher yields (Khush, 2001). About 30 years ago, the third generation of breeding was initiated, which is also known as the second green revolution or biotechnology-based breeding. In this generation, the application of transgenic technology along with the utilization of genome breeding technology, and genome-wide association studies (Risch and Merikangas, 1996, Fernie and Gutierrez-Marcos, 2019), marker-assisted breeding (Lander and Botstein, 1989) and genomic selection (Meuwissen et al., 2001) were harnessed for crop improvement. Now is the beginning of the fourth generation of breeding, which promotes design breeding (Figure 3.1). This includes the application of genome editing, precision breeding, and big data mining to tailor crops to demands.

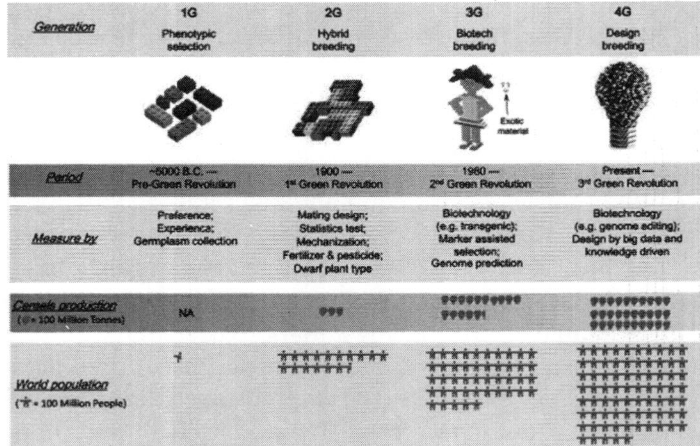

Figure 3.1. How Breeding History Feeds Human Civilization. The horizontal axis summarizes four milestone generations of breeding while the vertical axis presents specific characteristics of each generation. The first generation (1G) started ~10 000 years ago, focusing on simple phenotype-based collection and wild species domestication; it had a very limited production being able to feed about 50 million people. 2G breeding is characterized by hybrid and dwarf breeding and occurred during the early 20th century. In this generation people had already incorporated statistics and design into breeding practice and applied mechanization, fertilizers, and pesticides widely, resulting in the annual production of approximately 300 million tonnes of food to feed 1.65 billion people. Several modern biotechnologies have been applied from 1980 until now in 3G breeding, which was especially characterized by the application of transgenic technologies, whole-genome prediction, and marker-assisted selection using molecular markers. Between 1980 and today cereals production increased from 1.55 to 3 billion tonnes worldwide, increasing our capacity to feed from 4.5 to 7.5 billion people. Currently, 4G breeding is emerging as precision design agriculture as the integration of deep exploration on crop functional genomics, popular precision editing technologies, and big data mining are incorporated into breeding strategies. (Reproduced from Fernie and Yan 2019).

The Process of Domestication

Domesticated plant species span 160 taxonomic families (Meyer and Purugganan, 2013), with it being estimated that over 2500 species have undergone some extent of domestication (Meyer et al., 2012) and around 300 have been fully domesticated (Meyer and Purugganan, 2013, Salman-Minkov et al., 2016). Current models based on archaeological, genetic, and genomic evidence suggest that domestication is a multi-staged process that encompasses the onset of domestication, the increase in the frequency of desirable alleles, the formation of cultivated populations, and finally, deliberate breeding. However, the delineation of domestication is complicated in many species by multiple domestication events (Bitocchi et al., 2012, Dai et al., 2012, Schmutz et al., 2014, Fang et al., 2017, Liu and Yan, 2017, Wing et al., 2018), and often post-domestication genetic exchange with progenitor species occurs (Gross and Olsen, 2010). Three major routes have been taken toward crop speciation: selection, polyploidy, and introgression (Figure 3.2). A striking example is the speciation that occurred in the Brassicaceae family, which has involved favorable organ-shape selection from wild mustard, including terminal bud selection (for cabbage), flower selection (for broccoli), and stem selection (for kohlrabi) (Osnas, 2012), as well as allopolyploidy via hybridization with Chinese cabbage (for rutabaga) (Griffiths et al., 1999), and introgression from *Arabidopsis lyrata* transformed non-adapted *Arabidopsis arenosa* into adapted *A. arenosa* (Arnold et al., 2016) (Figure 2). Such diverse morphological and physiological changes that could be induced through simple genetic changes indicate that such approaches can be effectively aimed at tailoring crop production to a desired phenotype.

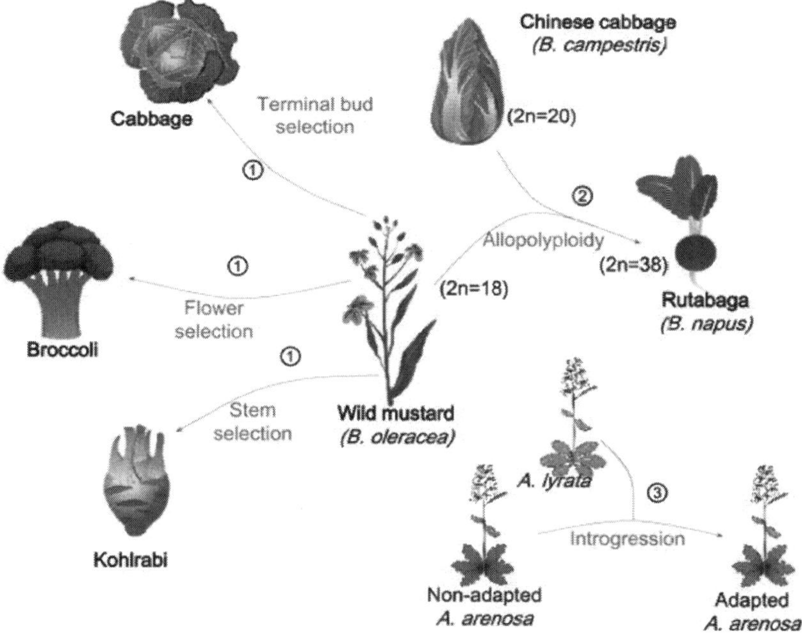

Figure 3.2. Routes for Crop Speciation. (1) A new species can be shaped in multiple ways including selection of specific organs (described in Osnas, 2012), (2) allopolyploidy among subspecies (described in Griffiths et al., 1999), or (3) absorbing new alleles through introgression from nearby species (described in Arnold et al., 2016). (Reproduced from Fernie and Yan 2019)

The complex of traits that commonly occurred upon domestication of diverse species has become known as the domestication syndrome, which has been extensively studied (Dillon et al., 2007, Alonso-Blanco et al., 2009, Gross and Olsen, 2010, He et al., 2011, Sakuma et al., 2011, Flint-Garcia, 2013, Abbo et al., 2014, Cornille et al., 2014, Dong et al., 2014, Avni et al., 2017, Che and Zhang, 2018, Hu et al., 2018). Many of these traits are common across different crop species, particularly within a crop type. For example, cereal species are characterized by the acquisition of modified seed shattering, seed size, and dormancy traits, while vegetable crops are characterized by the acquisition of modified fruit size and shape. Indeed, a mere 12 years ago the list of confirmed domestication genes numbered just over two dozen (Doebley et al., 2006). Since then, this number has increased about four-fold; recent examples of domestication genes include genes found to affect seed dormancy in rice, soybean, and tomato (Wang et al., 2018), or grain filling in maize (Sosso et al., 2015), and genes involved in the evolution of tuberous crops (Cheng et al., 2016, Hardigan et al., 2017). Indeed, there are many successful examples of this approach in a wide range of crops, including rice, maize, wheat, barley, sorghum, and tomato, and covering, but not limited to, many traits involved in plant and inflorescence architecture, yield, pigmentation, and ease of planting or harvesting (Table 1).

It has been observed that despite many species having similar domestication traits, either the same or different genes have been demonstrated to underlie their domestication (Gross and Olsen, 2010, Lenser and Theissen, 2013). For example, some quantitative trait loci (QTL) responsible for fruit weight in tomato and pepper, map on the same loci (Paran and van der Knaap, 2007), the stay green gene G confers control of seed dormancy in a range of species (Wang et al., 2018), and the glutinous grain improvement traits in rice, sorghum, barley, and millet are all controlled by orthologs of the Waxy gene (Meyer and Purugganan, 2013). Although initial studies suggested that the same genes were involved for all cereal domestication traits, fine mapping revealed that these were encoded by different genes (Li and Gill, 2006). For example, the canonical domestication gene tb1 of maize (Doebley et al., 1997) only has minor effects on branching in foxtail millet (Doust et al., 2004), while even within barley different genes are causal for changes in shattering within the two domesticated lineages (Azhaguvel and Komatsuda, 2007). These data suggest that both common and convergent mechanisms underlie the domestication syndrome.

While the domestication syndrome of some crops such as cereals has been characterized, those of vegetable crops are yet to be precisely characterized (Meyer et al., 2012). Similarly, other plants such as trees and rootstock cultivated plants have remained very poorly characterized (Breton et al., 2009, Warschefsky et al., 2016). This is due to the better conservation of fossil records of cereals, innovative research, and genetic tractability of the species. Further, genomics has revolutionized our understanding of the evolution of crops that have historically been characterized very poorly (Huang and Han, 2014, Sharma et al., 2018). Two factors have contributed significantly in this endeavour: the advent of next-generation sequencing and techniques that use broad natural allelic variance for assessing the genetic architecture of traits.

The Dynamics Of Crop Domestication

Understanding the dynamics of crop domestication is essential to discover the genomic signature of the phenotypic changes operated in the wild ancestral species. Until recently, it was widely believed that domestication was long and protracted, produced exclusively by human agency, and that convergent traits of the domestication syndrome would have a similar genetic basis (Purugganan, 2019). Growing evidence suggests that a more complex procedure is involved. First, although genomic analyses confirmed that in some cases, the domestication process took hundreds or even thousands of years, there are many examples of fairly recent and rapid domestication events. For example: sugar beet (*Beta vulgaris*) (Dohm et al., 2014), kiwifruit (*Actinidia chinensis*) (Ferguson, 2013), and African oil palm (*Elaeis guineensis*) (Zeven, 1972). There are also examples of "second cycle domestication", where existing domesticates were quickly adapted to new uses. For instance, changing soybean (*Glycine max*) from a vegetable to a protein and oil crop (Kofsky et al., 2018) and lupines (*Lupinus* species) from green manure to a seed crop (Gulisano et al., 2019). Second, operational selection, which occurs mostly because of human operations (hence the name) may have occurred after a first round of genetic changes created by the attraction of animals and plants to human habitations and their adaptation to this human-made environment (Spengler and Mueller, 2019). Pre-domestication of plants by grazing and foraging animals would also have contributed to the early selection of seed and horticultural crops (Zonneveld et al., 2018). Thus, it is possible that at least some traits of the domestication syndrome represent expectations ("pre-adaptations") for farming that were initially fixed by mutualism with animals (Spengler et al., 2021). Finally, studies involved in deciphering the function of "domestication genes", indicate a highly diversified genetic basis for convergent domestication traits in crops (Lai et al., 2018; Chen et al., 2021a). For example, the non-shattering trait selected during domestication was achieved through a widely divergent mechanisms in legumes and cereals. The mechanism of pod indehiscence in legumes and rachis flexibility in cereals, were found to be controlled by different developmental pathways and genes (Parker et al., 2021). Similarly, other fundamental domestication and breeding traits such as changes in growth habit, altered photoperiodic response for flowering or tuberization, reduced seed or fruit dispersal, and reduced seed dormancy are controlled by a multiplicity of genes from different families (Table 3.1).

Table 3.1 The diversified genetic basis of some relevant crop domestication and breeding traits.

Trait	Species	Gene	Type of variant	References
Changes in growth habit (dwarfism, determinate growth and/or side branching)	Wheat (*Triticum* spp.)	Rht genes (gibberellin insensitivity)	C	Peng et al., (1999)
	Rice (*Oryza* spp.)	sd1 (gibberellin biosynthesis)	C, ID	Sasaki et. al (2002)
	Common beans (*Phaseolus vulgaris*)	TFL1 (CETS)	C, ID, S, T	Kwak et.al (2012)
	Pea (*Pisum sativum*)	TFL1 (CETS) and Dt2 (MADSbox)	C (Dt1), R (Dt2)	Tian et. al (2010), Ping et. al. (2014)
	Tomato (*Solanum lycopersicum*)	SP/TFL1 (CETS)	C	Pnueli et. al (1998), Silva et. al (2018)
	Cotton (*Gossypium* spp.)	GbAF (CETS)	C	Si et. al ((2018)
Altered photoperiodic response	Barley (*Hordeum vulgare*)	Pseudo-response Regulator PpdH1	C	Turner et. al. (2005)
	Maize (*Zea mays*)	CMF transcription factor	R	Yang et. al. (2013), Huang et. al (2018)
	Sorghum (*Sorghum bicolor*)	Pseudo-Response Regulator 37 (PRR37)	C	Murphy et. al (2011)
	Sunflower (*Helianthus annuus*)	FT (CETS)	C	Blackman et. al. (2010)
	Soybean (*Glycine max*)	Pseudo-Response Regulator 37 (PRR37)	C	Lu et. al. (2020a), Wang et. al (2020)
	Common beans (*Phaseolus vulgaris*)	PHYTOCHROME A3 (PHYA3) and CONSTANSlike 2 C (PHYA3) C (PHYA3)	C, ID (COl-2)	Weller et. al. (2019), Gonza´lez et. al. (2021)
	Beet (*Beta vulgaris*)	Pseudo-Response Regulator BTC1	R	Pin et al., 2012
	Tomato (*Solanum lycopersicum*)	SP5G and SP11b (CETS) R(SP5G)	S (SP11b)	Soyk et al. (2017), Song et al. (2020b)
	Cassava (*Manihot esculenta*)	FT (CETS)	R	Adeyemo et. al. (2019)

Trait	Species	Gene	Type of variant	References
	Potato (*Solanum tuberosum*)	CYCLING DOF FACTOR	ID (CETS)	Kloosterman et al. (2013)
	Rose (*Rosa chinensis*)	CONSTANS and CONSTANS like 4	R	Lu et. al. (2020a)
Reduced seed or fruit dispersal	Barley (*Hordeum vulgare*)	Btr 1 and btr 2	C	Pourkheirandish et al. (2015)
	Foxtail millet (*Setaria italica*)	Les1 (MYB)	T	Mamidi et. al (2020)
	Rice (*Oryza* spp.)	qSH1 (homeodpmaimn) TF, sh4 (MYB3 binding domain), SH3 (YABBY)	C (qSH1, sh4) D (SH3)	Konishi et al. (2006), Li et al., (2006), Lv et al. (2018)
	Sorghum (*Sorghum bicolor*)	Sh1 (YABBY)	D, S	Lin et. al. (2012)
	Wheat (*Triticum* spp.)	Q (APETALA2)	C, R	Simons et. al. (2006)
	Soybean (*Glycine max*)	Shatt1-5; PDH1 (dirigent-like protein)	R (Shatt1-5), C (PDH1)	Dong et. al. (2014), Funatsuki et. al. (2014)
	Tomato (*Solanum lycopersicum*)	JOINTLESS-2 (MADS-box)	C, T	Roldan et. al (2017)
Reduced seed dormancy	Barley (*Hordeum vulgare*)	Qsd1 (alanine aminotransferase) Qsd2 (MKK3)	C	Sato et. al. (2016), Nakamura et. al. (2016)
	Rice (*Oryza* spp.)	Seed dormancy 4 (Zn-finger transcription factor), G (CAAX amino-terminal proteases	C, D	Sugimoto et. al. (2010), Wang et. al. (2018)
	Wheat (*Triticum* spp.)	Phs1 (MKK3)	C	Torada et. al. (2016)
	Common beans (*Phaseolus vulgaris*)	pectin acetylesterase 8	ID	Soltani et. al. (2021)
	Soybean (*Glycine max*)	G (CAAX amino-terminal protease)	C	Wang et. al. (2018)
	Cotton (*Gossypium* spp.)	GhMFT1 and GhMFT2 (CETS)	R	Yu et. al. (2019)

The type of variant refers to the allele changes found in domesticated versus wild relatives. C - non-synonymous coding sequence change, D - deletion, ID - insertion–deletion, R - regulatory variant (that alters gene expression), S - splicing variant, T - transposon insertion in coding sequence.

The diversified genetic basis for domestication traits in plants is quite unlike the findings for animals, in which strong loss-of-function mutations in pluripotent embryonic cells of the neural crest led to multiple traits of the domestication syndrome (Wilkins, 2020). This is not unexpected considering the marked differences in the general system of development between plants and animals (Meyerowitz, 2002). Most animals lack persistent stem cells, whereas plants retain embryonic cells in their meristems (Périlleux et al., 2019). Animals, therefore, have a closed developmental program and respond to environmental cues mainly through movement. Plants, on the other hand, are sessile, and their open and modular developmental pattern is highly flexible (Zsögön and Peres, 2018). The flexibility of the genetic circuitry controlling phenotypic outcomes is a consequence of the enormous environmental variability to which plants are subjected over the course of their lifetime. Thus, the plethora of genes underlying plant domestication traits, compared to the relatively few in animals, may reflect the higher developmental plasticity of plants. Plasticity itself is a quantitative trait strongly shaped by environmental conditions (Stotz et al., 2021). The initial degree of plasticity could have determined by which early domesticants were advanced and were subsequently abandoned (Vilela and González-Paleo, 2015; Piperno et al., 2019). The potential for domestication of a species may thus be directly related to its genomic (Dubcovsky and Dvorak, 2007), physiological (Matesanz and Milla, 2018), and developmental (Chen et al., 2021b) plasticity. Identification of the causal genes for plasticity and functional characterization of their roles in plant fitness will help in understanding future domestication processes and provide new targets for manipulation (Laitinen and Nikoloski, 2019; Liu et al. 2021b).

Identification of Domesticated Genes

The enhanced accessibility and affordability of next-generation sequencing technologies have allowed the identification of over 100 domestication genes (Fernie and Yan, 2019). Many of these, such as those associated with the loss of shattering, seed size, and dormancy, are conserved across various crops (Gross and Olsen, 2010; Lenser and Theissen, 2013). Other traits, such as fruit shape and evolution of tubers, seem to be specific to certain crops or crop types (Xiao et al., 2008, Cheng et al., 2016; Hardigan et al., 2017). Having identified the genes, they can now be used for de novo domestication of less cultivated or undomesticated species. Several reports on de novo domestication of less or undomesticated species indicate the possibility of application of this concept (Lemmon et al., 2018; Li et al., 2018; Zso¨go¨n et al., 2018; Hoyos et al., 2020; Yu et al., 2021). Initiation of projects such as The African Orphan Crops Consortium (Jamnadass et al., 2020) aims to use this approach at a very large scale.

Attempts have been made to use teosinte-maize (*Zea mays*) as a case-study to understand the following aspects: (a) how computational methods can be used to identify domestication genes and evaluate changes associated with these genes at both the phenotypic and molecular levels (Broman et al., 2019), (b) demonstrate the power of precision phenotyping to evaluate the potential phenotypic scope of such interventions (Yang et al., 2020), (c) describe the power of bioinformatics approaches and CRISPR/Cas libraries to streamline agricultural engineering strategies (Childs et al., 2012; Liu et al., 2020), (d) explain the benefit of harnessing recently acquired information concerning natural structural variations in crop and close-relative genomes (Alonge et al., 2020; Domı´nguez et al., 2020), (e) outline the potential role single cell sequencing for understanding application of these methods to polyploid species (Luo et al., 2020), and (f) discuss the role of molecular markers in metabolic models to improve genome selection (Tong et al., 2020).

Accelerating The Identification Of Domestication Genes

Rapid de novo domestication requires both genomic information and effective transformation procedures. Similarly, the alternative genetics-based approaches of crossing near relatives for introgressing a trait of interest or the adoption of mutagenesis-based strategies have been greatly facilitated by the generation of computational tools (Huang and George, 2011; Youens-Clark et al., 2011; Meng et al., 2015; Broman et al., 2019; Wei et al., 2021).

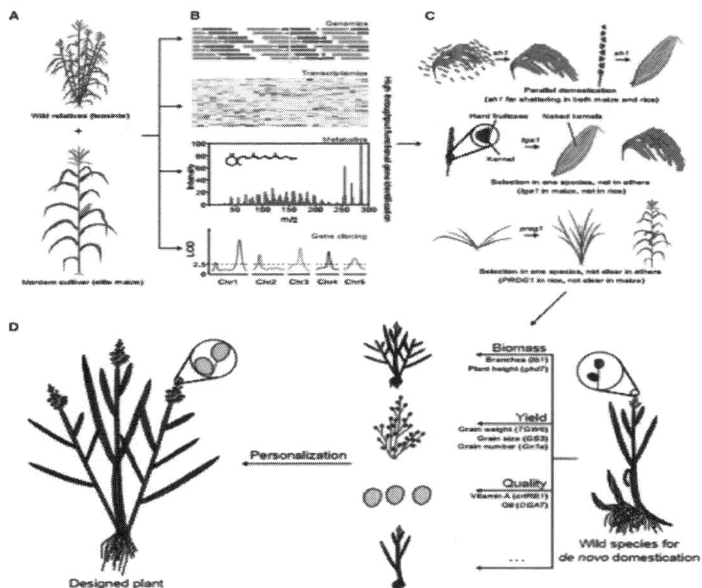

Figure 3.3. Overview of knowledge-driven de novo domestication. (A) the wild relatives and modern cultivars are vital germplasms (e.g. teosinte and maize) for tracing selection signals during plant domestication and also functional genomics research. (B) multiple-omics, including genomics, transcriptomics, metabolomics, and phenomics, are foundations for high-throughput gene cloning and gene function analysis. (C) and (D) wild species chosen for de novo domestication could be quickly shaped to meet different kinds of demands through selections on detected signals during previous domestications, especially parallel-selected genes. Ideal plants can thus be designed through the pyramiding and modification of genes of diverse functions. (Reproduced from Fernie et. al. 2021).

The use of multi-omics in de novo domestication is summarized in Figure 3.3 (Fernie et. al. 2021). Wild relatives and modern cultivars as exemplified by teosinte and maize (Figure 3.3, A). Comparisons of the progeny of a biparental cross between the wild and cultivated species are commonly made, subjected to metabolomics and transcriptomics, and the underlying quantitative trait loci including domestication genes are cloned (Figure 3.3, B). Literature information concerning the roles of these genes in other species can then be obtained. Previous studies have revealed that such genes are either subject to parallel domestication or species-specific domestication events (Figure 3.3, C). Following such assessments traits of interest, such as biomass, yield, and quality, can be modified via gene editing of the underlying genes allowing de novo domestication and personalization of the de novo domesticate (Figure 3.3, D).

The tracking of signals of selection is carried out by computational analyses known as selective sweeps, which are essentially a measure of the genetic diversity between cultivated and ancestral genotypes. Following this approach, it is assumed that those regions that have been selected during domestication harbour less diversity than those regions that are subject to less stringent selection (Zhu et al., 2018). When a genetic variant increases its frequency due to positive selection, the adjacent alleles also increase in their frequency in a process termed hitchhiking (Smith and Haigh, 2007). When the genetic variant under selection reaches high frequency (or even fixation), the hitchhiking effect reduces (or even eliminates), the genetic variation around the selected locus in an effect known as a selective sweep (Pavlidis and Alachiotis, 2017). Local FST outlier tests can be used to detect selective sweeps between populations of wild and domesticated taxa (Gepts, 2014). There are many different methods based on the FST statistic, which largely vary in the underlying model used to calculate the null distribution of FST values (Bonhomme et al., 2010; Lotterhos and Whitlock, 2015). Selecting a suitable method is complicated, with considerations including the sampling scheme taken, the total size of the dataset, the intensity of selective pressure, and the genetic structure of the population should be considered (BarreraRedondo et al., 2020). Some of the examples of the use of these tests in plants include the detection of domestication genes involved in apple (*Malus domestica*) fruit development, size, acidity, and sugar content (Khan et al., 2014), size, colour, and disease resistance in tomato (Zhu et al., 2018), and those for oil biosynthesis in sunflower (Baute et al., 2015).

With the availability of vast genome sequence information, it seems likely that the optimization of computational approaches will greatly accelerate the identification of domestication genes. If pedigree is established from genome information and this allows clear discrimination between wild relatives, landraces, and cultivars, similar experiments can be carried out to define domestication and improvement genes for many species. It is worth noting that a mere 15 years ago the number of confirmed domestication genes numbered just over two dozen (Doebley et al., 2006). However, there are now well over a hundred with notable recent genes being identified for dormancy in rice, soybean (*Glycine max*), and tomato

(Wang et al., 2018), the evolution of tubers in for example potato (Hardigan et al., 2017) and grain filling in maize (Sosso et al., 2015). Many of the genes involved are transcription factors (Fernie and Yan, 2019) as would perhaps be anticipated given the dramatic upregulation of gene expression on domestication evidenced in teosinte to maize comparisons (Swanson-Wagner et al., 2012). Beyond searching genomes for signatures of selection, detailed characterization of the transcriptome, metabolome, and phenome will represent important information when choosing which species is most appropriate for de novo domestication. The domestication genes already described are the likely tools for de novo domestication but deciding the likely species to domesticate first requires a better understanding of the physiology of a range of species. One could, for example, look for species that thrive in a particular environment, have a nutritionally excellent chemical composition, high nitrogen or water use efficiencies, or are naturally pest-resilient. Many domestication traits are shared by a very wide number of species, while others are more taxonomically restricted (Meyer et al., 2012). This clearly renders de novo domestication of completely wild species that are taxonomically distant from our major crops exceptionally hard. By contrast, those traits which are widely shared represent obvious targets.

Transcript profiles are tissue-specific and subject to temporal control, thus transcriptomics can reveal important loci involved in domestication traits such as suppression of secondary metabolites, changes in form, size, taste, absence of defense mechanisms, seed dormancy, and many other traits (Barrera-Redondo et al., 2020). For example, the assessment of altered gene expression on domestication was achieved in maize, tomato, lettuce (*Lactic sativa*), and common bean (*Phaseolus vulgaris*) by comparing the total gene expression of a range of wild species, progenitors, and cultivars, respectively (Swanson-Wagner et al., 2012; Koenig et al., 2013; Bellucci et al., 2014). A comparison of 38 diverse maize and 24 teosintes provided evidence for more than 600 genes having significantly different expression levels and almost twice as many altered co-expression patterns (Swanson-Wagner et al., 2012). Included among the 600 were a mere 46 that had been previously identified as targets of selection and a similar number presumed to result as an effect of inbreeding.

In tomato, both gene sequence and expression divergence between cultivated tomato and five related wild species was reported (Koenig et al., 2013). Based on sequence differences, Koenig et al. (2013) detected footprints of positive selection in over 50 genes and documented thousands of shifts at the gene-expression level. These rapidly evolving genes are commonly associated with environmental responses and stress tolerance. Similarly, decreased nucleotide and expression diversity and modified co-expression patterns characterized the domestication of the common bean (Bellucci et al., 2014). RNA sequencing of 240 accessions of lettuce revealed several regions as putative selective sweeps that occurred during domestication and divergence (Zhang et al., 2017). Genome-wide association studies (GWASs) further identified 5,311 expression quantitative trait loci (eQTL) regulating the expression of 4,105 genes, including nine eQTLs regulating genes associated with flavonoid biosynthesis and thereby underlying colour and nutritional content of the crop. The GWAS approach is highly useful alongside the selective sweep approach in characterizing changes in domestication. A list of GWAS databases and commonly used approaches is presented in Tables 3.2, and 3.3).

Table 3.2. List of databases and tools commonly used for GWAS and genomic studies.

Name	Link	References
easy GWAS: a cloud-based platform for comparing the results of genome-wide association studies	https://easygwas.ethz.ch/	Grimm et al. (2017)
Matapax: an online high-throughput genome-wide association study pipeline	https://matapax.mpimp-golm.mpg.de/	Childs et al. (2012)
GWAPP: a web application for genome-wide association mapping in Arabidopsis	https://gwas.gmi.oeaw.ac.at/	Seren et al. (2012)
GWAS Atlas: a curated resource of genome-wide variant–trait associations in plants and animals	https://bigd.big.ac.cn/gwas/	Ye et al. (2020)
PGSB Plants DB: updates to the database framework for comparative plant genome research	https://pgsb.helmholtz-muenchen.de/plant/index.jsp	Spannagl et al. (2016)
Phenotypic and genome-wide association with the local environment of Arabidopsis	http://www.personal.psu.edu/sma3/ CLIMtools.html	Ferrero-Serrano and Assmann (2019)
Rice Var Map: a comprehensive database of rice genomic Variations		Zhao et al. (2014)
GWASpro: a high-performance genome-wide association analysis Server	https://bioinfo.noble.org/GWASPRO/	Kim et al. (2018)
MaizeGDB 2018: the maize multi-genome genetics and genomics Database	https://www.maizegdb.org/	Kim et al. (2018)
TASUKE1: a web-based platform for exploring GWAS results and large-scale resequencing data	https://tasuke.dna.affrc.go.jp	Kumagai et al. (2019)
ZEAMAP, a comprehensive database adapted to the maize multi-omics era	http://www.zeamap.com	Gui et al. (2020)
MaizeCUBIC: a comprehensive variation database for a maize synthetic population	http://modem.hzau.edu.cn/	Luo et al. (2020)
CARMO: a comprehensive annotation platform for functional exploration of rice multi-omics data	http://bioinfo.sibs.ac.cn/carmo	Wang et al. (2015)
AraQTL—workbench and archive for systems genetics in Arabidopsis thaliana	http://www.bioinformatics.nl/AraQTL/	Nijveen et al. (2017)

Name	Link	References
WheatExp: an RNA-seq expression database for polyploid wheat	http://wheat.pw.usda.gov/WheatExp/	Pearce et al. (2015)
CerealsDB 2.0: an integrated resource for plant breeders and Scientists	https://www.cerealsdb.uk.net/ cerealgenomics/CerealsDB/ index NEW.php	Wilkinson et al. (2012)
The Triticeae Toolbox: combining phenotype and genotype data to advance small-grains breeding	http://triticeaetoolbox.org	Blake et al. (2016)
The AraGWAS Catalog: a curated and standardized Arabidopsis thaliana GWAS catalog	https://aragwas.1001genomes.org	TogninalLi et al. (2018)
Gramene: a resource for comparative analysis of plants genomes and pathway	http://www.gramene.org	Tello-Ruiz et al. (2018)
Ensembl Genomes 2020—enabling non-vertebrate genomic Research	http://www.ensemblgenomes.org	Howe et al. (2020)
SnpHub: an easy-to-set-up web server framework for exploring large-scale genomic variation data in the post-genomic era with applications in wheat	http://wheat.cau.edu.cn/Wheat_ SnpHub_Portal/	Guo et al. (2020)

Table 3.3. Commonly used packages for conducting GWASa.

Package	Description	Web page
TASSEL	Variety of algorithms MLM, GLM, weighted MLM, genome selection, fast association, supports P2D compression, can process GBS data, designed to determine dominance/additivity of effect, user-friendly GUI	http://www.maizegenetics.net/tassel
GAPIT	Package that can perform MLM and EMMA; supports P3D and EMMAx; works via R language	http://www.maizegenetics.net/gapit
A-DTest R	Package is ADGWAS: for GWAS https://github.com/maizego/A-D-testEMMAX Efficient mixed-model method for large genomic datasets; command-line interface only	http://genetics.cs.ucla.edu/emmax/index.html
GEMMA	Standard/multivariate/Bayesian linear mixed-model framework; estimates quantitative genetic traits and proportioning of variance; command line interface only	http://www.xzlab.org/software.html
ANGSD	Useful when genotypic states are not known with certainty; measures population genetic parameters; command-line interface only	http://www.popgen.dk/angsd/index.php/ ANGSD

Package	Description	Web page
Plink	Wide-ranging toolset for conducting GWAS; originally designed for human genome data; command-line interface only	http://pngu.mgh.harvard.edu/~purcell/ plink/
Lrgpr	Allows for testing of G 3 G and G 3 E; works via R language	http://lrgpr.r-forge.r-project.org/

A wide range of bitter-tasting compounds was selected against during the processes of domestication and improvement including the steroidal glycoalkaloids of tomato and potato (*Solanum tuberosum*) (Itkin et al., 2013; Schwahn et al., 2014), b-L-oxayl-2,3-diaminopropionic acid (b-LODAP) in grass pea (Emmrich et al., 2019), curcurbitadienol in cucumber (*Cucumis sativus*) (Shang et al., 2014; Zhou et al., 2016), glucosinolates in broccoli (*Brassica oleracea*) (Drewnowski and Gomez-Carneros, 2000), and the flavone-7- O-neohesperidoside in citrus (Frydman et al., 2013). While the decrease in the levels of these metabolites is rather predictable and easy to rationalize metabolome, studies in wheat (Beleggia et al., 2016), maize (Xu et al., 2019), rice (Deng et al., 2020), and tomato (Zhu et al., 2018), lettuce (Zhang et al., 2020), and tea (Zhang et al., 2020) revealed far more complex changes.

Beleggia et al. (2016) showed that the primary domestication of wheat was characterized by a reduction in unsaturated fatty acids in the primary domestication with altered amino acid content characterizing the secondary domestication. Maize, by contrast, was characterized by alkaloid, terpenoid, and lipid changes at the divergence between teosinte and tropical maize, whereas benzoxazinoid levels changed at the divergence between tropical and temperate maize (Xu et al., 2019). While rice displayed different changes again (Deng et al., 2020). Tomato domestication had highly diverse effects on the metabolome, with many metabolic changes being associated with the increase in size, others with breeding for colour preferences, and yet further by the introgression of disease resistance from wild relatives (Zhu et al., 2018). Likewise in lettuce, quinate and chlorogenic acid levels were strongly reduced on the domestication of lettuce, probably because of the desire to reduce bitterness (Zhang et al., 2020). Tea, on the other hand, did not display clear changes in its metabolite content across the domestication and improvement processes, probably due to the complex domestication process (Zhang et al., 2020).

While transcriptomics and metabolomics and corresponding GWAS analysis can clearly allow the computational discrimination of domestication genes, the number and scale of changes at the transcript and metabolite level suggest that transcriptomics and metabolomics will be necessary for the regulatory control of any future de novo domesticated product.

Molecular Mechanisms Underlying Crop Domestication

Plant breeding is basically the selection of optimal phenotypes, which are the best possible combinations of genotype and environment. The work of breeders began long before it was understood that genotypic variation is created by mutation, whereby new traits arise, and recombination, allows the stacking of useful traits into a single plant (Evans, 1998). Since both biological processes occur by chance, breeders create new varieties by crossing parents with traits of interest and selecting the best individuals among the progenies. However, breeders basically rely on a strong component of chance, as the phenotypes are not predictable and ought to be found among large numbers of variants. Thus, the first step to achieving full predictability in the generation of novel phenotypes is by controlling the basic processes that give rise to new genotypes: mutation and recombination.

Mutation is the ultimate source of genetic variation and occurs spontaneously at low frequencies. Soybean (Glycine max), for instance, has a genome of 1.15 Gbp and suffers a spontaneous mutation rate of one bp per ~5 × 10−8 bp per year. Assuming for simplicity that all mutations will be single-nucleotide polymorphisms (SNPs), one specific SNP of interest will occur in only one out of 100 million plants in one growing season. If the plants are grown at conventional density, this would represent a field area of 400 hectares. One of the major breakthroughs to accelerate the creation of genetic variation was the discovery of artificial mutagenesis by Hermann Müller (1890–1967). Application of physical or chemical treatments increases the mutation rate by many orders of magnitude, for instance, treating soybean seeds with ethyl methane sulfonate can generate one mutation per 74 kb of the genome (Tsuda et al., 2015). Thus, mutagenic treatments were used to create increased genetic variations in plants from the early 20th century in breeding programmes.

The first crop variety produced via induced mutagenesis was "Chlorina" tobacco (*Nicotiana tabacum*) in the 1930s (Jankowicz-Cieslak et al., 2017). Today, more than 3,200 mutant varieties in 170 species are grown commercially (https://mvd.iaea.org/). However, mutagenesis is an undirected and random process, and in many cases, the identity of the variants and their developmental and physiological bases remained largely unexplored. This lack of knowledge precluded targeted manipulation and extrapolation between different crop species. Furthermore, creating or discovering mutations that produce desirable agronomic traits is only the first step. The next challenge is to combine or stack different beneficial mutations into a single plant. This is achieved through crossing and selection in a segregating population, and the chances of finding the appropriate mutant combinations depend on the frequency of genetic recombination during meiosis (Choi and Henderson, 2015). Unlike the mutation rate, however, the rate of meiotic recombination is highly plastic and very difficult to determine accurately (Stumpf and McVean, 2003).

There is evidence that the domestication process itself may have favoured the selection of increased recombination rates (Ross-Ibarra, 2004). In the laboratory model Arabidopsis (*Arabidopsis thaliana*), crossovers occur in higher frequency at specific regions of the genome ("hotspots") (Choi and Henderson, 2015), and they can be influenced by environmental parameters, such as temperature (Lloyd et al.,

2018). Wide crosses to wild crop relatives have been instrumental in regaining genetic variation lost during the domestication process (Hajjar and Hodgkin, 2007). On the other hand, the introduction of foreign genome segments into crops may also reduce or altogether impair recombination, leading to breeding "blind spots" in the genome and potentially deleterious alleles may become fixed in these regions (Lenormand and Otto, 2000). A classic example is the introduction of root-knot nematode resistance in tomato from its wild relative *S. peruvianum* (controlled by the Mi-1 gene) and of resistance to yellow leaf curl virus from *S. chilense* (via the Ty-1 gene). The loci are located in close proximity on chromosome 6 in a region of suppressed recombination, so only commercial hybrid cultivars harbour both resistance traits (Lin et al., 2014).

The difficulty in combining favourable mutations in a single plant during the domestication process was partially bypassed by selecting genes with pleiotropic effects. For example, mutations of the Q gene in wheat allow free threshing (separation of the grain from the straw), a key domestication trait in cereals, but also simultaneously confers increased grain weight, roundness, and yield (Xie et al., 2018). The Q gene encodes an Apetala2-family transcription factor. The wild and domesticated alleles diverge by a nonsynonymous SNP that leads to an amino acid substitution and a synonymous SNP in a microRNA target site that leads to higher gene expression (Simons et al., 2006). Classical QTL mapping studies have helped to identify chromosome regions associated with specific domestication traits in crops (Paterson, 2002). However, both these approaches have failed to uncover pleiotropy in domesticated crops. Most studies revealed the effect of a single, or various very similar phenotypes at a time. The development of multi-phenotype databases such as those developed for *Arabidopsis* expected to overcome this issue (Togninalli et al., 2020). Similar databases for other crops are expected to be developed through High-throughput phenomics platforms, which have the potential to discover gene at an increased rate.

Regulating Mutation and Recombination to Produce Predictable Phenotypes

Understanding the genetic and physiological basis of advantageous phenotypes has remained a challenge to modern plant breeders (Korte and Farlow, 2013). However, the improved understanding of plant function generated in the last decades constitutes a valuable repertoire that can inform a new, knowledge-based breeding proposition (Li and Yan, 2020). Traits can now be designed, modelled, engineered, and stacked using a priori knowledge by targeted manipulation of the fundamental processes of mutation and, potentially, very soon also through recombination (Taagen et al., 2020). This implies that the mechanistic understanding of the connection between genotype and phenotype will increasingly be used for de novo domestication (Figure 3.4).

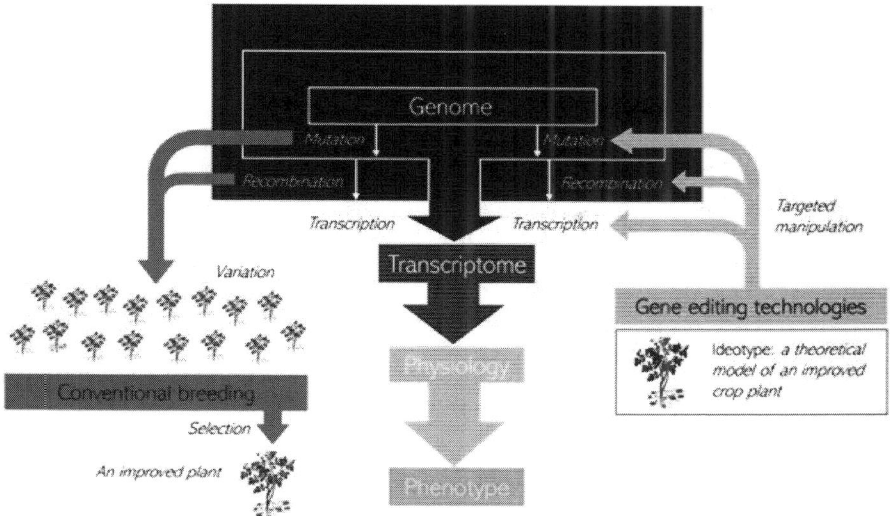

Figure 3.4. A pipeline that combines recently developed gene-editing technologies and conventional breeding to allow domestication and breeding of orphan crops and crop wild relatives. (Reproduced from Curtin et. al. 2022).

Targeted DNA modifications with a level of precision that was not thought possible even a few years ago are now routinely carried out in laboratories around the world (Gao, 2021; Pan et al., 2021a). Tools that modify DNA sequences with specific and predictable changes have undergone rapid improvements over the past decade. Beginning with customized zinc-finger nucleases and closely followed by TAL-effector nucleases, the technology has quickly expanded to multiple CRISPR/Cas platforms (Čermák, 2021). These Cas systems include four variants: nucleases, transposases/recombinases, base editors (BEs), and prime editors (PEs) (Anzalone et al., 2020). Both BE and PE can be used for the genetic manipulation of wild relatives of crops (Molla et al., 2021). They can perform a variety of tasks including the disruption of specific enhancer or transcription factor motif sequences, microRNA (miRNA) target sites, and precise modification of coding sequences to alter specific amino acid residues of proteins, (for example, those associated with pathogen resistance. BE works by enzymatically converting DNA nucleotide bases to generate point mutations without generating a double-strand break or requiring a DNA repair template (Koblan et al., 2021). PE is the most recent addition to the gene-editing tool kit and is essentially a *Streptococcus pyogenes* Cas9 nickase (SpCas9n). A nickase is an enzyme capable of introducing a single-strand cut in the DNA molecule) protein fused to a reverse transcriptase (RT) module (Anzalone et al., 2019). An editing template from a modified guide RNA (gRNA) called "pegRNA" (prime-editing guide RNA) is reverse transcribed by the SpCas9n–RT complex and incorporated into the nicked target locus. Unlike the BE machinery, PE reagents can introduce most types of edits, including short insertions and deletions (Anzalone et. al. 2019). However, the use of this platform in plants has remained limited (Lin et al., 2020), and further refinement is required before it can be adopted as a routine tool. Recently encouraging results have been reported in maize (*Zea mays*) with the efficiency of PEs up to 71.7% by increasing pegRNA expression with a composite 35S:CmYLCV:U6 promoter (Jiang et al., 2020).

The main bottlenecks preventing the widespread implementation of a platform for de novo domestication of wild species are the lack of methods for efficient delivery of the gene-editing reagents into the plant cells and efficient protocols for *in vitro* plant regeneration (Atkins and Voytas, 2020). Highly efficient DNA-free systems to deliver gene-editing reagents using RNA viruses have recently been described and hold promise to avoid the need for transgenes entirely (Ellison et al., 2020; Ma et al., 2020). However, these methods still require plant regeneration from explants using protocols that are time-consuming and labour-intensive and are feasible only for a handful of species (Anjanappa and Gruissem, 2021).

Research that focuses on improved transformation methods, especially for wild and recalcitrant exotic cultivars of certain crops, is paramount for effective gene editing. Two alternative approaches are emerging to overcome this hurdle. The first is the optimization of existing plant transformation and regeneration protocols. For example, using the embryonic axis as explant tissue has seen improved cultivar-independent transformation of legumes including soybean (*Glycine max*), cowpea (*Vigna unguiculata*), and common bean (*Phaseolus vulgaris*) (Paes de Melo et al., 2020; Song et al., 2020a; Che et al., 2021). In addition, regenerated shoots can be produced rapidly in 3–6 weeks and screened for targeted edits. These protocols are likely to become increasingly more efficient with the introduction of developmental regulator technology that has recently transformed monocot transformation (Lowe et al., 2018). The second aims at avoiding *in vitro* regeneration steps altogether by delivering gene-editing reagents to somatic cells of whole plants: Fast-Treated *Agrobacterium* Co-Culture (Fast-TrACC) followed by de novo meristem induction. Fast-TrACC is a protocol whereby an *Agrobacterium* culture harbouring a plasmid with plant developmental regulators (e.g. WUSCHEL, Shootmeristemless) that induce meristem formation and a reporter gene (e.g. luciferase, GUS) that may be included to monitor and calibrate the efficiency of reagent delivery (Nasti et al., 2021). The method was demonstrated in *Nicotiana benthamiana*, and required only 5–15 plants to create gene-edited shoots (Maher et al., 2020). When optimized and extended to other species, this platform will allow gene-edited somatic cells to form meristems that yield seed-producing shoots, increasing throughput and shrinking timescales for creating edited plants.

The best example of the potential of gene editing for crop breeding is canola, the improved varieties of rapeseed (*Brassica napus*), and turnip rape (*B. rapa*) that revolutionized Canadian agriculture in the 1960s. Prior to the 1970s, the high level of toxic erucic acid and glucosinolates hampered the use of rapeseed oils and meals for human and animal consumption, respectively (Khachatourians et al., 2001). Low erucic acid varieties of rapeseed and turnip rape were identified in Germany and Poland in 1960 and were used as parental lines to introduce the trait into commercial varieties that were only released in 1968 (*B. napus* "Oro") and 1971 (*B. rapa* "Span"). Today, a thorough understanding of erucic acid biosynthesis in canola and its genetic basis (Chiron et al., 2015) allows the creation of varieties with reduced erucic acid content via targeted gene editing (Okuzaki et al., 2018). Furthermore, altered fatty acid profiles in the seeds can be generated by precise mutation at a specific region of a gene in a much shorter timeframe via targeted gene manipulation (Huang et al., 2020).

Role of Genome Sequencing on De Novo Domestication

While detecting the phenotypic variance between a crop plant and its wild relatives is generally straightforward, the identification of the genetic basis for these differences has proved much more difficult. The most common approach has been the construction of segregating populations from a cross between a cultivated accession and a wild relative, followed by QTL analysis. Subsequently, fine mapping of the respective domestication gene should be carried out. Alternatively, the population genetic analysis should be carried out using whole-genome profiling by direct sequencing/re-sequencing of diverse varieties of a domesticated crop and its wild progenitor. This method is largely coupled with the genomic screening of artificial selection signatures to detect selective sweeps (Huang and Han, 2014).

Next-generation sequencing has facilitated the identification of domestication genes (Jackson et al., 2011). The availability of whole-genome sequences for a large number of crops and their wild species has facilitated this approach. The application of high-quality sequencing provides important information on the following features of domestication: genomic structural variation, transposable elements, and gene and whole-genome duplications (Olsen and Wendel, 2013, Scossa et al., 2016).

Genomic structural variation has been reported to have a linkage to some important agricultural traits. The copy-number variation is linked to variation in wheat vernalization and flowering time (Diaz et al., 2012), and barley freezing tolerance (Knox et al., 2010). Moreover, transposable elements are present at high frequency throughout crop genomes (Morrell et al., 2011). A Hopscotch retroelement insertion in the canonical domestication gene tb1 of maize has been identified as the cause of the single-stemmed maize phenotype (Studer et al., 2011). Similarly, the white berry phenotype of grapes (Walker et al., 2007) and blood orange colour (Butelli et al., 2012) are both the consequence of retrotransposon insertions. The role of whole-genome duplication in the domestication of crops has been well documented. A global survey of over 200 crops and over 2000 wild species revealing that polyploids are over-represented in crop species (Salman-Minkov et al., 2016). In addition to genome sequencing, considerable insight into domestication has also been provided by transcript profiling, epigenetics studies, and the use of metabolomics (Olsen and Wendel, 2013).

Role of Transcriptomics, Epigenetics, and Metabolomics on De Novo Domestication

In maize, a surprisingly large number of genomic regions are involved in domestication and improvement. These genes exhibit lower expression variability in maize than its progenitors (Hufford et al., 2012, Swanson-Wagner et al., 2012). Such large-scale reprogramming of transcription has also been observed in cotton (Rapp et al., 2010), tomato (Koenig et al., 2013), and common bean (Bellucci et al., 2014). This could be the consequence of the fact that many of the domestication genes encode transcription factors that may have additional target genes apart from those targeted by artificial selection. A more recent genome-wide study revealed the important contribution that alternative splicing makes to regulating phenotypic variance in maize (Liu et al., 2015, Chen et al., 2018).

Several epigenetic variations have been noted to contribute to crop domestication. For example, the colourless non-ripening locus in tomatoes results from the repression of an epiallele of a SQUAMOSA promoter binding protein-like transcription factor (Manning et al., 2006), and epialleles have been reported to associate with vitamin E accumulation in tomatoes. Roles for epialleles have also been reported in sex determination in melon (Martin et al., 2009), plant stature in rice (Miura et al., 2009), photoperiod sensitivity in cotton (Song et al., 2017), and soma clonal variation in oil palm (Ong-Abdullah et al., 2015). Furthermore, genome-wide studies investigating the methylome of important crop species suggest that the majority of these modifications are highly conserved within a species (Kim et al., 2015, Springer and Schmitz, 2017).

The application of metabolomics to study domestication is considerably recent. Although, most of the studies are restricted to QTLs (Fernie and Tohge, 2017, Fang et al., 2019), it is now becoming popular for studying the genetics of quality aspects of several crops (Klee and Tieman, 2018). The following studies have explicitly addressed how domestication and crop improvement affect the metabolome. First, a deep evaluation of changes in primary metabolism in cultivated wheat and its progenitor species demonstrated that changes in unsaturated fatty acid and amino acid content mark the domestication of emmer and durum wheat, respectively (Beleggia et al., 2016). Second, a multiomics approach studying fruit from several hundred tomato genotypes has illustrated how features of domestication have altered the metabolite content of this fruit (Zhu et al., 2018).

Role of Precision Phenotyping for De Novo Domestication

The strategy for identification of target genes for de novo domestication became relatively easy after analysis of the current crops and their wild relatives. Several key genes for improvement in orphan crops have been identified, including genes involved in plant architecture (barley Semi-Dwarf1, Teosinte Branched1, Deeper Rooting1, Phosphorus-Starvation Tolerance1, PROG1) (Jia et al., 2009; Uga et al., 2013; Mai et al., 2014; Shang et al., 2014; Studer et al., 2017), seasonal flowering time (Photoperiod-H1, Centroradialis) (Turner et al., 2005; Comadran et al., 2012), light competition (PHYB1 and PHYB2; Sheehan et al., 2007), seed or fruit retention (Shattering1 and others) (Lin et al., 2012; Meyer and Purugganan, 2013), fruit size (Fruit Weight2.2) (Frary et al., 2000), and length of the juvenile stage (Terminal Flower1, Flowering Locus T) (Bergonzi and Albani, 2011; Yamagishi et al., 2014). Similarly, targeted manipulation of multiple genes can also be readily carried out. However, selecting which relative is the best choice to target for the process is considerably more difficult.

This question is best resolved by comparative phenotyping of a range of species of local wild species against the major cultivated crop as well as select genotyping of the level of allelic variance in the domestication gene in question. However, it is important to note that these species are under-utilized and under-studied. The same could be held true until recently even for major crops such as cassava, sweet potato, and yam as well as nutritionally important crops such as quinoa. The recent publication of the genomes of these species (Wang et al., 2014; Jarvis et al., 2017; Yang et al., 2017; Scarcelli et al., 2019) alongside considerably better characterization of them at the physiological and metabolic levels (Obata et al., 2020; Price et al., 2020; Sonnewald et al., 2020) provides an effective blueprint as to how candidates for de novo domestication should be assessed.

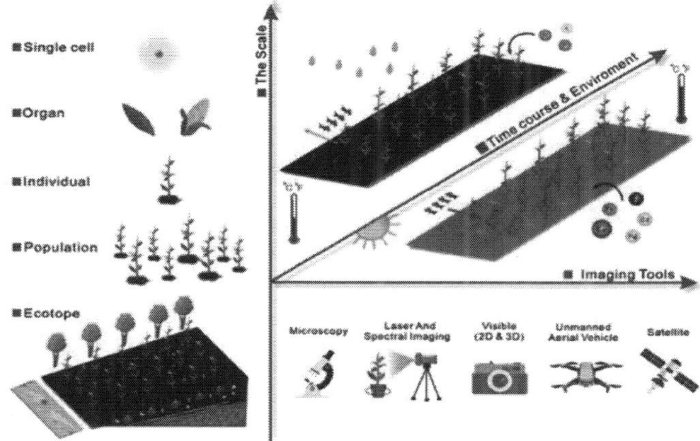

Figure 3.5. Utilizing multiple crop phenotyping tools for accurate high-throughput acquisition and analysis of multidimensional phenotypes on the organism-wide scale (from a single cell to ecotype). Examples range from micro to macro scale, through different environments (abiotic stress, biotic stress, etc.) and across the entire crop developmental process. (Reproduced from Fernie et. al 2021)

Fernie et. al. (2021) proposed a set of tools that should be brought to field trials for comparing the effects of established crops and candidates for de novo domestication is presented in Figure 3.5. It will be important to use contemporary technology to consider all levels of the spatial hierarchy from single cells (Luo et al., 2020) to ecosystem models (Tian et al., 2020), using a broad range of imaging tools to record and also infer trait variances (Yang et al., 2020). Recently the power of inferential data has been provided by a proof-of-concept study comparing hyperspectral imaging of metabolite content to measurements made by mass-spectrometry (Vergara-Diaz et al., 2020). This implies that it may be a reliable technique for spectral measurements of photosynthesis from unmanned aerial vehicles (Gago et al., 2020; Yang et al., 2020). The data emanating from such imaging platforms (Figure 3.6) will certainly form an essential component of deciding which species would be ideal for any given environment for de novo domestication. As an extension to this strategy, the use of the reciprocal transplant strategy (Hereford, 2009; A°gren and Schemske, 2012; Sork, 2018), whereby species are planted and monitored in the environment of one another in addition to their own, would allow the adaptation of exotic species to be quantified in a human-controlled environment as a first indicator that aids in the selection of the best candidates for de novo domestication.

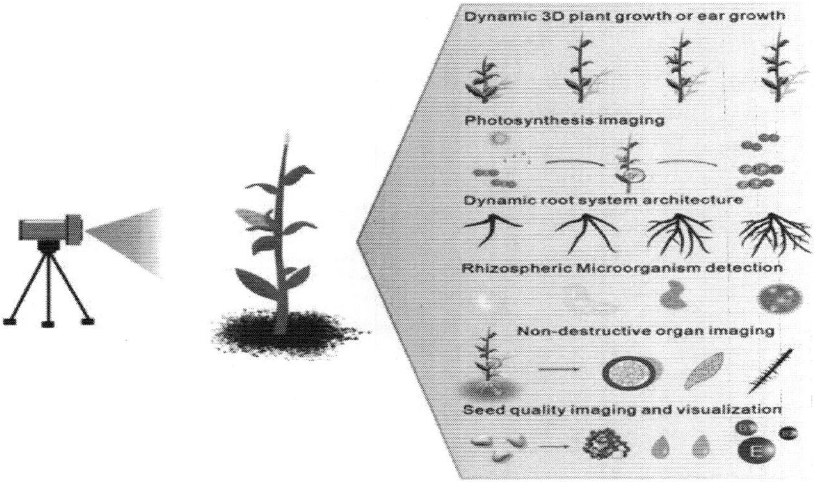

Figure 3.6. The utilization of crop phenotyping techniques. They can be applied widely for example in dynamic measurements of 3D plant growth, ear growth, photosynthesis, root system architecture, rhizospheric microorganism detection, non-destructive organ imaging, and the visualization of seed quality distribution. (Reproduced from Fernie et. al. 2021).

Application of Crispr/Cas Technology for De Novo Domestication

After the selection of the candidate species for de novo domestication, it is important to overcome two hurdles to accomplish the goal. First, it would be prudent to have the genome of the species to be domesticated in hand such that one could be sure that the genetic interventions planned were sensible. Next-generation sequencing technologies have rendered this relatively facile and even large polyploidy genomes can be tackled without trepidation. Second, identification of the genetic targets to modify a route for genetic manipulation. For this purpose, two clear possibilities exist. Using classical breeding, genes of interest can be introgressed into the candidate for de novo domestication, keeping in view that the source of these genes is closely related genetically to the recipient. Alternatively, chemical mutagenesis or the genome editing approach can be used. Since results of chemical mutagenesis is unpredictable, genome editing is preferred. Although, Zinc Finger Nucleases and Transcription Activator like effector nucleases are also utilized, Clustered Regularly Interspaced Short Palindromic Repeats/CRISPR-associated protein (CRISPR-Cas) has been used more frequently due to its simplicity and efficiency. A great advantage of the CRISPR/Cas system is that it can be used to edit multiple genes simultaneously (Xie and Yang, 2013; Zhang et al., 2014; Ma et al., 2015; Qi et al., 2016), and can generate large mutant libraries in a scalable high throughput manner (Lu et al., 2017; Meng et al., 2017; Liu et al., 2020).

CRISPR-Cas systems have already shown their superiority in precision breeding through editing coding regions and knocking out large numbers of genes of interest (Chen et al., 2019; Liu et al., 2021). Reports on fine-tuning of the expression of target genes via editing cis-regulatory sequences or changing the status of epigenetic marks is also very promising for future breeding designs (Liu et al., 2021). Application of big data analysis methods utilizing machine-learning approaches may help to understand and predict the function of each gene, a segment of sequence, or even each base. With CRISPR technology, it is not only possible to modify target sequences, but also can create new sequences, thereby altering their function(s). This will help to create more accurate and efficient de novo domestication in future (Liu et al., 2020). Rodrı´guez-Leal et al. (2017) provided a seminal example of obtaining quantitative variation in the important agronomic traits of fruit size, inflorescence branching, and plant architecture via gene editing. While this was carried out in a domesticated species as a proof-of-concept, it is clearly conceivable that introducing such common domestication traits to underutilized species may well improve their prospects as agricultural products.

Considerable improvements in the efficiency of plant regeneration from tissue culture are achieved by overexpression of plant developmental regulators including Leafy Cotyledon1 or Leafy Cotyledon2 (Lotan et al., 1998; Stone et al., 2001), WUSCHEL (Zuo et al., 2002), or Baby Boom (Boutilier et al., 2002). Moreover, it has been reported that a fusion protein combining wheat Growth-Regulating Factor4 and its cofactor Grfinteracting Factor1 substantially increases the efficiency and speed of regeneration in wheat, triticale, and rice and also increases the number of transformable wheat genotypes (Debernardi et al., 2020). These examples, alongside that of the transformation of the wild rice *O. alta* would appear to offer great promise for this approach. However, the ability to introduce the planned modification into the genome of the potential novel crop, either through classical, mutagenesis, or gene editing approaches, should not be underestimated as it is currently the major challenge facing the de novo domestication approach. A couple of recent developments have been found to be promising. The first is the development of efficient grafting methods that would allow wild species to act as root-stocks as a route to novel agricultural products. This approach has often been discarded due to graft incompatibilities, but this challenge has been overcome in many crops (Notaguchi, 2020). The second is the recent finding that gene-edited dicots can be generated via de novo meristem induction (Maher et al., 2020). In this method, developmental regulators and gene editing reagents can be delivered to somatic cells in order to generate inheritable changes in sequence via a route that bypasses tissue culture. The development of such approaches thus provides optimism that in the future the challenge of modifying lesser-grown plants could prove less daunting.

Several crop traits such as yield and drought tolerance are under complex, polygenic control. Thus, to achieve a desirable phenotype during crop improvement it is imperative to regulate the expression of multiple genes. Repurposing of the CRISPR toolkit now allows manipulation of gene expression using synthetic transcriptional activators or repressors. CRISPR activation (CRISPRa) and CRISPR interference (CRISPRi) are powerful technologies that allow for the simultaneous fine-tuning expression of multiple genes in plants (Lowder et al., 2015; Zhang et al., 2019; Pan et al., 2021a). For instance, with an improved CRISPRa system, termed CRISPR-Act3.0, multiple genes in the same metabolic pathway or in different biological processes could be upregulated at the same time (Pan et al., 2021b). Another promising application of CRISPRa and CRISPRi is their use in large-scale screening of a pool of genes or the whole genome for identifying a causal relationship between gene expression and a desirable phenotype. Once the gene target is identified, CRISPR-based genome editing tools can be used to edit cis-elements of the gene to achieve the expected expression level and create a predictable phenotype. However, it is yet to be fully understand the operating mechanism of cis-regulatory sequences, epigenetic modifications, and chromatin structure mediate transcription (Cramer, 2019).

Despite the immense interest in these approaches, there are some concerns about the application of the CRISPR-Cas system in crop breeding, the biggest of these being the possibility of deleterious effects caused by the integration of transgenic constructs or off-target mutations. Several studies have documented the off-target effects of the CRISPR-Cas system in plants (Xie and Yang, 2013; Zhang et al., 2014; Endo et al., 2015; Jacobs et al., 2015; Jin et al., 2019). Computational analysis of the likelihood of off-target effects is a prerequisite for precise de novo domestication approaches. Other methods of assessing genome-edited crops at the metabolome level have been proposed, which can assess the unintended effects of the editing and, if found to be negligible, such crops can be regarded as safe (Fraser et al., 2020). Such analyses are important for all new crops particularly for the do novo domesticates.

Integration Of Molecular Markers And Metabolites To Improve Phenotype Prediction

Several computational tools are being utilized for predicting phenotypes in de novo domestication. Recently, a gene expression model of the phenylpropanoid pathway in a wide range of wild species tomatoes based on metabolome data has been proposed (Tohge et al., 2020). Being a complex process, metabolomics data usually cannot be used to predict yield. Likewise, it is difficult to find a single metabolite whose level is predictive of yield. However, several studies have revealed metabolic signatures for yield in a range of species including *Arabidopsis*, maize, wheat, and tomato (Schauer et al., 2006; Meyer et al., 2012; Riedelsheimer et al., 2012; Obata et al., 2015; Vergara-Diaz et al., 2020).

Omics-level data are now being utilized for genomic selection approaches that utilize molecular markers and machine learning to identify superior genotypes with improved traits such as growth. Results from recent studies suggest that integrating molecular markers into metabolic models can dramatically improve the prediction accuracy of genomic selection strategies (Tong et al., 2020).

Manipulation Of Gene Expression To Create Predictable Phenotypes

The manipulation of cis-regulatory regions that control gene expression is increasingly gaining importance as an alternative to create desirable agronomic traits that closely resemble those produced by crop domestication and improvement (Rodríguez-Leal et al., 2017). Regulatory variants are the basis of many important domestication traits in crops such as maize (Chen et al., 2021a) and tomato (Alonge et al., 2020). However, two major hurdles exist for the application of this approach. First, in many cases, the controlling element(s) reside in distant regions of functional genes, which poses a challenge to effective genome manipulation. For instance, reduced branching in the maize plant is controlled by a transposon insertion 60-kb upstream of teosinte branched 1 (tb1), that enhances its expression in leaf axils and prevents outgrowth of axillary buds (Studer et al., 2011). The product of tb1 is a basic helix–loop–helix-containing transcription factor of the Teosinte Branched 1, Cycloidea, PCF family. Second, the lack of sufficient knowledge about the relationship between the promoter sequence and function to allow deliberate manipulation with predictable phenotypic outcomes. For example, variation in fruit size is known to be due to expression pattern of the gene fw2.2. Although the gene was identified more than two decades ago, the controlling element upstream of the coding sequence remains unknown, precluding its manipulation for breeding (Beauchet et al., 2021). Targeted dissection of promoter regions is starting to reveal the complex relationships between cis-regulatory mutations, gene expression patterns, and phenotypic outcomes (Hendelman et al., 2021; Liu et al., 2021a; Wang et al., 2021). Application of machine learning on these large datasets (Vo ngoc et al., 2020), coupled with genome-wide analysis through technologies such as ATAC-seq, which allow assaying chromatin accessibility (Buenrostro et al., 2015), make it likely that cis-element editing will soon be better designed to yield more predictable phenotypes (Figure 3.7).

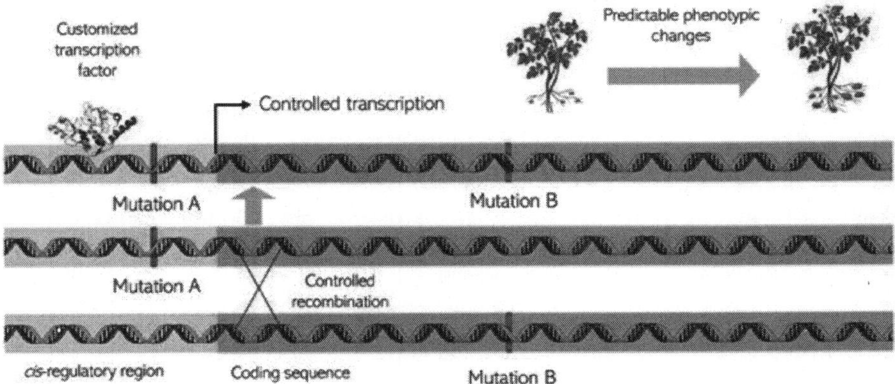

Figure 3.7. Schematic representation showing how manipulating mutation and recombination can create new traits predictably and reliably. (Reproduced from Curtin et. al. 2022).

The Status Of De Novo Domestication

Collective evidence suggests that there are many ways to domesticate a plant. Results of the limited number of domestication experiments published to date suggest that artificial selection needs in excess of 20 generations to change the phenotypes of wild or crop wild hybrids (Hilu and Wet, 1980, Hillman and Davies, 1990, Hillman and Davis, 1990). Furthermore, studies on the genetic architecture of domestication suggest that domestication phenotypes under simple genetic control can be achieved more quickly than traits with a complex genetic basis. More problematic than the timescale of domestication is the phenomenon that has become known as the cost of domestication (Charlesworth and Willis, 2009, Gepts, 2014, Gaut et al., 2018), that is, the selection can drag linked deleterious mutants to high frequency (Chun and Fay, 2011, Zhou et al., 2017). Most crops examined to date contain considerably more deleterious allelic variants than their wild relatives (Renaut and Rieseberg, 2015, Kono et al., 2016, Liu et al., 2017, Ramu et al., 2017, Wang et al., 2017, Zhou et al., 2017), with species that are clonally propagated being particularly susceptible to accrue such mutations (Gaut et al., 2018). This phenomenon clearly needs to be carefully considered when planning future de novo domestication strategies.

The productivity of plant species varies greatly across the globe, some are even poorly suited to their area of origin and cultivation. For example, the yields of cassava (*Manihot esculenta*) are three times higher in South East Asia than in Africa (Sonnewald et al., 2020). This instability aside, only 150 plant species are cultivated around the globe to a large extent and 70 per cent of the calories consumed by humans come from only 15 of these species (Fernie and Yan, 2019). More than 7,000 of the 400,000 plant species are regarded as semi-cultivated (Smykal et al., 2018) and could represent important germplasm for the design of future crops.

Recent studies have demonstrated the potential of gene editing for improving domestication traits. One of the studies is based on the orphan crop groundcherry (*Physalis pruinosa*) (Lemmon et al., 2018), while the other two describe de novo domestication of the wild tomato *Solanum pimpinellifolium* (Li et al., 2018b, Zsögön et al., 2018). The study on groundcherry focused on its undesirable characteristics for a crop, that is, its sprawling growth habit and small fruits, which drop to the ground due to strong stem abscission. It included the development of a transformation procedure for this species alongside whole-genome sequencing and RNA sequencing in order to provide much-needed genome resources. Thereafter, they examined homologs of the florigen suppressors and known tomato improvement genes, SELF-PRUNING (SP) and SELF-PRUNING 5G (SP-5G), with the knockout of the first of these proving too severe and causing extreme compactness but knockout of the second resulting in increased auxiliary flowering without changes in the primary shoot, which nevertheless resulted in enhanced fruit density (Lemmon et al., 2018). Similarly, targeting of the CLAVATA (CLV) pathway, which regulates shoot apical meristem size, resulted in increased floral meristem size, additional flower organs, and conversion from a two-locular to a larger three-locular fruit (Lemmon et al., 2018). The analogous studies in *S pimpinellifolium* targeted SP, SP-5G, CLV, and a total of eight further domestication genes, looking to improve architectural traits, day-length insensitivity, fruit size and shape, and vitamin A and C content (Li et al., 2018b, Zsögön et al., 2018). Both studies were highly successful in terms of the specific traits that they targeted, doubling the yield of *S pimpinellifolium* and increasing lycopene contents to 500% of control levels (Li et al., 2018b, Zsögön et al., 2018). The power of gene editing in breeding is provided by demonstrating that knocking out of the self-incompatibility gene S-RNase allows redomestication of potato into an inbred-line based diploid crop, representing a promising alternative to traditional clonal propagation of tetraploid potato (Ye et al., 2018).

Although the early examples of de novo domestication were restricted to the Solanaceae, the approach has been extended to pennycress (*Thlaspi arvense* L.) (McGinn et al., 2019), sunflower (*Helianthus annuus*) (Ekar et al., 2019; Van Tassel et al., 2020), and *Vigna stipulacea* (Takahashi et al., 2019). Pennycress has been de novo domesticated as a seed cover crop for the winter fallow period. This was carried out through CRISPR to produce insertion/deletion (indel) mutations in the Fatty Acid Elongation1 (FAE1) gene, thereby abolishing erucic acid production and creating an edible seed oil comparable to that of canola (McGinn et al., 2019). On the other hand, conventional breeding has been used to domesticate the oilseed Silphium, resulting in increased above ground biomass at the seedling and adult stages leading to increase in seed yield and modest improvement in harvest index (Van Tassel et al., 2020). A similar approach was followed in sunflower domestication to develop a perennial crop that can produce both high-value vegetable oil and continuous ground cover (Ekar et al., 2019).

The silencing of different genes (namely LABA1 and RAE2) via RNA interference has been used to shorten the awns of a rice introgression with long barber awns (Hua et al., 2015, Bessho-Uehara et al., 2016). Similarly, the silencing of genes of fatty acid biosynthesis (FAD2 and FAE1) improved the seed oil quality of the wild plant field cress (*Lepidum campestre*) (Ivarson et al., 2016). Mimicry of domestication has additionally been achieved by random mutagenesis followed by sequence evaluation of the "targeted" domestication gene. Examples of this approach include the shatter-proofing of weeping rice grass (*Microlaena stipoides*), an Australian wild relative of rice, via mutagenesis of the orthologs of qSH1 and sh4 genes (Shapter et al., 2013), while crop productivity in tomato has been improved by combining induced mutations in the florigen pathway (Park et al., 2014). Several genome-sequencing studies have been initiated for species that can be regarded as currently in the process of being domesticated, including intermediate wheatgrass (*Thinopyrum intermedium*), field pennycress (*Thlaspi arvense*), tree tobacco (Nicotiana glauca), and apiois (Apiose americana) (Dorn et al., 2015, Belamkar et al., 2016, Kantarski et al., 2017, Usadel et al., 2018). These studies provide clear proof of principle that the alteration of single domestication genes can dramatically improve the potential of little- or non-cultivated species.

Takahashi et al. (2019) carried out ethyl methanesulfonate (EMS) mutagenesis in a *V. stipulacea* population, and isolated mutants with reduced seed dormancy and shattering. Vigna represents an interesting legume species for de novo domestication given that it exhibits rapid growth, a short vegetative stage, and broad resistance to pests and diseases. Although it will be quite time-consuming, the authors postulate that by pyramiding these mutant phenotypes, they will be able to generate a primitive crop that can be cultivated without pesticides. In sorghum (*Sorghum bicolor*) and sugarcane (*Saccharum officinarum*), enhanced stress tolerance lines have been convincingly made (Zhang et al., 2018; Fernie and Yan, 2019). Yu et. al. (2021) in order to harness the advantages of polyploidization for genome buffering, vigorousness, and environmental robustness in rice, used the allotetraploid wild rice *Oryza alta*, for de novo domestication. The authors developed an efficient transformation system, thus facilitating gene editing and a high-quality genome assembly of *O. alta*. Following this, six agronomically important traits, namely shattering, awn length, hull colour, pericarp colour, panicle shape, and grain width, were rapidly improved in *O. alta*, thereby demonstrating the feasibility of its de novo domestication (Figure 3.8). This study is arguably the first true de novo domestication and illustrates the importance of developing transformation systems and genome sequences as enabling steps for this process.

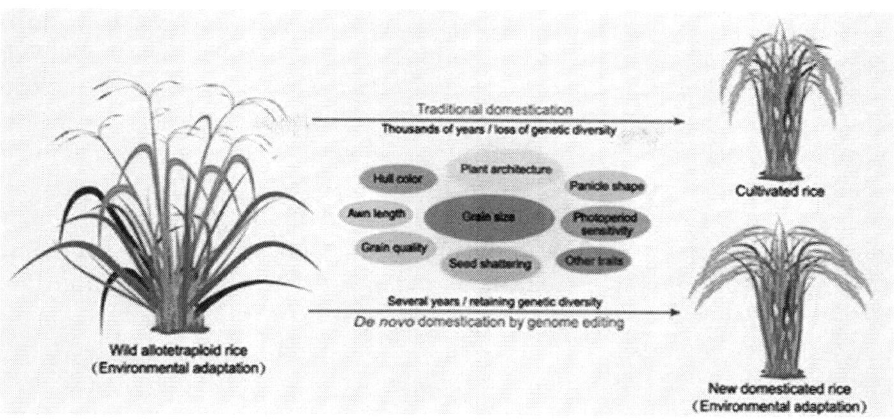

Figure 3.8. De novo domestication of wild plants towards new crops. By genome editing of key genes for domestication and agronomic traits, wild allotetraploid rice can be rapidly domesticated into new-type rice cultivars that have strong environmental adaptability and high grain yield. (Reproduced from Xie and Liu, 2021).

Two parallel approaches have been suggested for the de novo domestication of wild plants: traditional breeding approaches (Runck et al., 2014, DeHaan et al., 2016) and gene editing (Altpeter et al., 2016, Kantar et al., 2016, Zsögön et al., 2017). For gene editing, the CRISPR/Cas9 approach has become the method of choice (Altpeter et al., 2016, Pacher and Puchta, 2017, Scheben et al., 2017). This genome-editing tool, which is modified from a prokaryotic immune system, induces double-stranded DNA breaks by the action of Cas9 nuclease at a genome location corresponding to a designed guide RNA (Altpeter et al., 2016). Gene editing can, at least in theory, be used to: activate or suspend the function of any gene (Qi et al., 2013); create multiple different alleles of any gene (Rodriguez-Leal et al., 2017); achieve any base substitution (Gaudelli et al., 2017); add genes that do not exist in the original genome (Park et al., 2017); and delete any sequence including large chromosomal fragments or even the entire chromosome (Xiao et al., 2013). However, to date, prerequisites for the use of CRISPR/Cas9 are that the genome of the plant to be edited must be sequenced in order to identify known orthologs of domestication genes. As such, its use is currently largely restricted to cases wherein a loss-of-function mutation will lead to the domestication phenotype. That said, the number of species whose genomes have been edited is rapidly expanding (Osterberg et al., 2017), and important examples of loss-of-function mutants with beneficial effects include those enhancing rice and maize yields (Belamkar et al., 2016, Shi et al., 2017), the generation of virus-resistant cucumber (Chandrasekaran et al., 2016) and mildew-resistant wheat (Wang et al., 2014), improved soybean oil content (Haun et al., 2014), and potato storage ability (Clasen et al., 2016), as well as potentiating rubber production in dandelion (Iaffaidano et al., 2016).

Zsögön et al. (2017) recently assessed the domestication of the world's six major crops and postulated that gene-editing techniques could be used to introduce key monogenic traits into crop wild relatives as a means of de novo domestication (Table 3.4). They observed that most of the genes identified represent modifications that could readily be targeted by available gene-editing technology and noted that one of the targets had indeed already been modified by such techniques in tomato (Livne et al., 2015). Zsögön et al. (2017) also surveyed two of the three next most important world crops, cassava, and potato, arguing that potato and cassava were selected to accumulate biomass in underground sinks by alterations in the levels of cytokinins and the mobile signal SP6A, which are responsible for tuber initiation (Peres et al., 2005, Eviatar-Ribak et al., 2013). They suggested that the disease-resistant *Solanu demissum* and *Solanum stoloniferum* as potential substitute crops for potato, and the large-rooted and apoximic *Manihot glaziobii* and *Manihot neosana* (Nassar et al., 2008) as substitutes for cassava.

Table 3.4. Selected domestication genes spanning the range of traits that have commonly been described as domestication traits.

Trait	Crop	Domesticated gene	Function	References
Fruit size and weight	Tomato	SUN	Unknown protein	Xiao et. al. 2008
	Tomato	Fasciated	TF	Cong et. al. 2008
	Tomato	Fw2.2	Similar to human RAS	Frary et. al. 2000
	Rice	qSW5	Unknown protein	Shomura et. al. 2008
Grain filling	Rice	GIF1	Cell wall invertase	Wang et. al. 2008
	Maize	ZmSWEET4c	Hexose transporter	Sosso et. al. 2015
Growth habit	Soybean	Dt1	Signaling protein	Tian et. al. 2010
Plant/ inflorescence architecture	Barley	INT-C (HvTB1)	TF	Ramsay et. al. 2011
	Barley	Vrs1	TF	Komastsuda et. al. 2007
	Maize	ra1	TF	Sigmon and Vollbrecht 2010
	Wheat	Q	TF	Simons et. al. 2006
	Maize	tb1	TF	Doebley et. al. 1997
	Rice	PROG1	TF	Jin et. al. 2008, Tan et. al. 2008
	Rice	SD1	Gibberellin biosynthetic enzyme	Asano et. al. 2011

Trait	Crop	Domesticated gene	Function	References
Seed casing	Barley	Nud	TF	Taketa et. al. 2008
	Maize	tga1	TF	Wang et. al. 2005
Seed colour	Rice	Bh4	Amino acid transporter	Zhu et. al. 2011
	Rice	Rc	TF	Sweeney et. al. 2006
Seed dormancy	Rice	Sdr4	Zinc finger protein	Sugimoto et. al. 2010
	Rice	OsG	Protease	Wang et. al. 2018
	Soybean	G	Protease	Wang et. al. 2018
	Tomato	SolyG	Protease	Wang et. al. 2018
Shattering	Rice	qSH1	TF	Konishi et. al. 2006
	Rice	sh4	TF	Li et. al. 2006
	Sorghum	Sh1	TF	Lin et. al. 2012
Style length	Tomato	Style2.1	TF	Chen et. al. 2007
Flowering	Wheat	Vrn1	TF	Yan et. al. 2003
	Wheat	Vrn2	TF	Yan et. al. 2004
	Sunflower	HaFT1	TF	Blackman et. al. 2010

As an alternative, as well as being a powerful strategy to determine the genetic architecture of traits (Fernie et al., 2006), introgression breeding has been a staple of crop improvement over many years, with considerable gains being made in yield (Ashikari et al., 2005) and disease resistance (Nelson et al., 2018). In practice, this occurs by creating populations of near-isogenic lines in which either a single locus is introgressed via backcrossing (Belcher et al., 2012), many loci can be introgressed (Liu et al., 2016), or lines at the end of the inbreeding process harbouring residual heterozygous regions can be self-pollinated to produce near-isogenic lines divergent at these regions (Tuinstra et al., 1997). Ashikari et al. (2005) developed a process called QTL pyramiding by combining the QTL Gn1a gene for grain number with other desirable traits such as plant height. Gur and Zamir (2004) demonstrated that by pyramiding three independent yield-promoting genomic regions from the drought-tolerant *Solanum pennellii,* it was possible to elevate tomato yield dramatically. Introgression has been commonly used in breeding late blight disease resistance in potato (Haverkort et al., 2016) and submergence tolerance in rice (Bailey-Serres et al., 2010). However, relatively few studies have performed introgressions in the opposite direction, that is, the introgression of domesticated alleles into exotic backgrounds. Nevertheless, introgression breeding has two clear disadvantages with respect to gene editing. First, availability of limited range of phenotypic variance that can be introduced into species that are closely related and crossable. Second, incorporation of deleterious genes linked to the desirable gene and thereby accumulation of too many linked deleterious mutations in the recipient.

The increasing demand for biofuels has driven increased interest in the next wave of de novo domestications. For example, scientific breeding of leading candidates such as *Miscanthus* (Heaton et al., 2008) is underway, while research on close relatives of sorghum and sugarcane (Paterson et al., 1995a, Paterson et al., 1995b, Zhang et al., 2018b). and tree tobacco and *J. curcas* (Montes and Melchinger, 2016, Usadel et al., 2018) have been initiated. Moreover, farming of aquatic plants such as duckweed (Cheng and Stomp, 2009) and seaweed (Loureiro et al., 2015) has also been reported.

Conclusions

The ability to control mutation and recombination, the two fundamental biological processes underlying crop domestication and breeding, may soon make it possible to engineer novel plant phenotypes with agronomic potential. Technical advances in gene editing promise novel tools that will allow the manipulation, of not only gene coding sequences and their combination into a single plant but also cis-regulatory elements, epigenetic regulation, and quantitative genetic variation. Simultaneously editing multigene families and controlling epistatic interactions between many genes should then be feasible. Systems to customize gene expression under different environments will allow specific genetic control of developmental and physiological variation. However, it is important to bear in mind that all of these advances will ultimately rely on fundamental knowledge about the relationship between genotype and phenotype in crops and their wild relatives. Thus, the expanding knowledge about the genetic basis of complex agronomic traits in crops will represent a key asset to allow the improvement of orphan crops and the domestication of crop wild relatives.

Multiple roles for computation can be envisaged in the process of de novo domestication from comparative phenotypic evaluation of potential starting species, through genome analysis and population genomics to the comparative evaluation of de novo domesticates and their progenitors. The need for computation in the design of CRISPR/Cas strategies, the evaluation of GWASs, multi-omics integration, genomic prediction, and deep learning is equally pervasive. Despite its great promise, several important and even fundamental questions remain as yet unanswered. Among these the genetic architecture underlying traits, the level of standing genetic variation for domestication traits, and the accumulation of genetic load (i.e. deleterious genetic variants) have been noted as important factors determining the extent of domestication of various species. Less domesticated crops are often well adapted to diverse environments and of high nutritional value but need improvement in key domestication traits to render them serious alternatives to our existing crops. The key steps in de novo domestication are: (a) the comprehensive evaluation of a wide range of lesser-grown crops alongside the acquisition of (b) high quality genome sequences, (iii) knowledge concerning functional genes, and (c) the competence to generate variance in key genes of interest. A synergistic combination of conventional breeding and gene-editing technology will contribute to food security in the face of uncertain future conditions.

References

Abbo, S., Pinhasi van-Oss, R., Gopher, A., Saranga, Y., Ofner, I., & Peleg, Z. (2014). Plant domestication versus crop evolution: a conceptual framework for cereals and grain legumes. *Trends Plant Sci. 19*:351–360.

Abrouk, M., Ahmed, H.I., Cubry, P., Simonı´kova´, D., ... Cauet, S. (2020). Fonio millet genome unlocks African orphan crop diversity for agriculture in a changing climate. *Nat. Comm. 11*: 4488 A

Adeyemo, O.S., Hyde, P.T., & Setter, T.L. (2019). Identification of FT family genes that respond to photoperiod, temperature and genotype in relation to flowering in cassava (*Manihot esculenta*, Crantz). *Plant Reprod. 32*:181–191.

Alonge, M., Wang, X., Benoit, M., Soyk, S., Pereira, L., ... Zhang, L. (2020). Major impacts of widespread structural variation on gene expression and crop improvement in tomato. *Cell. 182*: 145–161.

Alonso-Blanco, C., Aarts, M.G., Bentsink, L., ... Keurentjes, J.J. (2009). What has natural variation taught us about plant development, physiology, and adaptation? *Plant Cell. 21*:1877–1896.

Alseekh, S., Kostova, D., Bulut, M., & Fernie, A.R. (2021). Genome-wide association studies: assessing trait characteristics in model and crop plants. *Cell. Mol. Life Sci. 78*: 5743–5754.

Alseekh, S., Scossa, F., & Fernie, A.R. (2020). Mobile transposable elements shape plant genome diversity. *Trends Plant Sci. 25*:1062–1064.

Altpeter, F., Springer, N.M., Bartley, L.E., ... Blechl, A.E. (2016). Advancing crop transformation in the era of genome editing. *Plant Cell. 28*:1510–1520.

Anjanappa, R.B., & Gruissem, W. (2021). Current progress and challenges in crop genetic transformation. *J. Plant Physiol. 261*: 153411.

Anzalone, A.V., Koblan, L.W, & Liu, D.R. (2020). Genome editing with CRISPR–Cas nucleases, base editors, transposases and prime editors. *Nat. Biotechnol. 38*: 824–844.

Anzalone, A.V., Randolph, P.B., Davis, J.R., ... Sousa, A.A. (2019). Search-and-replace genome editing without double-strand breaks or donor DNA. *Nature. 576*: 149–157.

Araus, J.L., & Kefauver, S.C. (2018). Breeding to adapt agriculture to climate change: affordable phenotyping solutions. *Curr. Opin. Plant Biol. 45*:237–247.

Arnold, B.J., Lahner, B., DaCosta, J.M., ... Weisman, C.M. (2016). Borrowed alleles and convergence in serpentine adaptation. *Proc. Natl. Acad. Sci. U S A 113*:8320–8325.

Asano, K., Yamasaki, M., Takuno, S., Miura, K., ... Katagiri, S. (2011). Artificial selection for a green revolution gene during japonica rice domestication. *Proc. Natl. Acad. Sci. USA. 108*:11034–11039.

Ashikari, M., Sakakibara, H., Lin, S., ... Yamamoto, T. (2005). Cytokinin oxidase regulates rice grain production. *Science. 309*:741–745.

Atefi, A., Ge, Y., Pitla, S., & Schnable, J. (2021). Robotic technologies for high-throughput plant phenotyping: contemporary reviews and future perspectives. *Front. Plant Sci.12*: 1082.

Atkins, P.A., & Voytas, D.F. (2020). Overcoming bottlenecks in plant gene editing. *Curr. Opin. Plant Biol. 54*: 79–84.

Avni, R., Nave, M., Barad, O., Baruch, K., ... Twardziok, S.O. (2017). Wild emmer genome architecture and diversity elucidate wheat evolution and domestication. *Science. 357*:93–97.

Azhaguvel, P., & Komatsuda, T. (2007). A phylogenetic analysis based on nucleotide sequence of a marker linked to the brittle rachis locus indicates a diphyletic origin of barley. *Ann. Bot. 100*:1009–1015.

Bailey-Serres, J., Fukao, T., Ronald, P., ... Ismail, A. (2010). Submergence tolerant rice: SUB1's journey from Landrace to Modern Cultivar. *Rice 3*:138–147.

Barrera-Redondo, J., Pinero, D., & Eguiarte, L.E. (2020). Genomic, transcriptomic and epigenomic tools to study the domestication of plants and animals: a field guide for beginners. *Front. Genet. 11*: 742.

Baute, G.J., Kane, N.C., Grassa, C.J., Lai, Z., & Rieseberg, L.H. (2015). Genome scans reveal candidate domestication and improvement genes in cultivated sunflower, as well as post-domestication introgression with wild relatives. *New Phytol. 206*: 830–838

Beauchet, A, Gévaudant, F., Gonzalez, N., & Chevalier, C. (2021). In search of the still unknown function of FW2.2/cell number regulator, a major regulator of fruit size in tomato. *J. Exp. Bot. 72*: 5300–5311.

Belamkar, V., Farmer, A.D., Weeks, N.T., Kalberer, S.R., Blackmon, W.J., & Cannon, S.B. (2016). Genomics-assisted characterization of a breeding collection of *Apios americana*, an edible tuberous legume. *Sci. Rep. 6*. https://doi.org/10.1038/srep34908.

Belcher, A.R., Zwonitzer, J.C., Santa Cruz, J., ... Krakowsky, M.D. (2012). Analysis of quantitative disease resistance to southern leaf blight and of multiple disease resistance in maize, using near-isogenic lines. *Theor. Appl. Genet. 124*:433–445.

Beleggia, R., Rau, D., Laido, G., Platani, C., Nigro, F., ... Fragasso, M. (2016). Evolutionary metabolomics reveals domestication-associated changes in tetraploid wheat kernels. *Mol. Biol. Evol. 33*: 1740–1753

Beleggia, R., Rau, D., Laido, G., Platani, C., ... Nigro, F. (2016). Evolutionary metabolomics reveals domestication-associated changes in tetraploid wheat kernels. *Mol. Biol. Evol. 33*:1740–1753.

Bellucci, E., Bitocchi, E., Ferrarini, A., Benazzo, A., ... Biagetti, E. (2014). Decreased nucleotide and expression diversity and modified coexpression patterns characterize domestication in the common bean. *Plant Cell. 26*: 1901–1912

Bergonzi, S., & Albani, M.C. (2011). Reproductive competence from an annual and a perennial perspective. *J. Exp. Bot. 62*: 4415–4422

Bessho-Uehara, K., Wang, D.R., Furuta, T., ... Minami, A. (2016). Loss of function at RAE2, a previously unidentified EPFL, is required for awnlessness in cultivated Asian rice. *Proc. Natl. Acad. Sci. USA .113*:8969–8974.

Bevan, M.W., Uauy, C., Wulff, B.B.H., Zhou, J., Krasileva, K., & Clark, M.D. (2017). Genomic innovation for crop improvement. *Nature. 543*: 346–354.

Beyer, P. (2010). Golden Rice and 'Golden' crops for human nutrition. *Nat. Biotechnol. 27*:478–481.

Beying, N., Schmidt, C., Pacher, M., Houben, A., & Puchta, H. (2020). CRISPR-Cas9-mediated induction of heritable chromosomal translocations in *Arabidopsis. Nat. Plants 6*: 638–645

Bitocchi, E., Nanni, L., Bellucci, E., Rossi, M., ... Giardini, A. (2012). Mesoamerican origin of the common bean (*Phaseolus vulgaris* L.) is revealed by sequence data. *Proc. Natl. Acad. Sci. USA. 109*: E788–E796.

Blackman, B.K., Strasburg, J.L., Raduski, A.R., Michaels, S.D., & Rieseberg, L.H. (2010). The role of recently derived FT paralogs in sunflower domestication. *Curr. Biol. 20*:629–635.

Blake, V.C., Birkett, C., Matthews, D.E, Hane, D.L., Bradbury, P., & Jannink, J.L. (2016). The Triticeae toolbox: combining phenotype and genotype data to advance small-grains breeding. *Plant Genome. 9*: 386.

Bonhomme, M., Chevalet, C., Servin, B., Boitard, S., Abdallah, J., Blott, S., & Sancristobal, M. (2010). Detecting selection in population trees: the Lewontin and Krakauer test extended. *Genetics 186*: 241–262.

Boutilier, K., Offringa, R., Sharma, V.K., Kieft, H., ... Ouellet, T. (2002). Ectopic expression of BABY BOOM triggers a conversion from vegetative to embryonic growth. *Plant Cell 14*: 1737–1749

Breton, C., Terral, J.F., Pinatel, C., Medail, F., Bonhomme, F., & Berville, A. (2009). The origins of the domestication of the olive tree. *C. R. Biol. 332*:1059–1064.

Broman, K.W., Gatti, D.M., Simecek, P., Furlotte, N.A., Prins, P., Sen, S., Yandell, B.S., & Churchill, G.A. (2019). R/qtl2: software for mapping quantitative trait loci with high-dimensional data and multiparent populations. *Genetics 211*: 495–502

Buenrostro, J.D., Wu, B., Chang, H.Y., & Greenleaf, W.J. (2015). ATAC-seq: a method for assaying chromatin accessibility genome-wide. *Curr. Protoc. Mol. Biol. 109*: 21.29.1–21.29.9.

Burghardt, L.T., Young, N.D., & Tiffin, P. (2017). A guide to genome-wide association mapping in plants. *Curr. Protoc. Plant Biol. 2*: 22–38.

Butelli, E., Licciardello, C., Zhang, Y., Liu, J., ... Mackay, S. (2012). Retrotransposons control fruit-specific, cold-dependent accumulation of anthocyanins in blood oranges. *Plant Cell. 24*:1242–1255.

Butelli, E., Titta, L., Giorgio, M., Mock, H.P., ... Matros, A. (2008). Enrichment of tomato fruit with health-promoting anthocyanins by expression of select transcription factors. *Nat. Biotechnol. 26*:1301–1308.

Callaway, E. (2018). CRISPR plants now subject to tough GM laws in European Union. *Nature. 560*:16.

Carvalho, F.P. (2017) Pesticides, environment, and food safety. *Food Energy Secur. 6:*48–60.

Čermák. T. (2021). Sequence modification on demand: search and replace tools for precise gene editing in plants. *Transgenic Res. 30*: 353–379.

Chandrasekaran, J., Brumin, M., Wolf, D., ...Leibman, D. (2016). Development of broad virus resistance in non-transgenic cucumber using CRISPR/Cas9 technology. *Mol. Plant Pathol. 17*:1140–1153.

Charlesworth, D., & Willis, J.H. (2009). The genetics of inbreeding depression. *Nat. Rev. Genet. 10*:783–796.

Che, P., Chang, S., Simon, M.K., Zhang, Z., Shaharyar, A., ... Ourada, J. (2021). Developing a rapid and highly efficient cowpea regeneration, transformation and genome editing system using embryonic axis explants. *Plant J. 106*: 817–830.

Che, G., & Zhang, X. (2018). Molecular basis of cucumber fruit domestication. *Curr. Opin. Plant Biol. 47*:38–46.

Chen, K., Wang, Y., Zhang, R., Zhang, H., & Gao, C. (2019). CRISPR/Cas genome editing and precision plant breeding in agriculture. *Ann. Rev. Plant Biol. 70*: 667–697

Chen, Q., Li, W., Tan, L., & Tian, F. (2021a). Harnessing knowledge from maize and rice domestication for new crop breeding. *Mol. Plant.14*: 9–26.

Chen, Z., Sun, J., Li, D., Li, P., He, K., Ali, F., Pan, Q., Mi, G., Chen, F., & Yuan, L. (2021b). Plasticity and domestication of root anatomy in maize-teosinte derived population. *J. Exp. Bot.* doi: 10.1093/jxb/erab406

Chen, K.Y., Cong, B., Wing, R., Vrebalov, J., & Tanksley, S.D. (2007). Changes in regulation of a transcription factor lead to autogamy in cultivated tomatoes. *Science. 318*:643–645.

Chen, Q., Han, Y., Liu, H., Wang, X., Sun, J., ... Zhao, B. (2018). Genome-wide association analyses reveal the importance of alternative splicing in diversifying gene function and regulating phenotypic variation in maize. *Plant Cell. 30*:1404–1423.

Cheng, F., Sun, R., Hou, X., Zheng, H., Zhang, F., ... Zhang, Y. (2016). Subgenome parallel selection is associated with morphotype diversification and convergent crop domestication in *Brassica rapa* and *Brassica oleracea*. *Nat. Genet. 48*:1218–1224

Cheng, J.J., & Stomp, A.M. (2009). Growing duckweed to recover nutrients from waste waters and for production of fuel ethanol and animal feed. Clean. *Soil Air Water. 37*:17.26.

Childs, L.H., Lisec, J., & Walther, D. (2012). Matapax: an online high-- throughput genome-wide association study pipeline. *Plant Physiol.158*:1534–1541

Chiron, H., Wilmer, J., Lucas, M.O., Nesi, N., Delseny, M., Devic, M., & Roscoe, T.J. (2015). Regulation of fatty acid elongation 1 expression in embryonic and vascular tissues of *Brassica napus*. *Plant Mol. Biol. 88*:65–83.

Choi, K., & Henderson, I.R. (2015). Meiotic recombination hotspots: a comparative view. *Plant J. 83*:52–61.

Chopra, R., Johnson, E.B., Emenecker, R., Cahoon, E.B., ... Lyons, J. (2020). Identification and stacking of crucial traits required for the domestication of pennycress. *Nat. Food.1*: 84–91.

Chun, S., & Fay, J.C. (2011). Evidence for hitchhiking of deleterious mutations within the human genome. *PLoS Genet. 7*:e1002240.

Clasen, B.M., Stoddard, T.J., Luo, S., ... Demorest, Z.L. (2016). Improving cold storage and processing traits in potato through targeted gene knockout. *Plant Biotechnol. J. 14*:169–176.

Comadran, J., Kilian, B., Russell, J., Ramsay, L., Stein, N., ... Ganal, M. (2012). Natural variation in a homolog of *Antirrhinum centroradials* contributed to spring growth habit and environmental adaptation in cultivated barley. *Nat. Genet. 44*:1388–1392

Comai, L. (2018). The taming of the shrub. *Nat. Plants 4*:742–743.

Cong, B., Barrero, L.S., & Tanksley, S.D. (2008). Regulatory change in YABBY-like transcription factor led to evolution of extreme fruit size during tomato domestication. *Nat. Genet. 40*:800–804.

Cornille, A., Giraud, T., Smulders, M.J., Roldan-Ruiz, I., & Gladieux, P. (2014). The domestication and evolutionary ecology of apples. *Trends Genet. 30*:57–65.

Council, N.R. (1996). Lost Crops of Africa: Volume I: Grains. Washington, DC: The National Academies Press.

Cramer, P. (2019). Organization and regulation of gene transcription. *Nature. 573*: 45–54

Cui, Z., Zhang, H., Chen, X., Zhang, C., Ma, W., ... Huang, C. (2018). Pursuing sustainable productivity with millions of smallholder farmers. *Nature. 555*:363–366.

Curtin, S., Qi, Y., Peres, L.E.P., Fernie, A.R., & Zsogon, A. (2022). Pathway de novo domestication of crop wild relatives. *Plant Physiology. 188*. 1746-1756.

Dai, F., Nevo, E., Wu, D., Comadran, J., ... Zhou, M. (2012). Tibet is one of the centers of domestication of cultivated barley. *Proc. Natl. Acad. Sci. USA. 109*:16969–16973.

Debernardi, J.M., Tricoli, D.M., Ercoli, M.F., Hayta, S., Ronald, P., Palatnik, J.F., & Dubcovsky, J. (2020). A GRF-GIF chimeric protein improves the regeneration efficiency of transgenic plants. *Natl. Biotechnol. 38*: 1274–1279.

DeHaan, L.R., Tassel, D.L., Anderson, J.A., ... Burton, E. (2016). A pipeline strategy for grain crop domestication. *Crop Sci. 56*:917–930.

Deng, M., Zhang, X., Luo, J., Liu, H., Wen, W., Luo, H., Yan, J., & Xiao, Y. (2020). Metabolomic analysis reveals differences in evolution between maize and rice. *Plant J. 103*:1710–1722

Denham, T., Barton, H., Castillo, C., ... Crowther, A. (2020). The domestication syndrome in vegetatively propagated field crops. *Ann. Bot. 125*: 581–597.

Diaz, A., Zikhali, M., Turner, A.S., Isaac, P., & Laurie, D.A. (2012). Copy number variation affecting the Photoperiod-B1 and Vernalization-A1 genes is associated with altered flowering time in wheat (*Triticum aestivum*). *PLoS One. 7*:e33234.

Dillon, S.L., Shapter, F.M., Henry, R.J., Cordeiro, G., Izquierdo, L., & Lee, L.S. (2007). Domestication to crop improvement: genetic resources for Sorghum and Saccharum (Andropogoneae). *Ann. Bot. 100*:975–989.

Dirzo, R., & Raven, P.H. (2003). Global state of biodiversity and loss. *Annu. Rev. Environ. Resour. 28*:137–167.

Doebley, J.F., Gaut, B.S., & Smith, B.D. (2006). The molecular genetics of crop domestication. *Cell. 127*: 1309–1321.

Doebley, J., Stec, A., & Hubbard, L. (1997). The evolution of apical dominance in maize. *Nature. 386*:485–488.

Doebley, J.F., Gaut, B.S., & Smith, B.D. (2006). The molecular genetics of crop domestication. *Cell. 127*:1309–1321.

Dohm, J.C., Minoche, A.E., Holtgräwe, D., ... Capella-Gutiérrez, S. (2014). The genome of the recently domesticated crop plant sugar beet (*Beta vulgaris*). *Nature. 505*: 546–549.

Domı´nguez, M., Dugas, E., Benchouaia, M., ... Leduque, B. (2020). The impact of transposable elements on tomato diversity. *Nat. Commun. 11*: 4058

Donald, C.M. (1968). Breeding of crop ideotypes. *Euphytica.17*:385–403.

Dong, Y., Yang, X., Liu, J., Wang, B.H., Liu, B.L., & Wang, Y.Z. (2014). Pod shattering resistance associated with domestication is mediated by a NAC gene in soybean. *Nat. Comm. 5*:3352-3357.

Dorn, K.M., Fankhauser, J.D., Wyse, D.L., & Marks, M.D. (2015). A draft genome of field pennycress (Thlaspi arvense) provides tools for the domestication of a new winter biofuel crop. *DNA Res. 22*:121–131.

Doust, A.N., Devos., K.M., Gadberry, M.D., Gale, M.D., & Kellogg, E.A. (2004). Genetic control of branching in foxtail millet. *Proc. Natl. Acad. Sci. USA. 101*:9045–9050.

Drewnowski, A., & Gomez-Carneros, C. (2000). Bitter taste, phytonutrients, and the consumer: a review. *Am. J. Clin. Nutr. 72*: 1424–1435

Drewnowski, A., & Popkin, B.M. (1997). The nutrition transition: new trends in the global diet. *Nutr. Rev. 55*:31–43.

Dubcovsky, J., & Dvorak, J. (2007). Genome plasticity a key factor in the success of polyploid wheat under domestication. *Science. 316*:1862–1866.

Ekar, J.M., Betts, K.J., Herman, A.C., Stupar, R.M., Wyse, D.L., Brandvain, Y., & Kantar, M.B. (2019). Domestication in real time: the curious case of a trigenomic sunflower population. *Agronomy. (Basel) 9*:785-790.

Ellison, E.E., Nagalakshmi, U., Gamo, M.E., Huang, P., Dinesh-Kumar, S., & Voytas, D.F. (2020). Multiplexed heritable gene editing using RNA viruses and mobile single guide RNAs. *Nat. Plants. 6*: 620–624.

Emmrich, P.M.F., Rejzek, M., Hill, L., Brett, P., ... Edwards, A. (2019). Linking a rapid throughput plate-assay with high-sensitivity stable-isotope label LCMS quantification permits the identification and characterisation of low b-L-ODAP grass pea lines. *BMC Plant Biol. 19*: 489

Endo, M., Mikami, M., & Toki, S. (2015). Multigene knockout utilizing off-- target mutations of the CRISPR/Cas9 system in rice. *Plant Cell Physiol 56*: 41–47

Evans, L.T. (1998). Feeding the ten billion: plants and population growth. Cambridge University Press, Cambridge, UK.

Eviatar-Ribak, T., Shalit-Kaneh, A., Chappell-Maor, L. .... Mu;;er, J. (2013). A cytokinin-activating enzyme promotes tuber formation in tomato. *Curr. Biol. 23*:1057–1064.

Fang, C., Fernie, AR., & Luo, J. (2019). Exploring the diversity of plant metabolism. trends in plant science. *Trends Plant Sci. 24*:83–98.

Fang, L., Gong, H., Hu, Y., Liu, C., Zhou, B., ... Huang, T. (2017). Genomic insights into divergence and dual domestication of cultivated allotetraploid cotton. *Genome Biol.18*:33.

Farnie, A.R., & Yan, J. (2019). De novo domestication: an alternate route towards new crops in the future. *Molecular Plant. 12*: 615-631.

Fedorova, M., & Herman, R.A. (2020). Obligatory metabolomic profiling of gene-edited crops is risk disproportionate. *Plant J.103*: 1985–1988.

Ferguson, A.R. (2013). Kiwifruit: the wild and the cultivated plants. *Adv. Food Nutr. Res. 68*: 15–32.

Fernie, A.R., Alsekh, S., Lin, J., & Yan, J. (2021). Using precision phenotyping to inform de novo domestication. *Plant Physiology. 186*: 1397-1411.

Fernie, A.R., & Gutierrez-Marcos, J. (2019). From genome to phenome: genome-wide association studies and other approaches that bridge the genotype to phenotype gap. *Plant J. 97*: 5–7.

Fernie, A.R., & Yan J. (2019). De novo domestication: an alternative route toward new crops for the future. *Mol. Plant.12*: 615–631.

Fernie, A.R., & Gutierrez-Marcos, J.F. (2019). From genome to phenome: genome-wide association studies and other approaches that bridge the genotype to phenotype gap. *Plant J. 97*:5–7.

Fernie, A.R., & Tohge, T. (2017). The genetics of plant metabolism. *Annu. Rev. Genet. 51*:287–310.

Fernie, A.R., Tadmor, Y., & Zamir, D. (2006). Natural genetic variation for improving crop quality. *Curr. Opin. Plant Biol. 9*:196–202.

Ferrero-Serrano, A., & Assmann, S.M. (2019). Phenotypic and genome-wide association with the local environment of Arabidopsis. *Nat. Ecol. Evol. 3*: 274–285

Fischer, A., Byerlee, D., & Edmeades, G. (2014). Crop yields and global food security: will yield increase continue to feed the world? *Australian Centre for International Agricultural Research, Canberra, Australia.*

Fisher, R.A. (1919). The correlation between relatives on the supposition of Mendelian inheritance. Earth Environ. Sci. Trans. R. Soc. Edinburg.

Flint-Garcia, S.A. (2013). Genetics and consequences of crop domestication. *J. Agric. Food Chem. 61*: 8267–8276.

Foucher, F., Morin, J., Courtiade, J., Cadioux, S., Ellis, N., Banfield, M.J., & Rameau, C. (2003). Determinate and late flowering are two terminal flower1/ centroradialis homologs that control two distinct phases of flowering initiation and development in pea. *Plant Cell. 15*: 2742–2754.

Francis, D., Finer, J.J., & Grotewold, E. (2017). Challenges and opportunities for improving food quality and nutrition through plant biotechnology. *Curr. Opin. Biotechnol. 44*:124–129.

Frary, A., Nesbitt, T.C., Grandillo, S., Knaap, E., ... Cong, B. (2000). fw2.2: a quantitative trait locus key to the evolution of tomato fruit size. *Science. 289*:85–88.

Fraser, P.D., Aharoni, A., Hall, R.D., Huang, S., Giovannoni, J.J., Sonnewald, U., & Fernie, A.R. (2020). Metabolomics should be deployed in the identification and characterization of gene-edited crops. *Plant J. 102*: 897–902.

Fridman, E., Carrari, F.F., Liu, Y.S., Fernie, A.R., & Zamir, D. (2004). Zooming in on a quantitative trait for tomato yield using interspecific introgressions. *Science. 305*:1786–1789.

Frydman, A., Liberman, R., Huhman, D.V., ... Carmeli-Weissberg, M. (2013). The molecular and enzymatic basis of bitter/non-bitter flavor of citrus fruit: evolution of branch-forming rhamnosyltransferases under domestication. *Plant J. 73*: 166–178

Fudge, J., Mangel, N., Gruissem, W., Vanderschuren, H., & Fitzpatrick, T.B. (2017). Rationalising vitamin B6 biofortification in crop plants. *Curr. Opin. Biotechnol. 44*:130–137.

Fuentes, P., Zhou, F., Erban, A., Karcher, D., Kopka, J., & Bock, R. (2016). A new synthetic biology approach allows transfer of an entire metabolic pathway from a medicinal plant to a biomass crop. *Elife. 5*. https://doi.org/10.7554/eLife.13664.

Funatsuki, H., Suzuki, M., Hirose, A., Inaba, H., ... Yamada, T. (2014). Molecular basis of a shattering resistance boosting global dissemination of soybean. *Proc. Natl. Acad. Sci. USA. 111*: 17797–17802.

Gago, J., Estrany, J., Estes, L., Fernie, A.R., ... Alorda, B. (2020). Nano and micro unmanned aerial vehicles (UAVs): a new grand challenge for precision agriculture? *Curr. Protoc. Plant Biol. 5*: e20103.

Gao, C. (2021). Genome engineering for crop improvement and future agriculture. *Cell. 184*: 1621–1635.

Gasparini, K., Moreira, J. dos. R., Peres L.E.P., & Zsögön, A. (2021). De novo domestication of wild species to create crops with increased resilience and nutritional value. *Curr. Opin. Plant Biol. 60*: 102-106.

Gaudelli, N.M., Komor, A.C., Rees, H.A., ... Packer, M.S. (2017). Programmable base editing of A*T to G*C in genomic DNA without DNA cleavage. *Nature. 551*:464–471.

Gaut, B.S., Seymour, D.K., Liu, Q., & Zhou, Y. (2018). Demography and its effects on genomic variation in crop domestication. *Nat. Plants. 4*:512–520.

Gepts, P. (2014). The contribution of genetic and genomic approaches to plant domestication studies. *Curr. Opin. Plant Biol.18*: 51–59.

Ghosh, S., Watson, A., Gonzalez-Navarro, O.E., ... Ramirez-Gonzalez, R.H. (2018). Speed breeding in growth chambers and glasshouses for crop breeding and model plant research. *Nat. Protocols. 13*: 2944–2963.

González, A.M, Vander Schoor, J.K., Fang, C., Kong, F., Wu, J., Weller, J.L., & Santalla M. (2021). Ancient relaxation of an obligate short-day requirement in common bean through loss of constans-like gene function. *Curr. Biol.* doi: 10.1016/j.cub.2021.01.075.

Griffiths, A.J.F., Gelbart, W.M., Miller, J.H., & Lewontin, R.C. (1999). Modern Genetic Analysis (New York: W.H. Freeman).

Grimm, D.G., Roqueiro, D., Salome,´ P.A, Kleeberger, S., ... Greshake, B. (2017). easyGWAS: a cloud-based platform for comparing the results of genome-wide association studies. *Plant Cell. 29*: 5–19.

Gross, B.L, & Olsen, K.M. (2010). Genetic perspectives on crop domestication. *Trends Plant Sci 15*: 529–537.

Gross, B.L., & Olsen, K.M. (2010). Genetic perspectives on crop domestication. *Trends Plant Sci. 15*:529–537.

Gui, S., Yang, L., Li, J., Luo, J., Xu, X., Yuan, J., Chen, L., Li, W., Yang, X., ... Wu, S. (2020). ZEAMAP, a comprehensive database adapted to the maize multi-omics era. *iScience. 23*:1012-1020.

Gulisano, A., Alves, S., Martins, J.N., & Trindade, L.M. (2019). Genetics and breeding of *Lupinus mutabilis*: an emerging protein crop. *Front. Plant Sci. 10*: 1385.

Guo, J., Li, C., Zhang, X., Li, Y., Zhang, D., Shi, Y., Song, Y., Li, Y., Yang, D., & Wang, T. (2020). Transcriptome and GWAS analyses reveal candidate gene for seminal root length of maize seedlings under drought stress. *Plant Sci. 292*: 1103-1109.

Gur, A., & Zamir, D. (2004). Unused natural variation can lift yield barriers in plant breeding. *PLoS Biol. 2*:e245.

Hajjar, R., & Hodgkin, T. (2007). The use of wild relatives in crop improvement: a survey of developments over the last 20 years. *Euphytica. 156*: 1–13.

Hardigan, M.A., Laimbeer, F.P.E., Newton, L., ... Crisovan, E. (2017). Genome diversity of tuber-bearing Solanum uncovers complex evolutionary history and targets of domestication in the cultivated potato. *Proc. Natl. Acad. Sci. USA .114*: E9999–E10008.

Harjes, C.E., Rocheford, T.R., Bai, L., ... Brutnell, T.P. (2008). Natural genetic variation in lycopene epsilon cyclase tapped for maize biofortification. *Science. 319:*330–333.

Haun, W., Coffman, A., Clasen, B.M., ... Demorest, Z.L., (2014). Improved soybean oil quality by targeted mutagenesis of the fatty acid desaturase 2 gene family. *Plant Biotechnol. J. 12*:934–940.

Haverkort, A.J., Boonekamp, P.M., Hutten, R., ... Jacobsen, E. (2016). Durable late blight resistance in potato through dynamic varieties obtained by cisgenesis: scientific and societal advances in the DuRPh project. *Potato Res. 59*:35–66.

He, Z., Zhai, W., Wen, H., Tang, T., Wang, Y., ... Lu, X. (2011). Two evolutionary histories in the genome of rice: the roles of domestication genes. *PLoS Genet. 7*: e1002100.

Heaton, E.A., Doehleman, F.G., & Long, S.P. (2008). Meeting US biofuel goals with less land: the potential of Miscanthus. *Glob. Change Biol. 14*:2000–2014.

Hendelman, A., Zebell, S., Rodriguez-Leal, D., ... Dukler, N. (2021). Conserved pleiotropy of an ancient plant homeobox gene uncovered by cis-regulatory dissection. *Cell.184*: 1724–1739.e16.

Henry, R.J., Rangan, P., & Furtado, A. (2016). Functional cereals for production in new and variable climates. *Curr. Opin. Plant Biol. 30*:11–18.

Hereford, J. (2009). A quantitative survey of local adaptation and fitness trade-offs. *Am. Nat. 173*: 579–588.

Herna´ndez, B.J.E., & Leo´n, J. (1994). Neglected Crops 1492 from a different perspective. *Plant Production and Protection Series No, 26 (Rome, Italy: FAO).*

Hillman, G.C., & Davies, M.S. (1990a). Measured domestication rates in wild wheats and barley under primitive cultivation, and their archeological implications. *J. World Prehist. 4*:157–222.

Hillman, G.C., & Davis, M.S. (1990b). Domestication rates in wild type wheats and barley under primitive agriculture. *Biol. J. Linn. Soc. 39*:39–78.

Hilu, K.W., & Wet, J.M.J.D. (1980). Effect of artificial selection on grain dormancy in Eleusine (Gramineae). *Syst. Bot. 5*:54–60.

Howe, K.L, Contreras-Moreira, B., De Silva, N., ... Maslen, G. (2020). Ensembl genomes 2020-enabling non-vertebrate genomic research. *Nucleic Acids Res. 48*: D689–D695.

Hoyos, V., Plaza, G., Li, X., & Caicedo, A.L. (2020). Something old, something new: evolution of Colombian weedy rice (*Oryza* spp.) through de novo de-domestication, exotic gene flow, and hybridization. *Evol. Appl. 13*: 1968–1983

Hu, M., Lv, S., Wu, W., Fu, Y., Liu, F., Wang, B., ... Li, W. (2018). The domestication of plant architecture in African rice. *Plant J. 94*:661–669.

Hua, L., Wang, D.R., Tan, L., Fu, Y., Liu, F., ... Xiao, L. (2015). LABA1, a domestication gene associated with long, barbed awns in wild rice. *Plant Cell 27*:1875–1888.

Huang, B.E., & George, A.W. (2011). R/mpMap: a computational platform for the genetic analysis of multiparent recombinant inbred lines. *Bioinformatics. 27*: 727–729

Huang, C., Sun, H., Xu, D., Chen, Q., Liang, Y., ... Wang, X. (2018). ZmCCT9 enhances maize adaptation to higher latitudes. *Proc. Natl. Acad. Sci. USA. 115*: E334–E341.

Huang, H., Cui, T., Zhang, L., Yang, Q., Yang, Y., Xie, K., Fan, C., & Zhou, Y. (2020). Modifications of fatty acid profile through targeted mutation at BnaFAD2 gene with CRISPR/Cas9-mediated gene editing in *Brassica napus*. *Theor. Appl. Genet.133*: 2401–2411.

Huang, S., Weigel, D., Beachy, R.N., & Li, J. (2016). A proposed regulatory framework for genome-edited crops. *Nat. Genet. 48*:109–111.

Huang, X., & Han, B. (2014). Natural variations and genome-wide association studies in crop plants. *Annu. Rev. Plant Biol. 65*:531–551.

Hufford, M.B., Xu, X., van Heerwaarden, J., ... Pyhajarvi, T. (2012). Comparative population genomics of maize domestication and improvement. *Nat. Genet. 44*:808–811.

Iaffaidano, B., Zhang, Y., & Cornish, K. (2016). CRSPR/Cas9 genome editing of rubber producing dandelions Taraxacum kok-saghyz using Agrobacterium rhizogenes without selection. *Ind. Crops Prod. 89*:356–362.

Itkin, M., Heinig, U., Tzfadia, O., Bhide, A.J, ... Shinde, B. (2013). Biosynthesis of antinutritional alkaloids in solanaceous crops is mediated by clustered genes. *Science. 341*: 175–179.

Ivarson, E., Ahlman, A., Lager, I., & Zhu, L.H. (2016). Significant increase of oleic acid level in the wild species Lepidium campestre through direct gene silencing. *Plant Cell Rep. 35:*2055–2063.

Jackson, S.A., Iwata, A., Lee, S.H., Schmutz, J., & Shoemaker, R. (2011). Sequencing crop genomes: approaches and applications. *New Phytol. 191*:915–925.

Jacobs, T.B., LaFayette, P.R., Schmitz, R.J., & Parrott, W.A. (2015). Targeted genome modifications in soybean with CRISPR/Cas9. *BMC Biotechnol. 15*: 16.

Jamnadass, R., Mumm, R.H., Hale, I., Hendre, P., ... Muchugi, A. (2020). Enhancing African orphan crops with genomics. *Nat. Genet. 52*: 356–360.

Jankowicz-Cieslak, J., Mba, C, & Till, B.J. (2017). Mutagenesis for crop breeding and functional genomics. In: Jankowicz-Cieslak, J., Tai, T.H, Kumlehn, J., & Till BJ (Eds), Biotechnologies for Plant Mutation Breeding: Protocols. *Springer International Publishing, Cham,* p. 3–18.

Jarvis, D.E., Ho, Y.S., Lightfoot, D.J., Schmo¨ckel, S.M., ... Li, B. (2017). The genome of Chenopodium quinoa. *Nature. 542*: 307–312.

Jia, Q, Zhang, J., Westcott, S., Zhang, X.Q., Bellgard, M., Lance, R., & Li, C. (2009). GA-20 oxidase as a candidate for the semidwarf gene sdw1/denso in barley. *Funct. Integr. Genomics. 9*: 255–262.

Jiang, Y.Y., Chai, Y.P., Lu, M.H., Han, X.L., Lin, Q., ... Zhang, Y. (2020). Prime editing efficiently generates W542L and S621I double mutations in two ALS genes in maize. *Genome Biol. 21*: 257.

Jin, S., Zong, Y., Gao, Q., Zhu, Z., Wang, Y., ... Qin, P. (2019). Cytosine, but not adenine, base editors induce genome-wide off-target mutations in rice. *Science. 364*: 292–295.

Jin, J., Huang, W., Gao, J.P., Yang, J., Shi, M., Zhu, M.Z., Luo, D., & Lin, H.X. (2008). Genetic control of rice plant architecture under domestication. *Nat. Genet. 40*:1365–1369.

Kantar, M.B, Nashoba, A.R, Anderson, J.E, Blackman, B.K, & Rieseberg, L.H. (2017). The genetics and genomics of plant domestication. *BioScience. 67*: 971–982.

Kantar, M.B., Tyl, C.E., Dorn, K.M., Zhang, X., ... Jungers, J.M. (2016). Perennial grain and oilseed crops. *Ann. Rev. Plant Biol. 67*:703–729.

Kantarski, T., Larson, S., Zhang, X., ... DeHaan, L. (2017). Development of the first consensus genetic map of intermediate wheatgrass (*Thinopyrum intermedium*) using genotyping-by-sequencing. *Theor. Appl. Genet. 130*: 137-150.

Khachatourians, G.G., Sumner, A.K., & Phillips, P.W.B. (2001). The biotechnology revolution in global agriculture: innovation, invention and investment in the canola industry. In: Phillips, P.W.B., & Khachatourians, G.G. eds, An introduction to the history of canola and the scientific basis for innovation, *CABI, Wallingford, UK*, p. 33–47.

Khan, M.A., Olsen, K.M., Sovero, V., Kushad, M.M., & Korban, S.S. (2014). Fruit quality traits have played critical roles in domestication of the apple. *Plant Genome. 7*: 204-208.

Khush, G.S. (2001). Green revolution: the way forward. *Nat. Rev. Genet. 2*:815–822.

Kim, B., Dai, X., Zhang, W., Zhuang, Z., ... Sanchez, D.L. (2018). GWASpro: a high-performance genome-wide association analysis server. *Bioinformatics. 35*: 2512–2514.

Kim, K.D., El Baidouri, M., Abernathy, B., ... Iwata-Otsubo, A. (2015). A comparative epigenomic analysis of polyploidy-derived genes in soybean and common bean. *Plant Physiol. 168*:1433–1447.

Klee, H.J., & Tieman, D.M. (2018). The genetics of fruit flavour preferences. *Nat. Rev. Genet. 19*:347–356.

Kloosterman, B., Abelenda, J.A., Gomez, M. del. M.C., ... Oortwijn M. (2013). Naturally occurring allele diversity allows potato cultivation in northern latitudes. *Nature. 495*: 246–250.

Knox, A.K., Dhillon, T., Cheng, H., Tondelli, A., Pecchioni, N., & Stockinger, E.J. (2010). CBF gene copy number variation at Frost Resistance-2 is associated with levels of freezing tolerance in temperate-climate cereals. *Theor. Appl. Genet. 121*:21–35.

Koblan, L.W, Arbab M., Shen, M.W., ... Hussmann, J.A. (2021). Efficient C•G-to-G•C base editors developed using CRISPRi screens, target-library analysis, and machine learning. *Nat. Biotechnol. 39*: 1–12.

Koenig, D., Jimenez-Gomez, J.M., Kimura, S., ... Fulop, D. (2013). Comparative transcriptomics reveals patterns of selection in domesticated and wild tomato. *Proc. Natl. Acad. Sci. USA. 110*: E2655–E2662.

Kofsky, J., Zhang, H., & Song, B.H. (2018). The untapped genetic reservoir: the past, current, and future applications of the wild soybean (*Glycine soja*). *Front. Plant Sci.* doi: 10.3389/fpls.2018.00949

Komatsuda, T., Pourkheirandish, M., He, C., ... Azhaguvel, P. (2007). Six-rowed barley originated from a mutation in a homeodomain-leucine zipper I-class homeobox gene. *Proc. Natl. Acad. Sci. USA. 104*:1424–1429.

Konishi, S., Izawa, T., Lin, S.Y., Ebana, K., Fukuta, Y., Sasaki, T., & Yano, M. (2006). An SNP caused loss of seed shattering during rice domestication. *Science. 312*:1392–1396.

Kono, T.J., Fu, F., Mohammadi, M., ... Hoffman, P.J. (2016). The role of deleterious substitutions in crop genomes. *Mol. Biol. Evol. 33*:2307–2317.

Korte, A., & Farlow, A. (2013). The advantages and limitations of trait analysis with GWAS: a review. *Plant Methods. 9:* 29.

Kumagai, M., Nishikawa, D., Kawahara, Y., ... Wakimoto, H. (2019). TASUKE1: a web-based platform for exploring GWAS results and large-scale resequencing data. *DNA Res. 26*: 445–452.

Kwak, M., Toro, O., Debouck, D.G., & Gepts, P. (2012). Multiple origins of the determinate growth habit in domesticated common bean (*Phaseolus vulgaris*). *Ann. Bot.110*:1573–1580.

Kwon. C.T., Heo, J., Lemmon, Z.H., Capua, Y., Hutton, S.F., Van Eck, J., Park, S.J., & Lippman, Z.B. (2020). Rapid customization of Solanaceae fruit crops for urban agriculture. *Nat. Biotechnol. 38*: 182–188.

Lai, X., Yan, L., Lu, Y., Schnable, J.C. (2018). Largely unlinked gene sets targeted by selection for domestication syndrome phenotypes in maize and sorghum. *Plant J. 93*: 843–855.

Laitinen, R.A.E., & Nikoloski, Z. (2019). Genetic basis of plasticity in plants. *J. Exp. Bot. 70*: 739–745.

Lander, E.S., & Botstein, D. (1989). Mapping mendelian factors underlying quantitative traits using RFLP linkage maps. *Genetics. 121*:185–199.

Ledford, H. (2013). US regulation misses some GM crops. *Nature. 500*:389–390.

Lemmon, Z.H., Reem, N.T., Dalrymple, J., Soyk, S., ... Swartwood, K.E. (2018). Rapid improvement of domestication traits in an orphan crop by genome editing. *Nat. Plants. 4*: 766–770.

Lenormand, T, & Otto, S.P. (2000). The evolution of recombination in a heterogeneous environment. *Genetics. 156*: 423–438.

Lenser, T., & Theissen, G. (2013). Molecular mechanisms involved in convergent crop domestication. *Trends Plant Sci. 18*: 704–714.

Li, S., & Xia, L. (2020). Precise gene replacement in plants through CRISPR/ Cas genome editing technology: current status and future perspectives. *aBIOTECH.1*: 58–73

Li, C., Zhou, A., & Sang, T. (2006). Rice domestication by reducing shattering. *Science. 311*: 1936–1939.

Li, Q., & Yan, J. (2020). Sustainable agriculture in the era of omics: knowledge-driven crop breeding. *Genome Biol. 21*: 154.

Li, T., Yang, X., Yu, Y., Si, X., Zhai, X., Zhang, H., Dong. W., Gao, C., & Xu, C. (2018). Domestication of wild tomato is accelerated by genome editing. *Nature Biotechnol. 36*:1160–1163.

Li, C., Zhou, A., & Sang, T. (2006). Rice domestication by reducing shattering. *Science. 311*:1936–1939.

Li, S., Tian, Y., Wu, K., Ye, Y., Yu, J., Zhang, J., ... Liu, Q. (2018a). Modulating plant growth-metabolism coordination for sustainable agriculture. *Nature. 560*:595–600.

Li, T., Yang, X., Yu, Y., Si, X., Zhai, X., ... Zhang, H. (2018b). Domestication of wild tomato is accelerated by genome editing. *Nat. Biotechnol.* https://doi.org/10.1038/nbt.4273.

Li, W., & Gill, B.S. (2006). Multiple genetic pathways for seed shattering in the grasses. *Funct. Integr. Genomics. 6*:300–309.

Lin, Q., Zong, Y., Xue, C., Wang, S., Jin, S., Zhu, Z., ... Wang, Y. (2020). Prime genome editing in rice and wheat. *Nat. Biotechnol. 38*: 582–585.

Lin, T., Zhu, G., Zhang, J., Xu, X., Yu, Q., Zheng, Z., ... Zhang, Z. (2014). Genomic analyses provide insights into the history of tomato breeding. *Nat. Genet. 46*: 1220–1226.

Lin, Z., Li, X., Shannon, L.M., Yeh, C.T., Wang, M.L., Bai, G., Peng, Z., Li, J., Trick, H.N., ... Clemente, T.E. (2012). Parallel domestication of the Shattering1 genes in cereals. *Nat. Genet. 44*: 720–U154.

Lin, Z., Li, X., Shannon, L.M., Yeh, C.T., Wang, M.L., ... Bai, G. (2012). Parallel domestication of the Shattering1 genes in cereals. *Nat. Genet. 44*:720–724.

Liu, H., & Yan, J. (2017). Rice domestication: an imperfect African solution. *Nat. Plants. 3*:17083.

Liu, H., Wang, X., Warburton, M.L., Wen, W., Jin, M., ... Deng, M. (2015). Genomic, transcriptomic, and phenomic variation reveals the complex adaptation of modern maize breeding. *Mol. Plant. 8*:871–884.

Liu, H.J., Jian, L., Xu, J., Zhang, Q., Zhang, M., Jin, M., ... Peng, Y. (2020). High-throughput CRISPR/Cas9 mutagenesis streamlines trait gene identification in maize. *Plant Cell. 32*: 1397–1413.

Liu, H.J., & Yan J. (2019). Crop genome-wide association study: a harvest of biological relevance. *Plant J. 97*: 8–18.

Liu, J., Fernie, A.R., & Yan, J. (2021). Crop breeding from experience-based selection to precision design. *J. Plant Physiol. 256*: 1533-1538.

Liu, J., Li, J., Wang, H., & Yan, J. (2020). Application of deep learning in genomics. *Sci. China Life Sci. 63*:1860–1878.

Liu, L., Gallagher, J., Arevalo, E.D., Chen, R., Skopelitis, T., Wu, Q., Bartlett, M., & Jackson, D. (2021a). Enhancing grain-yield-related traits by CRISPR–Cas9 promoter editing of maize CLE genes. *Nat. Plants. 7*: 287–294.

Liu, N., Du, Y., Warburton, M.L., Xiao, Y., & Yan, J. (2021b). Phenotypic plasticity contributes to maize adaptation and heterosis. *Mol. Biol. Evol. 38*:1262–1275.

Liu, Q., Zhou, Y., Morrell, P.L., & Gaut, B.S. (2017). Deleterious variants in Asian rice and the potential cost of domestication. *Mol. Biol. Evol. 34*:908–924.

Liu, Z., Cook, J., Melia-Hancock, S., Guill, K., ... Bottoms, C. (2016). Expanding maize genetic resources with predomestication alleles: maize-teosinte introgression populations. *Plant Genome. 9*. https:// doi.org/10.3835/ plantgenome2015.07.0053.

Livne, S., Lor, V.S., Nir, I., Eliaz, N., Aharoni, A., Olszewski, N.E., Eshed, Y., & Weiss, D. (2015). Uncovering DELLA-Independent gibberellin responses by characterizing new tomato procera mutants. *Plant Cell. 27*:1579–1594.

Lloyd, A., Morgan, C., H., Franklin, F.C., & Bomblies, K. (2018). Plasticity of meiotic recombination rates in response to temperature in *Arabidopsis*. *Genetics. 208*: 1409–1420.

Lobell, D.B., & Gourdji, S.M. (2012). The influence of climate change on global crop productivity. *Plant Physiol.160*:1686–1697.

Lotan, T., Ohto, M., Yee, K.M., West, M.A., Lo, R., ... Kwong, R.W. (1998). *Arabidopsis* LEAFY COTYLEDON1 is sufficient to induce embryo development in vegetative cells. *Cell.93*:1195–1205.

Lotterhos, K.E., & Whitlock, M.C. (2015). The relative power of genome scans to detect local adaptation depends on sampling design and statistical method. *Mol. Ecol. 24*: 1031–1046.

Loureiro, R., Gachon, C.M., & Rebours, C. (2015). Seaweed cultivation: potential and challenges of crop domestication at an unprecedented pace. *New Phytol. 206*:489–492.

Lowder, L.G., Zhang, D., Baltes, N.J., Paul, J.W., & Tang, X. (2015). A CRISPR/Cas9 toolbox for multiplexed plant genome editing and transcriptional regulation. *Plant Physiol.169*: 971–985.

Lowe, K., La Rota, M., Hoerster, G., Hastings, C., ... Wang, N. (2018). Rapid genotype "independent" *Zea mays* L. (maize) transformation via direct somatic embryogenesis. *In Vitro Cell Dev. Biol. Plant. 54*: 240–252.

Lowe, K., Wu, E., Wang, N., Hoerster, G., ... Hastings, C. (2016). Morphogenic regulators baby boom and wuschel improve monocot transformation. *Plant Cell. 28*:1998–2015.

Lu, J., Sun, J., Jiang, A., Bai, M., Fan, C., Liu, J., Ning, G., & Wang, C. (2020). Alternate expression of constant-like 4 in short days and constant in long days facilitates day-neutral response in Rosa chinensis. *J. Exp. Bot. 71*: 4057–4068.

Lu, Y., Ye, X., Guo, R., Huang, J., Wang, W., Tang, J., Tan, L., Zhu, J.K., Chu, C., & Qian, Y. (2017). Genome-wide targeted mutagenesis in rice using the CRISPR/Cas9 system. *Mol. Plant. 10*: 1242–1245

Luo, C., Fernie, A.R., & Yan, J. (2020). Single-cell genomics and epigenomics: technologies and applications in plants. *Trends Plant Sci. 25*: 1030–1040.

Luo, J., Wei, C., Liu, H., Cheng, S., Xiao, Y., Wang, X., Yan, J., & Liu J. (2020). MaizeCUBIC: a comprehensive variation database for a maize synthetic population. Database (Oxford).

Lv, S., Wu, W., Wang, M., Meyer, R.S., Ndjiondjop, M.N., ... Tan, L. (2018). Genetic control of seed shattering during African rice domestication. *Nat. Plants. 4*: 331–337.

Ma, X., Zhang, Q., Zhu, Q., Liu, W., Chen, Y., Qiu, R., ... Wang, B. (2015). A robust CRISPR/Cas9 system for convenient, high-efficiency multiplex genome editing in monocot and dicot plants. *Mol. Plant. 8*: 1274–1284.

Ma, X., Zhang, X., Liu, H., & Li, Z. (2020). Highly efficient DNA-free plant genome editing using virally delivered CRISPR–Cas9. *Nat. Plants. 6*: 773–779.

Maher, M.F., Nasti, R.A., Vollbrecht, M., Starker, C.G., Clark, M.D, & Voytas, D.F. (2020). Plant gene editing through de novo induction of meristems. *Nat. Biotechnol. 38*: 84–89.

Mai, C.D., Phung, N.T., To, H.T., Gonin, M., ... Hoang, G.T. (2014). Genes controlling root development in rice. *Rice (NY). 7*: 30.

Mamidi, S., Healey, A., Huang, P., Grimwood, J., ... Jenkins, J. (2020). A genome resource for green millet *Setaria viridis* enables discovery of agronomically valuable loci. *Nat. Biotechnol. 38*: 1203–1210.

Manning, K., Tor, M., Poole, M., Hong, Y., ... Thompson, A.J. (2006). A naturally occurring epigenetic mutation in a gene encoding an SBP-box transcription factor inhibits tomato fruit ripening. *Nat. Genet. 38*:948–952.

Martin, A., Troadec, C., Boualem, A., ... Rajab, M. (2009). A transposon-induced epigenetic change leads to sex determination in melon. *Nature. 461*:1135–1138.

Martin, C., & Li, J. (2017). Medicine is not health care, food is health care: plant metabolic engineering, diet and human health. *New Phytol. 216*:699–719.

Matesanz, S., & Milla, R. (2018). Differential plasticity to water and nutrients between crops and their wild progenitors. *Environ. Exp. Bot. 145*: 54–63.

McGinn, M., Phippen, W.B., Chopra, R., Bansal, S., ... Jarvis, B.A. (2019). Molecular tools enabling pennycress (*Thlaspi arvense*) as a model plant and oilseed cash cover crop. *Plant Biotechnol. J. 17*: 776–788.

Meng, L., Li, H., Zhang, L., & Wang, J. (2015). QTL IciMapping: integrated software for genetic linkage map construction and quantitative trait locus mapping in biparental populations. *Crop J.* 3: 269–283.

Meng, X., Yu, H., Zhang, Y., Zhuang, F., Song, X., Gao, S., Gao, C., & Li, J. (2017). Construction of a genome-wide mutant library in rice using CRISPR/Cas9. *Mol. Plant. 10*:1238–1241.

Meuwissen, T.H., Hayes, B.J., & Goddard, M.E. (2001). Prediction of total genetic value using genome-wide dense marker maps. *Genetics.157*:1819–1829.

Meyer, R.C., Witucka-Wall, H., Becher, M., ... Blacha, A. (2012). Heterosis manifestation during early *Arabidopsis* seedling development is characterized by intermediate gene expression and enhanced metabolic activity in the hybrids. *Plant J. 71*: 669–683.

Meyer, R.S., DuVal, A.E., & Jensen, H.R. (2012). Patterns and processes in crop domestication: an historical review and quantitative analysis of 203 global food crops. *New Phytol.196*: 29–48

Meyer, R.S., & Purugganan, M.D. (2013). Evolution of crop species: genetics of domestication and diversification. *Nat. Rev. Genet.14*: 840–852

Meyer, R.S., DuVal, A.E., & Jensen, H.R. (2012). Patterns and processes in crop domestication: an historical review and quantitative analysis of 203 global food crops. *New Phytol.196*:29–48.

Meyerowitz, E.M. (2002). Plants compared to animals: the broadest comparative study of development. *Science. 295*: 1482–1485.

Milla, R, & Osborne C.P. 2021. Crop origins explain variation in global agricultural relevance. Nat. Plants. 7: 598–607.

Miura, K., Agetsuma, M., Kitano, H., ... Yoshimura, A. (2009). A metastable DWARF1 epigenetic mutant affecting plant stature in rice. *Proc. Natl. Acad. Sci. USA. 106*:11218–11223.

Molla, K.A., Sretenovic, S., Bansal, K.C., & Qi, Y. (2021). Precise plant genome editing using base editors and prime editors. *Nat. Plants. 7*: 1166–1187.

Montes, J.M., & Melchinger, A.E. (2016). Domestication and breeding of Jatropha curcas L. *Trends Plant Sci. 21*:1045–1057.

Morrell, P.L., Buckler, E.S., & Ross-Ibarra, J. (2011). Crop genomics: advances and applications. *Nat. Rev. Genet.13*:85–96.

Mueller, N.G., Fritz, G.J., Patton, P., Carmody, S., & Horton, E.T. (2017). Growing the lost crops of eastern North America's original agricultural system. *Nat. Plants. 3*:17092.

Murphy, R.L, Klein, R.R., Morishige, D.T, ... Brady, J.A. (2011). Coincident light and clock regulation of pseudoresponse regulator protein 37 (PRR37) controls photoperiodic flowering in sorghum. *Proc. Natl. Acad. Sci. USA.108*: 16469–16474.

Nakamura, S., Pourkheirandish, M., Morishige, H., ... Kubo, Y. (2016). Mitogen-activated protein kinase 3 regulates seed dormancy in barley. *Curr. Biol. 26*: 775–781.

Napier, J.A., Olsen, R.E., & Tocher, D.R. (2018). Update on GM canola crops as novel sources of omega-3 fish oils. *Plant Biotechnol. J.*17:703–705.

Nassar, N.M.A., Hashimoto, D.Y.C., & Fernandes, S.D.C. (2008). Wild Manihot species: botanical aspects, geographic distribution and economic value. *Genet. Mol. Res.* 7:16–28.

Nasti, R.A., Zinselmeier, M.H., Vollbrecht, M., Maher, M.F., & Voytas, D.F. (2021). Fast-TrACC: a rapid method for delivering and testing gene editing reagents in somatic plant cells. *Front Genome Edit. 2*: 32.

Nasti, R.A,, & Voytas D.F. (2021). Attaining the promise of plant gene editing at scale. *Proc. Natl. Acad. Sci. USA .118*: e2004846117.

Nelson, R., Wiesner-Hanks, T., Wisser, R., & Balint-Kurti, P. (2018). Navigating complexity to breed disease-resistant crops. *Nat. Rev. Genet.19*:21–33.

Nijveen, H., Ligterink, W., Keurentjes, J.J.B., Loudet, O., ... Long, J. (2017). AraQTL—workbench and archive for systems genetics in *Arabidopsis thaliana. Plant J. 89*: 1225–1235.

Notaguchi, M., Kurotani, K.I., Sato, Y., Tabata, R., ... Kawakatsu, Y. (2020). Cell–cell adhesion in plant grafting is facilitated by b-1,4-glucanases. *Science. 369*: 698–702.

Obata, T., Klemens, P.A.W., Rosado-Souza, L., ... Schlereth, A. (2020). Metabolic profiles of six African cultivars of cassava (*Manihot esculenta* Crantz) highlight bottlenecks of root yield. *Plant J.102*: 1202–1219.

Obata, T., Witt, S., Lisec, J., Palacios-Rojas, N., ... Florez-Sarasa, I. (2015). Metabolite profiles of maize leaves in drought, heat, and combined stress field trials reveal the relationship between metabolism and grain yield. *Plant Physiol.169*: 2665–2683.

Okuzaki, A., Ogawa, T., Koizuka, C., Kaneko, K., Inaba, M., Imamura, J., & Koizuka, N. (2018). CRISPR/Cas9-mediated genome editing of the fatty acid desaturase 2 gene in *Brassica napus. Plant Physiol. Biochem.131*: 63–69.

Olsen, K.M., & Wendel, J.F. (2013). A bountiful harvest: genomic insights into crop domestication phenotypes. *Annu. Rev. Plant Biol.64*:47–70.

Ong-Abdullah, M., Ordway, J.M., Jiang, N., Ooi, SE., ... Kok, S.Y. (2015). Loss of Karma transposon methylation underlies the mantled somaclonal variant of oil palm. *Nature. 525*:533–537.

Osnas, J.L.D. (2012). The extraordinary diversity of *Brassica oleracea*. https://botanistinthekitchen.blog/2012/11/05/the-extraordinary-diversityof-brassica-oleracea/.

Osterberg, J.T., Xiang, W., Olsen, L.I., Edenbrandt, A.K., ... Vedel, S.E. (2017). Accelerating the domestication of new crops: feasibility and approaches. *Trends Plant Sci. 22*:373–384.

Pacher, M., & Puchta, H. (2017). From classical mutagenesis to nuclease-based breeding: directing natural DNA repair for a natural end-product. *Plant J. 90*:819–833.

Paes de Melo, B., Lourenço-Tessutti, I.T., ... Morgante, C.V. (2020). Soybean embryonic axis transformation: combining biolistic and agrobacterium-mediated protocols to overcome typical complications of in vitro plant regeneration. *Front. Plant Sci.11*:1228.

Pan, C., Sretenovic, S., & Qi, Y. (2021a). CRISPR/dCas-mediated transcriptional and epigenetic regulation in plants. *Curr. Opin. Plant Biol. 60*:101980.

Pan, C., Wu, X., Markel, K., Malzahn, A.A., ... Kundagrami, N. (2021b). CRISPR–Act3.0 for highly efficient multiplexed gene activation in plants. *Nat. Plants.7*: 942–953.

Pandelakis, M., Delgado, E., & Ebrahimkhani, M.R. (2020). CRISPR-based synthetic transcription factors in vivo: the future of therapeutic cellular programming. *Cell Sys. 10*:1–14.

Paran, I., & van der Knaap, E. (2007). Genetic and molecular regulation of fruit and plant domestication traits in tomato and pepper. *J. Exp. Bot. 58*:3841–3852.

Park, K.E., Powell, A., Sandmaier, S.E., ... Kim, C.M. (2017). Targeted gene knock-in by CRISPR/Cas ribonucleoproteins in porcine zygotes. *Sci. Rep.7*:42458.

Park, S.J., Jiang, K., Tal, L., Yichie, Y., ... Gar, O. (2014). Optimization of crop productivity in tomato using induced mutations in the florigen pathway. *Nat. Genet. 46*:1337–1342.

Parker, T.A., Lo, S., & Gepts, P. (2021). Pod shattering in grain legumes: emerging genetic and environment-related patterns. *Plant Cell. 33*: 179–199.

Paterson, A.H. (2002). What has QTL mapping taught us about plant domestication? *New Phytol.154*: 591–608.

Paterson, A.H., Lin, Y.R., Li, Z.K., Schertz, K.F., ... Doebley, J.F. (1995a). Convergent domestication of cereal crops by independent mutations at corresponding genetic-loci. *Science. 269*: 1714–1718.

Paterson, A.H., Schertz, K.F., Lin, Y.R., Liu, S.C., & Chang, Y.L. (1995b). The weediness of wild plants: molecular analysis of genes influencing dispersal and persistence of johnsongrass, *Sorghum halepense* (L.) Pers. *Proc. Natl. Acad. Sci. USA.92*:6127–6131.

Patton, P., & Williams, M. (2016). Chenopodium berlandieri: a potential appalachian Food? http://www.southeasternarchaeology.org/wp-content/uploads/SEAC2016 Bulletin_

Pavlidis, P., & Alachiotis, N. (2017). A survey of methods and tools to detect recent and strong positive selection. *J. Biol. Res. (Thessalon) 24*: 7.

Pearce, S., Vazquez-Gross, H., Herin, S.Y., Hane, D., Wang, Y., Gu, Y.Q, & Dubcovsky, J. (2015). WheatExp: an RNA-seq expression database for polyploid wheat. *BMC Plant Biol.15*: 299.

Peng, J., Richards, D.E., Hartley, N.M., Murphy, G.P., ... Devos, K.M. (1999). 'Green revolution' genes encode mutant gibberellin response modulators. *Nature. 400*: 256–261.

Peres, L.E.P., Carvalho, R.F., ... Zsog on, A. (2005). Grafting of tomato mutants onto potato rootstocks: an approach to study leaf-derived signaling on tuberization. *Plant Sci.169*:680–688.

Périlleux, C., Bouché, F., Randoux, M., & Orman-Ligeza, B. (2019). Turning meristems into fortresses. *Trends Plant Sci. 24*: 431–442.

Pimmental, D., Zuniga, R., & Morrison, D. (2005). Update on the environmental and economic costs associated with alien-invasive species in the United States. *Ecol. Econ. 52*:273–288.

Pin, P.A., Zhang, W., Vogt, S.H., Dally, N., ... Büttner, B. (2012). The role of a pseudo-response regulator gene in life cycle adaptation and domestication of beet. *Curr. Biol. 22*: 1095–1101.

Ping, J., Liu, Y., Sun, L., Zhao, M., Li, Y., She, M., Sui, Y., Lin, F., Liu, X., ... Tang, Z. (2014). Dt2 Is a gain-of-function MADS-domain factor gene that specifies semi-determinacy in soybean. *Plant Cell. 26*: 2831–2842.

Piperno, D.R., Holst, I., Moreno, J.E., & Winter, K. (2019). Experimenting with domestication: understanding macro- and micro-phenotypes and developmental plasticity in teosinte in its ancestral Pleistocene and early Holocene environments. *J. Archaeol. Sci.108*: 1049-1057.

Pnueli, L., Carmel-Goren, L., Hareven, D., ... Gutfinger, T. (1998). The self-pruning gene of tomato regulates vegetative to reproductive switching of sympodial meristems and is the ortholog of CEN and TFL1. *Development.125*:1979–1989.

Pourkheirandish, M., Hensel, G., Kilian, B., Senthil, N., ... Chen, G. (2015). Evolution of the grain dispersal system in barley. *Cell.162*: 527–539.

Prewitt, S.F., Ayre, B.G., & McGarry, R.C. (2018). Cotton centroradialis/terminal flower 1/self-pruning genes functionally diverged to differentially impact plant architecture. *J. Exp. Bot.69*: 5403–5417.

Price, E.J., Drapal, M., Perez-Fons, L., Amah, D., ... Bhattacharjee, R. (2020). Metabolite database for root, tuber, and banana crops to facilitate modern breeding in understudied crops. *Plant J. 101*: 1258–1268.

Purugganan, M.D. (2019). Evolutionary insights into the nature of plant domestication. *Curr. Biol. 29*: R705–R714.

Purugganan, M.D., & Fuller, D.Q. (2009). The nature of selection during plant domestication. *Nature. 457*:843–848.

Purugganan, M.D., & Jackson, S.A. (2021). Advancing crop genomics from lab to field. *Nat. Genet. 53*: 595–601.

Qi, L.S., Larson, M.H., Gilbert, L.A., ... Doudna, J.A. (2013). Repurposing CRISPR as an RNA-guided platform for sequence-specific control of gene expression. *Cell. 152*:1173–1183.

Qi, W., Zhu, T., Tian, Z., Li, C., Zhang, W., & Song, R. (2016). High-efficiency CRISPR/Cas9 multiplex gene editing using the glycine tRNA processing system-based strategy in maize. *BMC Biotechnol.16*: 58.

Ramsay, L., Comadran, J., Druka, A., ... Marshall, D.F. (2011). INTERMEDIUM-C, a modifier of lateral spikelet fertility in barley, is an ortholog of the maize domestication gene TEOSINTE BRANCHED 1. *Nat. Genet. 43*:169–172.

Ramu, P., Esuma, W., Kawuki, R., Rabbi, I.Y., ... Egesi, C. (2017). Cassava haplotype map highlights fixation of deleterious mutations during clonal propagation. *Nat. Genet. 49*:959–963.

Rapp, R.A., Haigler, C.H., Flagel, L., Hovav, R.H., Udall, J.A., & Wendel, J.F. (2010). Gene expression in developing fibres of upland cotton (*Gossypium hirsutum* L.) was massively altered by domestication. *BMC Biol. 8*:139.

Renaut, S., & Rieseberg, L.H. (2015). The accumulation of deleterious mutations as a consequence of domestication and improvement in sunflowers and other compositae crops. *Mol. Biol. Evol. 32*:2273–2283.

Riedelsheimer, C, Czedik-Eysenberg, A., Grieder, C., ... Brown, G.. (2012). Genomic and metabolic prediction of complex heterotic traits in hybrid maize. *Nat. Genet. 44*: 217–220.

Risch, N., & Merikangas, K. (1996). The future of genetic studies of complex human diseases. *Science. 273*:1516–1517.

Rodriguez-Leal, D., Lemmon, Z.H., Man, J., Bartlett, M.E., & Lippman, Z.B. (2017). Engineering quantitative trait variation for crop improvement by genome editing. *Cell. 171*:470–480.e8.

Roldan, M.V.G., Périlleux, C., Morin, H., ... Huerga-Fernandez, S. (2017). Natural and induced loss of function mutations in SlMBP21 MADS-box gene led to jointless-2 phenotype in tomato. *Sci. Rep. 7*: 4402.

Rosado-Souza, L., Scossa, F., Chaves, I,S., ... Kleessen, S. (2015). Exploring natural variation of photosynthetic, primary metabolism and growth parameters in a large panel of *Capsicum chinense* accessions. *Plants. 242*: 677–691.

Ross-Ibarra, J. (2004). The evolution of recombination under domestication: a test of two hypotheses. *Am. Nat.163*: 105–112.

Ross-Ibarra, J., Morrell, P.L., & Gaut, B.S. (2007). Plant domestication, a unique opportunity to identify the genetic basis of adaptation. *Proc. Natl. Acad. Sci. USA. 104*:8641–8648.

Runck, B.C., Kantar, M.B., Jordan, N.R., ... Anderson, J.A. (2014). Reflective plant breeding paradigm: a robust system of germplasm development to support strategic diversification of agroecosystems. *Crop Sci. 54*:1939–1948.

Sakuma, S., Salomon, B., & Komatsuda, T. (2011). The domestication syndrome genes responsible for the major changes in plant form in the *Triticeae* crops. *Plant Cell Physiol. 52*:738–749.

Salman-Minkov, A., Sabath, N., & Mayrose, I. (2016). Whole-genome duplication as a key factor in crop domestication. *Nat. Plants. 2*:16115.

Sasaki, A., Ashikari, M., Ueguchi-Tanaka, M., ... Itoh, H. (2002). Green revolution: a mutant gibberellin-synthesis gene in rice. *Nature. 416*: 701–702.

Sato, K., Yamane, M., Yamaji, N., Kanamori, H., ... Tagiri, A. (2016). Alanine aminotransferase controls seed dormancy in barley. *Nat. Comm.7*: 11625.

Scarcelli, N., Cubry, P., Akakpo, R., Thuillet, A.C., ... Obidiegwu, J. (2019). Yam genomics supports West Africa as a major cradle of crop domestication. *Sci. Adv. 5*: eaaw1947

Schauer, N., Semel, Y., Roessner, U., Gur, A., ... Balbo, I. (2006). Comprehensive metabolic profiling and phenotyping of interspecific introgression lines for tomato improvement. *Nat. Biotechnol. 24*: 447–454.

Scheben, A., Wolter, F., Batley, J., Puchta, H., & Edwards, D. (2017). Towards CRISPR/Cas crops: bringing together genomics and genome editing. *New Phytol. 216*:682–698.

Schemske, D.W. (2012). Reciprocal transplants demonstrate strong adaptive differentiation of the model organism *Arabidopsis thaliana* in its native range. *New Phytol.194*:1112–1122.

Schmidt, C., Fransz, P., Ro¨nspies, M., ... Dreissig, S. (2020). Changing local recombination patterns in *Arabidopsis* by CRISPR/Cas mediated chromosome engineering. *Nat. Comm.11*: 4418.

Schmutz, J., McClean, P.E., Mamidi, S., ... Wu, G.A. (2014). A reference genome for common bean and genome-wide analysis of dual domestications. *Nat. Genet. 46*:707–713.

Schneeberger K, & Weigel D. (2011). Fast-forward genetics enabled by new sequencing technologies. *Trends Plant Sci.16*: 282–288.

Schwahn, K., de Souza, L.P., Fernie, A.R., & Tohge T. (2014). Metabolomics-assisted refinement of the pathways of steroidal glycoalkaloid biosynthesis in the tomato clade. *J. Integr. Plant Biol. 56*: 864–875.

Schwartz, C., Lenderts, B., Feigenbutz, L., ... Barone, P. (2020). CRISPR-Cas9-mediated 75.5-Mb inversion in maize. *Nat. Plants. 6*: 1427–1431.

Scossa, F., Brotman, Y., de Abreu E Lima, F., ... Willmitzer, L. (2016). Genomics-based strategies for the use of natural variation in the improvement of crop metabolism. *Plant Sci. 242*:47–64.

Seren, U., Vilhja´lmsson, B.J., Horton, M.W, ... Meng, D. (2012). GWAPP: a web application for genome-wide association mapping in *Arabidopsis. Plant Cell. 24*: 4793–4805

Setter, T.L., Yan, J., Warburton, M., ... Ribaut, J.M. (2011). Genetic association mapping identifies single nucleotide polymorphisms in genes that affect abscisic acid levels in maize floral tissues during drought. *J. Exp. Bot. 62*:701–716.

Shang, Y., Ma, Y, Zhou, Y., Zhang, H., ... Duan, L. (2014). Plant science. Biosynthesis, regulation, and domestication of bitterness in cucumber. *Science. 346*: 1084–1088.

Shapter, F.M., Cross, M., Ablett, G., ... Malory, S. (2013). High-throughput sequencing and mutagenesis to accelerate the domestication of *Microlaena stipoides* as a new food crop. *PLoS One. 8*:e82641.

Sharma, T.R., Devanna, B.N., Kiran, K., ... Singh, P.K. (2018). Status and prospects of next generation sequencing technologies in crop plants. *Curr. Issues Mol. Biol. 27*:1–36.

Sheehan, M.J., Kennedy, L.M., Costich, D.E., & Brutnell, T.P. (2007). Subfunctionalization of PhyB1 and PhyB2 in the control of seedling and mature plant traits in maize. *Plant J. 49*: 338–353.

Shelef, O., Weisberg, PJ., & Provenza, F.D. (2017). The value of native plants and local production in an era of global agriculture. *Front. Plant Sci.8*:2069.

Shi, J., Gao, H., Wang, H., Lafitte, H.R., ... Archibald, R.L. (2017). ARGOS8 variants generated by CRISPR-Cas9 improve maize grain yield under field drought stress conditions. *Plant Biotechnol. J. 15*:207–216.

Shomura, A., Izawa, T., Ebana, K., ... Ebitani, T. (2008). Deletion in a gene associated with grain size increased yields during rice domestication. *Nat. Genet. 40*:1023–1028.

Si, Z., Liu, H., Zhu, J., Chen, J., Wang, Q., Fang, L., ... Gao, F. (2018). Mutation of SELF-PRUNING homologs in cotton promotes short-branching plant architecture. *J. Exp. Bot. 69*: 2543–2553.

Si, L., Chen, J., Huang, X., Gong, H., Luo, J., ... Hou, Q. (2016). OsSPL13 controls grain size in cultivated rice. *Nat. Genet. 48*:447–456.

Sigmon, B., & Vollbrecht, E. (2010). Evidence of selection at the ramosa1 locus during maize domestication. *Mol. Ecol.19*:1296–1311.

Silva, W.B., Vicente, M.H., Robledo, J.M., ... Reartes, D.S. (2018). Self-pruning acts synergistically with diageotropic to guide auxin responses and proper growth form. *Plant Physiol.176*: 2904–2916.

Simons, K.J., Fellers, J.P., Trick, H.N., Zhang, Z., Tai, Y.S., Gill, B.S., & Faris JD. (2006). Molecular characterization of the major wheat domestication gene Q. *Genetics.172*: 547–555.

Smith, J.M., & Haigh, J. (2007). The hitch-hiking effect of a favourable gene. *Genet. Res.89*: 391–403.

Smith, B.D. (2006). Eastern North America as an independent center of plant domestication. *Proc. Natl. Acad. Sci. USA.103*:12223–12228.

Smith, B.D., & Yarnell, R.A. (2009). Initial formation of an indigenous crop complex in eastern North America at 3800 B.P. *Proc. Natl. Acad. Sci. USA.106*:6561–6566.

Smykal, P., Trneny, O., Brus, J., Hana´cek, P., ... Rathore, A. (2018). Genetic structure of wild pea (*Pisum sativum* subsp. elatius) populations in the northern part of the Fertile Crescent reflects moderate cross-pollination and strong effect of geographic but not environmental distance. *PLoS One.13*: e0194056.

Smykal, P., Nelson, M.N., Berger, J.D., & von Wettberg, E.J.B. (2018). The impact of genetic changes during crop domestication. *Agronomy. 8*:119.

Soltani, A., Walter, K.A., Wiersma, A.T., ... Santiago, J.P. (2021). The genetics and physiology of seed dormancy, a crucial trait in common bean domestication. *BMC Plant Biol. 21*: 58.

Song, G., Han, X., Wiersma, A.T., Zong, X., Awale, H.E., & Kelly, J.D. (2020a). Induction of competent cells for Agrobacterium tumefaciens-mediated stable transformation of common bean (*Phaseolus vulgaris* L.). *PLOS One.15*: e0229909.

Song, J., Zhang, S., Wang, X., Sun, S., Liu, Z., ... Wang, K. (2020b). Variations in both FTL1 and SP5G, two tomato FT paralogs, control day-neutral flowering. Mol. Plant.13: 939–942.

Song, Q., Zhang, T., Stelly, D.M., & Chen, Z.J. (2017). Epigenomic and functional analyses reveal roles of epialleles in the loss of photoperiod sensitivity during domestication of allotetraploid cottons. *Genome Biol.18*:99-106.

Sonnewald, U., Fernie, A.R., Gruissem, W., ... Schla¨pfer, P. (2020). The Cassava Source-Sink project: opportunities and challenges for crop improvement by metabolic engineering. *Plant J. 103*:1655–1665.

Sork, V.S. (2018). Genomic studies of local adaptation in natural plant populations. *J. Hered.109*: 3–15.

Sosso, D., Luo, D., Li, Q.B., Sasse, J., Yang, J., ... Gendrot, G. (2015). Seed filling in domesticated maize and rice depends on SWEET-mediated hexose transport. *Nat. Genet.47*:1489–1493.

Soyk, S., Benoit, M., & Lippman, Z.B. (2020). New horizons for dissecting epistasis in crop quantitative trait variation. *Ann. Rev. Genetics. 54*: 287–307.

Soyk, S., Müller, N.A., Park, S.J., Schmalenbach, I., ... Jiang, K. (2017). Variation in the flowering gene self-pruning 5G promotes day-neutrality and early yield in tomato. *Nat. Genet. 49*: 162–168.

Spannagl, M., Nussbaumer, T., Bader, K.C., ... Martis, M.M. (2016). PGSB PlantsDB: updates to the database framework for comparative plant genome research. *Nucleic Acids Res. 44*: D1141–1147.

Spengler, R.N., & Mueller, N.G. (2019). Grazing animals drove domestication of grain crops. *Nat. Plants. 5*: 656–662.

Spengler, R.N., Petraglia, M., Roberts, P., ... Ashastina, K. (2021). Exaptation traits for megafaunal mutualisms as a factor in plant domestication. *Front. Plant Sci. 12*: 649394.

Springer, N.M., & Schmitz, R.J. (2017). Exploiting induced and natural epigenetic variation for crop improvement. *Nat. Rev. Genet.18*:563–575.

Stetter, M.G., Gates, D.J., Mei, W., & Ross-Ibarra, J. (2018). How to make a domesticate. *Curr. Biol. 27*: R896–R900.

Stetter, M.G., Vidal-Villarejo, M., & Schmid, K.J. (2020). Parallel seed color adaptation during multiple domestication attempts of an ancient new world grain. *Mol. Biol. Evol. 37*:1407–1419.

Stetter, M.G. (2020). Limits and constraints to crop domestication. *Am. J. Bot. 107*: 1617–1621.

Stone, S.L., Kwong, L.W., Yee, K.M., Pelletier, J., ... Lepiniec, L. (2001). Leafy cotyledons encodes a B3 domain transcription factor that induces embryo development. *Proc. Natl. Acad. Sci. USA. 98*:11806–11811.

Stotz, G.C., Salgado-Luarte, C., Escobedo, V.M., Valladares, F., & Gianoli, E. (2021). Global trends in phenotypic plasticity of plants. *Ecol. Lett. 24*: 2267–2281.

Studer, A.J., Wang, H., & Doebley, J.F. (2017). Selection during maize domestication targeted a gene network controlling plant and inflorescence architecture. *Genetics. 207*:755–765.

Studer, A., Zhao, Q., Ross-Ibarra, J., & Doebley, J. (2011). Identification of a functional transposon insertion in the maize domestication gene tb1. *Nat. Genet. 43*:1160–1163.

Stumpf, M.P.H., & McVean, G.A.T. (2003). Estimating recombination rates from population-genetic data. *Nat. Rev. Genet. 4*: 959–968.

Sugimoto, K., Takeuchi, Y., Ebana, K., Miyao, A., ... Hirochika, H. (2010). Molecular cloning of Sdr4, a regulator involved in seed dormancy and domestication of rice. *Proc. Natl. Acad. Sci. USA.107*:5792–5797.

Swanson-Wagner, R., Briskine, R., Schaefer, R., ... Hufford, M.B. (2012). Reshaping of the maize transcriptome by domestication. *Proc. Natl. Acad. Sci. USA. 109*:11878–11883.

Sweeney, M.T., Thomson, M.J., Pfeil, B.E., & McCouch, S. (2006). Caught red-handed: Rc encodes a basic helix-loop-helix protein conditioning red pericarp in rice. *Plant Cell.18*:283–294.

Sweetlove, L.J., Nielsen, J., & Fernie, A.R. (2017). Engineering central metabolism: a grand challenge for plant biologists. *Plant J. 90*:749–763.

Taagen, E., Bogdanove, A.J., & Sorrells, M.E. (2020). Counting on crossovers: controlled recombination for plant breeding. *Trends Plant Sci. 25*: 455–465.

Tadele, Z. (2019). Orphan crops: their importance and the urgency of improvement. *Planta. 250*: 677–694.

Takahashi, Y., Sakai, H., Yoshitsu, Y., Muto, C., ... Anai, T. (2019). Domesticating *Vigna Stipulacea*: A potential legume crop with broad resistance to biotic stresses. *Front. Plant. Sci.10*:1607.

Taketa, S., Amano, S., Tsujino, Y., Sato, T., ... Saisho, D. (2008). Barley grain with adhering hulls is controlled by an ERF family transcription factor gene regulating a lipid biosynthesis pathway. *Proc. Natl. Acad. Sci. USA.105*:4062–4067.

Tan, L., Li, X., Liu, F., Sun, X., Li, C., Zhu, Z., Fu, Y., ... Cai, H. (2008). Control of a key transition from prostrate to erect growth in rice domestication. *Nat. Genet. 40*:1360–1364.

Tello-Ruiz, M.K., Naithani, S., Stein, J.C., Gupta, P., ... Campbell, M. (2018). Unifying comparative genomics and pathway resources for plant research. *Nucleic Acids Res. 46*: D1181–D1189.

Tian, H., Xu, R., Canadell, J.G., ... Thompson, R.L. (2020). A comprehensive quantification of global nitrous oxide sources and sinks. *Nature. 586*: 248–256.

Tian, Z., Wang, X., Lee, R., Li, Y., ... Specht, J.E. (2010). Artificial selection for determinate growth habit in soybean. *Proc. Natl. Acad. Sci. U SA.107*:8563–8568.

Togninalli, M., Seren, Ü., Freudenthal, J.A., ... Monroe, J.G. (2020). AraPheno and the AraGWAS Catalog 2020: a major database update including RNA-Seq and knockout mutation data for *Arabidopsis thaliana. Nucleic Acids Res.48*: D1063–D1068.

Tohge, T., Scossa, F., & Fernie, A.R. (2015). Integrative approaches to enhance understanding of plant metabolic pathway structure and regulation. *Plant Physiol.169*: 1499–1511.

Tohge, T., Scossa, F., Wendenburg, R., Frasse, P., ... Balbo, I. (2020). Exploiting natural variation in tomato to define pathway structure and metabolic regulation of fruit polyphenolics in the *lycopersicum* complex. *Mol. Plant.13*: 1027–1036.

Tong, H., Ku¨ken, A., & Nikoloski, Z. (2020). Integrating molecular markers into metabolic models improves genomic selection for Arabidopsis growth. *Nat. Comm. 11*: 2410.

Tong, H., & Nikoloski, Z. (2020). Machine learning approaches for crop improvement: Leveraging phenotypic and genotypic big data. *J. Plant Physiol. 257*: 1533- 1538.

Torada, A., Koike, M., Ogawa, T., Takenouchi, Y., ... Tadamura, K. (2016). A causal gene for seed dormancy on wheat chromosome 4A encodes a MAP kinase. *Curr. Biol. 26*: 782–787.

Tsuda, M., Kaga, A., Anai, T., Shimizu, T., Sayama, T., ... Takagi, K. (2015). Construction of a high-density mutant library in soybean and development of a mutant retrieval method using amplicon sequencing. *BMC Genomics.16*: 1014.

Tuinstra, M.R., Ejeta, G., & Goldsbrough, P.B. (1997). Heterogeneous inbred family (HIF) analysis: a method for developing near-isogenic lines that differ at quantitative trait loci. *Theor. Appl. Genet. 95*:1005–1011.

Turner, A., Beales, J., Faure, S., Dunford, R.P., & Laurie, D.A. (2005). The pseudo-response regulator Ppd-H1 provides adaptation to photoperiod in barley. *Science. 310*: 1031–1034.

Uga, Y., Sugimoto, K., Ogawa, S., Rane, J., Ishitani, M., ... Hara, N. (2013). Control of root system architecture by deeper rooting 1 increases rice yield under drought conditions. *Nat. Genet. 45*:1097–1102.

Usade, B., Tohge, T., Scossa, F., Sierro, N., Schmidt, M., ... Vogel, A. (2018). The genome and metabolome of the tobacco tree, *Nicotiana glauca*: a potential renewable feedstock for the bioeconomy. *BioRxiv.* https:// doi.org/10.1101/351429.

Van Tassel, D.L., Tesdell, O., Schlautman, B., ... Rubin, M.J. (2020). New food crop domestication in the age of gene editing: genetic, agronomic and cultural change remain co-evolutionarily entangled. *Front. Plant Sci. 11*: 789.

Varshney, R.K., Singh, V.K., Kumar, A., Powell, W., & Sorrells, M.E. (2018). Can genomics deliver climate-change ready crops? Curr. Opin. Plant Biol. 45:205–211.

Vasconcelos, M.W., Gruissem, W., & Bhullar, N.K. (2017). Iron biofortification in the 21st century: setting realistic targets, overcoming obstacles, and new strategies for healthy nutrition. Curr. Opin. Biotechnol. 44:8–15.

Vergara-Diaz, O., Vatter, T., Kefauver, S.C., Obata, T., Fernie, A.R., & Araus, J.L. (2020). Assessing durum wheat ear and leaf metabolomes in the field through hyperspectral data. Plant J.102: 615–630.

Vergara-Diaz, O., Vatter, T., Vicente, R., Obata, T., Nieto-Taladriz, M.T., Aparicio, N., Carlisle Kefauver, S., Fernie, A., & Araus, J.L. (2020). Metabolome profiling supports the key role of the spike in wheat yield performance. Cell. 9:242.

Vilela, A.E., & González-Paleo, L. (2015). Changes in resource-use strategy and phenotypic plasticity associated with selection for yield in wild species native to arid environments. J. Arid Environ. 113: 51–58.

Vo ngoc, L., Huang, C.Y., Cassidy, C.J., Medrano, C., & Kadonaga, J.T. (2020). Identification of the human DPR core promoter element using machine learning. Nature.585: 459–463.

Walker, A.R., Lee, E., Bogs, J., McDavid, D.A., Thomas, M.R., & Robinson, S.P. (2007). White grapes arose through the mutation of two similar and adjacent regulatory genes. Plant J. 49:772–785.

Wallace, J.G., Rodgers-Melnick, E., & Buckler, E.S. (2019). On the road to breeding 4.0: unravelling the good, the bad and the boring of crop quantitative genomics. Annu. Rev. Genet. 52:421–444.

Wang, E., Wang, J., Zhu, X., Hao, W., Wang, L., ... Li, Q. (2008). Control of rice grain-filling and yield by a gene with a potential signature of domestication. Nat. Genet. 40:1370–1374.

Wang, H., Nussbaum-Wagler, T., Li, B., Zhao, Q., ... Vigouroux, Y. (2005). The origin of the naked grains of maize. Nature. 436:714–719.

Wang, J., Qi, M., Liu, J., & Zhang, Y. (2015). CARMO: a comprehensive annotation platform for functional exploration of rice multi-omics data. Plant J. 83: 359–374.

Wang, L., Beissinger, T.M., Lorant, A., ... Ross-Ibarra, C. (2017). The interplay of demography and selection during maize domestication and expansion. Genome Biol. 18:215.

Wang, L., Sun, S., Wu, T., Liu, L., Sun, X., Cai, Y., ... Li, J. (2020). Natural variation and CRISPR/Cas9-mediated mutation in GmPRR37 affect photoperiodic flowering and contribute to regional adaptation of soybean. Plant Biotechnol. J. 18: 1869–1881.

Wang, M., Li, W., Fang, C., Xu, F., Liu, Y., Wang, Z., ... Yang, R. (2018). Parallel selection on a dormancy gene during domestication of crops from multiple families. Nat. Genet. 50: 1435– 1441.

Wang, W., Feng, B., Xiao, J., Xia, Z., Zhou, X., Li, P., ... Zhang, W. (2014). Cassava genome from a wild ancestor to cultivated varieties. Nat. Comm. 5: 5110

Wang, X., Aguirre, L., Rodríguez-Leal, D., Hendelman, A., Benoit, M., & Lippman, Z.B. (2021). Dissecting cis-regulatory control of quantitative trait variation in a plant stem cell circuit. Nat. Plants. 7: 419–427.

Wang, Y., Cheng, X., Shan, Q., Zhang, Y., ... Liu, J. (2014). Simultaneous editing of three homoeoalleles in hexaploid bread wheat confers heritable resistance to powdery mildew. Nat. Biotechnol. 32:947–951.

Warschefsky, E.J., Klein, L.L., Frank, M.H., ... Chitwood, D.H. (2016). Rootstocks: diversity, domestication, and impacts on shoot phenotypes. Trends Plant Sci. 21:418–437.

Watt, M., Fiorani, F., Usadel, B., Rascher, U., Muller, O., & Schurr, U. (2020). Phenotyping: new windows into the plant for breeders. Ann. Rev. Plant Biol. 71: 689–712

Wei, X., Qui, J., Yong, K., Fan, J., Zhang, Q., ... Hua, H. (2021). A quantitative genomics map of rice provides genetic insights and guides breeding. Nat. Genet. 53: 243–253

Weller, J.L., Vander Schoor, J.K., Perez-Wright, E.C., ... Hecht, V. (2019). Parallel origins of photoperiod adaptation following dual domestications of common bean. *J. Exp. Bot.70*: 1209–1219.

Wilkins, A.S. (2020). A striking example of developmental bias in an evolutionary process: the "domestication syndrome. *Evol. Dev. 22*: 143–153.

Wilkinson, P.A., Winfield, M.O., Barker, G.L.A., ... Allen, A.M. (2012). Cereals 2.0: an integrated resource for plant breeders and scientists. *BMC Bioinformatics. 13*: 219

Wing, R.A., Purugganan, M.D., & Zhang, Q. (2018). The rice genome revolution: from an ancient grain to Green Super Rice. *Nat. Rev. Genet.19*:505–517.

Wright, S. (1921). Systems of mating I. biometric relations between parents and offspring. *Genetics. 6*:111–123.

Xiao, A., Wang, Z., Hu, Y., Wu, Y., Luo, Z., ... Yang, Z. (2013). Chromosomal deletions and inversions mediated by TALENs and CRISPR/Cas in zebrafish. *Nucleic Acids Res. 41*: e141.

Xiao, H., Jiang, N., Schaffner, E., Stockinger, E.J., & van der Knaap, E. (2008). A retrotransposon-mediated gene duplication underlies morphological variation of tomato fruit. *Science. 319*:1527–1530.

Xie, K., & Yang, Y. (2013). RNA-guided genome editing in plants using a CRISPR-Cas system. *Mol. Plant. 6*:1975–1983.

Xie, Q., Li, N., Yang, Y., Lv, Y., Yao, H., Wei, R., Sparkes, D.L., & Ma, Z. (2018). Pleiotropic effects of the wheat domestication gene Q on yield and grain morphology. *Planta. 247*:1089–1098.

Xie, X., & Liu, Y.G. (2021). De novo domestication towards new crops. *Natl. Sci. Rev. 8*:1-2.

Xu, G., Cao, J., Wang, X., Chen, Q., Jin, W., Li, Z., & Tian, F. (2019). Evolutionary metabolomics identifies substantial metabolic divergence between maize and its wild ancestor, Teosinte. *Plant Cell. 31*:1990–2009

Yamagishi, N., Kishigami, R., & Yoshikawa, N. (2014). Reduced generation time of apple seedlings to within a year by means of a plant virus vector: a new plant-breeding technique with no transmission of genetic modification to the next generation. *Plant Biotechnol. J.12*: 60–68.

Yan, J., Kandianis, C.B., Harjes, C.E., Bai, L., Kim, E.H., ... Yang, X. (2010). Rare genetic variation at *Zea mays* crtRB1 increases beta-carotene in maize grain. *Nat. Genet.42*:322–327.

Yan, L., Loukoianov, A., Blechl, A., Tranquilli, G., ... Ramakrishna, W. (2004). The wheat VRN2 gene is a flowering repressor downregulated by vernalization. *Science. 303*:1640–1644.

Yan, L., Loukoianov, A., Tranquilli, G., Helguera, M., Fahima, T., & Dubcovsky, J. (2003). Positional cloning of the wheat vernalization gene VRN1. *Proc. Natl. Acad. Sci. USA.100*:6263–6268.

Yang, J., Moeinzadeh, M.H., Kuhl, H., Helmuth, J., Xiao, P., . .. Haas, S. (2017). Haplotype-resolved sweet potato genome traces back its hexaploidization history. *Nat. Plants.3*: 696–703.

Yang, Q., Li, Z., Li, W., Ku, L., Wang, C., Ye, J., Li, K., Yang, N., Li, Y., ... Zhong, T. (2013). CACTA-like transposable element in ZmCCT attenuated photoperiod sensitivity and accelerated the postdomestication spread of maize. *Proc. Natl. Acad. Sci. USA .110*: 16969–16974.

Yang, W., Feng, H., Zhang, X., Zhang, J., ... Doonan, J.H. (2020). Crop phenomics and high-throughput phenotyping: past decades, current challenges, and future perspectives. *Mol. Plant.13*:187–214.

Ye, J., Tian, R., Meng, X., Tao, P., Li, C., Liu, G., Chen, W., Wang, Y., Li, H., ... Ye, Z. (2020). Tomato SD1, encoding a kinase interacting protein, is a major locus controlling stem development. *J. Exp. Bot.71*:3575–3587.

Ye, M., Peng, Z., Tang, D., Yang, Z., Li, D., Xu, Y., Zhang, C., & Huang, S. (2018). Generation of self-compatible diploid potato by knockout of S-RNase. *Nat. Plants. 4*:651–654.

Youens-Clark, K., Buckler, E., Casstevens, T., Chen, C., ... Declerck, G. (2011). Gramene database in 2010: updates and extensions. *Nucleic Acids Res. 39*: D1085–D1094.

Yu, H., Lin, T., Meng, X., Du, H., Zhang, J., Liu, G., ... Chen, M. (2021). A route to de novo domestication of wild allotetraploid rice. *Cell.184*:1156–1170.e14.

Yu, X., Liu, H., Sang, N., Li, Y., Zhang, T., Sun, J., & Huang, X. (2019). Identification of cotton mother of FT and TFL1 homologs, GhMFT1 and GhMFT2, involved in seed germination. *PLOS One.14*: e0215771.

Zamir, D. (2001). Improving plant breeding with exotic genetic libraries. *Nat. Rev. Genet.2*:983–989.

Zeven, A.C. (1972). The partial and complete domestication of the oil palm (*Elaeis guineensis*). *Econom. Bot. 26*: 274–279.

Zhang, H., Li, Y., & Zhu, J. K. (2018a). Developing naturally stress-resistant crops for a sustainable agriculture. *Nat. Plants.4*: 989–996.

Zhang, H., Zhang, J., Wei, P., Zhang, B., Gou, F., ... Feng, Z. (2014). The CRISPR/Cas9 system produces specific and homozygous targeted gene editing in rice in one generation. *Plant Biotechnol. J.12*: 797–807.

Zhang, J., Zhang, X., Tang, H., Zhang, Q., Hua, X., Ma, X., ... Zhu, F. (2018b). Allele-defined genome of the autopolyploid sugarcane *Saccharum spontaneum* L. *Nat. Genet.50*:1565–1573.

Zhang, L., Su, W., Tao, R., Zhang, W., Chen, J., ... Wu, P. (2017). RNA sequencing provides insights into the evolution of lettuce and the regulation of flavonoid biosynthesis. *Nat. Comm. 8*: 2264.

Zhang, W., Alseekh, S., Zhu, X., Zhang, Q., Fernie, A.R., Kuang, H., & Wen, W. (2020a). Dissection of the domestication-shaped genetic architecture of lettuce primary metabolism. *Plant J. 104*: 613–630

Zhang, W., Zhang, Y., Qiu, H., Guo, Y., Wan, H., ... Zhang, X. (2020b). Genome assembly of wild tea tree DASZ reveals pedigree and selection history of tea varieties. *Nat. Comm.11*: 3719.

Zhang, Y., Malzahn, A.A., Sretenovic, S., & Qi, Y. (2019). The emerging and uncultivated potential of CRISPR technology in plant science. *Nat. Plants.5*: 778–794.

Zhang, Y., Butelli, E., Alseekh, S., Tohge, T., Rallapalli, G., ... Luo, J. (2015). Multilevel engineering facilitates the production of phenylpropanoid compounds in tomato. *Nat. Commun.6*:8635-8638.

Zhang, Y., Massel, K., Godwin, ID., & Gao, C. (2018c). Applications and potential of genome editing in crop improvement. *Genome Biol.19*:210.

Zhao, H., Yao, W., Ouyang, Y., Yang, W., Wang, G., Lian, X., Xing, Y., Chen, L., & Xie, W. (2014). Rice Var Map: a comprehensive database of rice genomic variations. *Nucleic Acids Res. 43*: D1018–D1022.

Zhao, X., Meng, Z., Wang, Y., Chen, W., Sun, C., ... Cui, B. (2017). Pollen magnetofection for genetic modification with magnetic nanoparticles as gene carriers. *Nat. Plants.3*:956–964.

Zhou, Y., Ma, Y., Zeng, J., Duan, L., Xue, X., ... Wang, H. (2016). Convergence and divergence of bitterness biosynthesis and regulation in Cucurbitaceae. *Nat. Plants.2*:16183.

Zhou, Y., Massonnet, M., Sanjak, J.S., Cantu, D., & Gaut, B.S. (2017). Evolutionary genomics of grape (*Vitis vinifera* ssp. vinifera) domestication. *Proc. Natl. Acad. Sci. USA.114*:11715–11720.

Zhu, G., Wang, S., Huang, Z., Zhang, S., ... Liao, Q. (2018). Rewiring of the fruit metabolome in tomato breeding. *Cell.172*: 249–261.

Zhu, B.F., Si, L., Wang, Z., Zhou, Y., Zhu, J., ... Shangguan, Y. (2011). Genetic control of a transition from black to straw-white seed hull in rice domestication. *Plant Physiol.155*:1301–1311.

Zhu, G., Wang, S., Huang, Z., Zhang, S., ... Liao, Q. (2018). Rewiring of the fruit metabolome in tomato breeding. *Cell.172*:249–261.e12.

Zonneveld, M. van., Larranaga, N., Blonder, B., Coradin, L., Hormaza, J.I., & Hunter, D. (2018). Human diets drive range expansion of megafauna-dispersed fruit species. *Proc. Natl. Acad. Sci. USA.115*: 3326–3331.

Zsog, on. A., Cerma, T., Naves, E.R., ... Notini, M.M. (2018). De novo domestication of wild tomato using genome editing. *Nat. Biotechnol.36*:1211–1216.

Zsog, on. A., Cermak, T., Voytas, D., & Peres, L.E. (2017). Genome editing as a tool to achieve the crop ideotype and de novo domestication of wild relatives: case study in tomato. *Plant Sci.256*:120–130.

Zsögön, A., & Peres, L.E.P. (2018). Molecular control of plant shoot architecture. *Plant Cell.30*: tpc.118.tt1218–tpc.118.tt1218.

Zsögön, A., Čermák, T., Naves, E.R., Notini, M.M., ... Edel, K.H. (2018). De novo domestication of wild tomato using genome editing. *Nat. Biotech.36*:1211-1217.

Zuo, J., Niu, Q.W., Frugis, G., & Chua, N.H. (2002). The WUSCHEL gene promotes vegetative-to-embryonic transition in *Arabidopsis. Plant J. 30*: 349–359.

Chapter-4

High Throughput Phenotyping

Abstract

Since whole-genome sequencing of many crops has been achieved, crop functional genomics studies have stepped into the big-data and high-throughput era. However, the acquisition of large-scale phenotypic data has become one of the major bottlenecks hindering crop breeding and functional genomics studies. Nevertheless, recent technological advances provide great potential solutions to relieve this bottleneck and explore advanced methods for large-scale phenotyping data acquisition and processing in the coming years. The major progress on high-throughput phenotyping in controlled environments and field conditions as well as its use for post-harvest yield and quality assessment in the past decades have generated prospects to overcome this bottleneck. The multi-omics research combining high-throughput phenotyping with genetic studies has also contributed to this direction.

Keywords: *Phenotyping, high throughput phenotyping, multi omics, remote sensing, field phenotyping*

Introduction

Johannsen (1911) stated that "all types of organisms can be distinguished by direct inspection or with finer methods of measurement or description". Davis (1949) defined the genome as the material basis of the genotype, the term "phenome" was first defined as the sum total of extragenic, non-auto reproductive portions of the cell and represented the set of phenotypes. In the 1990s, in the context of the study of complex human disease and as the key complement of genomics, a new discipline, phenomics, emerged with the aim of bridging the gap between genes and clinical endpoints (Schork, 1997). Houle et al. (2010) defined phenomics as the acquisition of high-dimensional phenotypic data on an organism-wide scale. Fiorani and Schurr (2013) referred to plant phenotyping as the set of methodologies and protocols used to accurately measure plant growth, architecture, and composition at different scales. Yang et. al (2020) defined crop phenomics as the multidisciplinary study of high-throughput accurate acquisition and analysis of multidimensional phenotypes on an organism-wide scale through crop development.

High-throughput image-based plant phenotyping facilitates the computation of phenotypes by analyzing a large number of plants in short time intervals with precision, nullifying the need for time-consuming physical human labour (Das Choudhury et al., 2018). The process is generally non-destructive, allowing the same traits to be quantified repeatedly at multiple times and scales during a plant's life cycle. It is an interdisciplinary approach involving computer science, biology, remote sensing, statistics, and genomics.

Plants are not static but living organisms with constantly increasing complexity in shape, architecture, and appearance. Many plants alter leaf positioning (i.e., phyllotaxy) in response to light signals perceived through the photochrome in order to optimize light interception (Maddonni et al., 2002). In addition to variation in phyllotaxy, growth of individual leaves leads to self-occlusions and leaf crossovers over time posing challenges to image-based plant phenotyping analysis. Other challenges include variation in illumination, cluttered background, and change in zoom levels in the cameras often used to adjust for plant growth.

The state-of-the-art computer vision-based approaches mainly focus on the structural and morphological aspects of a plant for computing 2D and 3D phenotypes. However, quantification of physiological processes in plant components, i.e., leaves, stem segments, flowers, fruits, etc., may show differential behaviour as a function of environmental stress, and result in different leaf-level photosynthetic activity or carbohydrate content in a plant segment. Similarly, dynamic event-based phenotypes, i.e., identifying important events in the plant life cycle, e.g., emergence timing of a new leaf, automated germination detection, the timing of emergence detection of fruits and flowers, may provide important insights into the acclimation and adaptation strategies of plants.

Progress in sensors, aeronautics, and high-performance computing is paving the way. Recent advances in field HTPPs, with high capacity for data recording, scoring and processing, and non-invasive remote sensing methods, together with automated environmental data collection, and affordable cost has generated hope to overcome the issue. Laboratory analyses of key plant parts may complement direct phenotyping under field conditions. Improvements in user-friendly data management together with a more powerful interpretation of results should increase the use of field HTPPs.

Next-generation sequencing technology has greatly accelerated progress in functional genomics (Huang et al., 2010, Werner, 2010, Li et al., 2018), allowing quantitative trait locus (QTL) mapping and genome-wide association studies (GWASs) (Xiao et al., 2017) to become powerful tools for elucidating the genetic architecture of complex traits (Atwell et al., 2010, Huang et al., 2010, Tian et al., 2011, Wang et al., 2018a, Zhang et al., 2019a). Thereafter, many genes governing important agronomic traits have been identified (Zuo and Li, 2014, Yao et al., 2018, Fernie and Yan, 2019, Shi et al., 2019). For example, before 2000, only approximately 130 genes had been cloned in rice. By the end of 2017, more than 3000 genes were well characterized through various traditional phenotypic approaches (Yao et al., 2018). However, phenotypic data acquisition is still a bottleneck restricting crop breeding and functional genomics studies (Deery et al., 2016). Traditional crop phenotyping methods are labour-intensive, time-consuming, subjective, and often destructive (Furbank and Tester, 2011, Chen et al., 2014).

Advances in a range of technologies, from sensors to information technology (IT) and data extraction, combined with systems integration and decreasing costs, have led to the efficient assessment of the morphology and physiology, non-destructively and repeatedly across whole populations and throughout developmental stages of the plants (Ruiz-Garcia et al., 2009).

Importance of Phenotyping in Plant Breeding

It is expected that crop production will double by 2050 to feed the ever-increasing global population (Tilman and Cairns, 2014). However, achieving this goal will be a significant challenge for plant breeders as the yield of the crops would have to increase at a rate of 2.4% per year. The current average rate of increase is only 1.3%, with yields stagnating in up to 40% of land under cereal production (Ray et. al., 2012, 2013). Extensive breeding and agronomic efforts over the past 50 years have been responsible for tripling cereal yields (Pingali 2012). Continuing advances in the techniques available to breeders offer the potential to increase the rate of genetic improvement (Phillips, 2010). Attempts to exploit new molecular tools (Figure 4.1) to their full potential, particularly the ability to dissect the genetics of quantitative traits such as yield and stress tolerance (Blum, 2011, Cabrera-Bosquet et. al., 2012, Araus et. al., 2012, Tuberosa 2012, Cobb et. al., 2013), are limited by our ability to decipher the phenotypes of major crops (Araus and Cairns 2014).

Plant breeders and farmers have been making selections based on phenotypes long before the discovery of DNA and molecular markers. The development of improved varieties relies on the ability to identify the best genetic variation for advancement. Breeding is essentially a numbers game: the more crosses and environments used for selection, the greater the probability of identifying superior variation. Plant breeders want to be able to phenotype large numbers of lines rapidly and accurately identify the best progeny. To meet future needs there is a need to increase breeding efficiency. Advances in high-throughput genotyping have provided fast and inexpensive genomic information. Low-cost, high-throughput genotyping has paved the way for the development of large mapping populations and diversity panels of thousands of recombinant inbred lines for phenotyping (McMullen et. al, 2009). Although molecular breeding strategies, such as marker-assisted recurrent selection (MARS) and genomic selection, place greater focus on selections based on genotypic information, they still require phenotypic data (Jannick et. al. 2010). In genomic selection, phenotypes are not used for selection but are used to train a prediction model (Lorenz et. al., 2011); whereas in MARS, a single phenotyping cycle is used to identify markers for subsequent selection through generations. Similarly, phenotyping is necessary to identify promising events in transgenic studies (Gaudin et. al., 2012; Saint-Pierre et. al., 2012). Given that molecular breeding populations can include up to 5000 lines, the ability to accurately characterize all lines simultaneously is challenging (McMullen et. al., 2009). Advances in phenotyping are likely to be essential to capitalize on developments in conventional, molecular, and transgenic breeding and ensure the genetic improvement of crops for future food security.

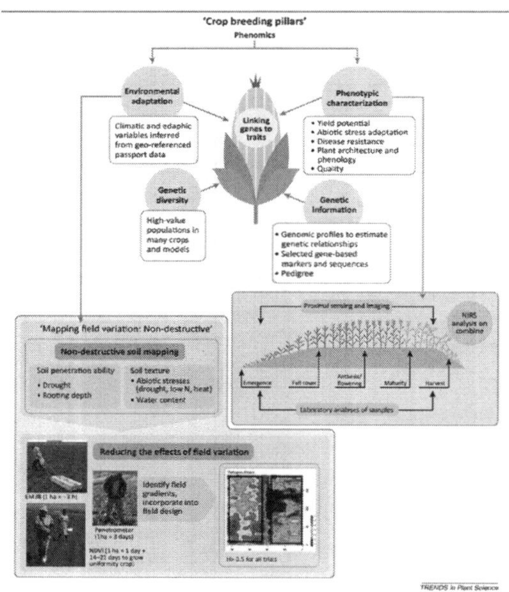

Figure 4.1. (Upper) The four pillars of the crop breeding pipeline (environmental adaptation, phenotypic characterization, genetic diversity, and genetic information) and the implications of phenotyping. The importance of phenotyping is highlighted by its involvement in two of these pillars. (Lower left) Mapping field variability in a non-destructive manner implies the use of different methodological alternatives and their further integration. (Lower right) Diagram of the main categories of phenotyping techniques deployed over the life cycle of an annual seed crop. Types of data acquisition include: proximal sensing and imaging at frequent intervals, laboratory analyses of samples taken at specific intervals, and near-infrared spectroscopy (NIRS) of seed for oil or protein content during combined harvesting. (Reproduced from Araus and Cairns 2014).

A Taxonomy For Plant Phenotyping

A plant phenotypic taxonomy that reflects plant phenotypes introduced in recent literature, as well as the new challenges that require future research attention is presented in Figure 4.2 (Deb Choudhury et al. 2019). The figure shows that the aboveground plant phenotypes are broadly classified into three categories, namely, structural, physiological, and temporal. The structural phenotypes refer to the morphological attributes of the plants, whereas the physiological phenotypes are related to traits that affect plant processes regulating growth and metabolism. Structural and physiological phenotypes are further divided into two groups: holistic and component. The holistic phenotypes consider the whole plant as a single object and compute its basic geometrical properties, e.g., height of the bounding rectangle to account for plant height, area of the convex hull to account for plant size (Das Choudhury et al., 2017, 2018). Component phenotypes are computed by considering individual components of the plants, i.e., leaves, stem, flower, and fruit. Examples of component phenotypes include leaf length, chlorophyll content of each leaf, stem angle, flower size, and fruit volume (Gage et al., 2017; He et al., 2017; Das Choudhury et al., 2018; Yin et al., 2018; Zhou et al., 2019).

The different parts of plants grow non-uniformly over space and time. This non-uniformity in growth is also regulated by the genotypes. The temporal phenotypes are computed by analyzing a sequence of images. We propose two different types of temporal phenotypes: trajectory-based and event-based. The structural and physiological phenotypes can be computed from a single image or a sequence of images to take into account of temporal phenotypic characteristics for better genetic variability, e.g., plant growth rate, leaf elongation rate, and trajectories of stem angle (Das Choudhury et al., 2016, 2018; Sun et al., 2018). It includes changes in phenotypic traits as a function of time, time rate of change in stem width, and relative growth rates of different leaves. Event-based phenotypes are based on specific events in a plant's life cycle, e.g., germination, emergence of a new leaf, flowering, etc. The timing of the events with reference to an origin (say planting date) is important in understanding a plant's behaviour (Agarwal, 2017). While the primary phenotypes refer to one specific characteristic of a plant, derived phenotypes are composed of two or more primary phenotypes, i.e., bi-angular convex-hull area ratio to account for phyllotaxy and plant aspect ratio to provide information on canopy architecture (Das Choudhury et al., 2016).

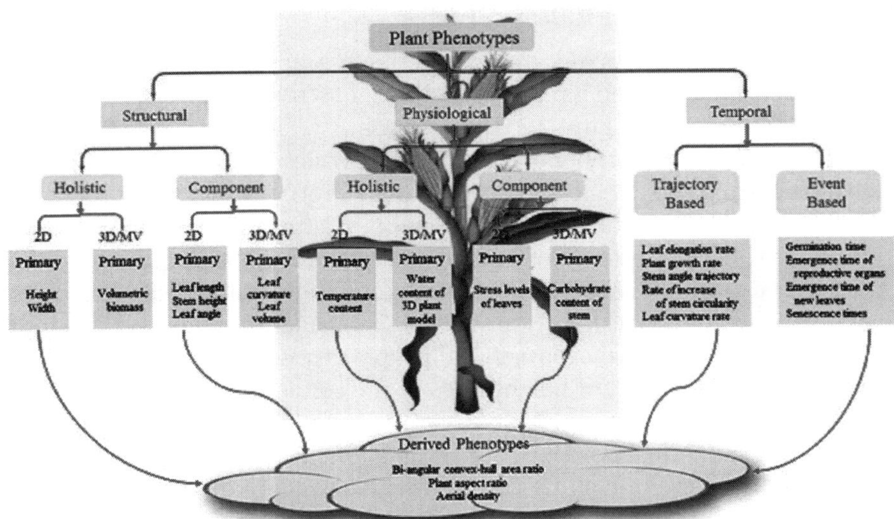

Figure 4.2. A taxonomy of phenotypes. (MV= Multiple View). (Reproduced from Das Choudhury et. al. 2019).

Advancements in Image-Based Plant Phenotyping

A brief description of the state-of-the-art methods to compute phenotypes from images organized using taxonomy, i.e., structural, physiological, and temporal is presented.

Structural Phenotypes

Different types of structural phenotypes are as follows.

2D phenotypes

Dellen et al. (2015) used a leaf-shaped model to detect each leaf of a tobacco plant. A graph-based tracking algorithm is used to track the detected leaves in a sequence to measure growth rate as a trajectory-based structural phenotype. Leaf alignment and tracking are formulated as two optimization problems by Yin et al. (2018) based on Chamfer matching and leaf template transformation from fluorescent videos for application in leaf-level photosynthetic capability estimation. Das Choudhury et al. (2018) introduced a set of new holistic and component phenotypes computed from 2D side view image sequences of maize plants, and demonstrated the temporal variations of these phenotypes regulated by genotypes. The method accepts plant image sequence as the input and produces a leaf status report containing the phenotypic information, i.e., the emergence timing, total number of leaves present at any point of time, total number of leaves emerged, the day on which a particular leaf stopped growing or lost, and the length and relative growth rate of individual leaves. The method in Das Choudhury et al. (2017)

introduces an algorithm to compute stem angle, a potential measure for plants' susceptibility to lodging, based on graph-based plant architecture determination. A time series clustering analysis is used to summarize the temporal patterns of the stem angles into different groups to provide further insight into genotype specific behavior of the plants. Unlike the methods by Das Choudhury et al. (2016, 2018) which focus on vegetative stage phenotyping analysis of maize, the method by Brichet et al. (2017) develops a robot-assisted imaging pipeline to track the growths of ear and silks based on an ear detection algorithm. The genotypic variation in silk growth rate under drought stress is experimentally demonstrated.

It has been suggested that future progress in image-based plant phenotyping will require a combined effort in the domains of image processing for feature extraction and machine learning for data analysis (Tsaftaris et al., 2016). In recent times, machine learning techniques have gained popularity in high-throughput 2D phenotyping, e.g., in the detection of branch-shaking locations for robotic cherry harvesting (Amatya et al., 2017) and plant growth analysis (Navarro et al., 2016). Deep learning is an emerging field that promises unparalleled results on many data analysis problems. Building on artificial neural networks, deep approaches have many more hidden layers in the network, and hence have greater discriminative and predictive power. A deconvolutional network is used for segmenting the rosette leaves from the background, and then a CNN is used for leaf counting in the method by Aich and Stavness (2017). The method by Atanbori et al. (2018) re-architects four existing deep neural networks to create "Lite" CNN models in an attempt to reduce their parameters while avoiding overfitting for cost-effective solutions in plant phenotyping. The method also introduces a global hyper-parameter for an efficient trade-off between parameter size and the accuracy of "Lite" CNN models. The method by Pound et al. (2017) uses CNN to identify quantitative trait loci by classifying biologically relevant features such as root tips, leaf and ear tips, and leaf bases to determine root and shoot architecture.

An open-source software tool based on CNN called Deep Plant Phenomics is proposed in Ubbens and Stavness (2017) to compute complex phenotypes from plant image sequences. ResNet50, a deep residual neural network, is used in Dobrescu et al. (2017) as a leaf prediction model to count the number of leaves of the rosette plants. In this method, leaf counting is modelled as a direct regression problem. A comprehensive summary of deep learning algorithms for the identification, classification, quantification, and prediction of plant stress phenotypes is presented by Singh et al. (2018). Deep learning has applications in a variety of plant phenotyping tasks, e.g., plant stalk count and stalk width (Baweja et al., 2017), leaf counting in rosette plants (Giuffrida et al., 2018; Ubbens et al., 2018), maize tassel counting (Lu et al., 2017), cotton bloom detection (Xu et al., 2018), wheat spikes detection (Hasan et al., 2018), and rice panicle segmentation (Xiong et al., 2017).

3D Phenotypes

The plants exhibit increasing architectural complexity with time due to self-occlusions and leaf crossovers, which pose challenges to the accurate estimation of 2D component phenotypes, e.g., leaf length, stem angle, and leaf curvature. To overcome these challenges, attempts have been made to reconstruct a 3D model of a plant for accurate estimation of phenotypes. A shape-from-silhouette method is used by Golbach et al. (2016) to reconstruct the 3D model of a tomato seedling from multi-view images. An algorithm for depth imaging-based detection of muskmelon plants for phenotyping in the greenhouse is proposed by Lei et al. (2014a). A detailed review of 3D reconstruction techniques for shoot topology for applications in phenotyping is provided by Gibbs et al. (2016). The method in Liu et al. (2017) uses the structure from motion method to reconstruct the 3D model of a plant using images from multiple side views. The method also deducted the optimal number of images needed for reconstructing a high-quality model. Most cereal crops, e.g., rice, wheat, and maize, have thin ribbon-like architectures with a lack of textural surfaces, which limits the success of 3D reconstruction to early growth stages where the distance between the camera and the plant is relatively small facilitating accurate camera calibration.

3D reconstruction of soybean canopies using multisource imaging is presented in Guan et al. (2018). A density-based spatial clustering of applications with noise removal is used to extract canopy information from the raw 3D point cloud. The method by Scharr et al. (2017) uses a voxel-based volume carving for 3D reconstruction of maize shoots to compute leaf-level traits. Depth cameras capture 3D information from plants to segment leaves and reconstruct plant models (Chéné et al., 2012). A semi-automated software pipeline is developed to reconstruct a 3D plant model from the images captured by a depth camera in McCormick et al. (2016). Standard shoot phenotypes such as shoot height, leaf angle, leaf length, and shoot compactness are measured from 3D plant reconstructions to characterize the shoot architecture. The Microsoft Kinect, originally designed for computer gaming environments, has found applications in 3D plant phenotyping in recent times (Chéné et al., 2012; Paulus et al., 2014; Polder and Hofstee, 2014.)

Polder and Hofstee (2014) used fused RGB and depth images captured by Microsoft Kinect sensor to segment muskmelon plants in the greenhouse with a cluttered background. Srivastava et al. (2017) provided an algorithm for 3D model reconstruction of wheat plants with occluded leaves, and used deep learning for drought stress characterization. The remote sensing technology using light detection and ranging (LiDAR) is hypothesized to dominate the future generation of plant phenotyping analysis in outdoor environments (Lin 2015). Sun et al. (2017, 2018) computed morphological traits of the cotton plants, e.g., canopy height, projected canopy area, and plant volume, based on 3D model reconstruction of the plants using top view images captured by LiDAR and real-time kinematic global positioning system (RTK-GPS).

Physiological Phenotypes

Water scarcity causes serious crop losses worldwide. Therefore, it is important to understand the plant resistance mechanisms to drought stress. Römer et al. (2012) used a simplex volume maximization technique for early drought stress detection using hyperspectral image sequences of barley plants. Hyperspectral image analysis has been used to investigate physiological processes, e.g., determination of salt stress in wheat based on vector-wise similarity measurement (Moghimi et al., 2018), early detection of abiotic stresses (Mohd et al., 2018), disease detection to prevent yield losses (Wahabzada et al., 2016), and early yield prediction (González-Sanchez et al., 2014).

Lobos and Poblete-Echevrria (2017) used a Matlab-based software called SK-UTALCA for applications in plant breeding, precision agriculture, crop protection, ecophysiology plant nutrition, and soil fertility by analyzing high-resolution spectral reflectances. A comprehensive overview of the machine learning techniques in the identification, classification, quantification, and prediction of various biotic and abiotic stress phenotypes is provided by Singh et al. (2016). Raza et al. (2014) combined information extracted from thermal and visible light images, and uses a support vector machine (SVM) and Gaussian processes (GP) to identify regions of spinach canopy showing a response to soil water deficit. Chlingaryan et. al. (2018) reviewed of machine learning approaches including back-propagation neural networks, GP, SVM, rotation forest, CNN, and LSTM, for crop yield prediction and nitrogen status estimation in precision agriculture.

Temporal Phenotypes

Das Choudhury et al. (2018) used line graphs to represent the trajectories of component phenotypes, i.e., leaf length, integral leaf-skeleton area, mid-leaf curvature, apex curvature, and stem angle, as a function of time in order to demonstrate the genetic influence on the temporal variations of these phenotypes. Agarwal (2017) used a method for plant emergence detection using adaptive hierarchical segmentation and optical flow-based tracking. The efficacy of the method was demonstrated based on experimental analysis on a dataset consisting of images captured at every 2–5 min interval starting immediately after planting the seeds.

Dynamic Plant Phenotyping

Multiple genes interact with the multiple environments (GxE) across the life cycle to affect performance (Orgogozo et al., 2015). The effect of the environment on plants can be particularly pronounced because their static lifestyle precludes the use of the avoidance strategies commonly used by animals. Sensor technologies now enable the detailed recording of the environmental history of plants as well as the dynamic response of plants or crops. Since the whole-genome sequencing of *Arabidopsis* and many other crops (https://genomevolution.org/wiki/index.php/Sequenced_plant_ genomes) has been achieved (Kersey, 2019), the next goal is to describe the whole phenotypes of important crops and to dissect the key functional loci (genes or QTLs) and better understand the crop genetic architectures. High-throughput phenotyping platforms will play a key role in achieving this goal. The high-throughput shoot phenotyping for *Arabidopsis* and larger crops and root phenotyping in the laboratory environment is presented in Table 4.1.

Table 4.1. High-throughput phenotyping platforms applied in different crops.

Applications	Platform	Advantages	Limitations	References
Shoot phenotyping for Arabidopsis in the laboratory	PHENOPSIS (DB); GROWSCREEN (FLUORO); Phenoscope; Phenovator; PlantScreen	Relatively affordable, rapid, and automatic measurements for large populations	Not suitable for larger crops	Fabre et al., 2011; Jansen et al., 2009; Tisné et al., 2013; Flood et al., 2016; Awlia et al., 2016
Shoot phenotyping for crops in the laboratory	TraitMill; Scanalyzer3D; PHENOARCH; HRPF	Dynamic and automatic obtain shoot growth, biomass, and lots of information for large populations	Costly; requires multidisciplinary experts to maintain and update	Reuzeau et al., 2005; Hairmansis et al., 2014; Brichet et al., 2017; Yang et al., 2014
	PhenoBox	Affordable; easy to maintain	Labour intensive for large screening	Czedik-Eysenberg et al., 2018
Root phenotyping in the laboratory	PlaRoM; Rhizoslides; Rhizoponics; RADIX; RhizoTubes	Affordable; obtain 2D root system architectures	Root growth in transparent media	Yazdanbakhsh and Fisahn, 2009; Le Marie et al., 2014; Mathieu et al., 2015; Le Marie et al., 2016; Jeudy et al., 2016
	GiARoots RootReader3D	Affordable; obtain 3D root system architectures	Root growth in transparent media	Galkovskyi et al., 2012; Clark et al., 2011
	GROWSCREEN-Rhizo	High-throughput to obtain shoot and root traits in soil-filled rhizotrons	Root growth is limited in 2D rhizotrons (rhizobox)	Nagel et al., 2012
	MRI–PET; PET-CT; MRI–CT	Obtain 3D root system architectures in soil-filled tubes	Costly; time consuming; lack of specialized prototype for crop study	Jahnke et al., 2009; Garbout et al., 2011; Metzner et al., 2015

Applications	Platform	Advantages	Limitations	References
Ground-based phenotyping in the field	CPRS, a fixed phenotyping tower	Easy to install and maintain	Limited crop information in fixed areas was obtained	Fukatsu et al., 2012
	Field Scanalyzer, a rail-based gantry phenotyping system	Integration of various optical sensors; high image resolution	Costly; limited image area; variable ambient light	Sadeghi-Tehran et al., 2017
	BreedVision, a self-propelled tractor equipped with multiple sensors mounted in a mobile dark chamber	Integration of various optical sensors; stable imaging conditions; non-restriction of image area	Restricted by wet soil and weather conditions (rainy and strong breeze)	Sadeghi-Tehran et al., 2017
Remote sensing in the field	Drones or UAVs equipped with multiple sensors	Integration of various sensors; non-restriction of imaging area; rapid measurements (crop growth, yield, stress response, etc.); flexible to install and use	Cannot obtain information below the canopy; strict operating and local flight laws should be noted to ensure flight safety	Maes and Steppe, 2019
Pocket phenotyping	PocketPlant3D, a smartphone equipped with APP to measure maize canopy and leaf traits canopy and leaf traits	Affordable; flexible to use; easy to popularize	Limited traits and single function; lack of robust models to face complex conditions in field	Confalonieri et al., 2017
	Seed Evaluation Accelerator (SEA)	Automatic threshing rice panicles and rapid measurements of yield-related traits	Cannot obtain 3D grain traits and panicle traits	Duan et al., 2011
	P-TRAP; PANorama	Quantify both rice panicle traits and grain traits; no need for threshing	Need to manually separate panicle branches; cannot obtain 3D grain traits	AL-Tam et al., 2013; Crowell et al., 2014

Applications	Platform	Advantages	Limitations	References
Post-harvest phenotyping	PhenoSeeder	Extract 3D traits from individual seeds with high accuracy	Low measuring speed; need for threshing	Jahnke et al., 2016
	X-ray CT	Extract 3D cereal grain traits and spike traits without threshing	Costly; time consuming; need to develop a bespoke image analysis pipeline for new species	Hughes et al., 2017 Hughes et al., 2019
	Hyperspectral imagings	Infer protein content, and other physiological or biochemical information	Costly; need bespoke image analysis and update model for new species, new physiological or biochemical indicators	Sun et al., 2019

High-Throughput Plant Phenotyping Platform

A schematic diagram for an image-based high-throughput plant phenotyping platform is shown in Figure 4.3. It analyses temporal image sequences of a group of plants (belonging to different genotypes) captured by multimodal cameras, i.e., visible light, fluorescent, near-infrared, infrared, and hyperspectral, from different viewing angles. The plants are imaged at regular intervals under various environmental conditions, e.g., drought, salinity, and thermal (Das Choudhury et al., 2016). Das Choudhury et. al. (2019) developed an advanced representation to explore computer vision-based plant phenotyping algorithms in multiple dimensions, (i.e., multimodal, Multiview, and temporal), regulated by genotypes under various environmental conditions. The image sequences captured by the visible light camera are often used to compute structural or morphological phenotypes. Fluorescent, near-infrared, and infrared images are respectively used to analyze the chlorophyll, water, and temperature content of the plants. Thermal infrared imaging is used as a proxy for a plant's temperature to detect differences in stomatal conductance as a measure of the plant response to the water status and transpiration rate for abiotic stress adaptation (Lei et al., 2014b). Hyperspectral imaging is uniquely suited to provide insights into the functional properties of plants, e.g., leaf tissue structure, leaf pigments, and water content (Mahlein et al., 2011) and stress resistance mechanisms (Wahabzada et al., 2016).

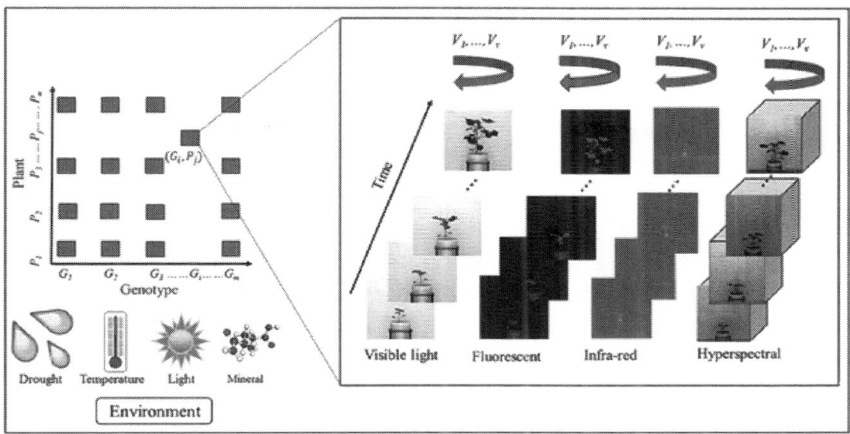

Figure 4.3. High-throughput plant phenotyping platform. (Reproduced from Das Choudhury et. al. 2019).

High-Throughput and Integrated Shoot Phenotyping

Over the past few decades, many intelligent and high-throughput phenotyping platforms have been developed for screening small plants (such as *Arabidopsis* and rosette species) as well as for larger plants (such as rice and maize). *Arabidopsis thaliana* is widely used as a model plant because of its short and rapid life cycle, relatively small genome, and wide geographic adaptation (Kaul et al., 2000), as well as its close relationship to crops in the Brassicaceae family. Automated and large-scale phenotyping under controlled environmental conditions is particularly attractive in such a species for forward genetics, reverse genetics, and quantitative genetics.

One of the first automated red-green-blue (RGB) imaging and weighing systems, PHENOPSIS, was developed to understand *Arabidopsis* responses to water deficit (Granier et al., 2006). However, it lacked integrated data management and had relatively simple environmental controls. Thereafter, the PHENOPSIS platforms and growth chambers were extended, and a database, PHENOPSIS DB, was developed to store hundreds of gigabytes of images and metadata and to provide data and image analysis modules. This database served as a template for other groups to develop a similar data management system (Fabre et al., 2011). The GROWSCREEN platform was developed to quantify the dynamics of seedling growth acclimation (total leaf area, relative growth rate, and root area) under altered light conditions (Walter et al., 2007). In addition, based on the linear displacement stages, the colour imaging device of GROWSCREEN could be replaced with a chlorophyll fluorescence imaging system, GROWSCREEN FLUORO, which allowed the simultaneous phenotyping of leaf growth and chlorophyll fluorescence in *Arabidopsis thaliana* and *Nicotiana tabacum* with a throughput of approximately 60 plants per hour (Jansen et al., 2009). Although GROWSCREEN had relatively limited capacity, it could screen about 200–500 plants along with the micro-environmental heterogeneity across the platform, thereby reducing the time for mapping for QTL analyses or GWAS.

A large-scale phenotyping platform PHENOSCOPE, was designed to obtain shoot growth and water content information for 735 pots, compensating for environmental heterogeneity by continuously rotating the 735 pots across the platform (Tisné et al., 2013). The PHENOVATOR utilized a high-speed x–y moving rail system to move a monochrome camera with eight specific band-pass filters above a platform carrying up to 1440 plants to obtain data on photosynthesis (ΦPSII, light-use efficiency of photosystem II electron transport) and plant growth traits (projected leaf area accumulation) (Flood et al., 2016). This configuration had a high acquisition speed, with the collection of leaf area in 20 min and PSII efficiency in less than an hour. Instead of moving sensors, the commercially available system PlantScreen transported trays (20 pots per tray) on conveyor belts from the growing area to a dark acclimation chamber and chlorophyll fluorescence (ChlF) and RGB imaging cabinets, with automatic weighing and watering (Awlia et al., 2016). Some photosynthetic parameters (such as FV/Fm) were robust in reflecting severe salt stress, while other parameters (ΦNPQ, qP, and ΦP) reflected early salt stress. However, the influence of the movement of potted plants and environmental variation (e.g., wind, temperature, and other microclimates (Brien et al., 2013) in a greenhouse may affect chlorophyll fluorescence parameter measurements (Haworth et al., 2018).

High-Throughput Phenotyping in Crop Breeding

Some of the first developments were aimed at the industrial-scale assessment of genetically modified crops. CropDesign (Belgium) developed TraitMill, comprising proprietary bioinformatics tools, a high-throughput gene engineering system, plant transformation methods, and a high-throughput phenotyping platform, which was used to determine morphometric traits (aboveground biomass, plant morphology, plant colour) that might be related to yield (Reuzeau et al., 2005). Unfortunately, the details of TraitMill, the experimental design, and outcomes were proprietary. LemnaTec GmbH (Germany) developed the Scanalyzer3D platform and installed versions around the world. The Plant Accelerator (Australian Plant Phenomics Facility, University of Adelaide, Adelaide, Australia), fitted with computer-controlled conveyor systems, automated weighing–watering devices, and imaging stations (RGB, near-infrared, fluorescence, and hyperspectral imaging), has a capacity of 2400 plants and has been used to assess the salinity tolerance of rice (Hairmansis et al., 2014), nitrogen/water deficiency in crops (Neilson et al., 2015), and the salinity tolerance of chickpea (Atieno et al., 2017). In addition, the total plant biomass is estimated from RGB images using the projected shoot area in three different view images and linear models, and the predication accuracy of the model can be improved by adding additional factors such as the growth date (Golzarian et al., 2011), or more structural or physiological features in a more sophisticated model (Chen et al., 2018). However, the commercial software of Scanalyzer3D, LenmanGrid, with its built-in algorithms, is not open source and cannot be easily modified to extract more traits. Therefore, its application to new species or experimental scenarios is often unsatisfactory (Hartmann et al., 2011).

The Leibniz Institute of Plant Genetics and Crop Plant Research developed HTPheno, a plugin for ImageJ, and other open-source image-processing software to extract standard traits from coloured images, such as plant height, width, and projected area (Hartmann et al., 2011). The Java-based open-source Integrated Analysis Platform (IAP) with broader functionalities and the flexibility to integrate third-party algorithms was developed (Klukas et al., 2014). They can support additional sensor types (fluorescence, near-infrared, infrared) and can cover a broader range of plant species (maize to *Arabidopsis*), and facilitates data management. IAP was used to extract features from images and model-derived parameters to determine the drought response (DR) of barley (Chen et al., 2014) and screen the genetic variation of growth dynamics in maize in combination with GWASs (Muraya et al., 2017).

The Bellwether Phenotyping Platform installed a Scanalyzer3D system within a climate-controlled growth facility at the Donald Danforth Plant Science Center (St. Louis, MO, USA). They developed PlantCV, platform-independent software based on open-source libraries (OpenCV, NumPy, and MatPlotLib) to process RGB, fluorescent, and NIR images and quantify different temporal responses to water availability in *Setaria* (Fahlgren et al., 2015).

Interestingly, the INRA group (PhenoArch Phenotyping Platform, INRA, Montpellier, France) developed an automatic platform to monitor silk growth dynamics in maize, by modifying two successive imaging cabins. The first cabin determines the spatial coordinates of potential ears and the second informs a robotic camera, thereby providing clearer images of maize ears and calculating silk traits across time (Brichet et al., 2017).

A multidisciplinary team from Huazhong Agricultural University and Huazhong University of Science and Technology (Wuhan, China) developed the High-throughput Rice Phenotyping Facility (HRPF). Yang et al. (2014) incorporated colour images, X-ray CT, automatic controls, and an image analysis pipeline that can monitor at least 15 agronomic traits in populations of up to 1920 rice plants, with a total greenhouse capacity of 5472 (Figure 4.4). Moreover, some image-based traits (i-traits) that are difficult to assess manually, such as leaf rolling and stay-green properties of drought-tolerant rice, can also be quantified (Duan et al., 2018). Together with GWASs and 51 i-traits, the HRPF can dissect the complex DR traits into heritable and simple i-traits and discover new genes for DR (Guo et al., 2018a). In addition, with minor adjustments of the image analysis pipeline, the HRPF can be extended to the phenotype of other species. These include 3D wheat plant architecture (Fang et al., 2016), the leaf traits of rape seedlings (Xiong et al., 2017), and the genetic architecture of variation in maize growth (Zhang et al., 2017). However, the establishment of high-throughput conveyor-based phenotyping systems is costly and time-consuming and requires in-depth knowledge of engineering and computational sciences to maintain function and flexibility. The unit cost depends on the throughput, so the implementation of such systems can only be justified at major research centers or companies.

Alternatively, Czedik-Eysenberg et al. (2018) developed a cheaper phenotyping platform "PhenoBox", which is a flexible open-hardware and open-source phenotyping system to extract shoot traits. Because of the low cost of materials, PhenoBox was widely applied to study infection of model grass, and salt stress in several crops, with additional sensors, such as hyperspectral imaging sensors.

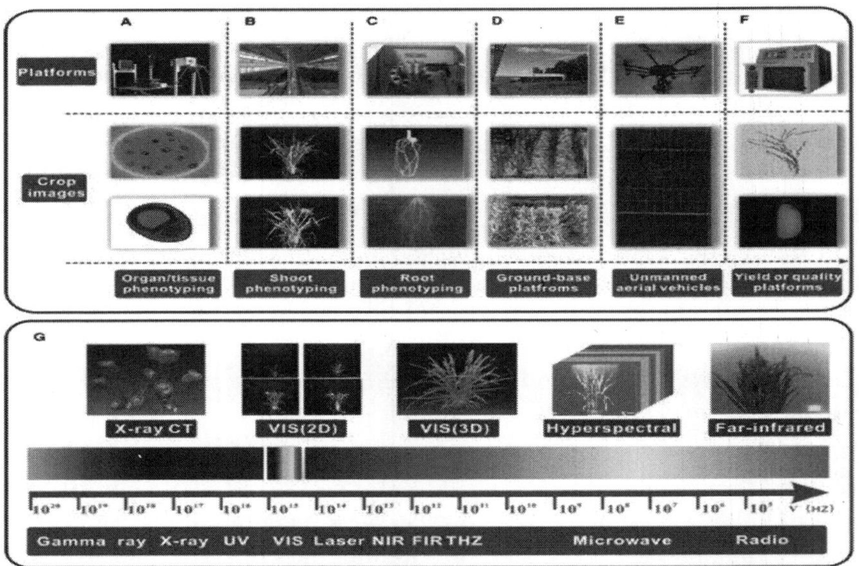

Figure 4.4. Schematic overview of phenotyping platforms across different scales. Organ/tissue (A), shoot (B), root (C), ground-base (D), and UAV I to yield/quality (F) phenotyping and exemplification of different spectra used in crop phenotyping (G). (Reproduced from Yang et. al 2020).

Choosing the appropriate imaging sensors and proper imaging-transfer mode are both vital in designing phenotyping facilities, which depend on the different experimental objectives. Different sensors have their pros and cons., which are as follows:

- RGB imaging: RGB imaging (also called visible light imaging) is cost-effective and is the most widely used to measure plant or organ morphological traits, biomass, and plant growth (Yang et al., 2014), but it cannot provide physiological information.
- Fluorescent imaging: Equipped with specific excitation illumination, fluorescent imaging can reflect the physiological signal, such as photosynthetic function and reactive oxygen species signal (Fichman et al., 2019).

- Thermal imaging: Thermal imaging (also called far-infrared thermal imaging) can obtain the plant or leaf temperature, which is also affected by environmental conditions (Zarco-Tejada et al., 2012). Moreover, both fluorescent and thermal imaging cannot provide sufficient spectral information.
- Hyperspectral imaging: Hyperspectral imaging can provide abundant spectral (visible and near-infrared) and spatial information simultaneously and can be assembled to assess disease severity, water status, and other physiological traits. However, the hyperspectral sensor is costly, and the processing of hyperspectral data (gigabyte per sample) can be complicated (Mahlein et al., 2018).
- 3D imaging: Compared with 2D imaging, 3D imaging techniques, which mainly include image-based (Fang et al., 2016) and laser scanning-based (Paulus et al., 2014) techniques, can generate 3D models and obtain more spatial and volumetric traits.

According to the merits of the different imaging technologies, the current trend is to combine multiple imaging techniques. Some studies have compared the different imaging techniques in plant phenotyping (Fiorani and Schurr, 2013, Zhao et al., 2019). Choosing the proper imaging-transfer mode (plant-to-sensor or sensor-to-plant) depends on the extracted traits, volume size of the measured species, greenhouse capacity, and other factors. CropDesign, Scanalyzer3D, and HRPF are all based on plant-to-sensor mode, which transfers the plants to imaging stations to extract biomass, shoot growth, and plant architectures. However, to measure certain physiological traits (such as canopy temperature) that are susceptible to environmental variation (air temperature and wind speed, etc.) (Ghanem et al., 2015), moving the camera and keeping the plants stationary would be better than transporting the plants to the inspection chamber via the long belt conveyor. In the case of Phenodyn (Montpellier, France; Avramova et al., 2019), which has displacement transducers and a balance recorder for each plant, can measures leaf elongation rate and transpiration rate with the capacity of 500 plants in the greenhouse. Apparently, combining both plant-to-sensor and sensor-to-plant modes would be interesting and have complementary advantages.

Phenotyping Plant Roots

Unlike most animals, individual land plants live at the interface between two or more distinct media. While the shoots usually develop in the air (and are relatively easily accessible for inspection, measurement, and sampling), the roots normally function within the soil, limiting direct observation. Roots are crucial organs that determine water and nutrient uptake in most crops, directly influencing their resilience to drought as well as affecting yield and quality.

Root system architecture (RSA) phenotyping in situ is challenging, especially in comparison with shoot phenotyping (Atkinson et al., 2019). A wide range of solutions have been evaluated and can provide useful information. The use of transparent (hydroponic, aeroponic, and gel-based) growth media, such as PlaRoM, permits the use of RGB cameras to acquire and extract growth dynamics and hair development, with a maximum capacity of 50 seedlings (Yazdanbakhsh and Fisahn, 2009). Other systems include Rhizoslides (a paper-based root growth and observation system) (Le Marie et al., 2014), Rhizoponics (a hydroponic rhizotron for the evaluation of the RSA of *Arabidopsis*) (Mathieu et al., 2015), RADIX (a rhizoslide platform used to screen the shoots and roots of 200 maize plants) (Le Marie et al., 2016), and RhizoTubes (an automatic "plant-to-sensor" platform comprising 1200 rhizotubes to obtain the RSA in approximately 6–8 weeks) (Jeudy et al., 2016). In addition, several open-source image analysis solutions evaluating the 2D RSA have been developed, including semiautomatic software (RootTrace, French et al., 2009; SmartRoot, Lobet et al., 2011; RootNav, Pound et al., 2013) and fully automatic software (EZ-Root-VIS, Shahzad et al., 2018).

Iyer-Pascuzzi et al. (2010) developed a 3D RSA imaging platform utilizing a gel-based growth cylinder to obtain 16 root traits. Other 3D image reconstruction and image analysis pipelines include GiARoots (Galkovskyi et al., 2012) and RootReader3D (Clark et al., 2011). Pineros et al., (2016) developed a plastic mesh vertical tower system to maintain the 3D root architecture and achieve 3D imaging in a hydroponic growth cylinder to obtain the RSA similar to that observed in the field.

However, it is not always apparent how results obtained from transparent media apply to normal growing conditions in the field. An intelligent mechanized root phenotyping platform, GROWSCREEN-Rhizo, was developed to image shoots and the roots simultaneously in transparent soil-filled rhizotrons. The throughput was 60 rhizotrons per hour out of a total capacity of 72 rhizotrons (Nagel et al., 2012). Moreover, additional sensors (i.e., hyperspectral) were integrated into this phenotyping system with the potential to dynamically reflect root chemical components, such as root water content and lignin change (Bodner et al., 2018).

X-ray CT, a 3D structural imaging application, can be used to visualize the inner 3D volume according to the differences in the X-ray attenuation of different materials, such as roots and soil. The first CT scanner (Cormack, 1963) was quickly followed by a clinically useful version (Hounsfield, 1976), for which the inventors received the 1979 Nobel Prize in Physiology and Medicine. The open-source software RooTrak was developed by the CPIB (Centre of Plant Integrative Biology, University of Nottingham) group (Mairhofer et al., 2012, Mairhofer et al., 2013) to exploit the differential X-ray attenuation between soil and roots and reconstruct the RSA in 3D. Multiple interacting root systems could also be extracted and separately analyzed based on a modified RooTrak (Mairhofer et al., 2015).

While using CT to scan the roots, it is important to take the following precautions:

- ❑ To ensure a better contrast of X-ray attenuation between soil and roots, soil preparation to maintain homogeneity, including sieving and drip,

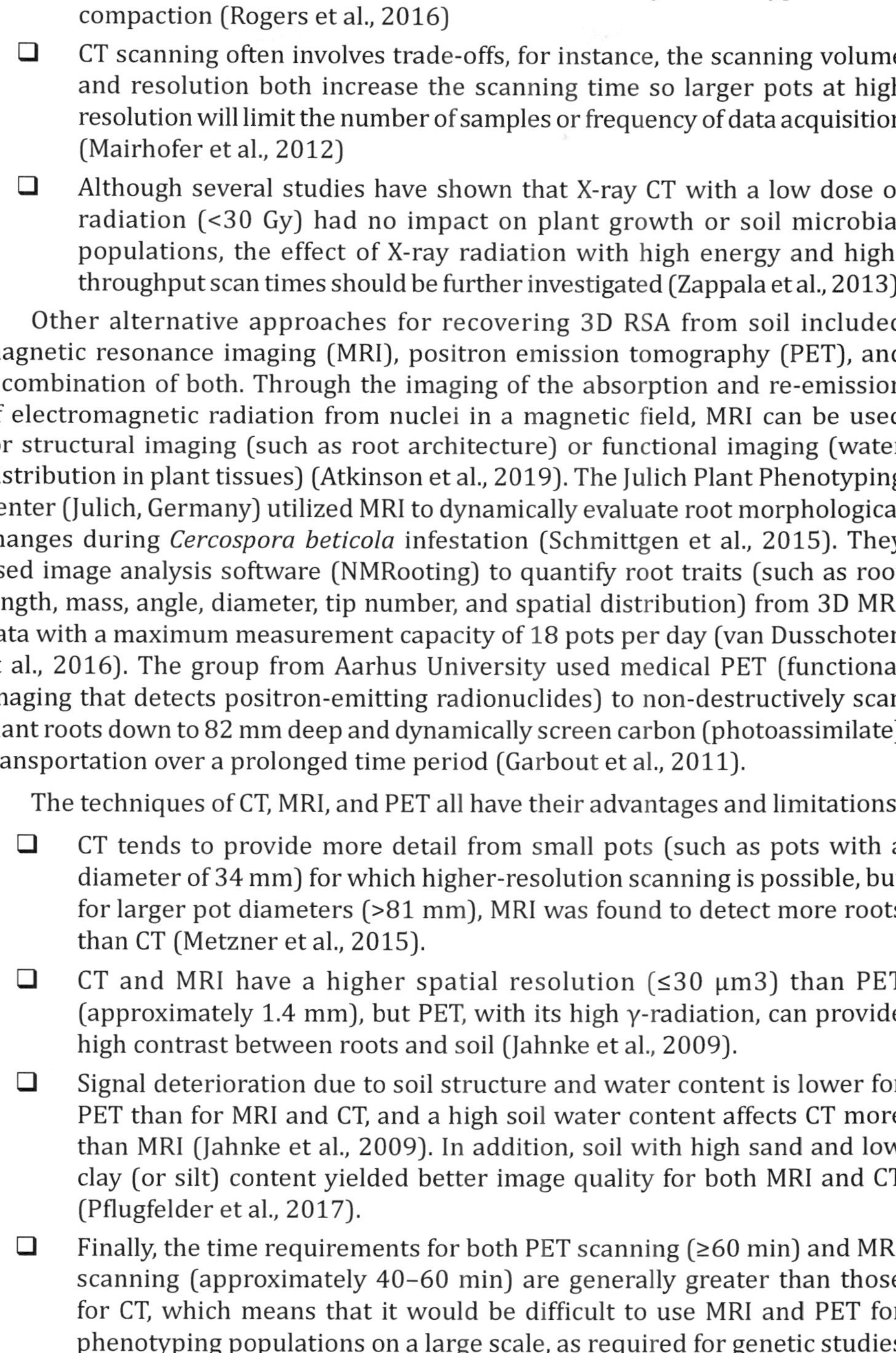

should be carried out. The RSA is also influenced by the soil type and soil compaction (Rogers et al., 2016)

- CT scanning often involves trade-offs, for instance, the scanning volume and resolution both increase the scanning time so larger pots at high resolution will limit the number of samples or frequency of data acquisition (Mairhofer et al., 2012)
- Although several studies have shown that X-ray CT with a low dose of radiation (<30 Gy) had no impact on plant growth or soil microbial populations, the effect of X-ray radiation with high energy and high-throughput scan times should be further investigated (Zappala et al., 2013).

Other alternative approaches for recovering 3D RSA from soil included magnetic resonance imaging (MRI), positron emission tomography (PET), and a combination of both. Through the imaging of the absorption and re-emission of electromagnetic radiation from nuclei in a magnetic field, MRI can be used for structural imaging (such as root architecture) or functional imaging (water distribution in plant tissues) (Atkinson et al., 2019). The Julich Plant Phenotyping Center (Julich, Germany) utilized MRI to dynamically evaluate root morphological changes during *Cercospora beticola* infestation (Schmittgen et al., 2015). They used image analysis software (NMRooting) to quantify root traits (such as root length, mass, angle, diameter, tip number, and spatial distribution) from 3D MRI data with a maximum measurement capacity of 18 pots per day (van Dusschoten et al., 2016). The group from Aarhus University used medical PET (functional imaging that detects positron-emitting radionuclides) to non-destructively scan plant roots down to 82 mm deep and dynamically screen carbon (photoassimilate) transportation over a prolonged time period (Garbout et al., 2011).

The techniques of CT, MRI, and PET all have their advantages and limitations:

- CT tends to provide more detail from small pots (such as pots with a diameter of 34 mm) for which higher-resolution scanning is possible, but for larger pot diameters (>81 mm), MRI was found to detect more roots than CT (Metzner et al., 2015).
- CT and MRI have a higher spatial resolution (≤ 30 μm3) than PET (approximately 1.4 mm), but PET, with its high γ-radiation, can provide high contrast between roots and soil (Jahnke et al., 2009).
- Signal deterioration due to soil structure and water content is lower for PET than for MRI and CT, and a high soil water content affects CT more than MRI (Jahnke et al., 2009). In addition, soil with high sand and low clay (or silt) content yielded better image quality for both MRI and CT (Pflugfelder et al., 2017).
- Finally, the time requirements for both PET scanning (≥ 60 min) and MRI scanning (approximately 40–60 min) are generally greater than those for CT, which means that it would be difficult to use MRI and PET for phenotyping populations on a large scale, as required for genetic studies (Jahnke et al., 2009, Metzner et al., 2015).

Thus, because of their different merits, the use of a combination of these tomography techniques, such as MRI–PET (Jahnke et al., 2009), PET–CT (Garbout et al., 2011), and MRI–CT (Metzner et al., 2015), is popular not only in medical imaging but also in root phenotyping for addressing focused biological questions. Currently, many important functional genes influencing RSA are likely to remain undiscovered due to the bottleneck associated with root phenotyping (Wing et al., 2018), necessitating the development of breakthrough technologies for below-ground imaging in the future.

Root Phenotyping In The Field

An effective method for assessing the RSA is still simply to dig up the root system and shake or wash off the soil; often termed "shovelomics" when scaled to the field level (Trachsel et al., 2010). Although this method is destructive and somewhat unpleasantly messy, image acquisition can be relatively rapid (up to 5000 images/h), and those images can be rich in trait information (39 root traits). Other root phenotyping methods include, for example, the core-break method, which was proposed to obtain the root depth (Wasson et al., 2017) and minirhizotrons with imaging sensors to observe root growth (Svane et al., 2019). But both the core-break and minirhizotrons methods can only detect limited roots, and the root measurements were subject to the sample positions. Some non-destructive detection methods, such as ground-penetrating radar (Delgado et al., 2017) and electrical impedance tomography (Corona-Lopez et al., 2019), could be used to phenotype root biomass and root development. However, the spatial resolution of both methods is much lower (~cm/pixel), and it is hard to detect individual fine roots.

The non-destructive root phenotyping in the field remains a challenge due to a lack of suitable technology. To solve this problem, the US Department of Energy funded the ROOTS project in 2017 (https://arpa-e.energy.gov/?q=programs/roots), which aims to evaluate and develop novel approaches, such as thermoacoustic tomography (TAT), associated particle imaging (API) using neutrons, tomographic electrical rhizosphere imaging (TERI), low-cost X-ray CT, and backscatter X-ray platforms, to dynamically image RSA. The long-term justification for the investment is associated with the production of crops that increase carbon uptake and decrease N2O emissions under field conditions. As the advancement of UAV and sensor techniques currently benefits canopy phenotyping in the field, we believe that breakthrough technologies could provide similar benefits below ground.

Field-Based High-Throughput Phenotyping Platforms

Combining advances in remote sensing, aeronautics, and high-performance computing is paving the way for the development of field-based HTPPs. Recently, several platforms have been developed, ranging from ground-based to aerial systems (e.g., the Australian Plant Phenomics Facility, http://www.plantphenomics.org/ hrppc/capabilities/technology). Ground-based HTPPs include modified vehicles equipped with a global positioning system (GPS) navigation device and sensors often referred to as 'phenomobiles'. Different 'phenomobiles' have been developed (http:// www.plant-phenotyping-network.eu/eppn/inra_diaphen). For example, in the case of cotton, a system has been developed that carries four sets of sensors to measure canopy height, reflectance, and temperature simultaneously on four adjacent rows, enabling the collection of phenotypic data at a rate of 0.84 ha h1. The mounting of sensors (such as NIRS) on agricultural harvesters may also be included within the category of 'phenomobiles' [61,62]. Ground-based HTPPs enable data to be captured at the plot level and require little post-processing. However, there are also limits on the scale at which ground-based HTPPs can be used. Moreover, simultaneous measurements of all plots within a trial are not possible with ground-based platforms. Aerial platforms are increasingly being considered as an alternative option to overcome limitations associated with ground-based HTPPs. Aerial HTPPs enable rapid characterization of many plots within minutes. Initial aerial HTPPs used small airplanes (e.g., crop-dusting airplanes). However, this is costly and it is difficult to safely achieve the low speeds required for high-definition images at low altitudes. The current generation of aerial HTPPs significantly varies in payload, initial costs, maintenance costs, and control. Recently developed alternatives include 'phenotowers', and blimps. However, this type of aerial HTPP has a maximum height of 50 m. Blimps are helium-filled balloons that can be held in a stationary position and have sensors mounted underneath. They can carry a heavy payload (several kilograms), enabling many sensors to be used concurrently. However, they require many people for control and considerable room for storage when inflated. Alternatives to blimps are unmanned aerial platforms (UAPs) such as "polycopters" (e.g., Ascending Technologies: http://www.asctec.de; MikroKopter: http://www. mikrokopter.de) and airplanes (see 'Innovative agricultural technologies for sustainable intensification' at the MAIZE Annual Report of the CGIAR Research Program: http://repository.cimmyt.org/xmlui/bitstream/handle/10883/3209/ 98018.pdf?sequence=1). Although the payload of UAPs is lower than blimps, they can generally carry up to 2 kg, enabling at least two sensors to be mounted for simultaneous image capture. UAPs enable greater flight control and autonomy and are becoming increasingly affordable. Unlike airplanes, "polycopters" can be maneuvered into a stationary position. However, advances in aeronautics and sensors now enable high-quality images to be obtained from unmanned airplanes. The autonomy and area covered by airplanes are larger and the risk of destruction by crashing is lower than "polycopters". Most UAPs carry an RGB/CIR camera, together with a thermal-imaging sensor (sometimes a conventional colour RGB digital camera). Replacing the RGB/CIR camera with a multispectral or hyperspectral imaging sensor increases the payload but opens a huge range of new possibilities. The software requirements for UAPs include programs to: (i) plan flight missions, (ii) gather the images, and (iii) extract the data for plots within the images. UAPs are controlled by an autopilot for autonomous flying; the ground control station and the UAP are radio-linked, transmitting position, altitude, and status. The imager is operated from the ground station.

Phenotyping Techniques in the Field

Most crops are grown outdoors, which is much more variable than the laboratory environment. Soil structure can vary within and among fields. Small differences in topology and aspect can alter the wind speed, the effect of solar radiation, evaporation rates, etc. Field crop breeding takes such variation into account through replicated multisite trials, but the application of appropriate phenotyping technologies remains a limitation for future genetic studies and breeding advances (Araus and Cairns, 2014). Field phenotyping can be broadly split into technological groups based on the sensor platform: ground-based, unmanned aerial vehicle (UAV), and handheld phenotyping (Table 1), which are discussed in the following sections.

Ground-Based Phenotyping

Ground-based sensor platforms can be either fixed or mobile. Towers and other fixed platforms are easy to construct and maintain and have been used to record stages of growth and development of crops, and monitor insect pest infestations. For example, digital cameras attached to a fixed phenotyping tower (Crop Phenology Recording System [CPRS]) were used to monitor rice growth and the occurrence of the rice bug (Sakamoto et al., 2011, Fukatsu et al., 2012). Shibayama et al. (2015a,b) used two digital cameras at a height of 12 m above a rice field to evaluate the nitrogen content and leaf area index of paddy rice. Zhou et al. (2017) developed a scalable and cost-effective field phenotyping platform (CropQuant), a web-based control system (CropMonitor), and a high-throughput trait analysis pipeline to measure dynamic crop growth, vegetative greenness, and plot orientation. The fixed phenotyping towers are easy to install and maintain, but can be used to obtain limited crop information in fixed areas. Costs for large-scale experiments increase in a near-linear fashion. In smart solution system, an ambient sensor cloud system is used, which employed open-source sensors and a free cloud data service. This allows long-time observations of individual plants (Hirafuji et al., 2011).

A rail-based gantry system was established by Rothamsted Research Laboratory for field phenotyping. The Field Scanalyzer measuring 125 m (length) × 15 m (width) × 6 m (height) was equipped with an ~300 kg sensor array, including a visible camera, a 3D laser scanner, a thermal infrared camera, a visible to near-infrared (VNIR) hyperspectral camera, a four-channel amplified radiometer, a chlorophyll fluorescence sensor, and a CO2 sensor (Sadeghi-Tehran et al., 2017, Virlet et al., 2017). The acquisition of data from all sensors took approximately 7 min for one plot (1.5 m2). This platform can be used to produce a detailed description of crop canopy development through all life stages. For example, to determine wheat ear emergence and flowering times (Sadeghi-Tehran et al., 2017). Similarly, Guo et. al (2018b) developed a high-throughput crop phenotyping platform (Crop3D) to measure 3D plant/leaf architectures and leaf temperature with the help of an integrated LiDAR sensor, RGB camera, thermal camera, and hyperspectral imager in a mobile gantry system. However, there are two challenges to its effective uses: First, the variable ambient light leads to differential images, which can be solved in part by night-time imaging or by the addition of an on-field reflectance standard (Virlet et al., 2017). Second, it was difficult to effectively analyze terabytes of hyperspectral images and laser-scanned images (Vadez et al., 2015), and interpret temporal series of thermal, multi- or hyperspectral-based features in changing environments (Virlet et al., 2017). In addition, such large rail-based systems can cover a limited area (e.g., 0.12 ha in the case of Rothamsted Research) are expensive in terms of both construction and maintenance.

Sensors mounted on manually operated carts or self-propelled tractors overcome many of these issues. A cart carrying multiple sensors, including an ultrasonic sensor, a normalized difference vegetation index (NDVI) sensor, a thermal infrared radiometer, a portable spectrometer, an RGB camera, and a proximity sensor, has been used to obtain soybean and wheat canopy traits (such as: height, NDVI, temperature, reflectance spectra, and RGB imagery) (Bai et al., 2016). Wang et al., (2018b) developed a ground-based high-throughput plant phenotyping system (HTPP) with various sensing technologies, to obtain 3D information and measure plant height. The system included an ultrasonic sensor, a LiDAR-Lite v2 sensor, a Kinect camera, and four digital single-lens reflex cameras. Young et. al. (2018) used stereo cameras, time-of-flight depth sensors, and infrared cameras to evaluate individual plant architecture traits. Nakarmi and Tang (2012) used a time of flight (TOF)-based 3D vision sensor to obtain an automated measurement of interplant distance.

Most imaging platforms in the open field are susceptible to variable environmental conditions. The BreedVision system resolved the problem of light intensities, by excluding ambient light and imaging within a mobile dark chamber (Busemeyer et al., 2013a). With the help of multiple sensors (3D time-of-flight camera, laser distance sensors, hyperspectral imaging, RGB sensors, and light curtain imaging) and specific trait calibrations, BreedVision can non-destructively measure plant traits, including plant height, tiller density, grain yield, moisture content, leaf colour, lodging, and dry biomass, at an operating speed of 0.5 ms−1. Remarkably, equipped with a pair of linear light barriers (one emitter and the other a receiver), light curtain imaging could be used to measure plant height, plant architecture, and biomass without the influence of indoor or field illumination conditions (Busemeyer et al., 2013a, Fanourakis et al., 2014). However, vehicle-based platforms have some limitations; they are restricted by soil and weather conditions and, in some cases, by topology. Specific crops, such as rice, require specialized vehicles.

Proximal (Remote) Sensing

Remote sensing phenotyping methods are non-destructive, non-invasive approaches (Cabrara Bosquet et. al 2011, Monneveux et. al. 2011, White et. al. 2012) based mostly on the information provided by visible/near-infrared (VIS-NIR) radiation reflected (or transmitted) and far-infrared (thermal) radiation emitted by the crop (Lobos et. al. 2013). These methods are termed proximal in the sense that information is gathered 'near' the crop. Remote sensing techniques may be deployed for screening in situ for a wide range of breeding objectives. These include: yield potential, adaptation to abiotic (water stress, extreme temperatures, salinity) and biotic (susceptibility to pests and diseases) limiting conditions, and quality traits. Many categories of traits may be measured under different environmental conditions using remote sensing approaches. These include: green biomass, photosynthetic transpiratory gas exchange, and quality traits or even to predict grain yield (Weber et. al. 2012). They working with a set of 300 maize test-crosses grown under different water and temperature regimes reported that partial least square regression (PLSR) models from hyperspectral reflectance of maize explained up to 40% of the variation in grain yield in each trial, with a relative efficiency of selection of 0.88 and 0.68 using leaf and canopy reflectance, respectively. Moreover, a strong agreement has been reported between remote sensing and imaging values of traits measured by HTPPs at the land and aerial levels, as well as through manual phenotyping approaches, with determination values ranging between 0.35 and 0.82 in the case of cotton under different water regimes (Androde-Sanchez et. al. 2013). These results confirmed the ability of the phenotyping system to measure multiple traits rapidly and accurately. Concerning NIRS, it is already routinely used in breeding for a wide array of food and feeds quality traits [43]. NIRS can be applied to drought or nutrient use efficiency screening, or other more general breeding/gene discovery objectives.

The implementation in imaging formats of proximal sensing with VIS-NIR and far-infrared radiation has enabled the process of taking measurements. This can be applied for measuring a single plot to the entire trial composed of different plots, provided that the image has enough resolution (pixels). The imaging speed is limited by post-processing, which includes image alignment, geometric and radiometric calibrations, atmospheric correction, automatic mosaicking, and algorithms for automatic image segmentation (Bernie et. al. 2009, Zarca-Tajada et. al. 2013).

Proximal (remote) sensing and imaging techniques include different approaches. The approaches that are most feasible for field phenotyping can be grouped into following categories:

- VIS-NIR spectroradiometry (including multispectral and the even less developed hyperspectral imaging),
- Infrared thermometry and thermal imaging, and
- Conventional digital photography (RGB colour cameras).

The first approach is undergoing rapid technical advances associated with the use of high-resolution full-range spectroradiometers that reach into the NIR region wavelengths around 2500 nm. This, together with the use of special adaptors with their own light source, enables such spectroradiometers to be used as field-portable

NIRS devices, generating large amounts of data with a wide range of potential applications. For example, development of empirical models to assess specific photosynthetic characteristics related to radiation use efficiency, or to predict a complex trait such as grain yield (Webner et. al. 2012), and the use of single spectroradiometrical indices to predict yield (Lobos et. al. 2013). At the imaging level, hyperspectral sensors are being developed along similar lines, although it is not possible for them to carry their own light source (specifically for large-scale images), thus this may represent a limitation. Nevertheless, hyperspectral imagery enables the assessment of complex traits, such as canopy photosynthesis and fluorescence under natural sunlight conditions (Zarco-Tejada et. al. 2013). Several software programs are available to process information to extract phenotype data from images, including free options (Cobb et. al. 2013).

Application of spectroradiometrical approaches has limitations due to the differences in plant architecture and developmental stage, and the anisotropic characteristics of the leaf (mostly due to the leaf surface: roughness, trichomes, epicuticular waxes). Further, spectroradiometrical approaches may negatively affect the estimation of biochemical traits (Cobb et. al. 2013). The effect of leaf anisotropy is more evident in the remote sensing-based approaches that use reflected instead of transmitted radiation. Moreover, care should be taken to standardize measures with plants at similar developmental stages and exhibiting a narrow range of variability in plant architecture. Elevation of the sun should also be taken into consideration because the majority of spectroradiometrical evaluations in the field use passive sensors. In the case of canopy temperature, variation in development and canopy architecture must be considered (e.g., plant height, soil covering, emerged spikes, leaf angle, and size) when interpreting performance and selecting cooler breeding lines (Maes and Steppe 2012, Rebetzke et. al. 2012). The environmental variability factors such as: light intensity, temperature, relative humidity, and wind speed) and timing of measurements, which affects the accuracy of thermal measurements, should also be considered (Costa et. al. 2013).

Digital photography is also a promising approach due to the low cost of the sensor (conventional camera) and availability of free software for data processing that enables estimation of green biomass, soil cover, plant colour (Casadesu et. al. 200, Mullan and Reynolda 2010), or even agronomical components such as plant or inflorescence density.

Other techniques such as laser image detection and ranging (Lidar) are also being adopted for field phenotyping. This is an active remote sensing technique that uses Lidar sensors to directly measure the 3D distribution of plant canopies as well as sub-canopy topography. Such techniques have enabled the development of high-resolution topographic maps with highly accurate estimates of vegetation height, cover, and canopy structure (Lefsky et. al. 2002, Omasa et. al. 2007, Hosoi and Omasa 2009). Furthermore, laser scanning and fluorescence enable the evaluation of photosynthetic performance and have the potential to apply in areas such as plant pathology (Romer et. al. 2011).

Different Types Of Cameras For Monitoring Crop Phenotypes

Different categories of imaging systems for remote sensing evaluation of vegetation are described below with examples of prototypes capable of being carried by UAPs of limited payload.

Rgb/Cir Cameras

The combination of colour infrared (CIR) and red, green, and blue light (called visible or RGB) imagery enables the estimation of green biomass (NDVI type of information). Miniaturization is advancing fast with cameras that weigh only 100 g and have spectral range bands in red, green, and NIR.

Multispectral Cameras

Multispectral imagers are widely used for crop monitoring via remote sensing. They can acquire a limited number of spectral bands at once in the VIS-NIR regions. Besides vegetation indices for evaluating green biomass, multispectral imagers can be formulated to other different spectral indices targeting senescence evaluation, nutrient status, pigment degradation, photosynthetic efficiency, or water content.

Hyperspectral Visible and Near-Infrared Imager

The hyperspectral visible and near-infrared imager enables the acquisition of hundreds of images at once, covering the entire electromagnetic spectrum between the visible and the near-infrared regions in a continuous mode (wavelengths ranging from 400 to 900 nm). Other configurations cover the range from 1000 to 2500 nm. Therefore, it is possible to run empirical calibrations (like in a 'NIRS mode') against a wide and miscellaneous set of traits.

Long-Wave Infrared Cameras (Thermal Imaging Cameras)

Long-wave infrared cameras and thermal imaging cameras render infrared radiation in the range of mm as visible light. Potential use of thermal imaging in phenotyping includes predicting water stress in crops, disease, and pathogen detection in plants, evaluating the maturing of fruits, and bruise detection in fruits and vegetables. The low resolution of the imaging (current cameras are in the range of 500 to 640 pixels) may limit the use of such cameras from aerial platforms. Thermal images obtained from a UAP using this type of camera have a resolution in the range of 20–40 cm.

Conventional Digital Cameras

Conventional digital cameras are low-cost instruments that enable, for example, plant cover (green biomass) or senescence to be estimated. Moreover, the software needed is usually freely available.

Remote Sensing with UAV

Drones (or UAVs) provide a flexible platform, quickly acquiring data over large areas and potentially providing high spatial resolution images (~1 mm per pixel).

Some advanced IT techniques, such as deep learning (DL), can handle millions of remote sensing images with high accuracy and high speed (Bauer et al., 2019). Thus, remote sensing has been widely used to monitor drought stress response, assess nutrient status and growth, detect weeds and pathogens, predict yield (Maes and Steppe, 2019), and identify QTLs (Wang et al., 2019).

Canopy colour and texture features obtained by UAV platforms at high spatial and temporal resolution facilitate phenotyping tasks (Yue et. al 2019), since the improvement of image quality and quantity provides detailed information for feature mining and analysis. Accordingly, high-resolution UAV imagery has been adopted for various phenotyping purposes, such as leaf area index estimation (Yao et al., 2017), wheat ear identification (Madec et al., 2019), weed detection (Hung et al., 2014), and seeding performance evaluation in rapeseed (Zhao et al., 2018). Madec et. al 2018) suggested that an optimal resolution of 0.3 mm can be used for high-throughput phenotyping in the field. Because of its combination of high spatial and temporal resolution, UAV-based phenotyping opens new possibilities in field research (Burkat et. al. 2018).

With the development of photogrammetric techniques, multiview stereopsis, and computer vision approaches, such as structure from motion (SFM), UAVs have emerged as a promising platform for obtaining 3D plant canopy structure information (Bendig et al., 2014, Hassan et al., 2019). SFM algorithms produce geometrically precise 3D point clouds by matching feature points from 2D images obtained by RGB sensors. SFM is effective for estimating plant height and biomass (Bendig et al., 2014, Malambo et al., 2018), but the penetration capability of photogrammetric techniques is limited compared with that of LiDAR. SFM can produce point clouds at the subpixel level with similar accuracy and quality to LiDAR (Hassan et al., 2019). Moreover, there is a strong correlation between the digital surface model (DSM, obtained by SFM algorithms using UAV data) and LiDAR estimations of plant height (Madec et al., 2017, Jimenez-Berni et al., 2018). Further, the SFM-UAV-based method is less costly and more flexible (Jimenez-Berni et al., 2018).

Different imaging sensors mounted on UAVs can be used to capture spectral information regarding visible or NIR bands for crop nutritional diagnosis and monitoring various types of stress. A consumer-grade RGB camera can be used to generate a range of RGB VIs, which have been found to have good performance in the assessment of stress conditions, such as low nitrogen (Elazab et al., 2016) and biotic stress (Vergara-Diaz et al., 2015). In addition, the NIR or IR bands can provide more spectral information, which improves the measuring accuracy of crop growth monitoring. Multispectral cameras can respond to spectral information from the red edge and NIR bands and have been utilized for chlorophyll-based diagnosis (Deng et al., 2018) and water stress monitoring (Zarco-Tejada et al., 2012). However, their spectral resolution is limited by the number of lenses on the sensors. Hyperspectral cameras cover a commonly used spectral region (400–1000 nm) for crop monitoring and can be used for assessing the leaf carotenoid content

(Zarco-Tejada et al., 2013), nitrogen concentration (Zheng et al., 2016), heat and drought stress (Trachsel et al., 2019). This type of sensor can capture the most comprehensive spectral information, but it is expensive, and data processing may be relatively complicated. In addition to imaging sensors, a VNIR non-imaging spectrometer can be mounted on a UAV to collect accurate spectral reflectance signals (Garzonio et al., 2017).

Remote sensing with UAVs has shown great potential for high-throughput phenotyping, which will enhance work in crop functional genomics and crop breeding. However, the following restrictions exist on the use of UAVs:

- The fly time and the load capacity are limited;
- Local flight laws and regulations may be a constraint
- Strict requirements for operating technicians should be implemented to ensure flight safety, and
- A low flight altitude will provide original images with higher quality but can also cause changes to the leaf architecture and physiology due to the strong wind produced by the UAV.

Moreover, lower flights will result in longer flight times, which will make it difficult to screen large populations with low-altitude drone flights. Thus, reasonable flight altitudes and other camera settings associated with different phenotyping goals should be thoroughly investigated.

Hand Held Phenotyping

Physiologists have long used specialized handheld portable tools to measure a range of functions (Kim et al., 2016). Many such instruments are necessarily complex, requiring training to obtain useful data. The challenges and opportunities related to portable phenotyping include:

- Decreasing the level of expertise to effectively handle the equipment,
- Integral data standardization (such as colour calibration and smart user guides),
- Integrating multiple sensors into one portable device with corresponding software for data analysis, and
- Exploiting the 5th generation mobile network and artificial intelligence techniques, such as DL. This could build a robust model to face complex conditions in the field.

Next-generation "pocket" or wearable phenotyping tools could provide the disruptive technology that will profoundly change and accelerate phenotyping.

The rapid development of smartphones with high-resolution RGB cameras and powerful computing has led to the creation of applications with ever-increasing utility. In addition, aspects of mobile phone technology have been incorporated into other bespoke portable instruments, increasing the range of optical and other sensors and enhancing the connectivity and portability of traditional phenotyping equipment. This is an emerging and rapidly evolving area, so only a few examples have been illustrated as future possibilities. PocketPlant3D measures maize canopy structure (Confalonieri et al., 2017). The smartphone is moved parallel with the leaf lamina from base to tip, and all the leaves can be scanned to obtain the whole leaf architecture. A machine learning-based app can diagnose *Cercospora* leaf spot and other sugar beet leaf diseases better than experts (Hallau et al., 2018). A smartphone application for Android devices, vitisBerry, can quantify the number of grapevine berries (Aquino et al., 2018).

Advances in Field Phenotyping

With the rapid development of UAV and peripheral equipment, many novel platforms for low-altitude remote sensing are rapidly emerging, such as mooring UAV platforms, cameras with 100 000 000 pixels, multisensory integrated platforms, miniaturization, and simplification UAV platforms. The UAV platform will partially replace the ground-based phenotyping platform and will likely be widely deployed based on new technologies in the future. In addition, integrated special UAV image processing and standardized analytical software will likely become available for a broader range of users, allowing nonexperts to process and analyze hyperspectral, thermal, or LiDAR data.

For ground-based phenotyping, is one of the trends in the development of field-intelligent robots, such as BreedVision (Busemeyer et al., 2013a). Equipped with multiple sensors integrated within an imaging chamber to ensure good image quality and automatic navigation, these types of robots are flexible to move to phenotype different fields and in the future could collaboratively work in a field.

In the case of handheld phenotyping tools, usually, attempts were made to integrate multiple sensors into one portable device and develop on-chip data analysis software to obtain and manage more traits together. Another trend is to decrease human intervention and add data standardization processes to improve measurement reliability and efficiency. With the advancements in artificial intelligence analysis techniques, fifth-generation mobile networks, and cloud-based technologies, smarter "pocket" phenotyping tools will likely be developed and change manual field phenotyping, which has been around for thousands of years. In addition, to achieve a real sense of "cost-effective phenotyping", the trade-off between investment in phenotyping techniques and manpower costs should be considered, which mainly depends on the different objectives (Reynolds et al., 2019b).

Data collection in the field using high-throughput technology has undergone rapid advances in recent years. However, the limiting factor now is how to manage and mine the vast amounts of data collected from fields through high-throughput technology. First, the biological objective of data collection should be clearly stated. Next, robust and user-friendly postprocessing and analysis tools for processing and interpreting raw data are urgently needed and should be improved (Araus and Cairns, 2014). For example, combining DL and multiple optical images obtained under different field conditions will create robust models for disease phenotyping and even the early detection of plant stress (Singh et al., 2018).

Each year, thousands of phenotyping experiments in environmentally controlled growth facilities or in the field will produce large amounts of phenotypic data. However, the replication of results by the same researcher or the reproducibility of results by different laboratories in independent experiments is often not satisfactory because of the unexplained variation of environments (Poorter et al., 2012). Thus, environmental factors are vital and should receive at least the same amount of attention as the traits that are measured, which leads to the next question: how to measure all the environmental impacts? Envirotyping, defined as a full set of next-generation high-throughput accurate envirotyping technologies, could help to address this issue (Xu, 2016). Moreover, by integrating multitype information, the genotyping × environment × management (G×E×M) interaction could also be investigated, and predictive phenomics would be possible (Xu, 2016, Araus et al., 2018). In recent years, crop yield growth (genetic gain) has been slowing down, affected by several factors, for example, population, genotype, heritability, GS model, and breeding scheme (Xu et al., 2020). Integrated with optimized experimental design, high-quality field trials, robust crop model, envirotyping, and other strategies, high-throughput and accurate phenotyping will improve the heritability and potential for genetic gain (Araus et al., 2018).

Post-Harvest Phenotyping

The harvested part of the crop is, generally, the most immediately economically relevant. Mechanized systems of harvesting already exist for handling and assessing the yield and quality, for both the harvesting and post-harvest processes. During the process of harvesting, the sensors and global positioning system (GPS) could also be integrated into the (combined) harvester to monitor, for example, the yield of blueberries (Farooque et al., 2013), grain yield (Li et al., 2016), and cotton yield (Pelletier et al., 2019). While several smart farming machines (e.g., advanced farming system, AFS, Case IH, US) have been commercialized in the harvesting process, post-harvest phenotyping is yet to take off on a commercial scale.

Quality assessment in the seed and milling industry uses imaging technologies for research applications. The relatively simple integration of 2D image capture, feature extraction, and data export has streamlined seed phenotyping to allow a wide, cost-effective survey of wheat germplasm, revealing the genetic architecture underlying grain shape and size variation in bread wheat (Gegas et al., 2010). This approach can be applied to almost any harvested part of a plant and can be further refined. For phenotyping rice spikelets, an integrated and fully automatic machine was developed (Duan et. al. 2011). The SEA unit can automatically thresh rice panicles, and score yield traits, with a capacity of up to 1440 plants per day, with a mean absolute percentage error of less than 5% (Duan et al., 2011). Similar approaches have been used for maize kernels (Miller et al., 2017), and stalks (Mazaheri et al., 2019), for which candidate genes for vascular bundle and rind traits were identified (Gage et al., 2017).

SmartGrain, a free-use software package, extracts seed size and shape (Tanabata et al., 2012), while GrainScan provides information on both size- and colour-related traits (Whan et al., 2014). With their user-friendly graphical user interfaces (GUIs), P-TRAP (AL-Tam et al., 2013) and PANorama (Crowell et al., 2014) are recommended as flexible tools to efficiently quantify both rice panicle traits and grain traits with good accuracy. Running on the Android operational system, a mobile application, SeedCounter, can measure grain number and grain size with high efficiency (20–60 s for 50 grains) (Komyshev et al., 2016).

Although they are robust and rapid, 2D imaging/scanning approaches fail to capture much of the morphological and other complexity commonly found in biological material. The phenoSeeder can extract 3D traits from individual seeds (Jahnke et al., 2016), which can then be traced through to early seedlings. Hyperspectral imaging can be used to infer grain cleanness (Wallays et al., 2009), insect-damaged wheat kernels (Singh et al., 2010), and grain protein content (Wang et al., 2004, Sun et al., 2019). An alternative approach to 3D morphometrics involves repurposing existing biomedical imaging techniques, such as X-ray microcomputed tomography (micro-CT) imaging (Hughes et al., 2017). The application of micro-CT to plant genetics will require the development of bespoke software as well as increased throughput (Hughes et al., 2019).

Ultimately, multidimensional seed traits must be combined with genetic analysis tools (such as GWAS or QTL) to dissect the genetic architecture of agronomic traits. Examples include sorghum panicle structure (Zhou et al., 2019) and the protein content of rice grains (Sun et al., 2019). Moreover, decreasing the cost of these novel photonics-based phenotyping tools and improving their reliability and extendibility could benefit crop grain research in the future.

Laboratory and Near Infra-Red Spectrum Analysis

Apart from proximal sensing approaches, the analysis of stable isotopes in plant samples (Yousfi et. al. 2012), may complement direct phenotyping under field conditions. When breeding for yield potential and adaption to abiotic stresses such as drought, application of carbon isotope discrimination (D13C) in dry matter is a promising tool. This frequently exhibits high heritability and genetic correlation with yield (Condon et. al. 2004, Condon et. a; 2012, Araus et. al. 2013), and has already been applied in breeding programs (Rebetzke et. al. 2008). The use of grain D13C relies on a good understanding of tested populations and assumes no confounding of genotypic differences in phenology, anthesis biomass, pre-anthesis water use, and remobilization of stem carbohydrates (Rebetzke et. al. 2008, Condon et. al. 2012, Araus et. al. 2018). Whereas D13C is still an expensive tool for use in large-scale phenotyping, low-cost surrogates are now available. In addition, if analyzed through the regular (mass spectrometry) technique, D13C can still be used to complement HTPP approaches.

Near Infra-Red Spectrum (NIRS) is regularly deployed not only at the laboratory level but also in harvesting machinery to analyze grain characteristics. When used in harvesting machinery, this technique secures a broad distribution of measurements within plots and covers substantially larger amounts of plot material than analyses at the laboratory level using conventional sample-based methods (Montes et. al. 2007, Mir et. al. 2012). The near-infrared spectrum captures the physical and chemical characteristics of the samples, either of vegetative plant tissues or harvested seeds. By using calibration models, several traits can be determined based on a single spectrum. Thus, NIRS is regularly used to analyze the protein, nitrogen, starch and oil content, grain texture, and grain weight of (intact) seeds (Montes et. al. 2007, Hacisalihoglu et. al. 2010, Mir et. al. 2012, White et. al. 2012). However, the same spectrum may be used to develop prediction models for analyzing traits of potential interest for phenotyping for stress adaptation, such as D13C, mineral content, or the composition of other stable isotopes (Ferrio et. al. 2001, Kleinebecker et. al. 2009, Cobrera-Bosquet et. al. 2011). Although the precision of these indirect estimations may be lower than those of direct analysis, the rapid, low-cost, and non-destructive nature of NIRS may justify its use, at least in the early generations of a breeding program, as a first screening approach when thousands of genotypes need to be evaluated.

Public Datasets

Development and public dissemination of datasets are vital for research in image-based plant phenotyping as it provides access to datasets, to many researchers, that they otherwise do not have the ability to generate. Standard datasets provide a common basis to compare the performance of plant phenotyping algorithms. Datasets that are publicly disseminated are summarized below.

- Leaf segmentation challenge (LSC) dataset: LSC dataset is released to advance the state-of-the-art in leaf segmentation, counting, and tracking of rosette plants. The dataset consists of images of two plant species, i.e., *Arabidopsis* (*Arabidopsis thaliana*) and tobacco (*Nicotiana tabacum*) organized into three subsets. Subsets A1 (Ara2012) and A2 (Ara2013) consist of top-view time-lapse images of *Arabidopsis thaliana* rosettes. The total number of images in Ara2012 and Ara2013 are 150 and 5048, respectively. Subset A3 (Tobacco) consists of top-view stereo image sequences of tobacco (*Nicotiana tabacum*) plants captured hourly for 30 days. The LSC dataset is publicly available from http://www.plant-phenotyping.org/CVPPP2014-challenge.
- Michigan State University Plant imagery dataset (MSU-PID): The MSU-PID dataset (Cruz et al., 2016) consists of images of *Arabidopsis* (total 2160 × 4 images) and bean (total 325 × 4 images) captured with four types of calibrated cameras, i.e., fluorescent, infrared, RGB colour and depth sensor to facilitate research in leaf segmentation, leaf counting, leaf alignment, leaf tracking, and 3D leaf reconstruction. A subset (576 × 4 Arabidopsis images and 175 × 2 bean images) is annotated to provide ground truth for leaf tip location, leaf segmentation, and leaf alignment. MSU-PID dataset is

publicly available from http://cvlab.cse.msu.edu/multi-modality-imagery-database-msu-pid.html

- Panicoid Phenomap-1: To stimulate the development and evaluation of holistic phenotypes of panicoid grain crops, a public dataset called Panicoid Phenomap-1 was introduced (Das Choudhury et al. 2016). It consists of visible light image sequences of 40 genotypes including at least one representative accession from five panicoid grain crops: maize, sorghum, pearl millet, proso millet, and foxtail millet. The images were captured by the Lemnatec scanalyzer high-throughput plant phenotyping facility at the University of Nebraska-Lincoln (UNL), USA.
- University of Nebraska-Lincoln Component Plant Phenotyping Dataset (UNL-CPPD): A benchmark dataset called UNL-CPPD (Das Choudhury et al. (2018), was developed to encourage research in leaf detection and tracking, leaf segmentation, evaluation of holistic, and component phenotypes for maize and cereal crops with similar architecture, e.g., sorghum. UNL-CPPD includes human-annotated ground-truth along with the original image sequences to facilitate image-based component phenotyping analysis.
- Komatsuna Dataset: Komatsuna Dataset (Uchiyama et al., 2017) contains images of the early growth stages of Komatsuna plants with a leaf annotation tool to facilitate 3D plant phenotyping analysis such as leaf segmentation, tracking, and reconstruction. A set of 5 Komatsuna plants are imaged every 4 h for 10 days using a RGB camera (Multiview dataset) and a RGB camera fitted with a structured light depth camera (RGB-D dataset). The dataset is freely available from http://limu.ait.kyushu-u.ac.jp/~agri/komatsuna/.
- University of Nebraska-Lincoln 3D Plant Phenotyping Dataset (UNL-3DPPD): This dataset consists of images of 20 maize and 20 sorghum plants for 10 side views to facilitate 3D plant phenotyping research. Plants were imaged once per day using the visible light camera of the UNL Lemnatec Scanalyzer 3D high-throughput phenotyping facility. Panicoid Phenomap-1, UNL-CPPD, and UNL-3DPPD and freely available from http://plantvision.unl.edu/.
- Deep Phenotyping Dataset: This dataset consists of 22 successive top-view image sequences of four accessions of *Arabidopsis*, i.e., Sf-2, Cvi, Landsberg, and Columbia, captured once daily to study temporal phenotypes for accession classification using convolutional neural network (CNN), recurrent neural network and long-short term memory (LSTM) (Taghavi Namin et al., 2018). The dataset is augmented by rotating each image by 90°, 180°, and 270° to avoid overfitting while training CNN, and freely available from https://figshare.com/s/e18a978267675059578f.

High-Throughput Phenotyping and Genetic Mapping

Genetic studies that have directly benefited from mechanized phenotyping platforms, both within the laboratory and in the field, are presented in Table 2. The phenotypic data listed in Table 2 can be categorized into organ-/tissue-related traits (Leiboff et al., 2015, Wu et al., 2019), plant morphological and leaf architecture traits (Bac-Molenaar et al., 2015, Yang et al., 2015, Condorelli et al., 2018), root anatomical traits (Courtois et al., 2013, Shi et al., 2013, Xie et al., 2017), biomass- or growth-related traits (Busemeyer et al., 2013b, Muraya et al., 2017, Zhang et al., 2017), drought- and salinity response-related traits (Honsdorf et al., 2014, Al-Tamimi et al., 2016, Guo et al., 2018a, Condorelli et al., 2018), and yield-related traits (Yang et al., 2014, Crowell et al., 2016, Zhou et al., 2019), which have been used in genetic mapping.

The size of shoot apical meristem (SAM) in the seedling stage correlates with early flowering phenotypes in maize that decrease the number of days to anthesis. Using a high-throughput image-processing pipeline, maize SAM morphological traits (shape and size) were extracted in an association panel and a backcross (BC) population. GWAS and QTL analyses demonstrated that the microscopic SAM morphology of seedlings is a predictor of adult phenotypes and that novel genes associated with SAM morphometric variation contribute to regulating SAM size (Leiboff et al., 2015, Leiboff et al., 2016). Combing a high-throughput micro-CT-RGB imaging system and GWASs, rice tiller traits, tiller growth traits, and plant traits of nine growth stages could be obtained. A total of 402 significantly associated loci were identified. In addition, two loci containing associations with both vigour-related traits and yield were also identified (Wu et al., 2019).

Leaves are primarily involved in photosynthesis and transpiration (Wang et al., 2011). The size, shape, degree of leaf greenness (chlorophyll content), and the number of leaves determine a plant's photosynthetic and yield potential (Pérez-Pérez et al., 2010, Wang et al., 2015). Two genetic mapping studies of leaf traits in rice and maize were performed at the HRPF (Yang et al., 2015). They conducted a GWAS of 29 leaf traits (i.e., leaf size, shape, and color) in a panel of 533 rice accessions at three growth stages using a self-designed high-throughput leaf scoring (HLS) system and detected many loci, including nine loci containing known leaf-related genes associated with leaf traits. In maize, 22 leaf architecture traits of a RIL population at 16-time points were obtained and used for QTL mapping. Moreover, some leaf traits (i.e., leaf angle and leaf length distribution) are predictive indicators of final yield. Interestingly, a QTL hotspot for SDLC on chromosome 10 overlapped with a QTL hotspot (Zhang et al., 2017) related to metabolic traits.

Root anatomical traits have important effects on the acquisition of nutrients and water from the soil and transportation to the aboveground parts of plants (Zhao et al., 2019). Uncovering the genetic basis of root traits will be helpful for improving the architecture of crop roots and promises to result in increases in water and nutrient use efficiency (Atkinson et al., 2019). Topp et al., (2013) dissected the genetic basis of 25 root-related traits obtained from semiautomatic in vivo 3D imaging and a digital phenotyping pipeline in a rice biparental mapping population. Courtois et al., (2013) performed GWASs of 15 rice root traits with a photography system, finding that most associations were identified for deep root mass and the number of deep roots, whereas no associations were detected for total root biomass or deep root proportions. In *Brassica napus*, the analysis of root architecture-related traits under high phosphate (Pi) and low Pi conditions resulted in the identification of 38 QTLs (Shi et al., 2013).

Drought and salinity are two important types of abiotic stress in many environments that can produce somewhat similar phenotypic effects (Munns and Tester, 2008, Munns et al., 2010). Drought stress responses interact with many other environmental parameters (e.g., temperature, relative air humidity, airflow, light, soil quality, and nutrient availability) (Granier et al., 2006) and are thus particularly challenging to replicate under field conditions. The quantitative measurement of drought resistance (DR) indicators (i.e., a leaf rolling indicator) combined with high-throughput phenotyping platforms provided an opportunity to measure both traditional and novel DR traits and mine for DR-related genes (Table 2). For example, based on LemnaTec's Scanalyzer3D, 44 and 21 DR QTLs were identified in a set of wild barley introgression lines (Honsdorf et al., 2014) and a wheat RIL population under water stress (Parent et al., 2015), respectively. In a rice study, 51 i-traits and traditional DR traits were identified in an association panel and a RIL population using a non-destructive phenotyping facility, and 93% of loci found in GWASs were colocalized with previously reported DR-related QTLs and some loci containing known DR-related genes. Sixty-nine i-trait–locus associations were identified by both GWAS and linkage analysis. The role of a DR gene, OsPP15, was confirmed by genetic transformation experiments, demonstrating that the combination of high-throughput phenotyping and genetic mapping is a promising novel approach for the discovery of causal genes for DR (Guo et al., 2018a) and other traits.

Many loci controlling yield and its components in different crops have been identified (Table 4.2) through the application of various phenotyping platforms, sequence technology, and GWASs (Huang et. al. 2010) and statistical methods for genetic mapping (van Eeuwijk et. al. 2010). However, most of the work listed in Table 2 is at the stage of QTL identification, and there are still several issues that need to be addressed. First, characterizing dynamic QTLs for a complex trait at multiple developmental stages (Li and Sillanpää, 2015). Second, comparing QTLs for the same traits across different large-scale species (e.g. rice and maize).

Table 4.2. Trait and Genotype Variation Discovery Platforms

Crops	Platform	Extracted traits/ features	Methods	References
Arabidopsis thaliana	An imaging system (11 cameras) acquiring root images in Petri plates (University of Wisconsin, USA)	Root tip angle, root gravitropism	LA	Moore et al., 2013
	PHENOPSIS (Fondettes, France)	Projected leaf area, fresh weight, growth traits, and growth model parameters	GWAS	Bac-Molenaar et al., 2015
	Near-infrared reflectance spectroscopic (NIRS, INRA, France)	Oil, protein, carbon, and nitrogen content	LA	Jasinski et al., 2016
Rice	Rhizoscope phenotyping platform (Montpellier, France)	Root system architecture, tiller number, and root/ shoot biomass	GWAS	Courtois et al., 2013
	3D root imaging, GiA Roots 2D and 3D image analysis platform	2D and 3D root system architecture (RSA) traits	LA	Topp et al., 2013
	High-throughput rice phenotyping facility (HRPF, Huazhong Agricultural University, China)	Plant morphological traits, biomass, and yield-related traits	GWAS	Yang et al., 2014
	LemnaTec3D Scanalyzer system (University of Nebraska, USA)	Salinity-responsive fluorescence colour classes	GWAS	Campbell et al., 2015
	High-throughput leaf scoring (HLS) (Huazhong Agricultural University, China)	Leaf size, shape, and colour traits	GWAS	Yang et al., 2015

Crops	Platform	Extracted traits/ features	Methods	References
	PANorama (Cornell University, USA)	Panicle phenotypes	GWAS and LA	Crowell et al., 2016
	Australian Plant Phenomics Facility, The Plant Accelerator, Australia	Relative plant growth rate, transpiration rate, and transpiration use efficiency, and select salinity tolerance traits	GWAS	Al-Tamimi et al., 2016
	High-throughput hyperspectral imaging system (HHIS, Huazhong Agricultural University, China)	Hyperspectral indices, chlorophyll content	GWAS	Feng et al., 2017
	Tractor-based high-throughput phenotyping (HTP, Colorado State University, USA)	Canopy height, canopy temperature depression, and three reflectance ratios	LA	Tanger et al., 2017
	HRPF and HLS	Image-based traits (i-traits), leaf colour-related traits, and yield traits to reflect drought response	GWAS and LA	Guo et al., 2018a
	High-throughput micro-CT-RGB imaging system (HCR, Huazhong Agricultural University, China)	Tiller traits, tiller growth traits, biomass, shoot morphological and shoot growth traits	GWAS	Wu et al., 2019
Triticale	Breedvision (University of Applied Sciences Osnabrück, Germany)	Biomass accumulation at three developmental stages	GWAS	Bumeyer et al., 2013b
Wild barley	Australian Plant Phenomics Facility, The Plant Accelerator, Australia	Traits to detect drought tolerance: shoot area, height, growth, colour, biomass, water-use efficiency, etc.	LA	Honsdorf et al., 2014

Crops	Platform	Extracted traits/ features	Methods	References
Wheat	Australian Plant Phenomics Facility, The Plant Accelerator, Australia	Biomass, leaf area, relative growth rate, transpiration, and water-use efficiency	LA	Parent et al., 2015
	A root imaging system and RootNav software (University of Nottingham, UK)	Root seedling traits: root angle, root length, root number, etc., mature plant height, grain yield	LA	Atkinson et al., 2015
	A germination paper-based "pouch and wick" phenotyping system, coupled with digital image analysis (University of Nottingham, UK)	Seedling root traits, yield, yield components, and phenology in field trials	LA	Xie et al. 2017
	Unmanned Aerial Vehicle (UAV) and a ground-based platform (University of Bologna, Italy)	Normalized Difference Vegetation Index, leaf chlorophyll content, phenology score, leaf rolling, and dry biomass, and select drought adaptive indicator	GWAS	Condorelli et al., 2018
	High-throughput plant phenotyping system in National Plant Phenomics Centre (Aberystwyth University, UK)	Plant area, height, water use, and senescence and fitted parameters	GWAS	Camargo et al., 2018
	A tractor-based semiautomatic phenotyping system (Beijing Forestry University, China)	Sensor-based traits, yield-related canopy architecture	GWAS	Jiang et al. 2019

Crops	Platform	Extracted traits/ features	Methods	References
Sorghum	Time-of-flight depth (Microsoft Kinect) sensor and open-source 3D image analysis pipeline (Texas A&M University, USA)	Shoot and leaf morphological traits	LA	McCormick et al., 2016
	Phenobot 1.0, an auto-steered and self-propelled field-based high-throughput phenotyping platform equipped with stereo RGB cameras (Iowa State University, USA)	Plant height and stem diameter	GWAS	Salas Fernandez et al., 2017
	RGB imaging box and semi-automated imaging analysis pipeline: Toolkit for Inflorescence (Iowa State University, USA)	Panicle shape and size	GWAS	Zhou et al. 2019
Maize	Axio Imager. Z10 (Carl Zeiss Microscopy, LLC, Thornwood, NY, USA) and image analysis using ImageJ (Cornell University, USA)	Shoot apical meristem (SAM) morphology	GWAS	Leiboff et al., 2015
	High-throughput plant phenotyping system (Leibniz Institute of Plant Genetics and Crop Plant Research, IPK, Germany)	Biomass, growth traits	GWAS	Muraya et al., 2017

Crops	Platform	Extracted traits/ features	Methods	References
	HRPF	Plant morphological traits, leaf architecture traits, colour traits, biomass-related traits, and growth traits	LA	Zhang et al., 2017
	PhenoArch phenotyping platform (INRA, France)	Plant leaf area, fresh biomass, transpiration rate	GWAS	Prado et al. 2018
	Unmanned aerial vehicle (Beijing Academy of Agriculture & Forestry Sciences, China)	Plant height and growth rate at four growth stages	GWAS	Wang et al., 2019
Large Pepper plant	SPICY: a sensor-to-plant phenotyping device equipped with RGB cameras and time-of-flight depth sensor (Wageningen University, The Netherlands)	Leaf size, leaf area, leaf angle, plant height	LA	Gerie et al., 2012
Brassica napus	Acquire root images using a flatbed scanner and image analysis using ImageJ (Huazhong Agricultural University, China)	Primary root length, lateral root length, lateral root number, lateral root density, and biomass traits under low Pi and high Pi conditions	LA	Shi et al. 2013
	High-throughput plant phenotyping system (Leibniz Institute of Plant Genetics and Crop Plant Research, IPK, Germany)	Estimated biovolume, projected leaf area, early plant height, colour uniformity	GWAS	Knoch et al., 2019

LA, linkage analysis; GWAS, genome-wide association study

Third, the measurements of some traits are subjective and error-prone (e.g., visual scoring of leaf rolling), or some traits cannot be manually measured non-destructively (e.g., digital biomass, root system architectures). Fourth, how to rapidly clone candidate genes from a large number of QTLs. One solution is to combine genetic transformation technology, CRISPR/Cas9 technology, other omics data, and statistical methods for speeding up QTL cloning (e.g., QTG-seq; Zhang et al., 2019a). This information could be used in omics-based strategies as well as in systemic and synthetic biology in molecular design breeding or molecular module programs and will greatly facilitate future agronomic improvements (Figure 4.5A) (Li et al., 2018, Wing et al., 2018, Zhang et al., 2019b).

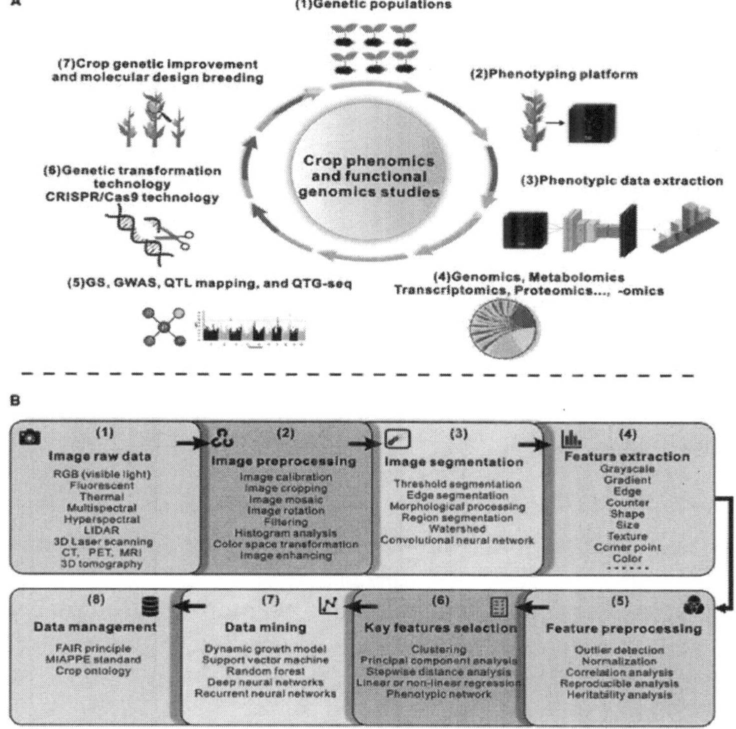

Figure 4.5. The Loop of Crop Phenotype-to-Genotype and Image Data Analysis. (A) Linking crop phenotyping to functional genomics studies and crop improvement with multi-omics. The high-throughput and multiscale phenotyping platform (2) can obtain the dynamic phenotypic traits of large crop genetic populations (1). The combination of phenotypic data (3) and other omics data (i.e., genomics, transcriptomics, proteomics, metabolomics) (4) can be used in mining QTL/genes via QTL mapping and GWASs (5), and combined with genetic transformation techniques to benefit crop genetic improvement (7). (B) The workflow of image data analysis: (1) image raw data obtained; (2) image preprocessing; (3) image segmentation; (4) feature extraction; (5) feature preprocessing; (6) key feature selection; (7) data mining; (8) data management. (Reproduced from Yang et. al. 2020).

High-Throughput Phenotyping and Genome Selection

Genome selection (GS) uses genome-wide markers and statistical modelling to select complex traits controlled by many alleles with small effects. It was first applied to cattle breeding (Meuwissen et al., 2001), but with the ever-decreasing cost of sequencing in recent years, GS is emerging as a powerful tool for estimating breeding values in crop breeding (Taylor, 2014). Its advantage is in predicting how crops will perform before a field test. To develop an accurate and robust predictive model, huge amounts of genotype and phenotype data from individuals or lines are necessary (Meuwissen et al., 2001). With the development of next-generation sequencing technology, markers can be easily and accurately obtained. However, the phenotyping of the plants remained a serious constraint.

High-throughput phenotyping platforms have been demonstrated to enhance GS in grain crops. For example, a UAV carrying a remote-sensing unit with either an RGB or near-infrared, green and blue (NIR-GB) camera has been used for the high-throughput phenotyping of sorghum plant height and for different genomic prediction models (Watanabe et al., 2017). UAV remote sensing will be an important and indispensable tool for high-throughput genomics-assisted crop breeding due to its relatively low cost and easy operation (Watanabe et al., 2017). Data on secondary traits (canopy temperature and green and red NDVI) obtained with remote sensing could be used in wheat GS to improve the prediction accuracy for grain yield (Rutkoski et al., 2016). The International Maize and Wheat Improvement Center (CIMMYT) found that GS strategies are a tractable way for crop breeders to increase the rate of genetic gain and select superior higher-yielding crop varieties more efficiently (Crain et al., 2018).

Sensor-Based Phenotyping-Assistance Breeding

Image-based phenotyping technologies can rapidly and accurately quantify biotic/abiotic resistance traits and grain yield and quality and have a high potential to accelerate crop breeding. A comparison of a traditional long-term breeding experiment with a more recent technology-supported example illustrates this. In Illinois in 1896, Hopkins began long-term directional recurrent selection for oil concentration in maize, increasing the oil content from 4.69% to 20.37% over approximately 100 generations (Dudley and Lambert, 2004), while Song et al., 1999 developed synthetic populations in China in which the oil content increased from 4.71% to 15.55% over just 18 cycles of selection (Song and Chen, 2004). This relative acceleration can be largely attributed to MRI and NIR spectroscopy.

Another example involves double haploid (DH) technology, which is already widely used in commercial crop breeding programs (Ishii et al., 2016). There is an ongoing need for better discrimination of haploid and hybrid kernels. An automatic screening system for maize haploid kernel identification based on the xenia effect value of the oil content and nuclear magnetic resonance (NMR) detection, developed by China Agricultural University, identified in vivo-induced haploid seeds with a mean accuracy of over 90% (Liu et al., 2012), which would accelerate the breeding process. Digital imaging can rapidly assess yield- and quality-related grain traits (Yang et al., 2014, Liang et al., 2016, Makanza et al., 2018). For example, a high-throughput maize kernel trait scorer had an efficiency that was seven times greater than that of manual operators (Liang et al., 2016). Furthermore, maize vascular bundles, as key transportation paths for delivering water, mineral nutrients, and organic substances, can be measured by X-ray micro-CT (Zhang et al., 2018).

Developing new varieties and focusing on cost-effective returns are key objectives of breeding companies and breeders. To improve the management of breeding data and optimize breeding programs, many commercial plant breeding software programs have been developed, such as AGROBASE Generation II (https://www.agronomix.com/) and Plant Research Information Sharing Manager (PRISM, http://www.sytseed.com/). However, to date, most commercial breeding software has limited functionality. Thus, the development of multifunctional breeding software that integrates genomes, phenotypes, environments, data management, and multiomics data analysis and can perform genetic mapping is urgently needed for crop breeding.

High Throughput Phenotyping for Abiotic Stress

The genetic architecture of abiotic tolerance, such as DR, is complex and influenced by many alleles with small effects (Fukao and Xiong, 2013). Thus, the search for generic drought tolerance using single major-effect genes have almost always been disappointing (Passioura, 2012). The following issues need to be addressed regarding high-throughput phenotyping platforms and high-throughput image analysis pipelines that can be used to extract an extensive list of 2D and 3D phenotypic traits related to stress and have the potential for understanding tolerance over time:

- Can the complexities of tolerance be decomposed into simple and more heritable traits? Using non-destructive phenotyping and image-based traits, this may be possible; complex DR, for instance, can be dissected into highly heritable and simple image-based traits (i-traits), including digital stay-green and digital leaf-rolling traits (Guo et al., 2018a). If the loci (genes or QTLs) identified by these simple traits can also affect yield under different stress scenarios, then the effects of the identified loci could benefit crop breeding (Tardieu and Tuberosa, 2010).
- How do root architectures contribute to yield and abiotic stress tolerance? The gene DRO1, for example, which controls the root growth angle and depth, has been used to increase rice yield under drought conditions by backcrossing with a shallow-rooting rice cultivar (Uga et al., 2013). However, variation in root architecture and function has not often been directly exploited in breeding, largely due to the huge challenge in assessing root traits.
- Can DL resolve the data analysis bottleneck? DL is a versatile tool that can be used to make sense of large datasets for crop stress ICQP (identification,

classification, quantification, and prediction) (Singh et al., 2018). However, due to the outstanding identification of diseases, most DL models have been applied in biotic stress and disease detection. In the future, with different sensor technologies (hyperspectral, thermal, CT, terahertz, MRI, radar, etc.) and large multiscale (lab-to-field) phenotypic datasets that include weather and other environmental data, DL could also be used to quantify abiotic stress and predict the loci that control stress tolerance through integration with genetic information and other omics information. This approach will likely work best with large and well-annotated datasets, making the case for open data.

Image Data Analysis and Big Data Organization

The imaging data formats vary widely depending on the different imaging sensors (e.g., RGB, thermal, hyperspectral, CT; Figure 4.5B). Therefore, it is challenging to conclude a common process of image analysis. Yang et. al. (2020) summarized a workflow of image data analysis in the following manner (Figure 4.5B):

- ❑ Image pre-processing to clip or merge raw images, enhance image contrast, remove the noise, and transform the space to get the image easily to handle in the next step;
- ❑ Image segmentation (e.g., threshold, morphological processing) to obtain the objects of interest from the background;
- ❑ Feature extraction to get the raw features according to the experimental targets, which mainly include grayscale, gradient, edge, counter, shape, size, texture, corner point, colour features, and so on;
- ❑ After extraction of huge amounts of raw features, feature pre-processing and traits selection should be adopted to filter and explore the key features;
- ❑ Data mining, for example, building dynamic growth using mechanistic models (Chen et al., 2014) or using DL (e.g., deep neural networks [DNNs] or recurrent neural networks [RNNs]); (Singh et al., 2018) to explore more spatial and temporal information.

Crop functional genomics has entered the large-scale and multi-omic stages. There is growing interest in combining phenomics data with genomic data and other omics data, such as metabolomics, proteomics, and transcriptomics. Data storage should be standardized to facilitate the joint analysis and reanalysis of data from independent experiments or interconnections with other available biological databases and resources (Billiau et al., 2012). Comprehensive metadata description and agreed-upon ontologies are therefore needed to construct a database of multi-omics information. However, the vast amounts of collected data are scattered as small pieces in many labs and vary in format and quality (Li et al., 2018). Data sharing and standardization across different communities remain a big challenge. Another challenge is the lack of funding and data infrastructure to handle these data (Bolger et al., 2019). To face the challenges, synergism efforts should be made and integrated resources and bioinformatics technologies will facilitate crop functional genomics and crop breeding efforts.

Challenges to Phenotyping

Most of the research in image-based plant phenotyping is focused on structural and mainly holistic physiological phenotypes. However, timing determination of important events in the life of a plant, e.g., germination, emergence and senescence of leaves, emergence of flowers and fruits, can provide crucial information about a plant's growth and response to biotic and abiotic stresses. Timing detection of such events using computer vision techniques remains an important open problem.

Most of the phenotypes have focused on either aboveground or belowground phenotypes. The third category of phenotype, called integrated phenotype, establishes a relationship between above- and belowground phenotypes affected by abiotic and biotic stresses. Primary root growth is inhibited during P-limitation, and its length determines the plant's capability to access stored water in the deeper layers of the soil substratum (Prasad et al., 2012; Paez-Garcia et al., 2015). Das Choudhury et. al. (2019) proposed a new integrated phenotype, e.g., the ratio of stem height to primary root length as a function of time, may be investigated for increased yield of stress-tolerant crops by enhancing the capacity of the plant for soil exploration and, thus, water and nutrient acquisition. Algorithms to compute them from plant imagery need to be developed Das Choudhury et. al 2019).

Efforts have been made in the early detection of stress, e.g., drought and minerals (Kim et al., 2011; Chen and Wang, 2014; van Maarschalkerweerd and Husted, 2015). However, future work is required in the investigation of phenotypes that characterize the prorogation of stress as a function of time and categorize stress into different stages, e.g., slight, moderate, extreme, and exceptional. The speed of recovery from the stresses regulated by genotypes is also an open challenge. Uga et al. (2013) demonstrated that controlling root growth angle contributes to drought tolerance.

Computing phenotypes from 2D images (e.g., stem angle) are inherently error-prone as they are dependent on accurate camera views. They are deficient too because the plant is a 3D structure and any projection onto 2D plane results in a loss of information. Thus, new and innovative 3D phenotypes, both holistic and component, based on an accurate reconstruction of the model of a plant must be developed to accurately characterize its properties. 3D plant model reconstruction has been successful for early growth stages with less architectural complexity (Golbach et al., 2016). Advanced 3D plant model reconstruction algorithms are yet to be developed for the entire life cycle of plants covering vegetative and reproductive stages for computation of derived phenotypes, e.g., stem cross-section area as a function of plant height at its different stages, 3D leaf area to leaf length ratio, 3D leaf curvature to leaf length ratio, carbohydrate content of stem with respect to stem volume and plant temperature to convex-hull volume ratio at different stages of stressed plants.

Conclusions

The capacity for undertaking precision phenotyping, particularly under repeatable and representative growing conditions in the field, is lagging far behind the capacity to generate genomic information. This is constraining breeding advances. Developments (e.g., sensors, platforms, and analytical capability) in HTPPs have enabled these measurements to be made in situ; however, the integration of data is yet to be commercialized. Field phenotyping of the appropriate traits, using low-cost, easy-to-handle tools, should become an integral and key component in the breeding pipeline. Using technological advances regarding phenotyping instrumentation should also go together with methods to characterize and control field site variation (for improving repeatability), adopt appropriate experimental designs, selection of the right traits, and, proper integration of heterogeneous datasets, analysis, and application, including prediction models. Efficient integration of all the components of the system is needed to pave the way for the adoption of field HTPPs in the future. This includes a more user-friendly data management system combined with data gathering and processing.

References

Agarwal, B. (2017). Detection of plant emergence based on spatio-temporal image sequence analysis. *Master's Thesis, University of Nebraska-Lincoln.*

Aich, S., & Stavness, I. (2017). Leaf counting with deep convolutional and deconvolutional networks. *The IEEE Intel. Conf. Computer Vision (ICCV) Workshops (Venice)*, 2080–2089.

AL-Tam, F., Adam, H., dos Anjos, A., Lorieux, M., Larmande, P., Ghesquiere, A., Jouannic, S., & Shahbazkia, H.R. (2013). PTRAP: a panicle trait phenotyping tool. *BMC Plant Biol. 13*:122.

Al-Tamimi, N., Brien, C., Oakey, H., Berger, B., Saade, S., Ho, Y.S., Schmockel, S.M., Tester, M., & Negrao, S. (2016). Salinity tolerance loci revealed in rice using high-throughput non-invasive phenotyping. *Nat. Commun. 7*:13342.

Amatya, S., Karkee, M., Zhang, Q., & Whiting, M.D. (2017). Automated detection of branch shaking locations for robotic cherry harvesting using machine vision. *Robotics 6*:31. doi: 10.3390/robotics6040031

Aquino, A., Barrio, I., Diago, MP., Millan, B., & Tardaguila, J. (2018). vitisBerry: an Android-smartphone application to early evaluate the number of grapevine berries by means of image analysis. *Comput. Electron. Agric. 148*:19–28.

Araus, J.L., & Cairns, J.E. (2014). Field high-throughput phenotyping: the new crop breeding frontier. *Trends Plant Sci. 19*, 52–61. doi: 10.1016/j.tplants.2013.09.008

Araus, J.L., Milton, D., King, S., ... Rastogi, L. (2012). Phenotyping maize for adaptation to drought. *Front. Physiol. 3*, 305.

Atanbori, J., Chen, F., French, A.P., & Pridmore, T.P. (2018). Towards low-cost image-based plant phenotyping using reduced-parameter CNN. *BMVC Workshop on Computer Vision Problems in Plant Phenotyping (Newcastle).*

Atieno, J., Li, Y., Langridge, P., Dowling, K., Brien, C., Berger, B., Varshney, R.K., & Sutton, T. (2017). Exploring genetic variation for salinity tolerance in chickpea using image-based phenotyping. *Sci. Rep. 7*:1300.

Atwell, S., Huang, Y.S., Vilhjalmsson, B.J., Willems, G., Horton, M., Li, Y., Meng, D., Platt, A., Tarone, AM., ... Hu, T.T. (2010). Genomewide association study of 107 phenotypes in Arabidopsis thaliana inbred lines. *Nature. 465*: 627–631.

Avramova, V., Meziane, A., Bauer, E., Blankenagel, S., Eggels, S., Gresset, S., Grill, E., Niculaes, C., Ouzunova, M., ... Poppenberger B. (2019). Carbon isotope composition, water use efficiency, and drought sensitivity are controlled by a common genomic segment in maize. *Theor. Appl. Genet. 132*:53–63.

Awlia, M., Nigro, A., Fajkus, J., Schmoeckel, S.M., Negrao, S., Santelia, D., Trtilek, M., Tester, M., Julkowska, M.M., & Panzarova, K. (2016). High-throughput non-destructive phenotyping of traits that contribute to salinity tolerance in Arabidopsis thaliana. *Front. Plant Sci. 7*:1414.

Bac-Molenaar, J.A., Vreugdenhil, D., Granier, C., & Keurentjes, J.J. (2015). Genome-wide association mapping of growth dynamics detects time-specific and general quantitative trait loci. *J. Exp. Bot. 66*:5567–5580.

Bai, G., Ge, Y., Hussain, W., Baenziger, P.S., & Graef, G. (2016). A multi-sensor system for high throughput field phenotyping in soybean and wheat breeding. C*omput. Electron. Agric. 128*:181–192.

Bauer, A., Bostrom, A.G., Ball, J., Applegate, C., Cheng, T., Laycock, S., Rojas, S.M., Kirwan, J., & Zhou, J. (2019). Combining computer vision and deep learning to enable ultra-scale aerial phenotyping and precision agriculture: a case study of lettuce production. *Hortic. Res. 6*:70.

Baweja, H.S., Parhar, T., Mirbod, O., & Nuske, S. (2017). Stalknet: a deep learning pipeline for high-throughput measurement of plant stalk count and stalk width. *Proc. Adv. Robotics (Zurich: Springer)*, 271–284.

Bendig, J., Bolten, A., Bennertz, S., Broscheit, J., Eichfuss, S., & Bareth, G. (2014). Estimating biomass of barley using crop surface models (CSMs) derived from UAV-based RGB imaging. *Rem. Sens. 6*:10395–10412.

Billiau, K., Sprenger, H., Schudoma, C., Walther, D., & Kohl, K.I. (2012). Data management pipeline for plant phenotyping in a multisite project. *Funct. Plant Biol. 39*:948–957.

Blum, A. (2011). Plant Breeding for Water-limited Environments. Springer

Bodner, G., Nakhforoosh, A., Arnold, T., & Leitner, D. (2018). Hyperspectral imaging: a novel approach for plant root phenotyping. *Plant Methods. 14*:84.

Brichet, N., Fournier, C., Turc, O., Strauss, O., Artzet, S., ... Pradal, C. (2017). A robot-assisted imaging pipeline for tracking the growths of maize ear and silks in a high-throughput phenotyping platform. *Plant Methods. 13*:96. doi: 10.1186/s13007-017-0246-7

Brien, C.J., Berger, B., Rabie, H., & Tester, M. (2013). Accounting for variation in designing greenhouse experiments with special reference to greenhouses containing plants on conveyor systems. *Plan Methods. 9*:5.

Burkart, A., Hecht, V.L., Kraska, T., & Rascher, U. (2018). Phenological analysis of unmanned aerial vehicle based time series of barley imagery with high temporal resolution. *Precis. Agr. 19*:134–146.

Busemeyer, L., Mentrup, D., Moller, K., Wunder, E., Alheit, K., Hahn, V., Maurer, H.P., Reif, J.C., Wurschum, T., ... Muller, J. (2013a). BreedVision–a multi-sensor platform for non-destructive field-based phenotyping in plant breeding. *Sensors. 13*:2830–2847.

Busemeyer, L., Ruckelshausen, A., Moller, K., Melchinger, A.E., Alheit, K.V., Maurer, H.P., Hahn, V., Weissmann, E.A., Reif, J.C., & Wurschum, T. (2013b). Precision phenotyping of biomass accumulation in triticale reveals temporal genetic patterns of regulation. *Sci. Rep. 3*:2442.

Cabrera-Bosquet, L., Livington, B., Hill, H., ... Brown, K. (2012). High-throughput phenotyping and genomic selection: the frontiers of crop breeding converge. *J. Integr. Plant Biol. 54*, 312–320.

Camargo, A.V., Mackay, I., Mott, R., Han, J., Doonan, J.H., Askew, K., Corke, F., Williams, K., & Bentley, A.R. (2018). Functional mapping of quantitative trait loci (QTLs) associated with plant performance in a wheat MAGIC mapping population. *Front. Plant Sci. 9*:887.

Campbell, M.T., Knecht, A.C., Berger, B., Brien, C.J., Wang, D., & Walia, H. (2015). Integrating image-based phenomics and association analysis to dissect the genetic architecture of temporal salinity responses in rice. *Plant Physiol. 168*:1476–1489.

Chen, D., Neumann, K., Friedel, S., Kilian, B., Chen, M., Altmann, T., & Klukas, C. (2014). Dissecting the phenotypic components of crop plant growth and drought responses based on high-throughput image analysis. *Plant Cell 26*:4636–4655.

Chen, D., Shi, R., Pape, J.M., Neumann, K., Arend, D., Graner, A., Chen, M., & Klukas, C. (2018). Predicting plant biomass accumulation from image-derived parameters. *Gigascience. 7*. https://doi.org/10.1093/gigascience/giy001

Chen, L., & Wang, K. (2014). Diagnosing of rice nitrogen stress based on static scanning technology and image information extraction. *J. Soil Sci. Plant Nutr. 14*, 382–393. doi: 10.4067/S0718-95162014005000030

Chéné, Y., Rousseau, D., Lucidarme, P., Bertheloot, J., Caffier, V., ... Morel, P. (2012). On the use of depth camera for 3D phenotyping of entire plants. *Comput. Electron. Agric. 82*, 122–127. doi: 10.1016/j.compag.2011.12.007

Chlingaryan, A., Sukkarieh, S., & Whelan, B. (2018). Machine learning approaches for crop yield prediction and nitrogen status estimation in precision agriculture: a review. *Comput. Electron. Agric. 151*, 61–69. doi: 10.1016/j.compag.2018.05.012

Clark, R.T., MacCurdy, R.B., Jung, J.K., Shaff, J.E., McCouch, S.R., Aneshansley, D.J., & Kochian, L.V. (2011). Three-dimensional root phenotyping with a novel imaging and software platform. *Plant Physiol. 156*:455–465.

Cobb, J.N., Cohn, B., Vavilov, G., ... Vinzel. P. (2013). Next-generation phenotyping: requirements and strategies for enhancing our understanding of genotype–phenotype relationships and its relevance to crop improvement. *Theor. Appl. Genet. 126*, 867–887.

Condorelli, G.E., Maccaferri, M., Newcomb, M., Andrade-Sanchez, P., White, J.W., French, A.N., Sciara, G., Ward, R., & Tuberosa, R. (2018). Comparative aerial and ground based high throughput phenotyping for the genetic dissection of NDVI as a proxy for drought adaptive traits in durum wheat. *Front. Plant Sci. 9*:893.

Confalonieri, R., Paleari, L., Foi, M., Movedi, E., Vesely, F.M., Thoelke, W., Agape, C., Borlini, G., Ferri, I., ... Massara, F. (2017). PocketPlant3D: analysing canopy structure using a smartphone. *Biosyst. Eng. 164*:1–12.

Cormack, A.M. (1963). Representation of a function by its line integrals with some radiological applications. *J. Appl. Phys. 34*:2722–2727.

Corona-Lopez, D.D.J., Sommer, S., Rolfe, S.A., Podd, F., & Grieve, B.D. (2019). Electrical impedance tomography as a tool for phenotyping plant roots. *Plant Methods 15:*49.

Courtois, B., Audebert, A., Dardou, A., Roques, S., Ghneim-Herrera, T., Droc, G., Frouin, J., Rouan, L., Goze, E., ... Kilian, A. (2013). Genome-wide association mapping of root traits in a japonica rice panel. *PLoS One 8*:e78037.

Crain, J., Mondal, S., Rutkoski, J., Singh, R.P., & Poland, J. (2018). Combining high-throughput phenotyping and genomic information to increase prediction and selection accuracy in wheat breeding. *Plant Genome 11*:170043.

Crowell, S., Falcao, A.X., Shah, A., Wilson, Z., Greenberg, A.J., & McCouch, S.R. (2014). High-resolution inflorescence phenotyping using a novel image-analysis pipeline, PANorama. *Plant Physiol. 165*:479–495.

Crowell, S., Korniliev, P., Falcao, A., Ismail, A., Gregorio, G., Mezey, J., & McCouch, S. (2016). Genome-wide association and highresolution phenotyping link Oryza sativa panicle traits to numerous trait-specific QTL clusters. *Nat. Commun. 7*:10527.

Cruz, J.A., Yin, X., Liu, X., Imran, S.M., Morris, D.D., ... Kramer, D.M. 2016. Multi-modality imagery database for plant phenotyping. Mach. Vis. Appl. 27, 735–749. doi: 10.1007/s00138-015-0734-6

Czedik-Eysenberg, A., Seitner, S., Guldener, U., Koemeda, S., Jez, J., Colombini, M., & Djamei, A. (2018). The 'PhenoBox', a flexible, automated, open-source plant phenotyping solution. *New Phytol. 219*:808–823.

Das Choudhury, S., Bashyam, S., Qiu, Y., Samal, A., & Awada, T. (2018). Holistic and component plant phenotyping using temporal image sequence. *Plant Methods. 14*:35. doi: 10.1186/s13007-018-0303-x

Das Choudhury, S., Goswami, S., Bashyam, S., Samal, A., & Awada, T. (2017). Automated stem angle determination for temporal plant phenotyping analysis. *ICCV Workshop on Computer Vision Problems in Plant Phenotyping (Venice: ACM Press)*, 41–50.

Das Choudhury, S., Samla, A., & Awada, T. (2019). Leveraging image analysis for high-throughput plant phenotyping. *Frontiers Plant Sci. 10*: 1-8.

Das Choudhury, S., Stoerger, V., Samal, A., Schnable, J.C., Liang, Z., & Yu, J.G. (2016). Automated vegetative stage phenotyping analysis of maize plants using visible light images. *KDD Workshop on Data Science for Food, Energy and Water (San Francisco, CA).*

Deery, DM., Rebetzke, G.J., Jimenez-Berni, J.A., James, R.A., Condon, A.G., Bovill, W.D., Hutchinson, P., Scarrow, J., Davy, R., & Furbank, R.T. (2016). Methodology for high-throughput field phenotyping of canopy temperature using airborne thermography. *Front. Plant Sci. 7*:1808.

Delgado, A., Hays, D.B., Bruton, R.K., Ceballos, H., Novo, A., Boi, E., & Selvaraj, M.G. 2017. Ground penetrating radar: a case study for estimating root bulking rate in cassava (*Manihot esculenta* Crantz). *Plant Methods. 13*:65.

Dellen, B., Scharr, H., & Torras, C. (2015). Growth signatures of rosette plants from time-lapse video. IEEE ACM Trans. *Comput. Biol. Bioinformatics 12*, 1470–1478. doi: 10.1109/TCBB.2015.2404810

Deng, L., Mao, Z.H., Li, X.J., Hu, Z.W., Duan, F.Z., & Yan, Y.N. (2018). UAV-based multispectral remote sensing for precision agriculture: a comparison between different cameras. *ISPRS J. Photogramm. Remote Sens. 146*:124–136.

Dobrescu, A., Valerio Giuffrida, M., & Tsaftaris, S.A. (2017). Leveraging multiple datasets for deep leaf counting. *The IEEE International Conference on Computer Vision (ICCV) Workshops (Venice)*, 2072–2079.

Duan, L., Han, J., Guo, Z., Tu, H., Yang, P., Zhang, D., Fan, Y., Chen, G., Xiong, L., ... Dai, M. (2018). Novel digital features discriminate between drought resistant and drought sensitive rice under controlled and field conditions. *Front. Plant Sci. 9*:492.

Duan, L., Yang, W., Huang, C., & Liu, Q. (2011). A novel machine vision-based facility for the automatic evaluation of yield-related traits in rice. *Plant Methods 7*:44.

Dudley, J.W., & Lambert, R.J. (2004). 100 generations of selection for oil and protein in corn. *Plant Breed. Rev. 24*:79–110.

Elazab, A., Ordonez, R.A., Savin, R., Slafer, G.A., & Araus, J.L. (2016). Detecting interactive effects of N fertilization and heat stress on maize productivity by remote sensing techniques. *Eur. J. Agron. 73*:11–24.

Fahlgren, N., Feldman, M., Gehan, M.A., Wilson, M.S., Shyu, C., Bryant, D.W., Hill, S.T., McEntee, C.J., Warnasooriya, S.N., ... Kumar, I. (2015). A versatile phenotyping system and analytics platform reveals diverse temporal responses to water availability in Setaria. *Mol. Plant 8*:1520–1535.

Fahlgren, N., Gehan, MA., & Baxter, I. (2015). Lights, camera, action: high-throughput plant phenotyping is ready for a close-up. *Curr. Opin. Plant Biol. 24*, 93–99. doi: 10.1016/j.pbi.2015.02.006

Fang, W., Feng, H., Yang, W., Duan, L., Chen, G., Xiong, L., & Liu, Q. (2016). High-throughput volumetric reconstruction for 3D wheat plant architecture studies. *J. Innov. Opt. Health Sci. 9*. https://doi.org/10.1142/S1793545816500371

Fanourakis, D., Briese, C., Max, J.F., Kleinen, S., Putz, A., Fiorani, F., Ulbrich, A., & Schurr, U. (2014). Rapid determination of leaf area and plant height by using light curtain arrays in four species with contrasting shoot architecture. *Plant Methods. 10*:9.

Farooque, A.A., Chang, Y.K., Zaman, Q.U., Groulx, D., Schumann, A.W., & Esau, T.J. (2013). Performance evaluation of multiple ground based sensors mounted on a commercial wild blueberry harvester to sense plant height, fruit yield and topographic features in real-time. *Comput. Electron. Agric. 91*:135–144.

Feng, H., Guo, Z., Yang, W., Huang, C., Chen, G., Fang, W., Xiong, X., Zhang, H., Wang, G., ... Xiong, L. (2017). An integrated hyperspectral imaging and genome-wide association analysis platform provides spectral and genetic insights into the natural variation in rice. *Sci. Rep. 7*:4401.

Fernie, A.R., & Yan, J. (2019). De novo domestication: an alternative route toward new crops for the future. *Mol. Plant. 12*:615–631.

Fichman, Y., Miller, G., & Mittler, R. (2019). Whole-plant live imaging of reactive oxygen species. *Mol. Plant. 12*:1203–1210.

Flood, P.J., Kruijer, W., Schnabel, S.K., van der Schoor, R., Jalink, H., Snel, J.F., Harbinson, J., & Aarts, M.G. (2016). Phenomics for photosynthesis, growth and reflectance in *Arabidopsis thaliana* reveals circadian and long-term fluctuations in heritability. *Plant Methods. 12*:14.

French, A., Ubeda-Tomas, S., Holman, T.J., Bennett, M.J., & Pridmore, T. (2009). High-throughput quantification of root growth using a novel image-analysis tool. *Plant Physiol. 150*:1784–1795.

Fukao, T., & Xiong, L.Z. (2013). Genetic mechanisms conferring adaptation to submergence and drought in rice: simple or complex? *Curr. Opin. Plant Biol. 16*:196–204.

Fukatsu, T., Watanabe, T., Hu, H., Yoichi, H., & Hirafuji, M. (2012). Field monitoring support system for the occurrence of *Leptocorisa chinensis* Dallas (Hemiptera: Alydidae) using synthetic attractants, field servers, and image analysis. *Comput. Electron. Agric. 80*:8–16.

Furbank, R.T., & Tester, M. (2011). Phenomics-technologies to relieve the phenotyping bottleneck. Trends Plant Sci. 16:635–644.

Gage, J.L., Miller, N.D., Spalding, E.P., Kaeppler, S.M., & de Leon, N. (2017). Tips: a system for automated image-based phenotyping of maize tassels. *Plant Methods. 13*:21. doi: 10.1186/s13007-017-0172-8

Galkovskyi, T., Mileyko, Y., Bucksch, A., Moore, B., Symonova, O., Price, C.A., Topp, C.N., Iyer-Pascuzzi, A.S., Zurek, P.R., ... Fang, S.Q. (2012). GiA Roots: software for the high throughput analysis of plant root system architecture. *BMC Plant Biol. 12*:116.

Garbout, A., Munkholm, L.J., Hansen, S.B., Petersen, B.M., Munk, O.L., & Pajor, R. (2011). The use of PET/CT scanning technique for 3D visualization and quantification of real-time soil/plant interactions. *Plant Soil. 352*:113–127.

Garzonio, R., Di Mauro, B., Colombo, R., & Cogliati, S. (2017). Surface reflectance and sun-Induced fluorescence spectroscopy measurements using a small hyperspectral UAS. *Rem. Sens. 9*:472.

Gaudin, A.C.M. Gill, M., Thomson, J., ... Green, S. (2013). Taking transgenic rice drought screening to the field. *J. Exp. Bot. 64*, 109–118.

Gegas, V.C., Nazari, A., Griffiths, S., Simmonds, J., Fish, L., Orford, S., Sayers, L., Doonan, J.H., & Snape, J.W. (2010). A genetic framework for grain size and shape variation in wheat. *Plant Cell. 22*:1046–1056.

Gerie, V.D.H., Song, Y., Horgan, G., Polder, G., Dieleman, A., Bink, M., Palloix, A., van Eeuwijk, F., & Glasbey, C. (2012). SPICY: towards automated phenotyping of large pepper plants in the greenhouse. *Funct. Plant Biol. 39*:870–877.

Ghanem, M.E., Marrou, H., & Sinclair, T.R. (2015). Physiological phenotyping of plants for crop improvement. *Trends Plant Sci. 20*:139–144.

Gibbs, J., Pound, M., French, A., Wells, D., Murchie, H.E., & Pridmore, T. (2016). Approaches to three-dimensional reconstruction of plant shoot topology and geometry. *Funct. Plant Biol. 44*, 62–75. doi: 10.1071/FP16167

Giuffrida, M.V., Doerner, P., & Tsaftaris, S.A. (2018). Pheno-deep counter: a unified and versatile deep learning architecture for leaf counting. *Plant J. 96*, 880–890. doi: 10.1111/tpj.14064

Golbach, F., Kootstra, G., Damjanovic, S., Otten, G., & van de Zedde, R. (2016). Validation of plant part measurements using a 3D reconstruction method suitable for high-throughput seedling phenotyping. *Mach. Vis. Appl. 27*, 663–680. doi: 10.1007/s00138-015-0727-5

Golzarian, M.R., Frick, R.A., Rajendran, K., Berger, B., Roy, S., Tester, M., & Lun, D.S. (2011). Accurate inference of shoot biomass from high-throughput images of cereal plants. *Plant Methods. 7*:2.

González-Sanchez, A., Frausto-Solis, J., & Ojeda, W. (2014). Predictive ability of machine learning methods for massive crop yield prediction. *Spanish J. Agric. Res. 12*, 313–328. doi: 10.5424/sjar/2014122-4439

Granier, C., Aguirrezabal, L., Chenu, K., Cookson, SJ., Dauzat, M., Hamard, P., Thioux, J.J., Rolland, G., Bouchier-Combaud, S., ... Lebaudy, A. (2006). PHENOPSIS, an automated platform for reproducible phenotyping of plant responses to soil water deficit in Arabidopsis thaliana permitted the identification of an accession with low sensitivity to soil water deficit. *New Phytol. 169*:623–635.

Guan, H., Liu, M., Ma, X., & Yu, S. (2018). Three-dimensional reconstruction of soybean canopies using multisource imaging for phenotyping analysis. *Remote Sens. 10:*1206. doi: 10.3390/rs10081206

Guo, Q., Wu, F., Pang, S., Zhao, X., Chen, L., Liu, J., Xue, B., Xu, G., Li, L., ... Jing, H. (2018a). Crop 3D—a LiDAR based platform for 3D high-throughput crop phenotyping. *Sci. China Life Sci. 61*:328–339.

Guo, Z., Yang, W., Chang, Y., Ma, X., Tu, H., Xiong, F., Jiang, N., Feng, H., Huang, C., ... Yang, P. (2018b). Genome-wide association studies of image traits reveal genetic architecture of drought resistance in rice. *Mol. Plant. 11*:789–805.

Hairmansis, A., Berger, B., Tester, M., & Roy, S.J. 2014. Imagebased phenotyping for non-destructive screening of different salinity tolerance traits in rice. *Rice. 7*:16.

Hallau, L., Neumann, M., Klatt, B., Kleinhenz, B., Klein, T., Kuhn, C., Rohrig, M., Bauckhage, C., Kersting, K., ... Mahlein, A.K. (2018). Automated identification of sugar beet diseases using smartphones. *Plant Pathol. 67*:399–410.

Hartmann, A., Czauderna, T., Hoffmann, R., Stein, N., & Schreiber, F. (2011). HTPheno: an image analysis pipeline for high-throughput plant phenotyping. *BMC Bioinformatics. 12*:148.

Hasan, M.M., Chopin, J.P., Laga, H., & Miklavcic, S.J. (2018). Detection and analysis of wheat spikes using convolutional neural networks. *Plant Methods. 14*:100. doi: 10.1186/s13007-018-0366-8

Hassan, M.A., Yang, MJ., Fu, L.P., Rasheed, A., Zheng, B.Y., Xia, X.C., Xiao, Y.G., & He, Z,H. (2019). Accuracy assessment of plant height using an unmanned aerial vehicle for quantitative genomic analysis in bread wheat. *Plant Methods. 15*:37.

Haworth, M., Marino, G., & Centritto, M. (2018). An introductory guide to gas exchange analysis of photosynthesis and its application to plant phenotyping and precision irrigation to enhance water use efficiency. *J. Water Clim. Change. 9*:786–808.

He, J.Q., Harrison, R.J., & Li, B. (2017). A novel 3d imaging system for strawberry phenotyping. *Plant Methods. 13*:93. doi: 10.1186/s13007-017-0243-x

Hirafuji, M., Yoichi, H., Kiura, T., Matsumoto, K., Fukatsu, T., Tanaka, K., Shibuya, Y., Itoh, A., Nesumi, H., ... Hoshi, N. (2011). Creating high-performance/low-cost ambient sensor cloud system using OpenFS (Open Field Server) for high-throughput phenotyping. *SICE Annual Conference.Tokyo*, p. 2090–2092.

Honsdorf, N., March, T.J., Berger, B., Tester, M., & Pillen, K. (2014). High-throughput phenotyping to detect drought tolerance QTL in wild barley introgression lines. *PLoS One. 9*:e97047..

Hounsfield, G.N. (1976). Historical notes on computerized axial tomography. *Can. Assoc. Radiol. J. 27*:135–142.

Huang, H., Liu, L., & Ngadi, M.O. (2014). Recent developments in hyperspectral imaging for assessment of food quality and safety. *Sensors. 14*, 7248–7276. doi: 10.3390/s140407248

Huang, X., Wei, X., Sang, T., Zhao, Q., Feng, Q., Zhao, Y., Li, C., Zhu, C., Lu, T., ... Zhang, Z. (2010). Genome-wide association studies of 14 agronomic traits in rice landraces. *Nat. Genet. 42*:961–967.

Hughes, N., Askew, K., Scotson, C.P., Williams, K., Sauze, C., Corke, F., Doonan, J.H., & Nibau, C. (2017). Non-destructive, high-content analysis of wheat grain traits using X-ray micro computed tomography. Plant Methods. 13:76.

Hughes, N., Oliveira, HR., Fradgley, N., Corke, F., Nibau, C., & Doonan, J.H. (2019). mCT trait analysis reveals morphometric differences between domesticated temperate small grain cereals and their wild relatives. *Plant J.* https://doi.org/10.1111/tpj.14312

Hung, C., Xu, Z., & Sukkarieh, S. (2014). Feature learning based approach for weed classification using high resolution aerial images from a digital camera mounted on a UAV. *Rem. Sens. 6*:12037–12054.

Ishii, T., Karimi-Ashtiyani, R., & Houben, A. (2016). Haploidization via chromosome elimination: means and mechanisms. *Annu. Rev. Plant Biol. 67*:421–438.

Iyer-Pascuzzi, A.S., Symonova, O., Mileyko, Y., Hao, Y., Belcher, H., Harer, J., Weitz, J.S., & Benfey, P.N. (2010). Imaging and analysis platform for automatic phenotyping and trait ranking of plant root systems. *Plant Physiol. 152*:1148–1157.

Jahnke, S., Menzel, M.I., van Dusschoten, D., Roeb, G.W., Buhler, J., Minwuyelet, S., Blumler, P., Temperton, V.M., Hombach, T., ... Streun, M. (2009). Combined MRI-PET dissects dynamic changes in plant structures and functions. *Plant J. 59*:634–644.

Jahnke, S., Roussel, J., Hombach, T., Kochs, J., Fischbach, A., Huber, G., & Scharr, H. (2016). PhenoSeeder - a robot system for automated handling and phenotyping of individual seeds. *Plant Physiol. 172*:1358–1370.

Jannick, J.L., Hendric, K., Omer, C., ... Jindel, R. (2010). Genomic selection in plant breeding: from theory to practice. *Brief. Funct. Genomics. 9*, 166–177.

Jansen, M., Gilmer, F., Biskup, B., Nagel, K.A., Rascher, U., Fischbach, A., Briem, S., Dreissen, G., Tittmann, S., ... Braun, S. (2009). Simultaneous phenotyping of leaf growth and chlorophyll fluorescence via GROWSCREEN FLUORO allows detection of stress tolerance in Arabidopsis thaliana and other rosette plants. *Funct. Plant Biol. 36*:902–914.

Jasinski, S., Lecureuil, A., Durandet, M., Bernard-Moulin, P., & Guerche, P. (2016). Arabidopsis seed content QTL mapping using high-throughput phenotyping: the assets of near infrared spectroscopy. *Front. Plant Sci. 7*:1682.

Jeudy, C., Adrian, M., Baussard, C., Bernard, C., Bernaud, E., Bourion, V., Busset, H., Cabrera-Bosquet, L., Cointault, F., ... Han, S. (2016). RhizoTubes as a new tool for high throughput imaging of plant root development and architecture: test, comparison with pot grown plants and validation. *Plant Methods. 12*:31.

Jiang, L., Sun, L., Ye, M., Wang, J., Wang, Y., Bogard, M., Lacaze, X., Fournier, A., Beauchene, K., ... Gouache, D. (2019). Functional mapping of N deficiency-induced response in wheat yield component traits by implementing high-throughput phenotyping. *Plant J. 97*:1105–1119.

Jimenez-Berni, J.A., Deery, D.M., Rozas-Larraondo, P., Condon, AG., Rebetzke, G.J., James, R.A., Bovill, W.D., Furbank, R.T., & Sirault, X.R.R. (2018). High throughput determination of plant height, ground cover, and above-ground biomass in wheat with LiDAR. *Front. Plant Sci. 9*:237.

Johannsen, W. (1911). The genotype conception of heredity. *Am. Nat. 45*:129–159.

Kaul, S., Koo, HL., Jenkins, J., Rizzo, M., Rooney, T., Tallon, L.J., Feldblyum, T., Nierman, W., Benito, M.I., ... Lin, X.Y. (2000). Analysis of the genome sequence of the flowering plant *Arabidopsis thaliana. Nature. 408*:796–815.

Kersey, P.J. (2019). Plant genome sequences: past, present, future. *Curr. Opin. Plant Biol. 48*:1–8.

Kim, T., Kim, J.I., Visbal-Onufrak, M.A., Chapple, C., & Kim, Y.L. (2016). Nonspectroscopic imaging for quantitative chlorophyll sensing. *J. Biomed. Opt. 21*:16008.

Kim, Y., Glenn, D.M., Park, J., Ngugi, H.K., & Lehman, B.L. (2011). Hyperspectral image analysis for water stress detection of apple trees. *Comput. Electron. Agric. 77*, 155–160. doi: 10.1016/j.compag.2011.04.008

Klukas, C., Chen, D., & Pape, J.M. (2014). Integrated analysis platform: an open-source information system for high-throughput plant phenotyping. *Plant Physiol. 165*:506–518.

Knoch, D., Abbadi, A., Grandke, F., Meyer, R.C., Samans, B., Werner, C.R., Snowdon, R.J., & Altmann, T. (2019). Strong temporal dynamics of QTL action on plant growth progression revealed through high-throughput phenotyping in canola. *Plant Biotechnol. J.* https://doi.org/10.1111/pbi.13171

Komyshev, E., Genaev, M., & Afonnikov, D. (2016). Evaluation of the SeedCounter, a mobile application for grain phenotyping. *Front. Plant Sci. 7*:1990.

Le Marie, C., Kirchgessner, N., Flutsch, P., Pfeifer, J., Walter, A., & Hund, A. (2016). RADIX: rhizoslide platform allowing high throughput digital image analysis of root system expansion. *Plant Methods. 12*:40.

Le Marie, C., Kirchgessner, N., Marschall, D., Walter, A., & Hund, A. (2014). Rhizoslides: paper-based growth system for non-destructive, high throughput phenotyping of root development by means of image analysis. *Plant Methods. 10*:13.

Lei, L., Qin, Z., & Danfeng, H. (2014a). Depth imaging-based detection of muskmelon plant for phenotyping in the greenhouse. *ASABE and CSBE/SCGAB Ann. Intel. Meeting (Montreal, QC)*, 153–159.

Lei, L., Qin, Z., & Danfeng, H. (2014b). A review of imaging techniques for plant phenotyping. *Sensors. 14*, 20078–20111. doi: 10.3390/s141120078

Leiboff, S., DeAllie, C.K., & Scanlon, M.J. (2016). Modeling the morphometric evolution of the maize shoot apical meristem. *Front. Plant Sci. 7*:1651.

Leiboff, S., Li, X., Hu, HC., Todt, N., Yang, J., Yu, X., Muehlbauer, G.J., Timmermans, M.C., Yu, J., ... Schnable, P.S. (2015). Genetic control of morphometric diversity in the maize shoot apical meristem. *Nat. Commun. 6*:8974.

Li, M, Xu, J., Zhang, N., Shan, J., & Yao, S. (2016). Study on the factors affecting grain yield measurement system. *Intel. Conf. Service Sci. Tech. Engg., 14-15 May, Suzhou, China*, p. 566–572.

Li, Y., Xiao, J., Chen, L., Huang, X., Cheng, Z., Han, B., Zhang, Q., & Wu, C. (2018). Rice functional genomics research: past decade and future. *Mol. Plant. 11*:359–380.

Li, Z., & Sillanpa€a, M.J. (2015). Dynamic quantitative trait locus analysis of plant phenomic data. *Trends Plant Sci. 20*:822–833.

Liang, X., Wang, K., Huang, C., Zhang, X., Yan, J., & Yang, W. (2016). A high-throughput maize kernel traits scorer based on line-scan imaging. *Measurement. 90*:453–460.

Lin, Y. (2015). Lidar: an important tool for next-generation phenotyping technology of high potential for plant phenomics? *Comput. Electron. Agric. 119*, 61–73. doi: 10.1016/j.compag.2015.10.011

Liu, J., Guo, T., Yang, P., Wang, H., Liu, L., Lu, Z., Xu, X., Hu, J., & Huang, Q. (2012). Development of automatic nuclear magnetic resonance screening system for haploid kernels in maize. *Trans. Chin. Soc. Agric. Eng. 28*:233–236.

Liu, S., Acosta-Gamboa, LM., Huang, X., & Lorence, A. (2017). Novel low cost 3D surface model reconstruction system for plant phenotyping. *J. Imaging. 3*:39. doi: 10.3390/jimaging3030039

Lobet, G., Pages, L., & Draye, X. (2011). A novel image-analysis toolbox enabling quantitative analysis of root system architecture. *Plant Physiol. 157*:29–39.

Lobos, GA., & Poblete-Echeverría, C. (2017). Spectral knowledge (sk-utalca): software for exploratory analysis of high-resolution spectral reflectance data on plant breeding. *Front. Plant Sci. 7*:1996. doi: 10.3389/fpls.2016.01996

Lobos, G.A., Camargo, A.V., del Pozo, A., Araus, J.L., Ortiz, R., & Doonan, J.H. (2017). Editorial: Plant phenotyping and phenomics for plant breeding. *Front. Plant Sci. 8*:2181. doi: 10.3389/fpls.2017.02181

Lorenz, A.J., Smith, K., James, B., ... Neil, D. (2011). Genomic selection in plant breeding: knowledge and prospects. *Adv. Agron. 110*, 77–123.

Lu, Y., Huang, Y., & Lu, R. (2017). Innovative hyperspectral imaging-based techniques for quality evaluation of fruits and vegetables: a review. *Appl. Sci. 7*:189. doi: 10.3390/app7020189

Maddonni, G.A., Otegui, M.E., Andrieu, B., Chelle, M., & Casal, J.J. (2002). Maize leaves turn away from neighbours. *Plant Physiol. 130*, 1181–1189. doi: 10.1104/pp.009738

Madec, S., Baret, F., de Solan, B., Thomas, S., Dutartre, D., Jezequel, S., Hemmerle, M., Colombeau, G., & Comar, A. (2017). Highthroughput phenotyping of plant height: comparing unmanned aerial vehicles and ground LiDAR estimates. *Front. Plant Sci. 8*:2002.

Madec, S., Jin, X., Lu, H., De Solan, B., Liu, S., Duyme, F., Heritier, E., & Baret, F. (2019). Ear density estimation from high resolution RGB imagery using deep learning technique. *Agric. For. Meteorol. 264*:225–234.

Maes, W.H., & Steppe, K. (2019). Perspectives for remote sensing with unmanned aerial vehicles in precision agriculture. *Trends Plant Sci. 24*:152–164.

Mahlein, A.K., Kuska, M.T., Behmann, J., Polder, G., & Walter, A. (2018). Hyperspectral sensors and imaging technologies in phytopathology: state of the art. *Annu. Rev. Phytopathol. 56*:535–558.

Mahlein, A.K., Oerke, E.C., Steiner, U., & Dehne, H.W. (2011). Recent advances in sensing plant diseases for precision crop protection. *Eur. J. Plant Pathol. 133*, 197–209. doi: 10.1007/s10658-011-9878-z

Maimaitijiang, M., Sagan, V., Sidike, P., Maimaitiyiming, M., Hartling, S., Peterson, K.T., Maw, M.J.W., Shakoor, N., Mockler, T., & Fritschi, F.B. (2019). Vegetation index weighted canopy volume model (CVMVI) for soybean biomass estimation from unmanned aerial system-based RGB imagery. *ISPRS J. Photogramm. Remote Sens. 151*:27–41.

Mairhofer, S., Sturrock, C.J., Bennett, M.J., Mooney, S.J., & Pridmore, T.P. (2015). Extracting multiple interacting root systems using X-ray microcomputed tomography. *Plant J. 84*:1034–1043.

Mairhofer, S., Zappala, S., Tracy, S., Sturrock, C., Bennett, MJ., Mooney, S.J., & Pridmore, T.P. (2013). Recovering complete plant root system architectures from soil via X-ray mu-computed tomography. *Plant Methods. 9*:8.

Mairhofer, S., Zappala, S., Tracy, S.R., Sturrock, C., Bennett, M., Mooney, S.J., & Pridmore, T. 2012. RooTrak: automated recovery of three-dimensional plant root architecture in soil from X-ray microcomputed tomography images using visual tracking. *Plant Physiol. 158*:561–569.

Makanza, R., Zaman-Allah, M., Cairns, J.E., Eyre, J., Burgueno, J., Pacheco, A., Diepenbrock, C., Magorokosho, C., Tarekegne, A., ... Olsen, M. (2018). High-throughput method for ear phenotyping and kernel weight estimation in maize using ear digital imaging. *Plant Methods 14*:49.

Malambo, L., Popescu, S.C., Murray, S.C., Putman, E., Pugh, N.A., Horne, D.W., Richardson, G., Sheridan, R., Rooney, W.L., ... Avant, R. (2018). Multitemporal field-based plant height estimation using 3D point clouds generated from small unmanned aerial systems high-resolution imagery. *Int. J. Appl. Earth Obs. Geoinf. 64*:31–42.

Mathieu, L., Lobet, G., Tocquin, P., & Perilleux, C. (2015). "Rhizoponics": a novel hydroponic rhizotron for root system analyses on mature Arabidopsis thaliana plants. *Plant Methods. 11*:3.

Mazaheri, M., Heckwolf, M., Vaillancourt, B., Gage, J.L., Burdo, B., Heckwolf, S., Barry, K., Lipzen, A., Ribeiro, C.B., ... Kono, T.J.Y. (2019). Genome-wide association analysis of stalk biomass and anatomical traits in maize. *BMC Plant Biol. 19*:45.

McCormick, R.F., Truong, S.K., & Mullet, J.E. (2016). 3D sorghum reconstructions from depth images identify QTL regulating shoot architecture. *Plant Physiol. 172*, 823–834. doi: 10.1104/pp.16.00948

McMullen, M.D., Muller, K., Vasu, D., ... Dutta, P. 2009. Genetic properties of the maize nested association mapping population. *Science. 325*, 737–740.

Metzner, R., Eggert, A., van Dusschoten, D., Pflugfelder, D., Gerth, S., Schurr, U., Uhlmann, N., & Jahnke, S. (2015). Direct comparison of MRI and X-ray CT technologies for 3D imaging of root systems in soil: potential and challenges for root trait quantification. *Plant Methods. 11*:17.

Meuwissen, T.H., Hayes, B.J., & Goddard, M.E. (2001). Prediction of total genetic value using genome-wide dense marker maps. *Genetics. 157*:1819–1829.

Miller, N.D., Haase, N.J., Lee, J., Kaeppler, S.M., de Leon, N., & Spalding, E.P. (2017). A robust, high-throughput method for computing maize ear, cob, and kernel attributes automatically from images. *Plant J. 89*:169–178.

Mishra, P., Asaari, M. S. M., Herrero-Langreo, A., Lohumi, S., Diezma, B., & Scheunders, P. (2017). Close range hyperspectral imaging of plants: a review. *Biosyst. Eng. 164*, 49–67. doi: 10.1016/j.biosystemseng.2017.09.009

Moghimi, A., Yang, C., Miller, M.E., Kianian, S.F., & Marchetto, P.M. (2018). A novel approach to assess salt stress tolerance in wheat using hyperspectral imaging. *Front. Plant Sci. 9*:1182. doi: 10.3389/fpls.2018.01182

Mohd., S., Mohd., A, Mohd S., Puneet, M., Stien, M., Stijn, D., Dirk, I., Nathalie, W., & Paul, S. (2018). Close-range hyperspectral image analysis for the early detection of stress responses in individual plants in a high-throughput phenotyping platform. *ISPRS J. Photogrammetr. Remote Sens. 138*, 121–138. doi: 10.1016/j. isprsjprs. 2018. 02.003

Moore, C.R., Johnson, L.S., Kwak, IY., Livny, M., Broman, K.W., & Spalding, E.P. (2013). High-throughput computer vision introduces the time axis to a quantitative trait map of a plant growth response. *Genetics. 195:*1077–1086.

Munns, R., James, R.A., Sirault, X.R.R., Furbank, R.T., & Jones, H.G. (2010). New phenotyping methods for screening wheat and barley for beneficial responses to water deficit. *J. Exp. Bot. 61*:3499–3507.

Munns, R., & Tester, M. (2008). Mechanisms of salinity tolerance. *Annu. Rev. Plant Biol. 59*:651–681.

Muraya, M.M., Chu, J., Zhao, Y., Junker, A., Klukas, C., Reif, JC., & Altmann, T. (2017). Genetic variation of growth dynamics in maize (*Zea mays* L.) revealed through automated non-invasive phenotyping. *Plant J. 89*:366–380.

Nagel, K.A., Putz, A., Gilmer, F., Heinz, K., Fischbach, A., Pfeifer, J., Faget, M., Blossfeld, S., Ernst, M., ... Dimaki, C. (2012). GROWSCREEN-Rhizo is a novel phenotyping robot enabling simultaneous measurements of root and shoot growth for plants grown in soil-filled rhizotrons. *Funct. Plant Biol. 39*:891–904.

Nakarmi, A.D., & Tang, L. (2012). Automatic inter-plant spacing sensing at early growth stages using a 3D vision sensor. *Comput. Electron. Agric. 82*:23–31.

Navarro, P J., Pérez, F., Weiss, J., & Egea-Cortines, M. (2016). Machine learning and computer vision system for phenotype data acquisition and analysis in plants. *Sensors. 16*:641. doi: 10.3390/s16050641

Neilson, E.H., Edwards, A.M., Blomstedt, C.K., Berger, B., Moller, B.L., & Gleadow, R.M. (2015). Utilization of a high-throughput shoot imaging system to examine the dynamic phenotypic responses of a C4 cereal crop plant to nitrogen and water deficiency over time. *J. Exp. Bot. 66*:1817–1832.

Orgogozo, V., Morizot, B., & Martin, A. (2015). The differential view of genotype–phenotype relationships. *Front. Genet. 6*:179.

Paez-Garcia, A., Motes, C.M., Scheible, W.R., Chen, R., Blancaflor, E.B., & Monteros, M.J. (2015). Root traits and phenotyping strategies for plant improvement. *Plants. 4*, 334–355. doi: 10.3390/plants4020334

Parent, B., Shahinnia, F., Maphosa, L., Berger, B., Rabie, H., Chalmers, K., Kovalchuk, A., Langridge, P., & Fleury, D. (2015). Combining field performance with controlled environment plant imaging to identify the genetic control of growth and transpiration underlying yield response to water-deficit stress in wheat. *J. Exp. Bot. 66*:5481–5492.

Passioura, J.B. (2012). Phenotyping for drought tolerance in grain crops: when is it useful to breeders? *Funct. Plant Biol. 39*:851–859.

Paulus, S., Behmann, J., Mahlein, AK., Plümer, L., & Kuhlmann, H. (2014a). Low-cost 3D systems: suitable tools for plant phenotyping. *Sensors. 14*, 3001–3018. doi: 10.3390/s140203001

Paulus, S., Schumann, H., Kuhlmann, H., & Leon, J. (2014b). High precision laser scanning system for capturing 3D plant architecture and analysing growth of cereal plants. *Biosyst. Eng. 121*:1–11.

Pe´rez-Pe´rez, J.M., Esteve-Bruna, D., & Micol, J.L. (2010). QTL analysis of leaf architecture. *J. Plant Res. 123*:15–23.

Pelletier, M.G., Wanjura, J.D., & Holt, G.A. (2019). Embedded microcontroller software design of a cotton harvester yield monitor calibration system. *Agri Engg. 1*:485–495.

Perez-Sanz, F., Navarro, PJ., & Egea-Cortines, M. (2017). Plant phenomics: an overview of image acquisition technologies and image data analysis algorithms. *Gigascience. 6*:1–18.

Pflugfelder, D., Metzner, R., van Dusschoten, D., Reichel, R., Jahnke, S., & Koller, R. (2017). Non-invasive imaging of plant roots in different soils using magnetic resonance imaging (MRI). *Plant Methods. 13*:102.

Phillips, R.L. (2010). Mobilizing science to break yield barriers. *Crop Sci. 50*, S99–S108

Pieruschka, R., & Schurr, U. (2019). Plant phenotyping: past, present, and future. *Plant Phenomics*. https://doi.org/10.34133/2019/7507131

Pineros, M.A., Larson, B.G., Shaff, J.E., Schneider, D.J., Falcao, A.X., Yuan, L., Clark, R.T., Craft, E.J., Davis, T.W., ... Pradier, P.L. (2016). Evolving technologies for growing, imaging and analyzing 3D root system architecture of crop plants. *J. Integr. Plant Biol. 58*:230–241.

Pingali, P.L. (2012). Green revolution: impacts, limits, and the path ahead. *Proc. Natl. Acad. Sci. U.S.A. 109*, 12302–12308.

Polder, G., & Hofstee, J. (2014). Phenotyping large tomato plants in the greenhouse using a 3D light-field camera. *ASABE and CSBE/SCGAB Ann. Intel. Meeting (Montreal, QC)*, 153–159.

Poorter, H., Fiorani, F., Stitt, M., Schurr, U., Finck, A., Gibon, Y., Usadel, B., Munns, R., Atkin, OK., ... Tardieu, F. (2012). The art of growing plants for experimental purposes: a practical guide for the plant biologist. *Funct. Plant Biol. 39*:821–838.

Pound, M.P., Atkinson, J.A., Townsend, A.J., Wilson, M.H., Griffiths, M., ... Jackson, A.S. (2017). Deep machine learning provides state-of-the-art performance in image-based plant phenotyping. *Gigascience. 6*, 1–10. doi: 10.1093/gigascience/gix083

Pound, M.P., French, A.P., Atkinson, J.A., Wells, D.M., Bennett, M.J., & Pridmore, T. (2013). RootNav: navigating images of complex root architectures. *Plant Physiol. 162*:1802–1814.

Prado, S.A., Cabrera-Bosquet, L., Grau, A., Coupel-Ledru, A., Millet, E.J., Welcker, C., & Tardieu, F. (2018). Phenomics allows identification of genomic regions affecting maize stomatal conductance with conditional effects of water deficit and evaporative demand. *Plant Cell Environ. 41*:314–326.

Prasad, S.S., Wasson, A., Rebetzke, G., Kirkegaard, J., Richards, R., ... Watt, M. (2012). Traits and selection strategies to improve root systems and water uptake in water-limited wheat crops. *J. Exp. Bot. 63*, 3485–3498. doi: 10.1093/jxb/ers111

Rahaman, M., Chen, D., Gillani, Z., Klukas, C., & Chen, M. (2015). Advanced phenotyping and phenotype data analysis for the study of plant growth and development. *Front. Plant Sci. 6*:619. doi: 10.3389/fpls.2015.00619

Ray, D.K., Dutta, P.K., Ghosh, P., ... Bhatt, G (2012). Recent patterns of crop yield growth and stagnation. *Nat. Commun. 3*, 1293.

Ray, D.K., Sen, K.L., Hazra, S.K., ... Gill, S. (2013). Yield trends are insufficient to double global crop production by 2050. *PLoS One. 8*, e66428.

Raza, S.E., Smith, H.K., Clarkson, G.J.J., Taylor, G., Thompson, A.J., ... Clarkson, J. (2014). Automatic detection of regions in spinach canopies responding to soil moisture deficit using combined visible and thermal imagery. *PLOS One. 9*:e97612. doi: 10.1371/journal.pone.0097612

Reuzeau, C., Pen, J., Frankard, V., de Wolf, J., Peerbolte, R., Broekaert, W., & van Camp, W. (2005). TraitMill: a discovery engine for identifying yield-enhancement genes in cereals. *Mol. Plant Breed. 3*:753–759.

Reynolds, D., Ball, J., Bauer, A., Davey, R., Griffiths, S., & Zhou, J. (2019a). CropSight: a scalable and open-source information management system for distributed plant phenotyping and IoTbased crop management. *Gigascience. 8*: giz009.

Reynolds, D., Baret, F., Welcker, C., Bostrom, A., Ball, J., Cellini, F., Lorence, A., Chawade, A., Khafif, M., ... Noshita, K. (2019b). What is cost-efficient phenotyping? Optimising costs for different scenarios. *Plant Sci. 282*:14–22.

Rogers, E.D., Monaenkova, D., Mijar, M., Nori, A., Goldman, D.I., & Benfey, P.N. (2016). X-Ray computed tomography reveals the response of root system architecture to soil texture. *Plant Physiol. 171*:2028–2040.

Römer, C., Wahabzada, M., Ballvora, A., Pinto, F., Rossini, M., ... Panigada, C. (2012). Early drought stress detection in cereals: simplex volume maximization for hyperspectral image analysis. *Funct. Plant Biol. 39*, 878–890. doi: 10.1071/FP12060

Ruiz-Garcia, L., Lunadei, L., Barreiro, P., & Robla, J.I. (2009). A review of wireless sensor technologies and applications in agriculture and food industry: state of the art and current trends. *Sensors. 9*:4728–4750.

Rungpichayapichet, P., Nagle, M., Yuwanbun, P., Khuwijitjaru, P., Mahayothee, B., & Müller, J. (2017). Prediction mapping of physicochemical properties in mango by hyperspectral imaging. *Biosyst. Engg. 159*, 109–120. doi: 10.1016/j. biosystemseng. 2017.04.006

Rutkoski, J., Poland, J., Mondal, S., Autrique, E., Perez, L.G., Crossa, J., Reynolds, M., & Singh, R. (2016). Canopy temperature and vegetation indices from high-throughput phenotyping improve accuracy of pedigree and genomic selection for grain yield in wheat. *Genes Genomes Genet. 6*:2799–2808.

Sadeghi-Tehran, P., Sabermanesh, K., Virlet, N., & Hawkesford, M.J. (2017). Automated method to determine two critical growth stages of wheat: heading and flowering. *Front. Plant Sci. 8*:252.

Saint-Pierre, C., Thomson, L., Ranger, P., ... Clerk, M. (2012). Phenotyping transgenic wheat for drought resistance. *J. Exp. Bot. 63*, 1799–1808.

Sakamoto, T., Shibayama, M., Kimura, A., & Takada, E. (2011). Assessment of digital camera-derived vegetation indices in quantitative monitoring of seasonal rice growth. *ISPRS J. Photogramm. Remote Sens. 66*:872–882.

Salas Fernandez, M.G., Bao, Y., Tang, L., & Schnable, P.S. (2017). A high-throughput, field-based phenotyping technology for tall biomass crops. *Plant Physiol. 174*:2008–2022.

Scharr, H., Briese, C., Embgenbroich, P., Fischbach, A., Fiorani, F., & Müller-Linow, M. (2017). Fast high resolution volume carving for 3D plant shoot reconstruction. *Front. Plant Sci. 8*:1680. doi: 10.3389/fpls.2017.01680

Schmittgen, S., Metzner, R., Van Dusschoten, D., Jansen, M., Fiorani, F., Jahnke, S., Rascher, U., & Schurr, U. (2015). Magnetic resonance imaging of sugar beet taproots in soil reveals growth reduction and morphological changes during foliar *Cercospora beticola* infestation. J. Exp. Bot. 66:5543–5553.

Schork, N.J. (1997). Genetics of complex disease - approaches, problems, and solutions. *Am. J. Respir. Crit. Care Med. 156*: S103–S109.

Selby, P., Abbeloos, R., Backlund, JE., Salido, M.B., Bauchet, G., Benites-Alfaro, O.E., Birkett, C., Calaminos, V.C., Carceller, P., ... Cornut, G. (2019). BrAPI: an application programming interface for plant breeding applications. *Bioinformatics. 35*:4147– 4155.

Shahzad, Z., Kellermeier, F., Armstrong, E.M., Rogers, S., Lobet, G., Amtmann, A., & Hills, A. (2018). EZ-Root-VIS: a software pipeline for the rapid analysis and visual reconstruction of root system architecture. *Plant Physiol. 177*:1368–1381.

Shi, C., Zhao, L., Zhang, X., Lv, G., Pan, Y., & Chen, F. (2019). Gene regulatory network and abundant genetic variation play critical roles in heading stage of polyploidy wheat. *BMC Plant Biol. 19*: 6.

Shi, L., Shi, T., Broadley, M.R., White, P.J., Long, Y., Meng, J., Xu, F., & Hammond, J.P. (2013). High-throughput root phenotyping screens identify genetic loci associated with root architectural traits in Brassica napus under contrasting phosphate availabilities. *Ann. Bot. 112*:381–389.

Shibayama, M., Sakamoto, T., Takada, E., Inoue, A., Morita, K., Takahashi, W., & Kimura, A. (2015a). Continuous monitoring of visible and near-infrared band reflectance from a rice paddy for determining nitrogen uptake using digital cameras. *Plant Prod. Sci. 12*:293–306.

Shibayama, M., Sakamoto, T., Takada, E., Inoue, A., Morita, K., Takahashi, W., & Kimura, A. (2015b). Estimating paddy rice leaf area index with fixed point continuous observation of near infrared reflectance using a calibrated digital camera. *Plant Prod. Sci. 14*:30–46.

Shrestha, R., Matteis, L., Skofic, M., Portugal, A., McLaren, G., Hyman, G., & Arnaud, E. (2012). Bridging the phenotypic and genetic data useful for integrated breeding through a data annotation using the Crop Ontology developed by the crop communities of practice. *Front. Physiol. 3*:326.

Singh, A.K., Ganapathysubramanian, B., Singh, A. K., & Sarkar, S. (2016). Machine learning for high-throughput stress phenotyping in plants. *Trends Plant Sci. 21*, 110–124. doi: 10.1016/j.tplants.2015.10.015

Singh, A.K., Ganapathysubramanian, B., Sarkar, S., & Singh, A. (2018). Deep learning for plant stress phenotyping: trends and future perspectives. *Trends Plant Sci. 23*:883–898.

Singh, C.B., Jayas, D.S., Paliwal, J., & White, N.D.G. (2010). Identification of insect-damaged wheat kernels using short-wave near-infrared hyperspectral and digital colour imaging. *Comput. Electron. Agric. 73*:118–125.

Song, T.M., & Chen, S.J. (2004). Long term selection for oil concentration in five maize populations. *Maydica. 49*:9–14.

Song, T.M., Kong, F., Li, C.J., & Song, G.H. (1999). Eleven cycles of single kernel phenotypic recurrent selection for percent oil in Zhongzong no. 2 maize synthetics. *J. Genet. Breed. 53*:31–35.

Srivastava, S., Bhugra, S., Lall, B., & Chaudhury, S. (2017). Drought stress classification using 3D plant models. *IEEE Int Conf. Computer Vision Workshops (ICCVW) (Venice)*, 2046–2054.

Sun, D., Cen, H., Weng, H., Wan, L., Abdalla, A., El-Manawy, A.I., Zhu, Y., Zhao, N., Fu, H., ... Tang, J. (2019). Using hyperspectral analysis as a potential high throughput phenotyping tool in GWAS for protein content of rice quality. *Plant Methods. 15*:54.

Sun, S., Li, C., & Paterson, A.H. (2017). In-field high-throughput phenotyping of cotton plant height using lidar. *Remote Sens. 9*: 377. doi: 10.3390/rs9040377

Sun, S., Li, C., Paterson, AH., Jiang, Y., Xu, R., ... Robertson, J.S. (2018). In-field high throughput phenotyping and cotton plant growth analysis using lidar. *Front. Plant Sci. 9*:16. doi: 10.3389/fpls.2018.00016

Svane, S.F., Jensen, C.S., & Thorup-Kristensen, K. (2019). Construction of a large-scale semi-field facility to study genotypic differences in deep root growth and resources acquisition. *Plant Methods. 15*:26.

Taghavi Namin, S., Esmaeilzadeh, M., Najafi, M., Brown, T.B., & Borevitz, J.O. (2018). Deep phenotyping: deep learning for temporal phenotype/genotype classification. *Plant Methods. 14*:66. doi: 10.1186/s13007-018-0333

Tanabata, T., Shibaya, T., Hori, K., Ebana, K., & Yano, M. (2012). SmartGrain: high-throughput phenotyping software for measuring seed shape through image analysis. *Plant Physiol. 160*:1871–1880.

Tanger, P., Klassen, S., Mojica, J.P., Lovell, J.T., Moyers, B.T., Baraoidan, M., Naredo, M.E., McNally, K.L., Poland, J., ... Bush, D.R. (2017). Field-based high throughput phenotyping rapidly identifies genomic regions controlling yield components in rice. *Sci. Rep. 7*:42839.

Tardieu, F., & Tuberosa, R. (2010). Dissection and modelling of abiotic stress tolerance in plants. *Curr. Opin. Plant Biol. 13*: 206–212.

Taylor, J.F. (2014). Implementation and accuracy of genomic selection. *Aquaculture. 420-421*: S8–S14.

Tian, F., Bradbury, P.J., Brown, P.J., Hung, H., Sun, Q., Flint-Garcia, S., Rocheford, T.R., McMullen, M.D., Holland, J.B., & Buckler, E.S. (2011). Genome-wide association study of leaf architecture in the maize nested association mapping population. *Nat. Genet. 43:*159–162.

Tilman, D., Kappa, U., Brown, J., ... Robertson, J. (2011). Global food demand and the sustainable intensification of agriculture. *Proc. Natl. Acad. Sci. U.S.A. 108,* 20260–20264

Tisne, S., Serrand, Y., Bach, L., Gilbault, E., Ben Ameur, R., Balasse, H., Voisin, R., Bouchez, D., Durand-Tardif, M., ... Guerche, P. (2013). Phenoscope: an automated large-scale phenotyping platform offering high spatial homogeneity. *Plant J. 74*: 534–544.

Topp, C.N., Iyer-Pascuzzi, A.S., Anderson, J.T., Lee, C.R., Zurek, P.R., Symonova, O., Zheng, Y., Bucksch, A., Mileyko, Y., ... Galkovskyi, T. (2013). 3D phenotyping and quantitative trait locus mapping identify core regions of the rice genome controlling root architecture. *Proc. Natl. Acad. Sci. U S A. 110*: E1695–E1704.

Trachsel, S., Dhliwayo, T., Perez, L.G., Lugo, J.A.M., & Trachsel, M. (2019). Estimation of physiological genomic estimated breeding values (PGEBV) combining full hyperspectral and marker data across environments for grain yield under combined heat and drought stress in tropical maize (*Zea mays* L.). *PLoS One. 14*: e0212200.

Trachsel, S., Kaeppler, S.M., Brown, K.M., & Lynch, J.P. (2010). Shovelomics: high throughput phenotyping of maize (*Zea mays* L.) root architecture in the field. *Plant Soil. 341*:75–87.

Tsaftaris, S.A., Minervini, M., & Scharr, H. (2016). Machine learning for plant phenotyping needs image processing. *Mach. Vis. Appl. 21*, 989–991. doi: 10.1016/j.tplants.2016.10.002

Tsaftaris, S.A., & Scharr, H. (2018). Sharing the right data right: a symbiosis with machine learning. *Trends Plant Sci. 24*: 99–102.

Tuberosa, R. (2012). Phenotyping for drought tolerance of crops in the genomics era. *Front. Physiol. 3*, 347.

Ubbens, J.R., & Stavness, I.K. (2017). Deep plant phenomics: a deep learning platform for complex plant phenotyping tasks. *Front. Plant Sci. 8:* 1190. doi: 10.3389/fpls.2017.01190

Ubbens, J.R., Cieslak, M., Prusinkiewicz, P., & Stavness, I. (2018). The use of plant models in deep learning: an application to leaf counting in rosette plants. *Plant Methods. 14*:6. doi: 10.1186/s13007-018-0273-z

Uchiyama, H., Sakurai, S., Mishima, M., Arita, D., Okayasu, T., ... Shimada, A. (2017). An easy-to-setup 3D phenotyping platform for komatsuna dataset. *The IEEE Int. Conf. Computer Vision (ICCV) Workshops (Venice)*, 2038–2045.

Uga, Y., Sugimoto, K., Ogawa, S., Rane, J., Ishitani, M., Hara, N., Kitomi, Y., Inukai, Y., Ono, K., ... Kanno, N. (2013). Control of root system architecture by DEEPER ROOTING 1 increases rice yield under drought conditions. *Nat. Genet. 45*:1097–1102.

Vadez, V., Kholova, J., Hummel, G., Zhokhavets, U., Gupta, S.K., & Hash, C.T. (2015). LeasyScan: a novel concept combining 3D imaging and lysimetry for high-throughput phenotyping of traits controlling plant water budget. *J. Exp. Bot. 66*:5581–5593.

van Dusschoten, D., Metzner, R., Kochs, J., Postma, J.A., Pflugfelder, D., Buhler, J., Schurr, U., & Jahnke, S. (2016). Quantitative 3D analysis of plant roots growing in soil using magnetic resonance imaging. *Plant Physiol. 170*:1176–1188.

van Eeuwijk, F.A., Bink, M.C., Chenu, K., & Chapman, S.C. (2010). Detection and use of QTL for complex traits in multiple environments. *Curr. Opin. Plant Biol. 13*:193–205.

van Maarschalkerweerd, M., & Husted, S. (2015). Recent developments in fast spectroscopy for plant mineral analysis. *Front. Plant Sci. 6*:169. doi: 10.3389/fpls.2015.00169

Vergara-Diaz, O., Kefauver, S.C., Elazab, A., Nieto-Taladriz, M.T., & Araus, J.L. (2015). Grain yield losses in yellow-rusted durum wheat estimated using digital and conventional parameters under field conditions. *Crop J. 3*: 200–210.

Virlet, N., Sabermanesh, K., Sadeghi-Tehran, P., & Hawkesford, M.J. (2017). Field Scanalyzer: an automated robotic field phenotyping platform for detailed crop monitoring. *Funct. Plant Biol. 44*:143–153.

Wahabzada, M., Mahlein, AK., Bauckhage, C., Steiner, U., Oerke, E.C., & Kersting, K. (2016). Plant phenotyping using probabilistic topic models: uncovering the hyperspectral language of plants. *Sci. Rep. 6*:22482. doi: 10.1038/srep22482

Wallays, C., Missotten, B., Baerdemaeker, J.D., & Saeys, W. (2009). Hyperspectral waveband selection for on-line measurement of grain cleanness. *Biosyst. Engg. 104*:1–7.

Walter, A., Scharr, H., Gilmer, F., Zierer, R., Nagel, K.A., Ernst, M., Wiese, A., Virnich, O., Christ, M.M., ... Uhlig, B. (2007). Dynamics of seedling growth acclimation towards altered light conditions can be quantified via GROWSCREEN: a setup and procedure designed for rapid optical phenotyping of different plant species. *New Phytol. 174*: 447–455.

Wang, H., Xu, S., Fan, Y., Liu, N., Zhan, W., Liu, H., Xiao, Y., Li, K., Pan, Q., ... Li, W. (2018a). Beyond pathways: genetic dissection of tocopherol content in maize kernels by combining linkage and association analyses. *Plant Biotechnol. J. 16*:1464–1475.

Wang, P., Zhou, G., Yu, H., & Yu, S. (2011). Fine mapping a major QTL for flag leaf size and yield-related traits in rice. *Theor. Appl. Genet. 123*:1319–1330.

Wang, Q., Xie, W., Xing, H., Yan, J., Meng, X., Li, X., Fu, X., Xu, J., Lian, X., ... Yu, S. (2015). Genetic architecture of natural variation in rice chlorophyll content revealed by a genome-wide association study. *Mol. Plant. 8*: 946–957.

Wang, X., Singh, D., Marla, S., Morris, G., & Poland, J. (2018b). Field based high-throughput phenotyping of plant height in sorghum using different sensing technologies. *Plant Methods. 14*: 53.

Wang, X.Q., Zhang, R.Y., Song, W., Han, L., Liu, X.L., Sun, X., Luo, M.J., Chen, K., Zhang, Y.X., ... Yang, H. (2019). Dynamic plant height QTL revealed in maize through remote sensing phenotyping using a high-throughput unmanned aerial vehicle (UAV). *Sci. Rep. 9*:3458.

Wang, Z., Wang, J., Liu, L., Huang, W., Zhao, C., & Wang, C. (2004). Prediction of grain protein content in winter wheat (Triticum aestivum L.) using plant pigment ratio (PPR). *Field Crops Res. 90*:311–321.

Wasson, A.P., Chiu, G.S., Zwart, A.B., & Binns, T.R. (2017). Differentiating wheat genotypes by Bayesian hierarchical nonlinear mixed modelling of wheat root density. *Front. Plant Sci. 8*:282.

Watanabe, K., Guo, W., Arai, K., Takanashi, H., Kajiya-Kanegae, H., Kobayashi, M., Yano, K., Tokunaga, T., Fujiwara, T., T.... sutsumi, N. (2017). High-throughput phenotyping of sorghum plant height using an unmanned aerial vehicle and its application to genomic prediction modeling. Front. Plant Sci. 8: 421.

Werner, T. (2010). Next generation sequencing in functional genomics. *Brief. Bioinform. 11*: 499–511.

Whan, A.P., Smith, A.B., Cavanagh, C.R., Ral, J.P., Shaw, L.M., Howitt, C.A., & Bischof, L. (2014). GrainScan: a low cost, fast method for grain size and colour measurements. *Plant Methods. 10*: 23.

Wing, R.A., Purugganan, M.D., & Zhang, Q. (2018). The rice genome revolution: from an ancient grain to Green Super Rice. *Nat. Rev. Genet. 19*. 505–517.

Wu, D., Guo, Z., Ye, J., Feng, H., Liu, J., Chen, G., Zheng, J., Yan, D., Yang, X., ... Xiong, X. (2019). Combining high-throughput microCT-RGB phenotyping and genome-wide association study to dissect the genetic architecture of tiller growth in rice. *J. Exp. Bot. 70*: 545–561.

Xiao, Y., Liu, H., Wu, L., Warburton, M., & Yan, J. (2017). Genomewide association studies in maize: praise and stargaze. *Mol. Plant. 10*: 359–374.

Xie, Q., Fernando, K.M.C., Mayes, S., & Sparkes, D.L. (2017). Identifying seedling root architectural traits associated with yield and yield components in wheat. *Ann. Bot. 119*:1115–1129.

Xiong, X., Duan, L., Liu, L., Tu, H., Yang, P., ... Wu, D. (2017). Panicle-seg: a robust image segmentation method for rice panicles in the field based on deep learning and superpixel optimization. *Plant Methods. 13*:104. doi: 10.1186/s13007-017-0254-7

Xiong, X., Yu, L., Yang, W., Liu, M., Jiang, N., Wu, D., Chen, G., Xiong, L., Liu, K., & Liu, Q. (2017). A high-throughput stereo-imaging system for quantifying rape leaf traits during the seedling stage. *Plant Methods. 13*: 7.

Xu, R., Li, C., Paterson, A.H., Jiang, Y., Sun, S., & Robertson, J.S. (2018). Aerial images and convolutional neural network for cotton bloom detection. *Front. Plant Sci. 8*: 2235. doi: 10.3389/fpls.2017.02235

Xu, Y. (2016). Envirotyping for deciphering environmental impacts on crop plants. *Theor. Appl. Genet. 129*: 653–673.

Xu, Y., Liu, X., Fu, J., Wang, H., Wang, J., Huang, C., Prasanna, B.M., Olsen, M.S., Wang, G., & Zhang, A. (2020). Enhancing genetic gain through genomic selection: from livestock to plants. *Plant Commun. 1*: 100-105.

Yang W., Feng, H., Zhang, X., Zhang, J., Doonan, J.H., Batchelor, W.D., Xiong, L., & Yan, J. (2020). Crop phenomics and high-throughput phenotyping: past decades, current challenges, and future perspectives. *Mol. Plant. 13*, 187–214.

Yang, W., Guo, Z., Huang, C., Duan, L., Chen, G., Jiang, N., Fang, W., Feng, H., Xie, W., ... Lian, X. (2014). Combining high-throughput phenotyping and genome-wide association studies to reveal natural genetic variation in rice. *Nat. Commun. 5*: 5087.

Yang, W., Guo, Z., Huang, C., Wang, K., Jiang, N., Feng, H., Chen, G., Liu, Q., & Xiong, L. (2015). Genome-wide association study of rice (Oryza sativa L.) leaf traits with a high-throughput leaf scorer. *J. Exp. Bot. 66*: 5605–5615.

Yano, K., Morinaka, Y., Wang, F., Huang, P., Takehara, S., Hirai, T., Ito, A., Koketsu, E., Kawamura, M., & Kotake, K. (2019). GWAS with principal component analysis identifies a gene comprehensively controlling rice architecture. *Proc. Natl. Acad. Sci. U S A. 116*: 21262–21267.

Yao, W., Li, G., Yu, Y., & Ouyang, Y. (2018). funRiceGenes dataset for comprehensive understanding and application of rice functional genes. *Gigascience. 7*: 1–9.

Yao, X., Wang, N., Liu, Y., Cheng, T., Tian, YC., Chen, Q., & Zhu, Y. (2017). Estimation of wheat LAI at middle to high levels using unmanned aerial vehicle narrowband multispectral imagery. *Rem. Sens. 9*: 1304.

Yazdanbakhsh, N., & Fisahn, J. (2009). High throughput phenotyping of root growth dynamics, lateral root formation, root architecture and root hair development enabled by PlaRoM. *Funct. Plant Biol. 36*: 938–946.

Yin, X., Liu, X., Chen, J., & Kramer, D.M. (2018). Joint multi-leaf segmentation, alignment, and tracking for fluorescence plant videos. IEEE Trans. *Pattern Anal. Mach. Intell. 40*, 1411–1423. doi: 10.1109/TPAMI.2017.2728065

Young, S.N., Kayacan, E., & Peschel, J.M. (2018). Design and field evaluation of a ground robot for high-throughput phenotyping of energy sorghum. *Precision Agric. 4*: 697–722.

Yue, J.B., Yang, G.J., Tian, Q.J., Feng, H.K., Xu, K.J., & Zhou, C.Q. (2019). Estimate of winter-wheat above-ground biomass based on UAV ultrahigh-ground-resolution image textures and vegetation indices. *ISPRS J. Photogramm. Remote Sens. 150*: 226–244.

Zappala, S., Helliwell, J.R., Tracy, S.R., Mairhofer, S., Sturrock, C.J., Pridmore, T., Bennett, M., & Mooney, S.J. (2013). Effects of X-ray dose on rhizosphere studies using X-ray computed tomography. *PLoS One. 8*: e67250.

Zarco-Tejada, P.J., Gonzalez-Dugo, V., & Berni, J.A.J. (2012). Fluorescence, temperature and narrow-band indices acquired from a UAV platform for water stress detection using a micro-hyperspectral imager and a thermal camera. *Remote Sens. Environ. 117*: 322–337.

Zarco-Tejada, P.J., Guillen-Climent, M.L., Hernandez-Clemente, R., Catalina, A., Gonzalez, M.R., & Martin, P. (2013). Estimating leaf carotenoid content in vineyards using high resolution hyperspectral imagery acquired from an unmanned aerial vehicle (UAV). *Agric. For. Meteorol. 171*: 281–294.

Zhang, H., Wang, X., Pan, Q., Li, P., Liu, Y., Lu, X., Zhong, W., Li, M., Han, L., ... Li, J. (2019a). QTG-Seq accelerates QTL fine mapping through QTL partitioning and whole-genome sequencing of bulked segregant samples. *Mol. Plant 12*: 426–437.

Zhang, J., Li, X.M., Lin, H.X., & Chong, K. (2019b). Crop improvement through temperature resilience. *Annu. Rev. Plant Biol. 70*: 753–780.

Zhang, X., Huang, C., Wu, D., Qiao, F., Li, W., Duan, L., Wang, K., Xiao, Y., Chen, G., ... Liu, Q. (2017). High-throughput phenotyping and QTL mapping reveals the genetic architecture of maize plant growth. *Plant Physiol. 173*: 1554–1564.

Zhang, Y., Ma, L., Pan, X., Wang, J., Guo, X., & Du, J. (2018). Micronscale phenotyping techniques of maize vascular bundles based on x-ray microcomputed tomography. *J. Vis. Exp. 140*: e58501.

Zhao, B., Zhang, J., Yang, C., Zhou, G., Ding, Y., Shi, Y., Zhang, D., Xie, J., & Liao, Q. (2018). Rapeseed seedling stand counting and seeding performance evaluation at two early growth stages based on unmanned aerial vehicle imagery. *Front. Plant Sci. 9*: 1362.

Zhao, C., Zhang, Y., Du, J., Guo, X., Wen, W., Gu, S., Wang, J., & Fan, J. (2019). Crop phenomics: current status and perspectives. *Front. Plant Sci. 10*: 714.

Zheng, H.B., Zhou, X., Cheng, T., Yao, X., Tian, Y.C., Cao, W.X., & Zhu, Y. (2016). Evaluation of a UAV-based hyperspectral frame camera for monitoring the leaf nitrogen concentration in rice. *IEEE Int. Geoscience Remote Sensing Symp., 10– 15 July, Beijing, China*, p. 7350–7353.

Zhou, J., Reynolds, D., Cornu, T.L., Websdale, D., ... Orford, S. (2017). CropQuant: an automated and scalable field phenotyping platform for crop monitoring and trait measurements to facilitate breeding and digital agriculture. *bioRxiv.* https://doi.org/10.1101/161547

Zhou, Y., Srinivasan, S., Mirnezami, S.V., Kusmec, A., ... Fu, Q. (2019). Semiautomated feature extraction from RGB images for sorghum panicle architecture GWAS. *Plant Physiol. 179*: 24–37.

Zuo, J., & Li, J. (2014). Molecular dissection of complex agronomic traits of rice: a team effort by Chinese scientists in recent years. *Natl. Sci. Rev. 1*: 253–276.

Chapter-5

Genome Editing for Crop Breeding

Abstract

Targeted editing of the genomes of living organisms not only permits investigations into the understanding of the fundamental basis of biological systems but also allows for improved productivity and quality of crops. This includes the creation of plants with valuable compositional properties and traits that confer resistance to various biotic and abiotic stresses. Recently, several novel genome editing systems have been developed, which include zinc finger nucleases (ZFNs), transcription activator-like effector nucleases (TALNEs), and clustered regularly interspersed short palindromic repeats/Cas9 (CRISPR/Cas9). These exciting new methods have proved themselves as effective and reliable tools for the genetic improvement of plants. The genome editing systems can also be used to exploit the genetic diversity present in the semi-domesticated and wild relatives of the cultivated crops by targeting homologous domesticated genes through allele mining. In this chapter, various tools available for gene editing, their merits, and demerits have been discussed.

Keywords: *Genome editing, CRISPR, base editing, gene expression, mutant screening, multiplexing.*

Introduction

Genome editing may be defined as a set of advanced molecular biological techniques that facilitates precise, efficient, and targeted modifications at specific loci in a genome. Use of transcription activators such as effector nucleases (TALENs), and zinc-finger nucleases (ZNFs) has been in use for such purposes, but came to the limelight recently due to the discovery of CRISPR (Clustered Regularly Interspersed

Short Palindromic Repeats) system, which is comparatively simple and easy to use for gene editing. The gene editing technologies basically use sequence-specific nucleases (SSNs) that can induce to identify specific DNA sequences and create double-stranded breaks (DSBs). Organisms internal repair systems fixes the DSBs either by homologous recombination (HR) or non-homologous end joining (NHEJ) mechanisms. The HR can cause gene replacements or insertions, while NHEJ can cause insertions or deletions of nucleotides, which leads to gene knockouts. Since gene editing technologies alter target design genetically they are considered to be different from genetic engineering technologies used for the development of genetically modified (GM) crops. After completion of editing the gene, it is not possible to distinguish between the edited gene and the naturally occurring mutants, in the segregating generations. Therefore, application of gene editing system should enhance the crop improvement process.

Zinc-Finger Nucleases

Zinc finger nucleases (ZNFs) is considered to be the first generation genome editing tools that was developed through chemically engineered nucleases, having folded up to ββα configuration. The enzyme is derived from the fusion of zinc-finger-based DNA-recognition modules and the DNA-cleavage domain of the *Fok*I restriction endonucleases (Figure 1a). The α-helix of the protein gets inserted into the DNA after binding of the protein into the major groove of the double helix of DNA. Each zinc finger recognizes and binds to a triplet nucleotide sequence and assembles into a group at the specific binding site(s). The monomer of ZFN has two different functional domains: an artificial ZF Cys2-His2 domain (N-terminal end) and a non-specific *Fok*I DNA cleavage domain (C-terminal end). Dimerization of the *Fol*I domain is critical for the enzymatic activity of ZEN. The individual zinc finger domains are interchangeable and according to the order of the domain, they can bind to specific site sequences in the genome. Several zinc finger domains capable of recognizing large numbers of triplet nucleotides have been generated to target large sequences of interest.

ZFNs have so far been used to modify rice, maize, soybean, rapeseed, tobacco, apple, fig petunia, and *Arabidopsis* (Bonawitz et al. 2018, Ran et al. 2018). Compared to other tools, ZFN is efficient, has high specificity, and minimal non-target effects. Although ZNFs have been used successfully for the development of herbicide tolerance, and stacking useful traits in maize and identification of safe regions for gene integration in rice, the technique has remained complicated and technically challenging for crop improvement. Therefore, the current focus is on improving the design and delivery system of ZNF technology for wider acceptability for crop improvement.

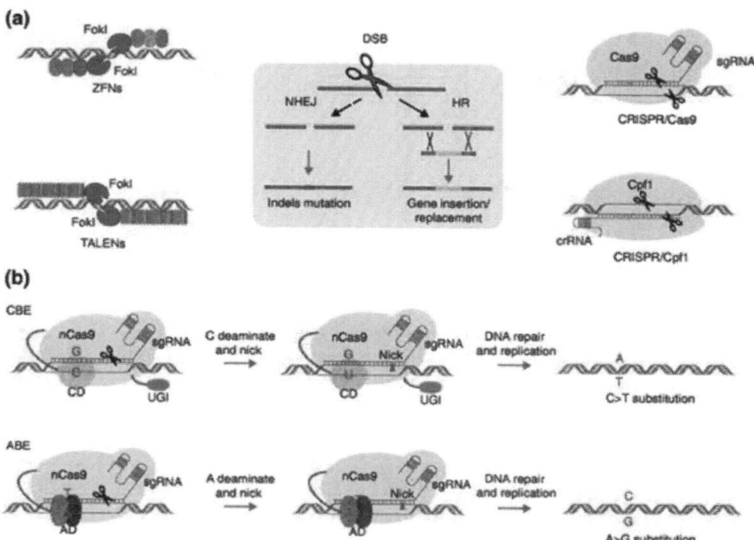

Figure 5.1. **(a) Genome editing tools and DNA repair mechanisms. ZFNs and TALENs on the left panel use FokI endonuclease to cut DNA double strands. Since FokI functions as a dimer, when two ZFNs or TALENs bind their targets and bring the FokI monomers into close proximity, cleavage occurs. CRISPR/Cas9 system on the right panel employs sgRNA for DNA binding and Cas9 protein for DNA cleavage. While CRISPR/Cpf1 system uses crRNA for DNA binding and Cpf1 protein for DNA cleavage. On the middle panel, when DSB was produced by genome editing techniques, the plant's endogenous repair systems fix the DSB by NHEJ or HR. NHEJ introduces small indels into the DSB and results in frame-shift mutations or premature stop codons. HR can cause gene replacements and insertions in the presence of a homologous donor DNA spanning the DSB. (b) Illustration of CRISPR/Cas9-mediated base editing. In the CBE system, nCas9 was fused to CD and UGI, and this complex could convert cytosine (C) in the targeting region to uracil (U), then U is changed to thymine (T) in DNA repair or replication processes, creating a C•G to T•A substitution. In the ABE system, nCas9 was fused to AD, and this system converts adenine (A) in the targeting region to inosine (I), which is treated as guanine (G) by polymerases, creating A•T to G•C substitutions. ABE adenine deaminases-mediated base editing, AD adenine deaminases, CBE cytidine deaminase-mediated base editing, CD cytidine deaminases, CRISPR clustered regularly interspaced short palindromic repeats, crRNA CRISPR RNA, DSB double-strand break, HR homologous recombination, nCas9 Cas9 nickase, NHEJ non-homologous end joining, sgRNA single-guide RNA, TALEN transcription activator-like effector nuclease, UGI uracil glycosylase inhibitor, ZFN zinc-finger nuclease. (Reproduced from Zhang et al. Genome Biology 2018, htttp;//creativecommons.org/ publication/zero/1.0/).**

Transcription Activator-Like Effector Nucleases

Transcription activator-like effector nucleases (TALENs) are a class of enzymes derived from the transcriptional activator-like effectors (TALE) repeats and the *Fok*I restriction enzyme (Figure 5.1a). The central domain of TALE protein is responsible for the binding of DNA and nuclear location signals. It also serves as the activator of transcription of the target gene. In the TALE monomer, the DNA-binding domain consists of a central repeat domain (CRD) which is responsible for the binding of DNA and host specificity. The CRD is made up of 34 amino acid residues that are arranged in tandem repeats. Each of the 34 amino acid long repeats binds to one nucleotide in the nucleotide sequence of the target gene. In the 34 amino acid repeats, two amino acids (located in 12 and 13 positions) are highly variable (called, repeat variable diresidue, RVD) and are responsible for the recognition of specific nucleotides. The last tandem repeat which binds to the nucleotide at the 3'-end of the recognition site has only 20 amino acid residues, and thus it is called half repeat. Although, TALE protein can be designated to bind any DNA sequence of interest, it is interesting to note that 5'-most nucleotide base of the DNA sequence should be thymidine, where TALE protein binds. Any deviation from this requirement can affect the efficacy of TALE recombinase (TALE-R), TALE transcription factor (TALE-TF), and transcription activator-like effector nucleases (TALENs). Since each TALE repeat recognizes and targets a single nucleotide, thereby increasing the number of potential targets and allowing more flexible target design, compared to ZFNs (Zhang et al., 2018).

For the creation of chimeric TALEN nuclease, plasmid vectors previously used the for creation of ZEN were used. The sequence encoding the DNA-binding TALE domain was inserted into such plasmid vectors, thereby creating a synthetic chimeric sequence-specific nuclease genetic construct. Such genetic construct has the DNA-binding domain of TALEs and the catalytic domain of restriction endonuclease *Fok*I. Through this process, artificial nucleases having DNA-binding domains and different repeat variable diresidues (RVDs) can be created, which can target any nucleotide sequence that may be of interest (Jung and Altpeter 2016).

The monomers with RVDs such as: Asparagine (Asn), and Isoleucine (IIe) (IN), Asparagine (Asn) and Glycine (Gly) (NG), Asparagine (Asn) and Asparagine (Asn) (NN), and Histidine (His) and Aspartic acid (Asp) (HD) bind to T, T, G and C bases, respectively. The first amino acid residue of all the RVDs, that is NI, NG, HD, and NN is involved in the stabilization of spatial conformation, although it does not directly bind to a nucleotide. The second amino acid residue binds to the nitrogenous bases of the nucleotide either through van der Waals force or through hydrogen bonding.

Through TALENs it is possible to introduce double stranded breaks in any location of the genome, provided that location carries the specific recognition sequence in the DNA-binding domain of TALEN. Another condition that has to be met is the presence of thymidine before 5'-end of the target sequence. However, it is possible to create mutants 5'-end thymidine constructs which can bind other nucleotides.

Genome of several plant species have been modified by using TALENs which include: rice, wheat, maize, barley, soybean, rapeseed, flex, sugarcane, tomato, tobacco, *Arabidopsis* and *Brachypodium* (Jung and Altpeter 2016, Kannan et al. 2018).

In 2012, the first TALEN-mediated application for genome editing was carried out in rice, where a susceptible gene (*OsSWEET14*) for bacterial blight was edited to produce a mutant rice cultivar having resistance to bacterial blight (Li et al. 2012). Thereafter, TALEN was used in wheat to edit three *TaMLO* homologue to produce a powdery mildew-resistant cultivar, and in maize by editing *GL2* gene to produce glossy phenotype with reduced epicuticular wax in the leaves making the plant with higher efficiency of scarification and cell wall composition. TALEN has been used to modify the nutritional profile of soybean, lower reducing sugar in potato, improve fragrance in rice, early flowering in *Brassica oleracea*, and high anthocyanin levels in tomato (Kannan et al. 2018). Although TALEN has the potential for its use in crop improvement, construction of TALE repeats has remained a challenge. Further, the efficiency of using TALEN for gene targeting remained variable.

Oligonucleotide-Directed Mutagenesis

In 2016, oligonucleotide-directed mutagenesis (ODM), a novel gene editing tool for plants has become a reality. In ODM, a specific 20 to 100-base-long oligonucleotide is used for targeted mutagenesis (Sauer et al. 2016). The sequence of this oligonucleotide is identical to the target sequence in the genome, except for a single base pair that is intended to e inserted in the genome to cause a mutation. Through this, the-directed editing of the gene (sequence) of interest is achieved. When the synthetic oligonucleotide having homology to a sequence of the targeted gene is exposed transiently in plant cells, it binds to the target sequence and activates the cells natural repair system. The repair system recognizes the mismatch (mutation) in the template DNA, and then copies the mismatch (mutation) into the target sequence. The desired targeted single nucleotide base editing is achieved through this process and the plant genome becomes capable of conferring novel functions or traits, while the synthetic oligonucleotide template is degraded. The plant cells with edited sequences were regenerated through tissue culture techniques, and varieties with desired traits are developed through traditional breeding methods.

CRISPR/Cas9 SYSTEM

CRISPR (Clustered regularly interspersed short palindromic repeats) were discovered in *E. coli* K12 in 1987 (Ishino et al. 1987). This unusual locus was found to include a set of short (29 nt) identical direct repeats that were randomly interspaced regularly by diverse short (32 nt) sequences. These loci were widespread in prokaryotes and often found to be adjacent to a set of CRISPR-associated sequences (*cas* genes).

Elements of CRISPR/Cas System

Two major sets of sequences constitute the CRISPR/Cas system: the repeat-spacer sequences and the *cas* genes (Figure 1b). The main defining feature of CRISPR locus is the short CRISPR repeat sequences. The length and sequences within the CRISPR locus are highly conserved. The sequence is partially palindromic and thus can form secondary hairpin structure.

The CRISPR repeats are separated by spacer sequences, the length of which are highly conserved, but the sequences are highly diverse. In fact, these sequences (called pro-spacers) are derived from foreign genetic elements such as viruses and plasmids, which invade the bacteria over time. Pro-spacer sequences are identified from the nucleic acids of the invaders that carry CRISPR-targeting short sequence motif called "pro-spacer adjacent motif" (PAM). These sequences are of about 2-5 nt recognition motif, essential for CRISPR-mediated interference. Pro-spacers are spliced from the invading nucleic acids and are specifically integrated at the "leader" end of the CRISPR-repeat sequence. The leader sequence is generally A-T rich and located upstream of the first CRISPR repeat, which has a promoter binding site for transcriptional regulators and can promote transcription of the repeat-spacer sequences.

In the majority of cases, CRISPR loci are relatively short, having about 30 repeats and 1.6 kb size. However, in some extreme cases, they may be quite large having hundreds of repeats. Usually, in most organisms, one or two CRISPR/Cas system exists. Although, in some exceptional cases more than two dozen CRISPR/Cas systems can exist.

In terms of occurrence, sequence, and number, *cas* genes are extremely diverse. Being a highly polymorphic protein family, they can carry out a wide variety of molecular functions. These include, recognition of RNA, binding with DNA, helicase, and nuclease activities.

Diversity of CRISPR/Cas System

Although CRISPR/Cas systems are widespread in bacteria and archaea, there exist extensive differences in their content and occurrences. Based on the phylogeny and functional roles at various stages of CRISPR-mediated immunity, three main types and ten sub-types of CRISPR/Cas systems have been identified (Table 5.1). The Type I system is found most frequently in bacteria, the Type II exclusively in bacteria, and Type III most frequently in archaea.

There exist two universally occurring *cas* genes, *cas*1 and *cas*2, which are found in three CRISPR/Cas types. In addition, each CRISPR/Cas type is associated with a unique gene that selectively occurs in a given type such as: *cas*3 with Type I, *cas*9 with Type II, and *cas*10 with Type III systems.

Table 5.1. Different types of CRISPR/Cas systems

Type	Gene	Sub-type	Host
I	Cas3	I-A	Sulfolobus solfataricus
		I-E	Escherichia coli
		I-F	Pseudomonas aeruginosa
II	Cas9	II-A	Streptococcus thermophilies
		II-B	Streptococcus pyogenes
III	Cas10	III-A	Staphylococcus epidermidis
		III-B	Pyrococcus furiosus

Mechanism of Action of Crispr-Mediated Interference

The CRISPR molecule is made up of short palindromic DNA sequences that are repeated along the molecule and are regularly spaced. Between these sequences are "spacers", foreign DNA sequences from organisms that have previously attacked the bacteria. The CRISPR molecule also includes CRISPR-associated genes or *cas* genes. These encode proteins that unwind and cut DNA, and are called helicase and nucleases, respectively.

The CRISPR immune system protects the bacteria from repeated virus attacks through three steps as described in the following section (Figure 5.2).

- **Adaptation**: when DNA from a virus invades the bacteria, the viral DNA is processed into short segments and integrated as a new spacer between the repeats. They serve as the genetic memory of the previous infections.
- **Production of CRISPR RNA (crRNA)**: The CRISPR sequence undergoes transcription, including spacers and *cas* genes, creating a single-stranded RNA. The resulting single-stranded RNA is called CRISPR RNA (crRNA), which contains copies of the invading viral DNA sequence in its spacers.
- **Targeting**: The CRISPR RNAs will identify viral DNA and guide the CRISPR-associated proteins (Cas nucleases) to them. The protein cleaves and destroys the targeted viral material.

It is possible to make use of the CRISPR/Cas9 systems recognition of specific DNA sequences in crop improvement programs. Instead of viral DNA as spacers, breeders can design their own sequences, based on their specific gene of interest. Once the sequence of the gene of interest is known, it can be easily used in CRISPR. It will then act just like a spacer for the system and guide the Cas9 protein (Cas nucleases) to a DNA matching sequence. The mechanism of gene-editing through CRISPR/Cas is described in Figure 5.3. Comparison of the parameters of different genome editing systems for plants is presented in Table 5.2.

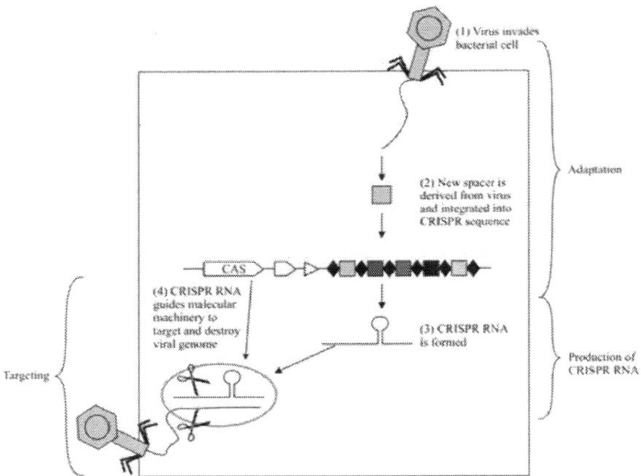

Figure 5.2. The steps of CRISPR-mediated immunity. **CRISPR loci contain clusters of repeats (black diamonds) and spacers (colored boxes) that are flanked by a "leader" sequence (L) and CRISPR-associated (cas) genes. During adaptation, new spacers derived from the genome of the invading virus are incorporated into the CRISPR array by an unknown mechanism. The synthesis of a new repeat is also required. During crRNA biogenesis a CRISPR precursor transcript is processed by Cas endoribonucleases within repeat sequences to generate small crRNAs. During targeting, the match between the crRNA spacer and target sequences (complementary protospacer) specifies the nucleolytic cleavage of the invading nucleic acid.**

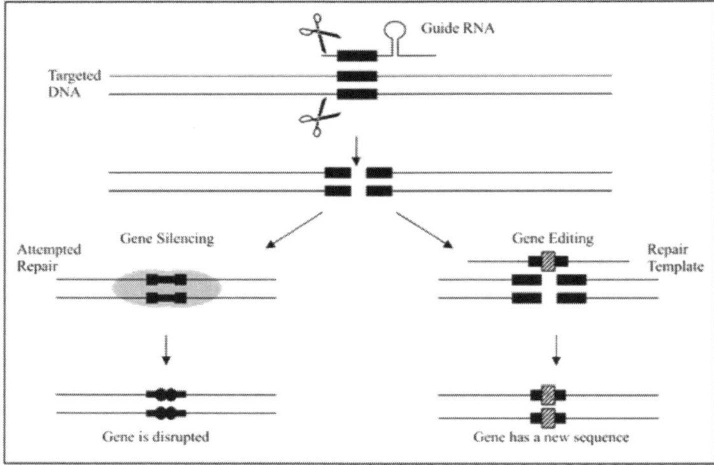

Figure 5.3. Mechanism of Gene Editing through CRISPR/Cas9 (For details see text)

Table 5.2. Comparison of the parameters of different genome editing techniques for plants

Parameters	ZFNs	TELENs	ODM	CRISPR/Cas9
Components	Nonspecific Fok1 nuclease, Zn finger domains	Nonspecific Fok1 nuclease domain, TALE DNA-binding domains	Exocuclease polynucleotide	crRNA, Cas9 protein
Catalytic domain	Restriction endonuclease Fok1	Restriction endonuclease Fok1	No catalytic domain	HNH and RUVC
Structural protein	Dimeric protein	Dimeric protein	Nonprotein in nature	Monomeric protein
Length of target sequence (bp)	24-36	24-39	64-88	20-22
Cloning	Required	Required	Not required	Not required
Protein engineering	Required	Required	Not required	Required to test gRNA
gRNA production	Not required	Not required	Not required	Required
Mode of action	DS-breaks in target DNA	DS-breaks in target DNA	Info. Strand directs conversion(s) within target region	DS-breaks or SS-nicks in target DNA
Target recognition efficiency	High	High	High	High
Mutation rate	High	Medium	Medium	Low
Large scale library creation	Not possible	Difficult	Difficult	Possible
Multiplexing	Difficult	Difficult	Difficult	Possible

Activities Carried Out Through CRISPR/Cas9

The following activities can be carried out through CRISPR.Cas9 system.

Gene Knock-Out

Gene silencing using CRISPR started with the use of a single guide RNA (sgRNA) to the target genes and initiated a double-stranded break using the Cas9 endonuclease. These breaks are then repaired by an innate DNA repair mechanism, the non-homologous end-joining (NHEJ). However, NHEJ is error-prone and results in genome deletions or insertions, which then translates into permanent silencing of the target gene.

DNA-Free Gene Editing

CRISPR can be used for DNA-free gene editing without the use of a DNA vector, requiring only RNA or protein components. A DNA-free gene editing system can be a good choice to avoid the possibility of undesirable genetic alternations due to the plasmid DNA integrating at the cut site or random vector integrations (Metje-Sprik et al. 2019).

Gene Insertion or "Knock-ins"

The CCRISPR-induced double-stranded break can also be used to create a gene "knock-ins" by exploiting the cells' homologous-directed repair. The precise insertion of a donor template can alter the coding region of a gene. Several studies have demonstrated that single-stranded DNA can be used to create precise insertions using CRas9 systCRISPRem.

Transient Gene Silencing

By modifying the Cas9 protein so that it cannot cut DNA, transient gene silencing or transcriptional repression can also be done. The modified Cas9, led by a guide RNA, targets the promoter region of a gene and reduces transcriptional activity and gene expression. Transient activation or up-regulation of specific genes can be effectively carried.

Technological Breakthroughs for Gene Editing

Genome editing technology has shown great potential for its use in crop improvement. However, the technology has several limitations such as low efficiency of HR, efforts on off-target, restrictive protospacer adjacent motif (PAM) sequences etc. To overcome these issues, several novel innovations have been introduced which are discussed in the following section.

Base Editing

In plants, the efficiency of homologous recombination (HR) of double-stranded breaks (DSBs) has been found to be much less template-free non-homologous end joining (NHEJ), thereby making it difficult to induce single nucleotide substitutions (Molla et al. 2008). On the other hand, studies on genome-wide association have shown that single-nucleotide changes are responsible for variations in elite traits in several crops. Therefore, the development of an efficient technique for precise point mutation is urgently required (Bharat et al. 2019).

Recently the CRISPR/Cas9-mediated gene editing technology has been developed, which can precisely convert one base to another, without using DNA repair template. In this technology, the Cas9 nickase (nCas9) or dead Cas9 (dCas9) fused to an enzyme having base conversion activity is used. For example, cytosine (C) gets converted to uracil (U) through cytidine deaminases, which can pair with adenine (A) and subsequently adenine (A) can pair with thymidine (T). Thus G:C gets converted to A:T (Figure 1b). Similarly, adenine (A) can be converted to inosine (I) through adenine deaminases, which can pair with cytosine (C). subsequently, C and pair with G, thereby convert A:T to G:C. Base editing through cytidine-deaminase-mediation (CBE) has been adopted in *Arabidopsis*, rice, wheat, maize, watermelon, and tomato to create among other things herbicide-resistant plants (Zong et al. 2017, Hu et al. 2018, Hua et al. 2018). Comparative representation of the mechanism of CRISPR/Cas9 and base-editing system is presented in Figure 5.4 (Mishra et al., 2020).

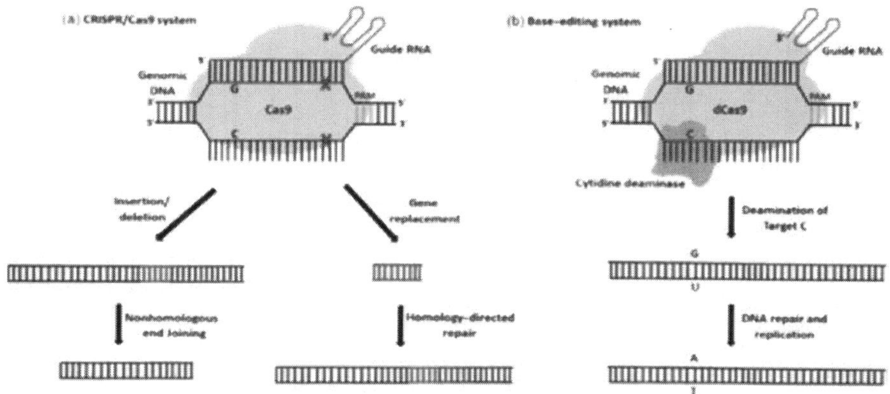

Figure 5.4. Comparative representation of the mechanism of CRISPR/Cas9 and a base-editing system. (a) In a CRISPR/Cas9 system, the Cas9-sgRNA complex moves along the DNA strand and makes a double-stranded break (DSB) where the Cas9 encounters the appropriate protospacer adjacent motif (PAM) and the sgRNA matches the target DNA sequence. These DSBs are subsequently repaired either by nonhomologous end-joining (NHEJ) or by homology-directed repair pathway (HDR). (b) In a base-editing system, a catalytically dead Cas9 endonuclease (dCas9) fused to a catalytic cytidine deaminase domain is guided by a sgRNA molecule to make single-base substitutions without creating a double-stranded break in the DNA. (Reproduced from Mishra et. al. Plant Biotechnology Journal, 18, 2020. doi:10.1111/pbi.13225).

Due to non-availability of naturally occurring enzyme which can catalyze adenine deamination in DNA, the adenine-deaminase-mediated base editing (ABE) is more complicated. In 2017, an efficient ABE technology has been developed by using several rounds of directed evolution and protein engineering. Thereafter, ABE has been used in *Arabidopsis*, rice and wheat to produce point mutations with altered desirable phenotypes.

CBE is also used to generate nonsense mutations to disrupt or knockout expression of gene of interest. CBE has been found to be more specific than conventional SSN-mediated knockout method. Thus, CBE has provided new dimension for gene editing and broaden its potential applications for solving various challenges of crop improvement (Yan et al. 2018). The characteristics, catalytic window and functions of base editors have been presented in Table 5.3.

Table 5.3. Characteristics, catalytic widow, and functions of base editors.

Base editors	Characteristics	Types of base substitutions	Catalytic window
DNA base editor			
BE1	(APOEC1-XTEN-dCas9): Composed of a cytidine deaminase enzyme APOBEC1 (from rats) linked to a catalytically dead Cas9 (dCas9) by a amino acid XTEN linker	C to T	-17 to -13
BE2	(APOBEC-XTEN-dCas9-UGI): UGI is fused to the C terminus of BE1	C to T	-17 to -13
BE3	(APOEC-XTEN-Cas9-GI): Rapobec1 fused to the N-terminus of nickase Cas9 D10A through a 16-amino acid XTEN linker and a UGI fused to the C-terminus by a 4-amino acid linker	C to T	-16 to -12
BE4	Composed of rAPOEC1 fused to Cas9D10A through a 32-aa linker and two UGI molecule are linked to both C and N terminal of Cas9 nickase by a 9-aa linker	C to T	-17 to -13
ABE	TadA is fused to a catalytically impaired CRISPER/Cas9 mutant	A to G	-17 to -14
CRISPR-X	dCas9 is used to target a hyperactive AID variant to induce localized, diverse point mutations. The sgRNA backbone contains two MS2 RNA hairpins that each recruit two MS2 proteins fused to AID	C to T	-50 to +50
TAM	dCas9 is fused to human AID, co-expressed with UGI	C to T	-16 to -12
S a B E 4 - GAM	Gam protein fused to Staphylococcus aurea Cas9-derived BE4	C to T	-19 to -9
YEE-BE3	(W90Y+R126E+R132E): triple mutant	C to T	-15 to -13
RNA base editor			
ADAR	Catalytically inactive Cas13 (dCas13) is fused to a naturally occurring ADAR (adenine deaminase acting on RNA)	A to I	-50 to +50

Cytosine Base Editors

Cytosine base editors are the vectors that catalyze the conversion of cytosine to thymine. The enzyme cytidine deaminase removes an amino group from cytosine thereby converting it to uracil, which results into a U-G mismatch. This mismatch gets resolved through a DNA repair mechanism to form U-A base pairs. Thereafter, a A-T base pair is formed in the newly synthesized strand. This results in the conversion of C-G to A-T in a programmable manner (Figure 5.5). The first-generation base editor (BE 1) was developed by David Liu and coworkers, in 2016. It was composed of an enzyme named cytidine deaminase APOBEC 1 (from rats) linked to a dCas9 by a 16 amino acid XTEN linker. The XTEN is a peptide that links the two proteins and maintains a balance between them. A group of naturally occurring cytidine deaminases such as, apolipoprotein B mRNA editing enzymes and catalytic polypeptide-like (APO-BEC) enzymes, found in vertebrates, acts to protect them from invading viruses. These enzymes act on single-stranded DNA/ RNA as substrates (Mishra et al., 2020).

One of the limitations in BE1 was the frequent removal of uracil by uracil DNA glycosylase (UDG) which results in low editing efficiency. To overcome this limitation, a series of improved base editors were developed. The second generation base editor -BE2 (APOBEC-XTEN-dcas9-UG) was developed by adding a uracil DNA glycosylase inhibitor (UGI) to the C terminal end of the DNA targeting module. Addition of UGI leads to inhibition of the activity of UDG that catalyzes the removal of U from DNA in cells and could initiate base excision repair (BER) pathway. UGI is an 83-residue protein derived from *Bacillus subtilis* bacteriophage PS1 that can block UDG activity in human cells. This inhibition of BER leads to a threefold increase in the editing efficiency in human cells. Subsequently, BE3 base editor was developed, which was composed of rAPOBC1 fused to the N-terminus of nickase cas9 D10A through a 16-amino acid XTEN linker and a UGI fused to the C terminus by a 4-amino acid linker (Figure 5.6). The major improvement in BE3 was the replacement of dCas9 with Cas9 nickage (nCas9), which nicks the strand opposite to the deaminated cytidine (Mishra et al., 2020).

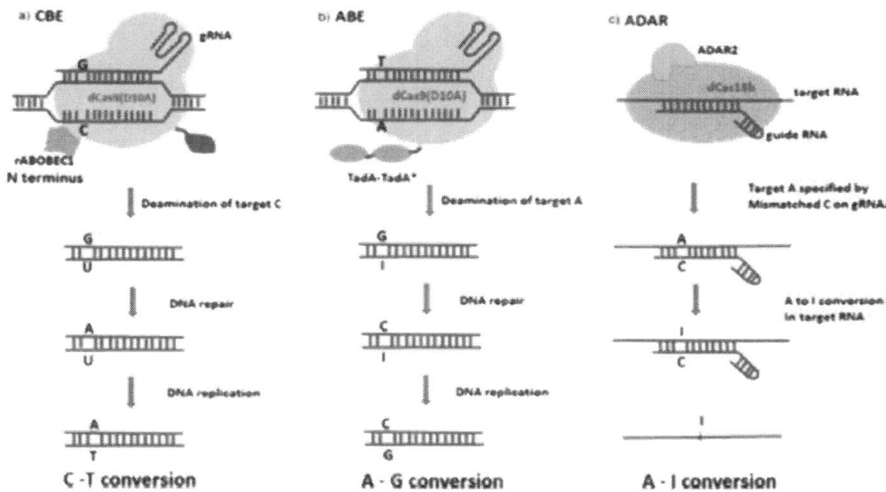

Figure 5.5. A comparison of three different approaches of base editing. (a) CBE – mediated base-editing strategy results in C-T conversions. (b) ABE –mediated base-editing strategy results in A-G conversions. (c) ADAR – mediated RNA base-editing results in A-1 conversion. (Reproduced from Mishra et. al. Plant Biotechnology Journal, 18, 2020. doi:10.1111/pbi.13225)

To increase the base-editing efficiency, fourth-generation base editors BE4 (derived from *S. pyogenes* Cas9p base editor) and SaBE4 (derived from *S. auteus* Cas9 base editor 4) were developed by linking rAPOBEC1 to Cas9D10A through a 32-aa linker and fusing two UGI molecules to both N and C terminals of Cas9 nickase through a 9-aa linker. By using the UGI it is possible to block the access of UNG to the uracil intermediate, inhibiting BER, and thereby minimizing the formation of undesired by-products. In addition, a double-stranded DNA end-binding protein Gam (derived from bacteriophage Mu), was fused to the N terminal end of a Cas9 nickage that binds to the free ends of DSB and reduces indel formation during base-editing process, which improves purity of the product. Both E4-Gem and SaE4-Gam showed decrease in non-T product formation and increase C to T editing efficiency compared to BE4 and SaBE4.

Apart from introducing point mutation in a precise and programmable manner, deaminases are used to create a diverse library of point mutations localized to a targeted region of the genome (Mishra et al. 2020).

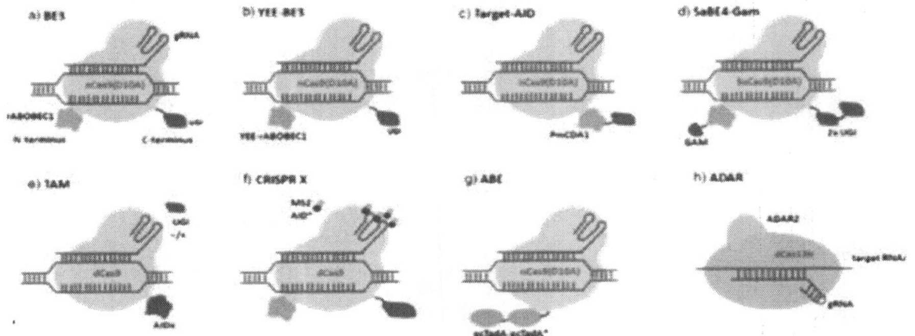

Figure 5.6. Structural representation of base-editing platforms: (a) BE3 employs Cas9 nickase (nCas9D10A) along with a cytidine deaminase rAPOBEC1(orange) and an uracil DNA glycosylase inhibitor (UGI) (Green). (b) YEE-BE3 employs YEE-rAPOBEC1. (c) Target-AID employ PmCDA1 (d) SaBE4-gam employs SaCas9D10A, 2 9 UGI and has a Gam protein (red) fused to its terminus. (e) and (f) The TAM and CRISPR-X systems used dCas9 to recruit variants of the deaminase AID (AIDx or MS2-AID*D). (g) ABE is composed of ecTadA (WT)-ecTadA* (7.10) heterodimer fused to Cas9n. (h) Catalytically inactive Cas13 (dCas13) is fused to a naturally occurring ADAR2 (adenosine deaminase acting on RNA). (Reproduced from Mishra et. al. Plant Biotechnology Journal, 18, 2020. doi:10.1111/pbi.13225)

Adenine Base Editors

Capabilities of base editing were further expanded by the development of a class of AEEs that could modify adenine bases. Adenine DNA deaminases do not occur in nature. Li et al (2017) developed ABEs by using *E. coli* TadA through protein engineering and directed evolution technique (Figure 6). The *E. coli* TadA is a tRNA adenine deaminase that can convert adenine to inosine in the single-stranded anticodon loop of tRNA Arg. It also shares homology with APOBEC enzyme. Development of the first-generation ABEs was made by fusing a TadA with a catalytically impaired CRISPR/Cas9 mutant. Seventh-generation ABEs (ABE7.10) were developed for conversion of A-T to G-C in a wide range of targets with increased efficiency and product quality.

RNA Base Editors (ADAR)

Zhang et al. (2017) developed RNA base editors by using a catalytically inactive Cas13 (dCas13) and a naturally occurring adenosine deaminase acting on RNA (ADAR) to direct the conversion of adenosine to inosine in mammalian cells. Cas13 is a type VI CRISPR-associated RNA-guided RNAs with abilities to bind RNA. The adenosine deaminase while acting on RNA (ADAR) family of enzymes, mediates endogenous editing of transcripts via hydrolytic deamination of adenosine to inosine. These enzymes can edit RNA precisely. The system used to edit RNA transcript is called RNA Editing for Programmable A to I Replacement (REPAIR). REPAIR is considered as a promising RNA editing platform having broad applicability for research, therapeutics and biotechnology (Figure 9.6).

Application of Base Editors in Crop Improvement

Cytosine and adenine base editors have been used successfully to edit specific genes in wade range of crop plants, including rice, maize, wheat, tomato, cotton, and watermelon (Table 5.4). In 2017, multiple herbicide resistance point mutations have been introduced through multiplex base editing, in rice (Yan et al. 2018). Genetic variations were also induced in rice by using a CRISPR/Cas9 toolkit comprised of rBE3 and rBE4 (rice base editors). Subsequently, rBE5 was used to target *Pi-d2* gene in rice, which harbor a point mutation modulating defense response to blast fungus. Recently, Zong et al. (2018) developed a new plant base editor, A3A-PBE by using human APOBEC3A, fused to Cas9 nickage to further enhance the base-editing efficiency in plants. The third-generation base editors, BE3, are used successfully to create C to A substitutions in various crop plants. However, the editing was possible for 5 nucleotide sequences only and thus editing activity was low. To overcome this, the previous base editor nCas9-PBE was modified to create A3A-PBE, in which the rat APOBEC1 was replaced with human APOBEC3A, and the codons were optimized for cereals. The efficiency of A3A-PBE (in terms of C to T conversion) was found to be increased by more than 13 percent compared nCas-PBE, in rice and wheat genes. Further, the editing was possible for 17 nucleotides and which generate a low frequency of undesirable on-target indels.

Like CBEs, ABEs have also been successfully used for base editing in different crop plants. The ABEs developed for mammalian cells have been well adapted and optimized for editing plant cells for creating point mutations. A modified version of ABE7-10 (for mammalian cells), named ABE-P1 (ABE plant version 1) has been found to be highly efficient ABE for conversion of A-T to G-C in a programmable manner in rice. In 2018, Yan and coworkers developed a fluorescence-tracking ABE by using *E. coli* TadA variants and Cas9 variants for the conversion of A to G. Application of this variant not only helped to induce A to G conversion but also made it convenient to select the base-edited plants through detection of fluorescence, in rice.

Li et al. (2019) synthesized rice codon-optimized ABE-nCas9 tool to induce targeted A-T to G-C point mutation in the rice genome. They cloned the rice-optimized ecTasA XTEB+N-TadA*7.10 into pHUN411 binary vector under the control of a maize ubiquitin promoter, and the rice amylose synthesis gene *Wx* was targeted by the vector. This system was found to be very efficient in converting A-T to G-C without any off-target mutations.

Table 5.4. Genes targeted by adenine and cytidine base editors in different crops.

Crops	Gene Targeted	Base Editors Used	Functions
Rice	ALS	CBE	Herbicide resistance
	C287	CBE	Herbicide resistance
	GL2IOsGRF4	ABE	Grain size and yield
	NRT1, 18, and SLR1	CBE	Enhance nitrogen use efficiency
	Pi-d2	CBE	Blast resistance
	OsACC-T1	ABE	Herbicide resistance
	OsCDC48	CBE	Regular Senescence and death
	OsMPK6	ABE	Pathogen response gene
	OsRLCK185, OsCERK1	CBE	Defense response
	OsPDS, OsSBEIIb	CBE	Nutritional improvement
	OsSPL14	CBE ABE	Herbicide resistance Plant architecture and grain yield
	SLR1	ABE	Dellla protein for plant height
	Wx	ABE	Amylose synthesis
Wheat	TaLOX2	CBE	Lipid metabolism
	TaDEP1, TaGW2	ABE	Panicle length and grain weight
Maize	ZmCENH3	CBE	Chromosomal segregation
Potato	StALS, StGSS	CBE	Herbicide resistance, Starch synthesis
Tomato	SLALS1	CBE	Herbicide resistance
Watermelon	ALS	CBE	Herbicide resistance
ABE= adenine base editor, CBE=cytidine base editor			

DNA-Free Genome Editing Systems

In a conventional genome editing system, the DNA cassette having edited genetic components is delivered and integrated into the host genome. Since integration of the transgene occurs at random, it may generate undesirable genetic changes. In case the DNA cassette is degraded, the resulting fragments may get integrated into the host genome thereby producing undesirable effects. Expression of genome editing tools for long periods results in off-target effects in plants, due to the presence of abundant nucleases. Integration of foreign DNA into the plant genome may also affect the regulatory mechanisms in GM plants. Therefore, DNA-free genome editing is a revolutionary technology, as the risk of producing undesirable off-target mutations can be reduced substantially.

Success in using a DNA-free genome editing system has been accomplished through both protoplast-mediated transformation and particle bombardment. In 2015, the first DNA-free genome editing in plants was achieved by Woo and his colleagues. They transferred CRISPR/Cas9 ribonucleoproteins (RNPs) into protoplasts of *Arabidopsis*, lettuce, rice, and tobacco. Thereafter, targeted mutations were produced in grapes and apples by delivering purified CRISPR/Cas9 RNPs into protoplasts. Since efficient, regenerable protoplast systems are not available for many important crops, the technology is not getting the required boost.

Particle bombardment-mediated DNA-free genome editing has been accomplished in maize and wheat. Genome-edited plants have been produced by delivering CRISPR/Cas9 RNA and CRISPR/Cas9 RNPs into wheat embryos by particle bombardment. Similarly, by delivering CRISPR/Cas9 RNPs through particle bombardment, knockout mutants were produced in maize. Application of CRISPR/Cas9 RNPs has been found to cause few off-target effects in plants compared to CRISPR/Cas9.

Zong et al. (2018) have described a gene editing system by combing base editing and DNA-free genomic editing in wheat. They achieved a base conversion frequency of 1.8% for C-to-T. Such developments generate a lot of promise for crop improvement and commercialization of edited plants.

CRISPR/Cpf1 System

The type II CRISPR/SpCas9 system although efficient and simple, it can only recognize DNA sequences upstream of the 5'-NGG-3' PAMS. Therefore, it restricts it application to any potential target sites. The type V CRISPR/Cpf1 system has demonstrated its potential to overcome this problem. Unlike SpCas9 which creates blunt ends, the Cpf1 recognizes T-rich PAMs and generates cohesive ends with 4 to 5 nucleotide overhangs (Figure 1a). Recently, Cpf1 derived from *Francisella novicida* (FnCpf1) has been used successfully for targeted mutagenesis in rice and tobacco, and the Cpf1 ortholog from *Lachnospiraceae* bacterium (LbCpf1) induced targeted mutations in rice.

The FnCpf1 and LbCpf1 nucleases can generate precise gene insertions at the target site in rice at a much higher frequency than most other genome-editing nucleases. To expand the scope of CRISPR/Cpf1-mediated genome editing system, a variant (LbCPF1-RR) has been developed that enables the multiplex editing of the target genes having TYCV PAMs.

The CRISPR/Cpf1-mediated DNA-free genome editing system can be combined with base editing and DNA-free genome editing systems to achieve specific objectives in crop improvement programs. Recently, CRISPR/Cpf1-mediation and DNA-free genome editing system has been combined to achieve specific objectives in rice.

A variant of Cpf1, named as Cas12a, has been developed which differs from classical CRISPR/Cas system in several aspects: (1) the nucleases are smaller ranging from 135 to 158 kDa, (2) a naturally occurring single guide RNA is present in the system, (3) cutting of Cas12a results into staggered cuts Cas9 cutting with blunt ends, (4) for Cas12a system the protospacer adjacent motifs has to be rich in thymidine, whereas it should be rich in guanine for Cas9, and (5) from the recognition site of Cas12a, the DNA is cut distal site, whereas DNA is cut at proximal site by Cas9.

Prime Editing Technology

Anzalone et al. (2019) from Brood Institute of MIT and Harvard have developed a new CRISPR genome editing system by combining CRISPR/Cas9 and a reverse transcriptase into a single system. They called this system "prime editing", which can edit the human genome very efficientiy with precision, in a highly versatile fashion. The technique is so efficient that it can rectify up to 89 percent of known disease-causing genetic variations.

Prime editing differs from earlier gene editing systems in that it uses RNA to direct the insertion of new DNA sequences into human genome. This new RNA is called prime editing guide (pegRNA), which directs a catalytically impaired Cas9 protein, fused to reverse transcriptase to the target site. Unlike double-stranded breaks created by CRISPR/Cas9 system, the modified Cas9 cuts the DNA only in one strand. The pegRNA contains additional RNA nucleotides which encodes the new edited sequence. The reverse transcriptase reads the RNA extensions and then incorporates the corresponding DNA nucleotides into the target site.

The "prime editing" has shown lot of promise and the technology is available freely to all the researchers and non-profit organizations. Thus it will be interesting to see how the will take advantage of this technology for crop improvement.

CRISPR/Cas9 for Crop Improvement

CRISPR/Cas9 has played a huge part in the increase in genome editing studies in recent years. The system has broad applications in plant and animal improvement, as well as in the medical field. As a relatively young technique, various discoveries and innovations for its efficient use in wider applications are in the offing. Researchers have found that the CRISPR/Cas9 system can be applied to nearly every organism.

CRISPR, derived from *Streptococcus pyogenes* has been developed as a versatile genome editing technology for a wide range of applications. Compared to ZENs and TALENs, the CRISPR system has been found to be simple, efficient, low cost, and capable of targeting multiple genes. Therefore, CRISPR has become the choice of breeders for crop improvement. Accordingly, CRISPR has been used to address a variety of issues in breeding several crop plants including rice, wheat, maize, barley, sorghum, soybean, potato, tomato, rapeseed, flex, cotton, cucumber, lettuce, grapes, grapefruit, apple, orange, watermelon and *Camelina* (Qi et al. 2016, Braatz et al. 2017, Jacob et al. 2017, Ali et al. 2018, Anderson et al. 2018, Curtin et al. 2018, He et al. 2018, Huang et al. 2018, Li et al. 2018a,b, Okuzaki et al. 2018, Zhang et al. 2019). The application of CRISPR has been mostly to induce null alleles by knocking the gene. This can be carried out by introducing small indels into the target gene to create frame-shift mutation or by introducing stop codons to interrupt mRNA prematurely (Figure 5.9.1a). Both gene knockout mutants and insertion and replacement mutants have been produced through genome-editing technologies in different plant species. Mutants thus produced are useful for the improvement of several crop plants (Table 5.5). CRISPR/Cas9 system has been adopted for gene editing for the improvement of several traits, such as yield, plant architecture, plant aesthetics, and disease resistance in important crop plants.

Table 5.5. Specific traits of crop plants improved by genome-editing technique

Plants	Target gene	GE Technique	Type of DNA repair	Trait modified
Brassica oleracea	FRIGIDA	TALENs	NHEJ	Early flowering
Camelina sativa	FAD2	CRISPR/Cas9	NHEJ	Decreased poly-unsaturated fatty acids
Cassava	EPSPS	CRISPR/Cas9	HR	Herbicide resistance
Citrus paradise	CsL0B1	CRISPR/Cas9	NHEJ	Citrus canker resistance
Cucumber	eiF4E	CRISPR/Cas9	NHEJ	Virus resistance
Cucumis sativus	EIF4E	CRISPR/Cas9	NHEJ	Broad virus resistance
Grape	VvWRKY52	CRISPR/Cas9	NHEJ	Botrytis cinerea resistance
Grapefruit	CsLOB1 promoter	CRISPR/Cas9	NHEJ	Alleviated citrus canker
Gossypium hirsutum	Gh14-3-3d	CRISPR/Cas9	NHEJ	Verticillium wil resistance
Grapefruit	CsLOB1	CRISPR/Cas9	NHEJ	Citrus canker resistance
Maize	ZmTLP	ZFNs and CRISPR/Cas9	HR	Herbicide tolerance
Maize	ZmTLP	ZFNs	HR	Trait stacking
Maize	ZmGL2	TALENs	NHEJ	Reduced epicuticular wax in leaves
Maize	ZmMTL	TALENS	NHEJ	Induction of haploid plant
Maize	Wx1	CRISPR/Cas9	NHEJ	High amylopectin content
Maize	TMSS	CRISPR/Cas9	NHEJ	Thermosensitive male-sterile
Maize	ARGOS8	CRISPR/Cas9	HR	Draught stress tolerance
Maize	ALS	CRISPR/Cas9	HR	Herbicide resistance
Maize	ZmHKT1	CRISPR/Cas9	NHEJ	Salinity tolerance
Maize	PPR, RPL	CRISPR/Cas9	NHEJ	Reduced zein protein
Mushroom	PRO	CRISPR/Cas9	NHEJ	Anti-browning
Orange	CsLOB1	CRISPR/Cas9	NHEJ	Citrus canker resistance
Potato	Vinv	TALENs	NHEJ	Minimizing reducing sugar

Plants	Target gene	GE Technique	Type of DNA repair	Trait modified
Potato	Wx1	CRISPR/Cas9	NHEJ	High amylopectin content
Potato	ALS	CRISPR/Cas9	HR	Herbicide resistance
Potato	CP and Rep sequences	CRISPR/Cas9	NHEJ	Yellow leaf curl resistance
Potato	SIJAZ2	CRISPR/Cas9	NHEJ	Bacterial speck resistance
Potato	SIMIo1	CRISPR/Cas9	NHEJ	Powdery mildew resistance
Potato	SINPR1	CRISPR/Cas9	NHEJ	Drought tolerance
Potato	GBSS	CRISPR/Cas9	NHEJ	Increased amylopectin/ amylose
Rice	OsQQR	ZFNs	HR	Trait stacking
Rice	OsSWEET14	TALENs	NHEJ	Bacterial blight resistance
Rice	OsBADH2	TALENs	NHEJ	Fragrant rice
Rice	LAZY1	CRISPR/CAS9	NHEJ	Tiller spreading
Rice	Gn1a, GS3,DEP1	CRISPR/Cas9	NHEJ	Enhance grain number. Large grain size, dense erect panicle
Rice	SEIIb	CRISPR/Cas9	NHEJ	High amylose content
Rice	OsERF922	CRISPR/Cas9	NHEJ	Rice blast resistance
Rice	OsSWEET13	CRISPR/Cas9	NHEJ	Bacterial blight resistance
Rice	OsMATL	CRISPR/Cas9	NHEJ	Induction of haploid plants
Rice	ALS	CRISPR/Cas9	HR	Herbicide resistance
Rice	EPSPS	CRISPR/Cas9	NHEJ	Herbicide resistance
Rice	eIF4G	CRISPR/Cas9	NHEJ	Tungro virus resistance
Rice	OsNAC041	CRISPR/Cas9	NHEJ	Salinity tolerance
Rice	OsOTS1	CRISPR/Cas9	NHEJ	Salinity tolerance
Rice	OsOTS2	CRISPR/Cas9	NHEJ	Salinity tolerance
Rice	OsAnn3	CRISPR/Cas9	NHEJ	Cold tolerance
Rice	SAPK2	CRISPR/Cas9	HDR	Drought and salinity tolerance
Rice	OsMPK2 OsPDS OsBADH2	CRISPR/Cas9	HDR	Multiple stress tolerance
Rice	OsAAP3	CRISPR/Cas9	NHEJ	Grain yield
Rice	OsCCD7	CRISPR/Cas9	NHEJ	High tillering
Rice	GW5	CRISPR/Cas9	NHEJ	Grain weight
Rice	Hd2, Hd4, Hd5	CRISPR.Cas9	NHEJ	Early heading
Rice	OsSWEET11	CRISPR/Cas9	NHEJ	Grain weight
Rice	OsGRF4	CRISPR/Cas9	NHEJ	Grain size
Rice	IPA, GS3, DEP1, Gn1a	CRISPR/Cas9	NHEJ	Improved yield
Rice	GS3, GW2, GW5, TGW6	CRISPR.Cas9	NHEJ	Grain weight
Rice	SBEIIb	CRISPR/Cas9	NHEJ	Amylose, starch resistance
Rice	OsNAC14	CRISPR/Cas9	NHEJ	Draught tolerance
Rice	SAPK1, SAPK2	CRISPR/Cas9	NHEJ	Salinity tolerance

Plants	Target gene	GE Technique	Type of DNA repair	Trait modified
Rice	Waxy	CRISPR/Cas9	NHEJ	Enhanced glutinosity
Sugarcane	COMT	TALENs	NHEJ	Improved cell wall composition
Sugarcane	COMT	TALENs	NHEJ	Improved saccharification efficiency
Soybean	FAD2-1A, FAD2-1B	TALENs	NHEJ	High oleic content
Soybean	FAD2-1A, FAD2-1B, FAD3A	TALENs	NHEJ	High oleic, low linoleic content
Soybean	ALS	CRISPR/Cas9	HR	Herbicide resistance
Soybean	Drb2a, Drb2	CRISPR/Cas9	NHEJ	Draught and salt tolerance
Soybean	GmFT2	CRISPR/Cas9	NHEJ	Delayed flowering
Soybean	FAD2-1A, FAD2-1B	CRISPR/Cas9	NHEJ	Improved oil quality
Tomato	ANT1	TALENs	HR	High anthocyanin
Tomato	SIMLO1	CRISPR/Cas9	NHEJ	Powdery mildew resistance
Tomato	SIJAZ2	CRISPR/Cas9	NHEJ	Bacterial speck resistance
Tomato	SP5G	CRISPR/Cas9	NHEJ	Early harvest time
Tomato	SIAGL6	CRISPR/Cas9	NHEJ	Parthenocarpy
Tomato	IncRNA1459	CRISPR/Cas9	NHEJ	Long self-life
Tomato	SO, SP5G, CLV3, WUS, GGP1	CRISPR/Cas9	NHEJ	Tomato domestication
Tomato	SIMAPK3	CRISPR/Cas9	NHEJ	Drought tolerance
Tomato	SICBF1	CRISPR/Cas9	NHEJ	Cold tolerance
Tomato	SGR1, LCY-E, BIc, LCY-B1	CRISPR/Cas9	NHEJ	Increased lycopene
Tomato	SIGAD2, SIGAD3	CRISPR/Cas9	NHEJ	Enhanced γ-aminobutyric acid
Wheat	EDR1	CRISPR/Cas9	NHEJ	Powdery mildew resistance
Wheat	GW2	CRISPR/Cas9	NHEJ	Increased grain weight and protein content
Wheat	TaDREB2, TaDREB3	CRISPR/Cas9	NHEJ	Draught tolerance
Wheat	TaGW2	CRISPR/Cas9	HR	Grain weight
Wheat	α-gliadin	CRISPR.Cas9	HR	Low gluten
GE Technique= Genome editing technique, HR= Homologous recombination, NHEJ= Non-homologous end joining, TALEN= Transcription activator-like effector nuclease, ZFN= Zinc-finger nuclease, CRISPR= Clustered regularly interspersed short palindromic repeats.				

Usually, cells grown in culture are used for genetic modification through the CRISPR system. The viral vector or plasmid having high and stable synthesis of CRISPR/Cas9 system elements was introduced into the plant cells. Alternatively, cultured protoplasts and a plasmid containing CRISPR/Cas9 elements are used to obtain genetically modified plants. The third approach could be use of *Agrobacterium* sp. having a plasmid that contains CRISPR/Cas9 system.

CRISPR has been used successively in rice for increasing grain number, dense erect panicle, large grain size, overall yield, resistance against blast and bacterial blight, and resistance to herbicide. Ying Wang and coworkers have designed several CRISPR sgRNAs and successfully deleted fragments of the dense and erect panicle1 (DEP1) gene in the *Indica* rice line IR58025B. Improvements in yield-related traits, such as dense and erect panicles and reduced plant height, were observed in the mutant plant produced.

CRISPR/Cas9-mediated gene editing can be used to disrupt disease-causing genes, known as "S-gene" for the development of disease-resistant plants. in rice, ethylene-responsive gene *OsERF922* was generated through CRISPR/Cas9 tool, which showed reduced blast lesions and increased resistance against blast disease. Similarly, bacterial blight-resistant plants were produced by targeted mutagenesis of *SWEET13* gene. Peng and coworkers have the edited effector binding elements (EBEs) by CRISPR/Cas9 system in the *CsLOB1* gene promoter region, thereby increased disease resistance in *Citrus sinensis* against *Xanthomonas citri*.

CRISPR technology was also used successively to develop resistance to powdery mildew in wheat and tomato, improved self-life and heat stability in soybean, high amylopectin in maize, resistance to ipomovirus (cucumber vein yellowing virus) and papaya ring spot mosaic virus-W in cucumber, draught resistance in maize, and herbicide resistance in flex.

Shah and coworkers edited the *TaMLO* gene through CRISPR/Cas technique and produced wheat lines resistant to powdery mildew disease caused by *Blumeria graminis.* Similarly, wheat and tomato lines having resistance against powdery mildew were developed through CRISPR/Cas9-mediated multiplex gene editing of *EDR1* gene and *MLO* gene, respectively.

In hexaploid wheat, geminiviral-based DNA replicons was utilized for transient expression of the CRISPR/Cas9 system against wheat dwarf virus (WDV), and 12 fold up-regulation was observed in ubiquitin gene expression. Stable over-expression of sgRNA and Cas9 that particularly target Gemini-virus genome to prevent its growth has been adopted to develop virus resistant plants. On the other hand, the CRISPR/Cas9 system has been used to mutate viral genomes to reduce infection caused by them. Recently, a new ortholog of Cas9 has been discovered in *Francisella novicida* (FnCas9) to edit RNA virus genome. FnCas9 can inhibit the replication of the tobacco mosaic virus and cucumber mosaic virus and provide immunity against them.

Cai et al. (2018) have used the CRISPR/Cas9 system to induce mutations on GmFT2a, an integrator in the photoperiod flowering pathway of soybean. The developed soybean plants showed late flowering, resulting in increased vegetative size. The mutation was also found to be stably inherited in the following generations.

Tain et al (2018) used the CRISPR/Cas9 system to target CIPDS, the phytoene desaturase in watermelon, to achieve the albino phenotype. All genome-edited watermelons harbor mutations in CIPDS and showed full or mosaic albino phenotype. This study served as a proof of concepts of using the CRISPR/Cas9 system in watermelon breeding.

Citrus plants resistant to citrus canker caused by *Xanthomonas citri* sub sps. Citri (Xcc), a serious disease of citrus, has been developed through CRISPR/Cas9. The promoter of the CsLOB1 gene, which promotes canker development, was targeted for the development of canker resistant citrus plants.

CRISPR/Cas9 was also used to generate mutants in the flowering suppressor SELF-PRUNING5G (*SP5G*) gene in tomato to manipulate photoperiod response. The mutations brought about by CRISPR/Cas9 caused rapid flowering and enhanced the compact growth habit of field tomatoes, resulting in a quick burst of flower production and early yield.

Crop yield is a complex, multi-genic, and quantitative character that is influences by several features. It has been demonstrated that the CRISPR/Cas9 technology can be effectively utilize to enhance crop yield.

The CRISPR/Cas9 technique has been used to knock out the genes that are known to negatively regulate yield-related traits such as, tiller number (*OsAAP3*), panicle size (*TaDEP1, OsDEP1*), grain weight (*TaGASR7, TaGW2*), and grain number (*OsGn1a*). In rice, by employing a multiplexing gene editing strategy, three genes including GS2, GS3, GA5, and TGWA6 were mutated simultaneously and headed towards trait pyramiding and enhancing grain size and weight. Similarly, Li et al (2018b) applied the CRISPR/Cas9 system to knockout three yield-related genes (*Hd2*, *Hd4*, and *Hd5*), which resulted in early heading in rice. In wheat, knockout of the gene *GASR7* via CRISPR/Cas9 technique increased kernel weight.

The CRISPR/Cas9 technology has also been used for quality improvement in crops such as storage quality, nutritional value, starch content, and fragrance. In rice, the cooking and eating quality has been improved by mutating the *Waxy* gene using CRISPR/Cas9 techniques. The nutritional value (high amylose content) of rice has also been improved by knocking out the *SBEllb* gene. In 2018, Sanchez and coworkers produced low-gluten wheat through CRISPR/Cas-mediated gene editing of the glutin –encoded gene family α-gliadin. High-yielding soybean plants with improved levels of oleic acid were developed by disrupting the *FAD2-1B* and *FAD2-1A* genes employing CRISPR/Cas9 system. Sorghum nourishment quality has been improved by employing CRISPR/Cas9 technique, targeting *k1C* genes which were responsible for poor digestibility and hindered production of improved amino acids.

Crispr/Cas9-Mediated Gene Editing In Plants

Several efforts have been successfully employed for target gene editing in many crop plants via CRISPR.Cas9-based genome editing tools. Factors that have been reported to affect the editing ability of the CRISPR/Cas9 system include targeted DNA, GC contents, Cas9 codons, sgRNA structure, and expression of Cas9 and sgRNA. Therefore, all these factors should be highly optimized to develop increased efficiency of CRISPER/Cas9 system.

Designing CRISPR/Cas9 Delivery System

Initial attempts to use CRISPR/Cas9 in plants were not encouraging due to low efficiency. With the improvement of technology, highly efficient vector delivery systems for CRISPR/Cas9 have been designed for genome editing in plants. Some of the important development includes, viral infection suppression, disruption of gene of cis-elements, genomic deletion, gene knockout, and multiplex genome editing. Discovery of new Cas9 variants, efficient screening methods for knockout mutants, vector selection, and construction and employment of the most appropriate delivery system for Cas9 expression cassette has led to more precise, accurate, and targeted delivery of the Cas9 system in plants. Construction, screening and delivery of the CRISPR/Cas system into plants are described in the following section

Cargo-Vector for CRISPR/Cas9 System

Both single-vector and binary-vector systems are used in CRISPR/Cas9-mediated gene editing. For plant transformation, any specific vector constructed with several gRNAs and Cas9 protein expression cassette, can be applied. Various structural constructs of gRNAs can be utilized for various Cas9 proteins to design a unique gRNA - Cas9 nuclease, for accuracy and easiness in experimental design. In a single vector system, RNA polymerase III-driven promoters (U6/U3) are designed for gRNA expression, whereas in the case of Cas9 gene expression ubiquitin and CaMV355 promoters based on RNA polymerase II are used. Recently, certain new adjustments are made in the single-vector system, which includes polymerase II and dual polymerase II promoters. Single polymerase II vectors are used to control the expression of gRNA and the Cas9 gene at the same time, while the dual polymerase II vectors utilize two different promoters to control the expression of gRNA and Cas9 gene. With the addition of these new technologies has helped to decrease the vector length and thereby improve the efficiency of the system.

Bioinformatics Tools for Designing CRISPR/Cas9 Construct

One of the most critical steps in developing a highly precise genome editing system is the designing of a sgRNA construct for CRISPR/Cas9. Several online tools with plant databases are available, which permit the designing of sgRNA for the identification of new target sites (Table 5.6). For example, Xie et al. (2014) developed a web tool named CRISPR-PLANT to design efficient sgRNA constructs for CRISPR/Cas9-based gene editing. Similarly, a novel web tool was developed by Michno et al (2015) for rapid detection of target loci in soybean for CRISPR/Cas9-mediated gene editing, and the CRISPR-P was developed by Lei et al. (2014) for designing sgRNA for every plant having an available sequenced genome and to help evaluate off-targets.

Construction of sgRNA Expression Cassette

The most important step in CRISPR/Cas9-mediated gene editing system is the construction of a unique sgRNA cassette (Chen et al. 2019). This works as a guide for the Cas9/sgRNA complex which comprises 98 nucleotides with 20 nucleotide target sequences. In plants, RNA polymerase III is used to transcribe sgRNA with the help of U3 or U6 promoters. Since the expression cassettes of sgRNA-U3/U6 promoters are of very small length (about 300-600 bp), adopter ligation or overlapping PCR can be applied to construct these expression cassettes. Ma et al. (2015) developed a robust cloning-free approach for sgRNA expression cassette based on PCR technique. Direct cloning of sgRNA expression cassette into binary vectors for the CRISPR/Cas9 system developed using Gibson assembly or Golden Gate cloning strategy. Gao and Zhao (2014) utilized a ribozyme mechanism to generate sgRNA by transcription of pre-RNA through RNA polymerase II, through which it is possible to ligate inducible or constitutive promoters to obtain the desired function of sgRNA (Sun et al. 2019).

Table 5.6. List of various sgRNA (single guide RNA) designing bioinformatics tools for the CRISPR/Cas system.

Tool Name	Year	Description and Function	Web Link
CRISPR Design	2013	Precise sgRNA construction for target sites, assess off-target sites	http://www.genome-engineering.org
sgRNAcas9	2014	Rapid design of sgRNA with less off-target	https://www.biootools.com/col.jsp? id=103/
CHOPCHOP	2014	Detect optimal target sites for sgRNA, produce potential scores for target sites	https://chopchop.cbu.uib.no/
CRISPR-P	2014	Generate synthetic sgRNA, predict potential sites for enzyme cut	https://www.cbi.hzau.cn/crisper
GPP Web Portal	2014	Produce potential sgRNA score	https://www.roadinstitute.org/rnai/public/analysis-tools/sgrnadesign
SSFinder	2014	High-throughput detection of target sites	https://code.google.com/p/ssinder
E-CRISP	2014	Potential target site evaluation	https://www.-crisp.org/E-CRISP/designcrispr.html
Cas-OFFinder	2014	Based on RNA-guided endonucleases, robust for detecting off-target sites	http://www.rgenome.net/cas.oinder/
CRISPRseek	2014	Screen sgRNA for targeted sequences, produce cleavage scores for predicted off- targets	https://www.bioconductor.org/packages/ release/bioc/html/CRISPERseek.html
CRISPR-PLANT	2014	Construct specific sgRNAs for particular plant species	https://www.genome.arizona.edu/crisper
CRISPRdirect	2014	Design sgRNA with minimum off-target	https://crisper.dbcls.jp

Tool Name	Year	Description and Function	Web Link
Azimuth	2015	Design sgRNA for both on-target and off-target models	https://research.microsoft.com/en-us/projects/azimuth/
CCTop	2015	Predict target sgRNA sequence based on possible off-target	https://crisper.cos.uni-heidelerg.de
Cas-Designer	2015	RNA-guided endo- nucleases, provides all information about off-targets and out-of frame scores	http://rgenome.net/csas-designer/
CRISPy	2016	Target prediction for sgRNA, geographical representation of results	http://crispy.secondarymetaolites.org
phytoCRISP-Ex	2016	UNIX-based standalone, Cas9 target prediction	http://www.phytocrispex.biologie.ens.fr/ CRISP-Ex/
CRISPR-DO	2016	Specific for both coding and non-coding targets, provides information regarding off-targeted sites and its functional conservation	http://cistrome.org/crisper/
CRISPRpred	2017	Efficient designing of sgRNA based on target in silico prediction	https://githubb.com/khaled-buet/CRISPRpred
CRISPR-P 2.0	2017	Predict on-target scores, analyze and detect guide sequence	http://cbi.hzau.edu.cn/CRISPR2
sgRNA Score 2.0	2017	Design sgRNA for several PAM sites	http://crispr.med.harvard.edu/ sgRNAScoreV2
CRISPR-Local	2018	Design sgRNA for non-reference cultivars, predict sgRNA that can target multiple genes	http://crisper.hzau.edu.cn/CRISPR-Local/
CRISPRInc	2019	Design sgRNA for lncRNAs, works for all species	http://www.crisprlnc.org

Construction of Cas9 Expression Cassette

In the case of Cas9 nuclear localization in eukaryotes, the coding sequence of Cas9 (4107 bp) must be fused with the nuclear localization signal. In plants, user-biased codons were used to design highly efficient and optimized Cas9 expression cassettes for improved gene editing. For example, in rice utilization of codon-optimized Cas9p has been improved by enhancing GC content. Usually, constitutive promoters such as 35S *Cauliflower mosaic virus* (CaMV) and ubiquitin from *Arabidopsis thaliana*, maize, and rice can be used to control the expression of Cas9 in monocot and dicot plants, for highly targeted gene editing, through callus-based transformation approaches.

Transformation Approaches for CRISPR/Cas9-Based Vector Delivery into Plants

In genome editing through CRISPR/Cas9-mediated system, the cargo-vector containing the expression cassette of both sgRNA and the *Cas9* gene must be directed to target sites in plant cells. For transformation of the cargo-vector, floral dip and biolistic approaches are generally adopted. Currently advanced strategies such as ribonucleoprotein complex, virus-mediated delivery, and plasmid delivery systems are used for the transformation of plants. PEG-mediated and biolistic transformation techniques are also being used for the direct delivery of *Cas9* gene expression cassettes. However, in certain cases, regeneration from protoplast may be difficult due to heritable targeted mutations. *Agrobacterium*-mediated transformation has been found to be the most efficient approach for the stable transformation of the CRISPR/Cas9 system in plants (Zhang et al. 2018, Zhang et al. 2019b).

Strategies for Mutant Screening

Numerous mutant libraries have been created by CRISPR/Cas9 system in several important crop plants such as rice and tomato. Scientists are required to screen huge number of mutants, using a strategy that includes the detection of off-target and on-target edits. To assist in screening and identifying the desired mutants, different techniques have been developed, which include high-resolution melting analysis (HRMA), annealing at critical temperature polymerase chain reaction (ACT-PCR), polyacrylamide gel electrophoresis (PAGE)-mediated genotyping, restriction enzyme site loss technique, and T7 endonucleases I (T7E1) approach. However, there exist certain advantages and limitations for each technique.

Rapid detection of the mutant is possible only when it manifests a clear phenotype. for example, in rice and tomato a visible albino phenotype was observed when a gene Phytoen Desaturate mutated via the CRISPR/Cas9 system, which was applied as a phenotype marker to detect CRISPR/Cas9 edited plants. In another approach, high-throughput sequencing can be used to screen all the mutants generated by the CRISPR/Cas9 system, with precision and accuracy (Du et al. 2016). Whole genome sequencing is quite beneficial and helpful for the detection of DNA-free plants edited by the CRISPR/Cas9 system.

Repair Of Cleaved Genome Sites

In any gene editing system, the repair of the DNA breaks created by the nucleases is an important step. Usually, the DNA breaks are repaired by the endogenous repair mechanism of the cells such as: homologue-dependent repair (HDR) or non-homologous end-joining (NHEJ) mechanisms. In HDR mechanism, a sequence having homology to target is used as a template to repair DNA breaks or lesions. The template that contains the desired sequence is flanked by sequences having homology to both sides of the breakpoint, thereby forcing the insertion of the desired sequence into the target site. Thus in HDR, homologous recombination enables insertion or gene recovery. In NHEL mechanism, the ends of the cleaved

DNA are joined together, which results into insertion or deletion of nucleotides, resulting shifting of the gene reading frame. This process may lead to gene knockout. Since the HDR mechanism is comparatively simple and efficient and it is preferred by molecular plant breeders.

Software Toolkit For Crispr-Based Genome Editing

Many steps are involved in executing CRISPR/Cas9 and Cpf1 systems in plants. These include:

- Selection of specific target sites(s) having no highly homologous sequences which may act as the off-target site(s) in the genome,
- Designing and synthesis of oligonucleotides similar to the target sequence,
- Preparation and expression of the cassette(s) for sgRNA (single guie RNA) for generating target-specific sequence specificity,
- Construction of plant-transformation/expression vector(s), and
- Evaluation of the outcome of gene (genome) editing approach.

Several software such as CRISPR-P, E-CRISPR, Breaking-Cas are available to carry out some of the activities mentioned above. However, most of the toolkits cannot carry out all the steps simultaneously. Xie et al. (2017) developed a web-based software package named CRISPR-GE (Genome Editing) which is capable to expedite all experimental designs and analyze the mutants for CRISPR/Cas9/Cpf1-based genome editing in plants.

Safety Aspects Of Genome Editing System

Genome editing systems are expected to preserve the native genomic structure. Therefore, they are considered to be safe technologies for crop improvement. However, there exist certain concerns related to the biosafety issues of the crops developed through such techniques. Some of the important aspects of genome editing system are discussed in the following section.

Non-Target Effects

Careful selection of sites for the creating double-stranded breaks in the genomic DNA is essential to minimize the non-target effects of the genome editing systems. This can be achieved through prior bioinformatics analysis. It is important to avoid sites with repeated sequences and sites having high homology with other regions of the genome. Several software is now available to assist in the selection of target sites and nuclease design and their validation.

Regulations on Plants Created by Genome Editing

Through genome editing technology it is possible to introduce stably inherited point mutations (modifications) into the plant genome and thereafter remove the transgenic region easily. This allows the creation of non-transgenic plants with improved traits. The technology is much faster than the conventional breeding methods. The plants developed through genome editing technology are nearly identical to the plants developed through classical breeding methods. However, their safety must be ascertained based on the developed product rather than the process involved. It is important to note that ODM-derived products plants are indistinguishable from the plants derived through conventional breeding methods. Therefore, it is apparent that plants derived through ODM should not be brought under any regulations. Through CRISPR/Cas9 technology it is possible to obtain marker-free (for example, antibiotic resistance) modified plants. Accordingly, the existing regulations on transgenic plants should also be not applicable to plants developed through CRISPR/Cas9 technology. Therefore, plants derived through genome editing systems should not be considered as genetically modified (GMO) plants.

FUTURE PROSPECTS

The following areas have great potential for the application of gene editing systems for crop improvement.

Multiplexing and Trait Stacking in Crop Breeding

In plants, manipulation of agronomic traits requires precise engineering of complex metabolic pathways, controlled by complex genetic networks. Therefore, it is important to develop molecular tools with the capability to manipulate multiple genes simultaneously. CRISPR and its variations have the capability to manipulate multiple genes simultaneously and therefore gain popularity among plant breeders.

Wang and his coworkers have exploited the multiplexing capability of CRISPR to create a system enabling clonal reproduction from F1 hybrids in rice. This helped to preserve the favorable high degree of heterozygosity. This was achieved by altering three meiotic genes to function as mitosis-like cell division, thereby producing diploid gametes and tetraploid seeds. An increase in additional ploidy levels was prevented by targeting a gene involved in fertilization.

Generation of Genomic Diversity for Crop Breeding

CRISPR/Cas has been demonstrated to have the potential to generate a broad range of allelic diversity at specific loci. Shen et al (2018) could simultaneously edit eight yield and quality traits genes in rice. They further could isolate homozygous mutated alleles of all the eight targeted genes. The mutants isolated represent octuple, septuple, and sixtuple mutants as well as heterozygous mutants for all the targeted genes. Thus they could generate ample genetic diversity for selection within one generation. It has also been shown that editing the same QTls can deliver different results on the basis of the genetic background.

High-Throughput Mutant Libraries

Complete genome sequencing for many crops has now been completed, but the functions of many gene sequences are unknown, which may control important agronomic characters. Therefore, the next step should be to systematically analyze the functions of the sequenced genes. Gene knockout is an effective strategy for determining gene functions. Hence, the construction of large-scale mutant libraries of the whole genomes of different crops shall provide the opportunity for the study of functional genomics of crop plants.

Gene Regulation

Gene editing systems can also be used to study the mechanism of regulation of gene expression in plants. Repression and activation of genes can be achieved by fusing transcriptional repressors or activators to the DNA-binding domains of genome-editing constructs (such as ZFP, TALE or dCas9), which target the regulatory regions of endogenous genes. Recently, CRISPR/Cas9 technology was used to alter the *cis*-regulatory control of quantitative trait loci in tomato.

Genome editing could also be used to reveal the function of many non-canonical RNAs that are linked to crop improvement. Since most no-coding transcripts are of nuclear origin, they lack open reading frames. The genome editing tools can modulate transcription directly and thus optimally suitable to interrogate the function of each RNAs.

Another application of CRISPR/Cas9 system could be in the formation of conditional alleles, to provide spatial and temporal control of gene expression, to understand the function of lethal genes. Application of inducible or tissue-specific promoters for Cas9/sgRNA expression can be used to understand the regulation of gene expression in a specific tissue, in development stages, and under varying environmental conditions.

It is now possible to label endogenous genes with fluorescent proteins and visualize their *in vivo* expression. Through fluorescently labeled dCas9, it is possible to study the changes in genome dynamics during the developmental stages of plants under different environmental conditions. Application of dCas9 can facilitate to understand the selection process of activation/repression of effector domains to specific genomic loci for regulation of endogenous gene expression in plants.

Genome editing technologies can be successfully used for epigenetic editing through selection of proteins responsible for the modification of histones and methylation of DNA. These technologies have emerged as a novel way to understand regulating cellular functions in plants. CRISPR/Cas9 system can be used to understand epigenetic regulation and for the identification of proteins attached to enriched chromatin carried out through the enrichment of chromatin target sites. It is also possible to use CRISPR/Cas9 to identify regulatory proteins which bind to specific DNA sequences to control the expression of genes.

Utilization of Genetic Diversity from Wild and Uncultured Species

Out of over 300,000 plant species, only about 200 are commercially used. Three major crops rice, wheat, and maize provide energy for human consumption.

Improvement in these three crops may not be possible forever to meet the increasing demand. Therefore, it is important to explore the possibilities of utilizing vast genetic diversity in the wild and uncultured plant species. Genome editing systems have the potential to resolve the issue. One of the approaches could be to target the so-called domestication genes in the wild and uncultured plant species and accelerate the domestication process through CRISPR. During the domestication process of crop plants, certain favorable traits such as favorable plant architecture, simultaneous flowering, easy harvesting, larger fruit size, high yield, etc. were given importance during the selection process. With the increase in our understanding of the genetic basis of these domestication traits, it has become much easier to identify several such domestication genes.

Zsogon et al. (2018) have demonstrated the de-novo domestication of *Solanum pimpinellifolium*, an ancient stress-tolerant tomato relative, through the application of CRISPR/Cas9 system. Similarly, Li et al. (2018a) have demonstrated the de-novo domestication of four wild tomato accessions having resistance against specific stress conditions through CRISPR. More recently, Lemmon and coworkers have achieved de-novo domestication in *Physalis pruinosa,* an orphan crop of the Solanaceae family, after the identification of domestication genes and their alteration through CRISPR technology.

Plant Breeding Innovations through CRISPR/Cas9

CRISPR-Cas9 has received a lot of attention in recent years due to its range of applications, including biological research, breeding and development of agricultural crops and animals, and human health applications. These include gene silencing, DNA-free CRISPR/Cas9 gene editing, homology-directed repair (HDR), and transient gene silencing or transcriptional repression (CRISPRi).

Traditional plant breeding methodologies have contributed towards advancement in agriculture during the past several decades. Traditional methods basically depend on the existence of genetic variability in the natural population and induced mutations thereof. Mutations are usually rare and occur at random. Moreover, desirable mutations may not occur in the elite varieties. The process of creating variability and their utilization is often time-consuming and laborious process. In contrast, genome editing is easier, faster, and can be used precisely to modify the targeted gene(s) in any crop for its improvement. A variety of genome editing techniques is now available for crop improvement (Figure 5.7). Thus it is important for the plant breeders to select the optimum system for a given plant species, based on the objective of the breeding program. After selecting the appropriate genome-editing tools, the target sequences are designed and introduced into the vector along with the appropriate genetic cargo (DNA, RNA, or RNPs). After the introduction of the genetic cargo into the plant cells, the target sequence shall get modified (Zhang et al. 2018). The calli developed from the edited cells should then be regenerated to produce the edited plants (Figure 5.7).

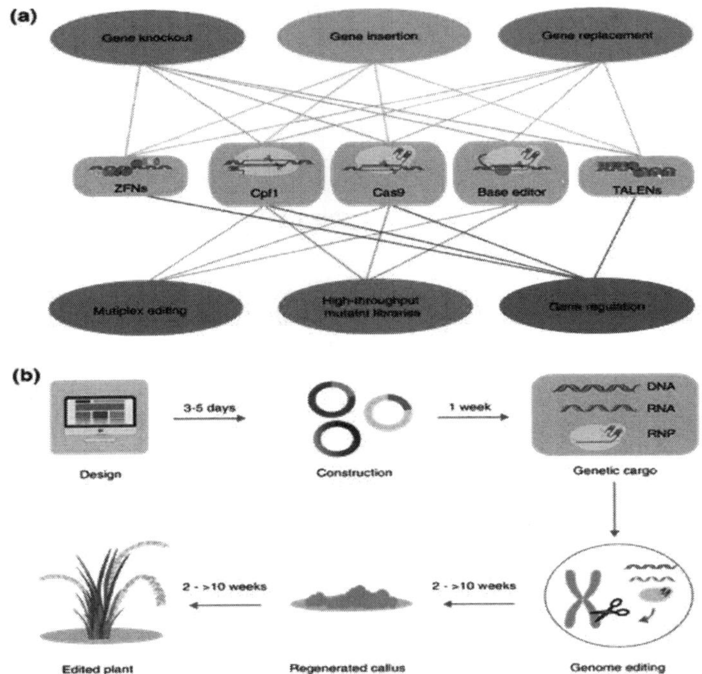

Fig. 5.7. (a) The network of genome editing methods and the corresponding genome editing tools. (b) Flow chart illustrating the successive steps in plant genome editing, and the estimated time needed for each step. RNP ribonucleoprotein, TALEN transcription activator-like effector nuclease, ZFN zinc-finger nuclease. (Reproduced from Zhang et al. Genome Biology 2018, htttp;// creativecommons.org/ publication/ zero/1.0/).

It is well known that a protoplast-based regeneration system is not available for many important crop plants. Therefore, alternative methods should be developed for gene-editing technology. Use of pollen or immature embryos that can coax to germinate *in vitro* could be a successful method for the application of gene-editing in plants. Since the regeneration process is not required, such an approach could provide better results. With the progress and development of genome-editing tools, it is expected to speed up the crop breeding process to meet the ever-increasing demand for food.

Application of Cas-CLOVER Technology

CRISPR is considered to be the most precise and efficient tool for creating genetic modifications in animals, plants, and microorganisms. However, it may also create some off-target mutations, because it uses a single RNA guide.

To overcome this problem, a new unique gene-editing tool called Cas-CLOVER was developed by Hera Biolabs in 2018. Technically it is fully dimeric Clo51-dCas9 genome editing technology, which is practically similar to CRISPR-Cas9. But it utilizes a different nuclease protein called Clo51 endonuclease. Cas-CLOVER is a combination of proteins that involves a nuclease-inactivated Cas9 protein melded to the Clo51 endonuclease. Cas-CLOVER tool requires two RNA guides and their activity dependents on specific spacer lengths. Unlike CRISPR, it is considered to have robust editing efficiency, with high fidelity. In addition, it creates no off-target mutation as demonstrated by Next Generation Sequencing (NGS). The basic features of Cas-CLOVER system are as follows (Li et al., 2018):

- ❑ Exclusively licensed technology and suv0licensable by Hera,
- ❑ Functionally similar other CRISPR/Cas9 technologies, but uses a different nuclease protein called C1o51, which is covered under a set of patents distinct from other CRISPR/Cas9 technologies,
- ❑ Cas-CLOVR is a fusion protein that comprises a nuclease-inactivation Cas9 protein fused to the C1o51 endonuclease,
- ❑ Cas-CLOVER achieves greater specificity through the utilization of two guide RNAs as well as a nuclease activity that requires dimerization of subunits associated with each guide RNA

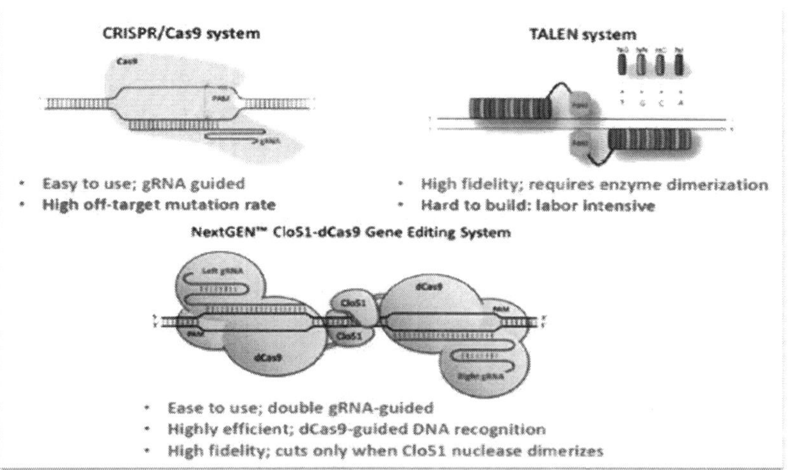

Figure 5.8. Comparative assessment of the CRISPR/Cas9, TALEN, and Cas-COVER systems of gene editing. For explanation see text. (Reproduced from: Li et. al. 2018. Precision CRISPR Congress Poster Presentation, Boston, MA)

Thus, Cas-CLOVER genome editing tool is considered not only to be novel but also as an advanced tool. It has been shown to be high specificity and efficiency in both proliferating and resting T cells. It has the capacity to overcome the various limitations of CRISPR and TALENS. The capability to provide higher fidelity genome editing proved to be safer in clinical applications. Cas-CLOVER is expected to open new doors of research in genome editing science, in various organisms including plants. Comparative assessments of CRISPR/Cas9, TALENs, and Cas-CLOVER technologies are presented in Figure 5.8.

CRISPR-like Gene Editing system in Eukaryotes

The CRISPR gene editing system originated in bacteria, and is part of a defense that bacterial cells use against viruses. In CRISPR gene editing, a guide RNA brings a genome-cutting enzyme (typically Cas9) to a specific place in the genome, where a cut is made. This system has been modified, adapted, and used in many ways that have been revolutionary, triggering advances in the research lab, and has even treated disease in a small number of people. A team of researchers led by Feng Zhang at the McGovern Institute for Brain Research at MIT and the Broad Institute of MIT and Harvard has uncovered the first programmable RNA-guided system in eukaryotes — organisms that include fungi, plants, and animals (Saito et al. 2023). It is based on a protein known as Fanzor, which also uses RNA as a guide that takes it to a specific place in the genome to make a cut (Saito et al. 2023).

The Fanzor proteins can be reprogrammed so that they can target the DNA in human cells. Fanzors are also smaller, and are likely easier to deliver to cells compared to the CRISPR/Cas9 method. Much like CRISPR, this new technique will probably be altered and improved in a variety of ways before it can be brought to the clinic, but it opens up many new gene editing possibilities. This research has also shown that mechanisms that cut DNA are not only found in bacterial or prokaryotic cells, all the kingdoms of life carry some system like it. This is the first time a DNA-cutting system has been discovered in eukaryotes.

Saltos et al. (2023) identified and isolated Fanzors from various species including fungi, algae, and amoeba. Characterization of the proteins revealed their endonuclease power - they can cut DNA. They also use RNA molecules that do not code for protein to target specific genomic sequences. Fanzor proteins sit within the eukaryotic genome, in regions known as transposable elements. They also may have migrated to eukaryotes from prokaryotes during horizontal gene transfer.

Saito et al (2023) led the biochemical characterization of the Fanzor proteins, showing that they are DNA-cutting endonuclease enzymes that use nearby non-coding RNAs known as ωRNAs to target particular sites in the genome. It is the first time this mechanism has been found in eukaryotes, such as animals.

Unlike CRISPR proteins, Fanzor enzymes are encoded in the eukaryotic genome within transposable elements, and the team's phylogenetic analysis suggests that the Fanzor genes have migrated from bacteria to eukaryotes through so-called horizontal gene transfer.

The investigators showed that Fanzors can create insertions and deletions at specific places in the human genome, though they were initially far less efficient at cutting DNA compared to CRISPR methods. With systematic testing of different mutations in the Fanzor proteins, they boosted Fanzor's activity ten times. Fanzor also appear to have low levels of off-target activity and do not seem to affect nearby DNA or RNA molecules.

To explore Fanzor's potential as a genome editing tool, the researchers demonstrated that it can generate insertions and deletions at targeted genome sites within human cells. The researchers found the Fanzor system to initially be less efficient at snipping DNA than CRISPR-Cas systems, but by systematic engineering, they introduced a combination of mutations into the protein that increased its activity 10-fold. Additionally, unlike some CRISPR systems and the OMEGA protein TnpB, the team found that a fungal-derived Fanzor protein did not exhibit "collateral activity," where an RNA-guided enzyme cleaves its DNA target as well as degrading nearby DNA or RNA. The results suggest that Fanzors could potentially be developed as efficient genome editors.

They analyzed the molecular structure of the Fanzor/ωRNA complex and illustrated how it latches onto DNA to cut it. Fanzor shares structural similarities with its prokaryotic counterpart CRISPR-Cas12 protein, but the interaction between the ωRNA and the catalytic domains of Fanzor is more extensive, suggesting that the ωRNA might play a role in the catalytic reactions.

Like CRISPR-based systems, the Fanzor system can be easily reprogrammed to target specific genome sites, and it could one day be developed into a powerful new genome editing technology for research and therapeutic applications. The abundance of RNA-guided endonucleases like Fanzors further expands the number of OMEGA systems known across kingdoms of life and suggests that there are more yet to be found.

Conclusions

The new CRISPR-based gene editing tools have become increasingly efficient and flexible in plant cells. Apart from targeted gene mutagenesis, other types of genetic modifications such as base substitution, gene knocking and replacement have now become available in several plant species. It is now theoretically possible to obtain precise gene editing at any locus without requiring any selection marker.

CRISPR-based gene editing technology can be classified into two categories; gene-mutagenesis and gene-correction tools. Gene mutagenesis is used to introduce full or partial loss-of-function mutations into the target gene (e.g. canonical CRISPR/Cas9 system and base editing system). On the other hand, for gene-correcting purposes, the modifications on the target gene have to be very precise. The base editing systems and fragment deletion, insertion, and replacement tools can be used for this purpose. Although much progress has been made, the low homology repair (HR) frequency in plant cells is still a problem for precise gene editing.

Base editing has become efficient means of precisely converting one base to another. Diverse base editing systems have been established successfully to alter/modify genes in plants for both biological functional analysis and crop improvement. Several variants of Cas9 and mutations of cytidine deaminase and adenine deaminase have been developed and incorporated into CBE and ABE base editors thereby expanding their editing scope and improving their specificity and editing efficiency in plants. With the introduction of new base editing technology and improvement in their specificity, base editing will have unparalleled potential for crop improvement.

One of the big challenges in crop breeding is to achieve efficient delivery of CRISPR components into the reproductive cells to generate heritable gene modifications. The second big challenge is to decide which gene(s) to edit in order to improve a particular character. It is expected that with the aid of CRISPR gene editing systems, the underlying genes for complex traits shall be identified and subsequently edited for crop improvement.

References

Ali, Z., Eid, a., Ali, S., & Mahfouz, M.M. (2018). Pea early-browning virus-mediated genome editing via the CRISPR/Cas9 system in *Nicotiana benthamiana* and *Arabidopsis. Virus Research, 244*, 333-337.

Anderson, M., Turesson, H., Olsson, N., Falt, , A.S., Ohlsson, P., Gonzalez, M.N., ... Hofvander, P. (2018). Genome editing in potato via CRISPER-Cas9 ribonucleoprotein delivery. *Physiologia Plantarum, 164*, 378-384.

Anzalone, A., Randolph, P., Davis, J., Sousa, A. A., Koblan, L. W., Levy, J. M., ... Li, D. R. (2019). Search-and replace genome editing without double-stranded breaks or donor DNA. *Nature*. Retrieved from: http://doi:1038/s41586-019-1711-4.

Barrangou, R., & Horvath, P. (2014). Functions and applications of RNA-guided immune system. In RA Mayers (Ed.) *Encyclopedia of molecular cell biology and molecular medicine: RNA biology* (pp. 1-24), New York, Wiley-VCH Verlag GmBh & Co.

Bharat, S.S., Li, S., Li, J. Yan, L., & Xia, L. (2019). Base editing in plants: current status and challenges, *The Crop Journal*, Retrieved from: https://doi.org/10.1016/j.cj.2019. 10.002.

Blin, K., Pedersen, L.E., Webbber, T., & Lee, S.Y. (2016). CRISPy-web: an online resource to design sgRNA for CRISPR applications. *Synthetic and Systems Biotechnology, 1*, 118-121.

Bonawitz, N. D., Ainley, W. M., Itaya, A., Chennareddy, S. R., Cicak, T., Effinger, K., ... Pareddy, D. R. (2018). Zinc finger nuclease-mediated targeting of multiple transgenes to an endogenous soybean genomic locus via non-homologous end joining. *Plant Biotechnology Journal, 17*, 750-761.

Braatz, J., Harloff, h.j., Mascher, M., Stein, N., Himmelbach, A., & Jung, C. (2017). CRISPR-Cas9 targeted mutagenesis leads to simultaneous modification of different homogenous gene copies in polyploidy oilseed rape (*Brassica napus*). *Plant Physiology, 174*, 935-942.

Butt, H., Jamil, M., Wang, J.Y., Al-Babili, S., & Mahfouz, M. (2018). Engineering plant architecture via CRISPR/Cas9-mediated alternation of strigolactone biosynthesis. *BMC plant Biology, 18,* 174-186.

Cai, Y., Chen, L., Shi Sun, S., Wu, C., Weiwei Yao, W., Jiang, B., Tianfu Han, T., & Hou, W.(2018). CRISPR/Cas9-Mediated Deletion of Large Genomic Fragments in Soybean. *International Journal of Molecular Sciences, 19,* 3835-3846. doi:10.3390/ijms19123835

Chari, R., Yeo, N.C., Chavez, A., & Church, G. (2017). Sigma score 2.0- a species independent model to predict CRISPR/Cas9 activity. *ACS Synthetic Biology., 6,* 902-904.

Chen, L., Li, W., Katin-Grazzini, L., Ding, J., Gu, X., LI, Y., ... Deng, Z.. (2018). A method for the production and expedient screening of CRISPER.Cas9-mediated non-transgenic mutant plants. *Horticultural Research, 5,* 13-18.

Chen, W., Zhang, G., Li, J., Zhang, X., Huang, S., Xiang, S., ... Liu, C. (2019). CRISPRInc: a manually curated database of validated sgRNAs for IncRNAs. *Nucleic Acid Research, 47,* D63-D68.

Curtin, S.J., Xiong, Y., Michoo, J.M., Campbell, B.W., Stec, A.O., Cermak, T., ... Stuper, R.M. (2018). CRISPR/Cas9 and TALENs generate heritable mutations for genes involved in small RNA processing of *Glycine max* and *Medicago truncatula. Plant Biotechnology Journal, 16,* 1125-1137.

Du, H., Zeng, X., Zhao, M., Cui, X., Wang, Q., Yang, H ... Yu, D. (2016). Efficient targeted mutagenesis in soybean by TALENs and CRISPR/Cas9. *Journal of Biotechnology, 217,* 90-97.

Feng, Z., Zhang, B., Ding, W., Lin, X., Yang, D. L., Wei, P., ... Zhu, KJ. K. (2013). Efficient genome editing in plants using a CRISPR.Cas system. *Cell Research, 23,* 1229-1232.

Gao, Y., & Zhao, Y. (2014). Self-processing of ribozyme-flanked RNAs into guide RNAs in vitro for CRISPR-mediated genome editing. *Journal of Integrated Plant Breeding, 56,* 343-349.

Gehrke, J.K., Cervantes, O., Clement, M.K., Wu, Y., Zeng, J., Bauer, D.E., ... Joung, J.K. (2019). An APOBEC3A-Cas9 base editor with minimized bystander and off-target activities. *Nature Biotechnology, 36,* 977-982.

Grechko, V. V. (2002). Molecular DNA markers in phylogeny and systematics. *Russian Journal of Genetics, 38,* 851-868.

Grohmann, L., Keilwagen, J., Duensing, N., Dogand, E., Harting, F., Withelm, R., ... Sprink, T. (2019). Detection and identification of genome editing in plants: challenges and opportunities. *Frontiers of Plant Science, 10,* 1-8.

He, J., Zhao, X., Larocu, A., Laroche, A., Lu, Z. X., Lin, H. K., & Li, Z. (2014). Genotyping-by-sequencing (GBS), an ultimate marker-assisted selection (MAS) tool to accelerate plant breeding. *Frontiers of Plant Breeding, 5,* 1-6.

He, Y., Zhu, M., Wang, L., Wu, J., Wang, Q., Wang, R., & Zhao, Y. (2018). Programmed self-elimination of the CRISPR/Cas9 construct greatly accelerates the isolation of edited and transgene-free rice plants. *Molecular Plants, 11*, 1210-1213.

Hess, G.T., Tycko, J., Yao, D., & Bassik, M.C. (2017). Methods and applications of CRISPER-mediated base editing in eukaryotic genomes. *Molecular Cell. 68*, 26-43.

Hsu, P. D., Lander, E., & Zhang, F. (2014). Development and applications of CRISPR-Cas9 for Genome Engineering. *Cell, 157*, 1262-1278.

Hsu, P.D., Scott, D.A., Weinstein, J.A., Ran, T.A., Konermann, S., Agarwal, V., ... Zhang, F. (2013). DNA targeting specificity of RNA-guided Cas9 nucleases. *Nature Biotechnology, 31*, 827-832.

Hu, X., Meng, X., Liu, Q., Li, J., & Wang, K. (2018). Increasing the efficiency of CRISPER-Cas9-VQR precise genome editing in rice. *Plant Biotechnology Journal, 16*, 292-297.

Hua, K., Tao, X., Yuan, F., Wang, D., & Zhu, J.K. (2018). Precise A-T to G-C base editing in the rice genome. *Molecular Plant, 11*, 627-630.

Hua, Y., Wang, C., Huang, J., & Wang, K. (2017). A simple and efficient method for CRISPER/Cas9-induced mutant screening. *Journal of Genetics and Genomics, 44*, 207-213.

Huang, J., Li, J., Zhou, J., Wang, L., Yang, S., Hurst, L.D., ... Tian, D. (2018). Identifying a large number of high yielding genes in rice by pedigree analysis, whole-genome sequencing, and CRISPR-Cas9 gene knockout. *Proceeding National Academy of Science, USA, 115*, E7559-E7567.

Jacob, T.B., Zhang, N., Patel, D., & Martin, G.B. (2017). Generation of a collection of mutant tomato lines using pooled CRISPR libraries. *Plant Physiology, 174*, 2023-2037.

Ji, X., Si, X., Zhang, Y., Zhang, H., Zhang, F., & Gao, C. (2018). Conferring DNA virus resistance with high specificity in plants using virus-inducible genome-editing system. *Genome Biology, 19*, 197-202.

Jiang, G. L. (2013). Molecular marker and marker-assisted breeding in plants. Retrieved from: http://dx.doi.org/10.5772/52583.

Jung, J. H., & Altpeter, F. (2016). TALEN mediated targeted mutagenesis of the caffeic acid O-methyltransferase in highly polyploidy sugarcane improves cell wall composition for production of bioethanol. *Plant Molecular Biology, 92*, 131-142.

Kamburova, V. S., Nikitina, E. V., Shermatov, S. E., Buriev, Z. T., Kumpatla, S. P., Emani, C., & Abdurakhmonov, Y. (2017). Genome editing in plants: an overview of tools and applications. *International Journal of Agronomy, Article ID7315351*, Retrieved from: https://doi:org./10.11 5 5/2017/7315351.

Kannan, B., Jung, J. H., Moxley, G. W., Lee, S., & Altpeter, F. (2018). TALEN-mediated targeted mutagenesis of more than 100 COMT copies/alleles in highly polyploidy sugarcane improves saccharification efficiency without compromising biomass yield. *Plant Biotechnology Journal, 16*, 856-866.

Kim, H., Kim, S.T., RYU, J., Kang, B.C., Kim, J.S., & Kim, S.G. (2017). CRISPER/CPF1-mediated DNA-free plant genome editing. *Nature Communication, 8*, 14406-14411.

Lei, Y., Lu, L., Lin, H.Y., Li, S., Xing, F., & Chen, L.L. (2014). CRISPR-P: a web tool for synthetic single-guide RNA design of CRISPER-system in plants. *Molecular Plants, 7*, 1494-1496.

Li, J., Sun, Y., Du, J., Zhao, Y., & Xia, L. (2017). Generation of targeted point mutations in rice by a modified CRISPR/Cas9 system. *Molecular Plants, 10*, 526-529.

Li, R., Liu, C., Zhao, R., Wang, L., Chen, L., Yu, W., ... Shen, L. (2019). CRISPER/Cas9-mediated SINPR1 mutagenesis reduces tomato plant drought tolerance. *BMC Plant Biology, 19*, 38-46.

Li, R., Zhang, L., Wang, L., Chen, L., Zhao, R., Sheng, J., & Shen, L. (2018). Reduction of tomato-plant chilling tolerance by CRISPER-CAS9-mediated SICBF1 mutagenesis. *Journal of Agriculture and Food Chemistry, 66*, 9042-9051.

Li, T., Yang, X., Yu, Y., Si, X., Zhai, X., Zhang, H. ..... Xu, C. (2018). Domestication of wild tomato is accelerated by genome editing. *Nature Biotechnology, 36*, 1160-1163.

Li, X., Wang, X., Tong, M., Tan, Y., Down, J.D., Shedlock, D.J., Ostertag, E.M. (2018). Cas-CLOVER™: A high-fidelity genome editing system for safe and efficient modification of cells for immunotherapy. *Precision CRISPR Congress Poster Presentation, Boston, MA.*

Li, X., Wang, Y., Chen, S., Tian, H., Fu, D., Zhu, B., ... Zhu, H. (2018). Lycopene is enriched in tomato fruit by CRISPR/Cas9-mediated multiplex genome editing. *Frontiers in Plant Sciences, 9*, 559-564.

Liang, Z., Chen, K., Zhang, Y., Lin, J., Yen, K., Qin, J.L. & Geo.,C. (2018). Genome editing of bread wheat using biolistic delivery of CRISPR/Cas9 in vitro transcripts or ribonucleoproteins. *Nature Protocols, 13*, 413-430.

Lin, C.S., Hsu, C.T., Yang, L.H., Lee, L.Y., Fu, J.Y., Chang, Q.W., ... Zhang, R. (2018). Application of protoplast technology to CRISPR/Cas9 mutagenesis: from single-cell mutation detection to mutant plant regeneration. *Plant Biotechnology, 16*, 1295-1310,

Liu, H., Ding, Y., Zhou, Y., Jin, W., Xie, K., & Chen, L.L. (2017). CRISPER-P 2.0: an improved CRISPR/Cas9 tool for genome editing in plants. *Molecular Plant, 10*. 530-532.

Lu, H.P., Lin, S.M., Xu, S.L., Chen, W.Y., Zhou, X., Tan, Y.Y. ... Shu, Q.Y. (2017). CRISPER-S: an active interference element for a rapid and inexpensive selection of genome-edited, transgene-free rice plants. *Plant Biotechnology Journal, 15*, 1371-1373.

Lu, Y., Ye, X., Gao, R., Huang, J., Wang, W., Tang, J., ... Qian, Y. (2017). Genome-wide targeted mutagenesis in rice using the CRISPER/Cas9 system. *Molecular Plant,*

10, 1242-1245.

Ma, J., Koster, J., Qin, Q., Hu, S., Li, W., Chen, C., ... Liu, X.S. (2016). CRISPR-DO for genome-wide CRISPR design and optimization. *Bioinformatics. 32*, 3336-3338.

Ma, X., Zhang, Q., Zhu, Q., Liu, W., Chen, Y., Qin, R., ... Liu, Y. G. (2015). A robust CRISPER/Cas9 system for convenient, high-efficient multiplex genome editing in monocot and dicot plants. *Millennium Plant, 8,* 1274-1284.

Makarova, K. S., Haft, D. H., Barrangou R., Brouns, S. J., Carpentier, E., Horvath, P., ... Koonin, E. V. (2011). Evolution and classification of the CRISPER-Cas system. *National Review of Microbiology, 9,* 467-477.

Mao, Y., Botella, J. R., Liu, Y., & Zhu, J. K. (2019). Gene editing in plants: progress and challenges. *National Science Review, 6,* 421-437.

Mao, Y., Botella, J.R., & Zhu, J. K. (2017). Heritability of targeted gene modifications induced by plant-optimised CRISPER systems. *Cellular and Molecular Life Sciences, 74*, 1075-1093.

Meng, X., Yu, H., Zhang, Y., Zhuang, F., Song, X., Gao, S., ... Li, J. (2017). Construction of a genome-wide mutant library in rice using CRISPER/Cas9. *Molecular Plant, 10,* 1238-1241.

Metje-Sprik, J., Menz, J., Modrzejewski, D., & Sprink, T. (2019). DNA-free genome editing: past, present and future. *Frontiers in Plant Science, 9,* 1-9. Retrieved from: http://doi:10.3389/fpls.2018.01957.

Michno, J.M., Wang, X., Liu, J., Curtin, S.J., Kono, T.J., Stupar, R.M. (2015). CRISPR/Cas mutagenesis of soybean and *Medicago truncatula* using a new web-tool and modified Cas9 enzyme. *GM Crops and Food, 6*, 243-252.

Mishra, R., Joshi, R. K., & Zhao, K. (2020). Base editing in crops: current advances, limitations and future implications. *Plant Biotechnology Journal. 18*, 20-31.

Molla, K.A., & Yang, Y. (2019). CRISPR/Cas-mediated base editing: technical considerations and practical applications. *Trends in Biotechnology, 37*, 1121-1142.

Moose, S. P., & Mumm, R. H. (2008). Molecular plant breeding as a foundation for 21st century crop improvement. *Plant Physiology, 147*, 969-977.

Nadeem, M. A., Nawaz, M. A., Shahid, M. Q., Dogan, Y., Comertpay, G., Yidiz, M., ... Chung, G. (2018). DNA molecular markers in plant breeding: current status and recent advances in genomic selection and genome editing. *Biotechnology and Biotechnological Equipment, 32,* 261-285.

Okuzaki, A., Ogawa, T., Koizuka, C., Kaneko, K., Inabas, M., Imamura, J., & Koizuka, N. (2018). CRISPER/Cas9-mediated genome editing of the fatty acid desaturase 2 gene in *Brassica napus. Plant Physiology and Biochemistry, 131*, 63-69.

Patwardhan, A., Ray, S., & Roy, A. (2014). Molecular marker in phylogenetic studies-a review. *Journal of Phytogenetics and Evolution Biology, 2,* 2-9.

Pawluk, A., Davidson, A.R., & Maxwelll, K.L. (2018). Anti-CRISPER: discovery, mechanism and function. *National Review of Microbiology. 16*, 12-17.

Perez-de-Castro, A. M., Vilanova, S., Canizares, J., Pascual, J. M., Blanca, M. J., Diez, J., Prohens, J., & Pico, B. (2012). Application of genomic tools in plant breeding. *Current Genomics, 13,* 179-195.

Qi, L.S., Larson, M.H., Gilbert, L.A., Doudna, J.A., Weissman, J.S., Arkin, A.P., ... Lim, W.A. (2013). Repurposing CRISPR as an RNA-guided platform for sequence-specific control of gene expression. *Cell, 152,* 1173-1183.

Qi, W., Zhu, T., Tian, Z., Li, C., Zhang, W., & Song, R. (2016). High-efficiency CRISPER/Cas9 multiplex gene editing using the glycine tRNA-processing system-based strategy in maize. *BMC Biotechnology, 16,* 58-73.

Qi, Y (Ed.) (2019). *Plant Genome Editing with CRISPR Systems, Methods and Protocols,* New York, Springer.

Rahman, M.K., & Rahman, M.S. (2017). CRISPRpred: a flexible and efficient tool for sgRNAs on-target activity prediction in CRISPR/Cas9 systems. Public Library of Science *(PLoS) ONE, 12,* e0181943.

Ran, Y., Patron, N., Kay, P., Wong, D., Buchanan, M., Cao, Y. Y., ... Webb, S. R., (2018). Zinc finger nuclease-mediated precision genome editing of an endangered gene in hexaploid bread wheat (*Triticum aestivum*) using a DNA repair template. *Plant Physiology Journal, 16*, 2088-2101.

Rastogi, A., Muruk, O., Bowler, C., & Tirichine, L. (2016). A web-based and stand-alone application to find specific target sequences for CRISPER/Cas editing. *BMC Bioinformatics, 17*, 261-278.

Razzaq, A., Saleem, F., Kanwal, M., Mustafa, G., Yousaf, H. M. I., Hameed, M. K., ... Joyia, F. A., (2019). Modern trends in plant genome editing: an inclusive review of the CRISPER/Cas9 toolbox. *International Journal of Molecular Sciences, 20,* 4045-4089.

Sauer, N. J., Mozoruk, J., Miller, R. B., Warburg, Z. J., Walker, K. ., Beetham, P. R., ... Gocal, G. F. (2016). Oligonucleotide-directed mutagenesis for precision gene editing. *Plant Biotechnological Journal, 14,* 496-502.

Schiml, S., Fauser, F., & Puchta, H. (2014). The CRISPR/Cas system can be used as nuclease for *in planta* gene targeting and as paired nickases for directed mutagenesis in *Arabidopsis* resulting in heritable progeny. *The Plant Journal, 80*, 1139-1150.

Shah, Q., Wang, Y., Li, J., Zhang, Y., Chen, K., Liang, Z., ... Gao, C. (2013). Targeted genome modification of crop plants using a CRISPER-Cas system. *Natural Biotechnology, 31,* 686-688.

Shen, L., Wang, C., Fu, Y., Wang, J., Lin Q., Zhang, X., ... Wang, K. (2018). QTL editing confers opposing yield performance in different rice varieties. *Journal of Integrated Plant Biology, 60*, 89-93.

Shimatani, Z., Ariizumi, T., Fujikura, U., Konodo, A., Ezura, H., & Nishida, K. (2019). Targeted base editing with CRISPER-deaminase in tomato. *Methods in Molecular Biology, 1917*, 297-307.

Stemmer, M., Thumberger, T., Dei Soi Keyer, M., Wittbrodt, J., & Mateo, J.L. (2015). CCTop: an intuitive, flexible and reliable CRISPER/Cas9 target prediction tool. Public Library of Science (*PLoS) ONE, 10*, e0124633.

Sun, J., Liu, H., Liu, J., Cheng, S., Peng, Y., Zhang, Q., ... Chen, L.L (2019). CRISPER-local: a local single-guide RNA (sgRNA) design tool for non-reference plant genome. *Bioinformatics, 35*, 2501-2503.

Tian, S., Jiang, L., Cui, X., Zhang, J., Guo, S., Li, M., ... Zong, M. (2018). Engineering herbicide-resistant watermelon variety through CRISPER/Cas9-mediated base-editing. *Plant Cell Report, 67*, 1068-1079.

Varshney, R., Nayak, S. N., Hoisington, D., & Graner, A. (2009) Molecular plant breeding: methodology and achievements. In D. J. Somers (Ed.) *Methods in molecular biology* (pp. 283-304). New York, Humana Press.

Wolter, F., & Puchta, H. (2017). Knocking out consumer concerns and regulator's rule: efficient use of CRISPR/Cas ribonucleoprotein complexes for genome editing in cereals. *Genome Biology, 18*, 682-698.

Wolter, F., Schindele, P., & Puchta, H. (2019). Plant breeding at the speed of light: the power of CRISPR/Cas to generate directed genetic diversity at multiple sites. *BMC Plant Biology, 19*,176-183.

Xie, X., Ma, X., & Zhu, Q. (2017). CRISPR-GE: a convenient software toolkit for CRISPR-based genome editing. *Molecular Plant, 10*, 1246-1249.

Xie. K., Zhang, J., & Yang, Y. (2014). Genome-wide prediction of highly specific guide RNA spacers for CRISPER-Cas-mediated genome editing in model plants and major crops. *Molecular Plant, 7*, 923-926.

Yan, F., Kuang, Y., Ren, B., Wang, J., Zhang, D., Lin, H., ... Zhou, H. (2018). Highly efficient A-T to G-C base editing by Cas9n-guided tRNA adenosine deaminase in rice. *Molecular Plant, 11*, 631-634.

Zhang, A., Liu, Y., Wang, f., Li, T., Chen, Z., Kong, D., ... Wang, J. (2019). Enhanced rice salinity tolerance via CRISPR/Cas9-targeted mutagenesis of the OsRR22 gene. *Molecular Breeding, 39*, 47-56.

Zhang, K., Raboanatahiry, N., Zhu, B., & Li, M. (2017). Progress in genome editing technology and its application in plants. *Front. Plant Sci. 8*, 177-184. doi: 10.3389/fpls.2017.00177

Zhang, S., Zhang, R., Song, G., Gao, J., Li, W., Han, X., ... Li, G. (2018). Targeted mutagenesis using the *Agrobacterium tumefaciens*-mediated CRISPR-Cas system in common wheat. *BMC Plant Biology, 18*, 302-312.

Zhang, Y., Massel, K., Godwin, I. D., & Gao, C. (2018). Applications and potential of genome editing in crop improvement. *Genome Biology, 19,* 201- 2011.

Zhang, Z., Hua, L., Gupta, A., Tricoli, D., Edwards, K. J., Yang, B., & Li, W. (2019). Development of an Agrobacterium-delivered CRISPR/Cas9 system for wheat genome editing. *Plant Biotechnology Journal, 17,* 1623-1635.

Zong, Y., Song, Q., Li, C., Jin, S., Zhang, D., Wang, Y., ... Gao, C. (2018). Efficient C-to-T base editing in plants using a fusion of ncas9 and human apobec3a. *Nature Biotechnology, 36,* 950-963.

Zong, Y., Wang, Y., Li, C., Zhang, R., Chen, K., Ran, Y., ... Gao, C. (2017). Precise base editing in rice, wheat and maize with a Cas9-cytidine deaminase fusion. *Nature Biotechnology, 35,* 438-440.

Zsogon, A., Cermak, T., Naves, E. R., Notini, M. M., Edel, S. W., Weinl, S., ... Peres, L.F.O. (2018). *De novo* domestication of wild tomato using genome editing. *Nature Biotechnology, 36,* 1211-1216.

Chapter-6

Speed Breeding

Abstract

Breeding crops in a conventional way demands considerable time, space, inputs for selection, and the subsequent crossing of desirable plants. The duration of the seed-to-seed cycle is one of the crucial bottlenecks in the progress of plant research and breeding. During the early phases of breeding a significant amount of time, space, and resources are invested in the selection, and genetic advancement stages after initial crosses are performed with parental genotypes. Speed breeding (SB), has the potential to reduce the time required for cultivar development, release, and commercialization. SB relies mainly on photoperiod extension, temperature control, and early seed harvest, and has the potential to accelerate the rate of plant improvement.

Well demonstrated in the case of long-day plants, the SB protocols are being extended to short-day plants to reduce the generation interval time. Flexibility in SB protocols allows them to align and integrate with diverse research purposes including population development, genomic selection, phenotyping, and genomic editing. Though SB has been extensively used in plant phenotyping and the pyramiding of multiple traits for the development of new crop varieties, certain challenges and limitations hamper its widespread application across diverse crops. However, the existing constraints can be resolved by further optimization of the SB protocols for critical food crops and their efficient integration in plant breeding programmes.

Keywords: *Speed breeding, rapid generation advancement, speed breeding techniques, genome selection*

Introduction

The conventional breeding procedures can take more than 10 years to develop and release an improved variety in the absence of an integrated pre-breeding programme (Ahmar et al., 2020; De La Fuente et al., 2013). Field selection processes take an entire season, and the slow rate of advancement of each generation during conventional breeding is attributable to the inherent nature of the crop cycles. In the common cereal and legume crops a crop cycle is typically 3 to 6 months per crop, per generation, and per year. In other crops such as cassava, it takes 15 to 18 months to complete one breeding generation. In some agro-ecologies, variable weather conditions such as extreme temperatures, erratic and poor rainfall distribution, day length, etc. allow for only one crop cycle per year. Under some tropical crop production conditions, two generations per year are attainable (Laux et al., 2010). Furthermore, continuous field selection and early rapid generation advancement require skill and expertise in phenotyping, as well as production resources to manage many segregating populations before yield and economic traits can be evaluated once homozygous or genetically stable breeding populations have been created (Mobini et al., 2015). The duration of each crop breeding cycle can be reduced using modern technologies such as doubled haploid breeding (Dwivedi et al., 2015) and speed breeding (Hickey et al., 2019).

Climatic fluctuations involving rising temperatures, more frequent floods, and drought are predicted to lead to novel diseases and more frequent pest outbreaks, requiring an agile plant breeding response. Lin et al. (2016) highlighted the urgent need for increasing the current rate of genetic gain of critical food crops to safeguard global food security. Improving the rate of genetic gain will rely on accelerated crop breeding pipelines in order to allow rapid delivery of improved crop varieties. As inferred by the breeder's equation (Moose and Munm 2008), plant breeding can be accelerated by improving factors that influence the genetic gain per unit time (Bohra et al 2020, Sinha et al. 2021, Varshney et al. 2021), crucially, the breeding cycle time (t) (Cobb et al. 2019).

Since the 1940s, the speed of plant lifecycle turnover has been manipulated in plant breeding using techniques such as single-seed descent (Brim 1996) and shuttle breeding (Borlaug 1968). More recently, researchers have manipulated controlled-environment (CE) growth conditions to further truncate plant lifecycle time. Techniques to enhance the cycle turnover are widely grouped under the term speed breeding (SB) (Ghosh et al. 2018) and include accelerated single-seed descent (aSSD: rapid development of homozygous lines), rapid generation cycling (RGC: more breeding cycles per year using DNA marker technology), fast generation cycling (FGC: more generations per year using stressed conditions and in vitro culture of immature embryos), and rapid generation turnover (RGT: an increasing number of generations per year using immature seed harvest and photoperiod response). Since the early 21st century, this suite of SB techniques has been applied across economically and scientifically important crops, and pasture families (e.g. Poaceae, Fabaceae, and Brassicaceae) to achieve up to three-fold improvement in annual generation turnover compared to conventional generation advancement systems.

Rapid Generation Advancement Through Speed Breeding

Speed breeding techniques can be used to expedite breeding outcomes including the generation of crosses, mapping populations, and evaluation of agronomic traits of interest. Plants are grown under controlled environment (CE) conditions, and researchers manipulate day/night temperature, available light spectrum and intensity, and photoperiod duration in order to reduce time to floral initiation and hasten embryo development and seed maturity (Mobini et al. 2016, Croser et al. 2016, Hickey et al. 2017, Cazzola et al. 2020). Emphasis is placed on light, with plants responding to changes in light duration and quality, by compressing their time to flowering. The use of artificial electric lamps for accelerating plant growth and development has been long established (Pfeiffer 1926, Wheeler 2008). Photoperiod extension has since been used widely in long-day species to manipulate the time to flowering (Sysoeva et al. 2010). The advent of advanced LED lighting systems complemented efforts to accelerate lifecycle turnover, enabling manipulation of wavelength composition to trigger light responses, such as shade avoidance, and encourage rapid progression to flowering (Ballare et al. 1996, Pazos-Navarro et al. 2017, Singh et al. 2015, Christopher et al. 2015, Ribalta et al. 2017).

Speed breeding is a suite of techniques that involves the manipulation of environmental conditions under which crop genotypes are grown, aiming to accelerate flowering and seed set, to advance to the next breeding generation as quickly as possible (Figure 6.1). The method saves breeding time and resources through rapid generation advancement (Table 6.1). Various selection methods can be integrated into speed breeding, such as the single seed descent (SSD), single pod descent (SPD), single plant selection (SPS), clonal selection, and marker-assisted selection (MAS) to shorten the breeding cycle and for efficient resource use (Hickey et al., 2017; Watson et. al., 2018; Samineni et al., 2019). Speed breeding results in 3 to 9 generations per year compared to 1 to 2 generations per year achieved with conventional selection approaches (Ghosh et al., 2018; Ochatt et al., 2002). Accordingly, speed breeding offers opportunities to rapidly develop homozygous and stable genotypes, and to facilitate rapid generation advancement, resulting in accelerated development and release of new cultivars (Watson et al., 2018). Speed breeding technology also fits well with MAS and high-throughput phenotyping methodologies for multiple trait selection.

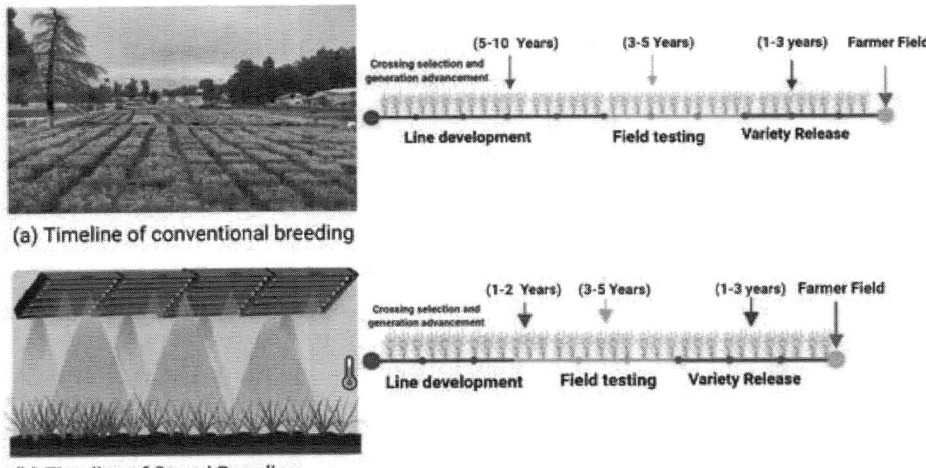

Figure 6.1. Timelines of varietal development with (a) conventional breeding and (b) speed breeding. The image was created using BioRender. (Reproduced from Samantara et al. 2022)

Table 6.1. List of crops where speed breeding has increased generation turnover.

Type of photoperiod	Family	Species	Generation/year	References
Long day	Poaceae	Oat (*Avena sativa*)	~ 7generations	Liu et al. 2017
		Barley (*Hordeum vulgare*)	~ 6 generations	Hicket et al. 2017
		Wheat (*Triticum aestivum*)	4 - 6 generations	Mukade et al. 1974
	Fabaceae	Clover (*Trifolium subterraneum*)	2.7 – 6.1	Pizos-Navarro et al. 2017
		Lentil (*Lens culinaris*)	~ 8 generations	Mobini et at. 2014
		Chickpea (*Cicer sativum*)	~ 6 generations	Watson et al. 2018, Atieno et al. 2017
		Pea (*Pisum sativum*)	6.8 generations	Ochatt et al. 2002, Ribalta et al. 2017
			5 generations	Mobini and Warkentin 2016
		Faba bean (*Vicia faba*)	7 generations	Mobini ert al. 2014
		Narrow-leaf lupin (*Lupinus angustifolius*)	5 generations	Croser et al. 2016
	Brassicaceae	Rapeseed (*Brassica napus*)	~ 5 generations	Watson et al. 2018
	Linaceae	Flex (*Linum usitatissimum*)	~ 3 generations	Sysoeva et al. 2010

Type of photoperiod	Family	Species	Generation/year	References
Short day	Poaceae	Rice (*Oryza sativa*)	~ 4 – 5 generations	Rana et al. 2019, Collard et al. 2017
		Sorghum (*Sorghum bicolor*)	4 generations	Foster 2014
	Fabaceae	Soybean (*Glycine max*)	~ 5 generations	Nagatoshi and Fujita 2019, Jahne et al. 2020
		Pigeonpea (*Cajanas cajan*)	~ 4 generations	Saxena et al. 2017
		Bambara groundnut (*Vigna subterranea*)	~ 4 generations	Ochatt et al. 2002
		Groundnut (*Arachis hypogaea*)	~ 4 generations	O'Connor et al. 2013
	Amaranthaceae	Grain amaranthus (*Amaranthus* spp,)	~ 6 generations	Stetter et al. 2016

The availability of a low-cost growth room design highlights the versatility of the SB 'recipe', which can be tailored according to local resources and purposes (Watson et al. 2018). SB technology has facilitated rapid phenotyping in wheat and the analysis of multiple disease-resistance traits in European two-rowed barley (Hickey et al. 2017, Atieno et al. 2017). A combination of SB technology and marker-assisted selection (MAS) has accelerated the development of herbicide-tolerant chickpea (Croser et al. 2021) and the introgression of valuable allelic variation from wild relatives in lentil (Lulsdorf and Bannize 2018). These practical breeding outcomes highlight the potential of the global suite of SB techniques to substantially accelerate genetic gain.

Table 6.2. Techniques for rapid generation advancement with corresponding days to flowering, number of generations achieved per year, and selection methods used in different crops.

Crops	Techniques	Days to flowering	No. of generations / year	Selection methods	References
Amaranth	Photoperiod and temp.	28	6	SSD	Stetter et al. 2016
Arabidopsis thaliana	Plant hormones, immature seed germination, and photoperiod	20-26	10	-	Ochatt and Sanwan 2008
Barley	Photoperiod, temp., immature seed germination, and embryo rescue	24-36	9	SSD	Zheng et al. 2013

Crops	Techniques	Days to flowering	No. of generations / year	Selection methods	References
Canola	Photoperiod, light intensity, temp., immature seed germination, embryo rescue, and soil moisture	73	4	SSD	Watson et al. 2018
Cheakpea	Photoperiod and immature seed germination	33	7	SPD	Samineni et al. 2019
Faba bean	Plant hormone, photoperiod, light intensity and immature seed germination	29-32	7	SPD	Mobini et al. 2015
Groundnut	Photoperiod and temp.	25-27	3	SPD	O'connor et al. 2013
Lentil	Plant hormone, photoperiod, light intensity, and immature seed germination	31-33	8	SPD	Mobini et al. 2015
Pea	Plant hormone, photoperiod, and immature seed germination	33	5	-	Mobini and Warkentin 2016
Pigeon pea	Photoperiod, temp., and high density planting	50-56	4	SPD	Saxena et al. 2019
Rice	Photoperiod, temp., and immature seed germination	75-85	4	SPD	Collard et al. 2017
Sorghum	Photoperiod, temp., and immature seed germination	40-50	6	SSD	Foster et al. 2014
Soybean	Photoperiod, temp., and immature seed germination	23	5	SSD	Jahne et al. 2020
Wheat	Photoperiod, temp., soil fertility, immature seed germination, and embryo rescue	28-41	7.6	SSD	Zheng et al. 2013
Abbreviations: SPD, single pod descent; SSD, single seed descent; SPS, single plant selection; —, not available.					

Speed Breeding Techniques

Various techniques used for speed breeding are described in the following section.

Rapid Development Of Homozygous Lines For Accelerated Breeding

Speed breeding techniques have been used on various crops to rapidly develop homozygous lines after initial crosses of selected parents with complementary traits. The technique depends on the manipulation of photoperiod, light intensity, temperature, soil moisture, soil nutrition, and high-density planting. These methods have been used to induce early flowering and seed set, reducing the time taken to generate each breeding generation (Table 6.2). By adopting this method, it was possible to produce 3 to 9 breeding generations per year. This is ideal for accelerated breeding and population evaluation across the target production environments involving various selection methods such as SSD, SPD, and SPS (El-Hashash and El-Absy, 2019). Speed breeding relies on deliberate manipulation of the following growing conditions.

Manipulation Of Photoperiod Regime

Photoperiod refers to the length of daily exposure of plants to scheduled light and dark regimes to enhance rapid growth, development, flowering, and seed set (Vince-Prue, 1994). Different crop species and genotypes within species have variable photoperiod requirements for flower induction and seed set (Kouressy et al., 2008; Saito et al., 2009). Thus, it is important to determine the optimum light quality, intensity, and photoperiod that trigger flower initiation in various crops and genotypes (Figure 6.2).

Light quality, which includes the instantaneous and cumulative amount, delivered per day has a direct effect on plant growth, net photosynthetic rate, stomatal conductance, intercellular CO2, and transpiration rate (Yang et al., 2017). Also, the daily light-to-dark hours influence flowering rate and maturity. Light sources emitting photosynthetic active radiation (PAR) within the range of 400–700 nm with an intensity of 360–650 μmol/m2/s have been successfully used in many crops, including wheat, barley, chickpea, pea, and canola to facilitate speed breeding (Ghosh et al., 2018; Watson et al., 2018) (Figure 6.3). Light-based speed breeding protocols provide a benefit for sustained photosynthesis for the entire light period, all year round (Bhatta et al., 2021). Dubcovsky et al. (2006) reported that a photoperiod of 22 hr light and 2 hr dark under PAR of 150–190 μE m−2s−1 reduced the total number of days to flowering by half when compared with the corresponding wheat genotypes grown with 12/12 hr light/dark. This procedure induced flowering in 35 and 39 days in several wheat genotypes, such as Paragon, Watkins landrace W352, and a late flowering Paragon x W352 F6 recombinant inbred line. The plants grown with 12/12 hr light/dark were still at the stem elongation growth stage when the corresponding plants grown with 22/2 hr light/dark had started flowering (Figure 6.4). In another study, when plants of a sensitive winter wheat genotype, G3116 (AY485969) were grown under short day (SD) conditions (8/16 hr light/dark) for six weeks before being transferred to long day (LD) conditions (16/8 hr light/dark) at a light intensity of 200–270 µmol/m2/s, flowering was induced without the need for vernalization. The days to flowering of non-vernalized genotypes grown in SD-LD condition were comparable with genotypes vernalized for six weeks (Dubcovsky et. al., 2006).

Figure 6.2. Breeding for improved wheat lines under Speed Breeding conditions. (Reproduced from Mundia 2020).

Zheng et. al. (2013) reported that a photoperiod of 16/8 hr light/dark with a light intensity of 500 µmol/m2/s reportedly induced early flowering in seven barley genotypes: namely, Franklin (36 days), Gairdner (35 days), Gimmett (33 days), Commander (30 days), Fleet (29 days), Baudin (26 days), and Lockyer (25 days). This day-length regime was combined with the germination of immature seed, and moisture and nutrition management, resulting in an increase in the number of generations advanced per year reaching 6–8 and 7–9 in wheat and barley, respectively. Samineni et al. (2019) reported that a photoperiod length of 12/12 hr light/dark using a standard incandescent bulb of 60 W with a light intensity of 870 lm induced early flowering in chickpeas. This regime reportedly induced early flowering ranging from 37–38, 50–51, and 60–65 days in early-maturing (i.e., JG 11 and JG 14), medium-maturing (IGGV 10 and JG 16) and late-maturity (C 235 and CDC - Frontier) chickpea genotypes. Depending on the test genotypes, the number of days to flowering in early-, medium- and late-maturity genotypes were reduced by 8–19, 7–16, and 11–27 days, respectively. In grain amaranth, flowering was induced in about four weeks by growing plants for two weeks in LD conditions (16/8 hr light/dark) followed by SD conditions (8/16 hr light/dark) using a light intensity of 150 mmol (Stetter et al., 2016). Continuous exposure to light (24 hr light) with a 450 W PAR lamp induced relatively early flowering at 25 to 27 days after germination in groundnut (O'Connor et al., 2013). Adjusting the photoperiod is cost-effective when using low-energy, light-emitting diodes (LED), which can use battery-based inverter systems, charged from solar panels. The use of solar power systems is an effective and sustainable approach for indoor speed breeding in countries with unreliable electricity supplies.

Figure 6.3. Accelerated plant growth and development under speed breeding (22-hour photoperiod conditions) (left) compared to standard long-day conditions (16-hourphotoperiod) (right) in LED-supplemented glasshouses (Reproduced from Ghosh et al., 2018).

Figure 6.4. Speed breeding for improvement of wheat. Flowering and seed set was enhanced after exposing the plants to 22 hrs of photoperiod, resulting in growing 4-6 generations per year compared to the normal 2 generations, while exposed to 12 hrs of photoperiod. (Reproduced from Mundia 2020).

Regulation Of Temperature Regime

Adjustments to soil and air temperatures affect germination and growth responses, leading to rapid growth, flowering, seed set, and maturity. Low and high-temperature extremes activate a wide range of effects on the rate of plant development, including a transition from the vegetative to the reproductive stages (Hatfield and Prueger, 2015; McClung et al., 2016). The temperatures required for germination for most crops are between 12 and 30°C, whereas the optimum temperature for growth, flowering, and seed set varies from 25 to 30°C for most crops. Temperatures maintained at 25 ± 1 °C under 12/12 hr light/dark conditions were used for germination of direct sown immature seed in chickpeas (Samineni et al., 2019). Temperature regulation allowed the germination of immature seeds (harvested 16–24 days after flowering), freshly planted into pots, which allowed to produce 7 generations per year. In winter wheat, vernalization or cold temperature stress is required at the vegetative stage to accelerate the transition to the reproductive stage (Yan et al., 2004; Dubcovsky et al., 2006). Temperatures above 33°C can lead to decreased pollen viability and increased male sterility in rice, sorghum, and soybean (Wiebbecke et al., 2012; Hatfield and Prueger, 2015; Singh et al., 2015). Therefore, temperatures inside the critical limit can facilitate flowering, seed set, and maturity for speed breeding. Zheng et al, (2013) used temperatures of 20–22°C for the germination of immature seeds derived from embryo culture in wheat and barley. After germination, seedlings were transferred to a temperature regime of 25/22°C synchronized with a photoperiod of 16/8 hr light/dark for rapid plant growth and early flowering. O'Connor et al (2013) used a temperature regime of 17/32°C, to facilitate rapid plant growth, flower induction, and seed set under conditions of constant light (24/0 hr light /dark) in groundnuts. Photoperiod-sensitive genotypes of various crops have shown variable responses to temperature regimes that affect their transition from the vegetative to the reproductive stage. In sorghum, early flowering was induced by exposing sensitive genotypes (BTx642/Tx7000 RIL population and parental lines) to LD conditions (14/10 hr light/dark) or to SD conditions (10/14 hr light/dark) at 30/23°C day/night temperature (Yang et al., 2014). Solar battery-powered air-conditioning systems could provide a cost-effective, stable technology for indoor speed breeding programmes.

Regulation Of Soil Moisture

Soil moisture stresses can cause significant changes in plant growth and development processes affecting plant height, days to flowering, and seed set and maturity (Anjum et al., 2017; Hussain et al., 2018). Drought or flooding stress can trigger early flowering and maturation, which can be used in speed breeding. Drought stress is the most applied technique for crops such as wheat, barley, and pearl millet (Shavrukov et al., 2017). Drought causes early flowering in pearl millet, which may have evolved as an 'escape mechanism' to produce the next generation (Vadez et al., 2012). However, in no-tillering, late-flowering genotypes of pearl millet, drought stress can result in sterility, and extend the flowering period up to

18 days in high-tillering genotypes (De Rouw and Winkel, 1998). In cowpea, plants grown under drought stress flowered about 12 days earlier than those grown under well-watered conditions (Agbicodo et al., 2009; Goufo et al., 2017). In wheat and barley, irrigation when plants show wilt symptoms promotes plant growth and development. Zheng et al. (2013) combined watering regimes with embryo rescue, adjusted photoperiod, and adjusted temperatures to produce 8 and 9 generations per year in wheat and barley, respectively. In contrast, in sorghum, drought stress delayed flowering up to 9 days in cultivars Segaolane (Botswana 79), Mahube (SDS 2583), and Phofu (Macia) when compared with the same genotypes grown under well-watered conditions (Munamava and Riddoch, 2001). These studies highlighted the opportunity to optimize water supply in crop speed breeding facilities for more efficient generation turnover.

After flowering, rapid grain filling and maturation can be facilitated by gradually reducing soil moisture content. Reducing the watering frequency from daily to twice per week, from 4 to 6 weeks after flowering, followed by non-watering in the last week before harvesting has been used in several crops including wheat, barley, canola, and chickpea for speed breeding (Watson et al., 2018). Soil moisture management approaches are applicable in both field and indoor growing environments.

Plant Population Density

High-density planting implies growing plants at higher densities than the density required to produce maximum yield. Usually, high-density planting results in taller plants due to competition for competition, which leads to a rapid transition from the vegetative to the reproductive growth stages (Warnasooriya and Brutnell, 2014). This approach is useful to induce early flowering and maturity, increasing the number of generation cycles per year. In rice, up to four generations per year were achieved using a high-density planting of 400 plants m-2, with intra-row spacing of 5 cm and inter-row spacing of 5 cm (5 × 5 cm), compared to the conventional planting of 25 plants m-2 (20 × 20 cm) (Rahman et al., 2019). The length of a crop cycle in rice can be reduced by 15 to 40 days, (i.e. 90 days from 105 days, and 105 days from 145 days) (Rahman et al., 2019) using high-density planting. However, others have reported that high-density planting did not accelerate flowering in rice (Hayashi et al., 2006; Fukushima et al., 2011). Jones and Johnson (1991) reported that a plant density of between four and eight plants m-2 had non-significant effects on plant growth and grain yield in sorghum. Villar et al. (1989) reported that increasing plant density from 16 to 38 plant m-2 reduced the days to flowering from 59 to 50 days in sorghum. In cotton, high-density planting (11 plant m-2) slightly reduced flowering to 25–26 days from 26–31 days compared to lower-density planting (9 plant m-2) (Khan et al., 2017, 2019). Thus, it is apparent that genotype differences affect plant responses to high-density planting under various field conditions. Therefore, for speed breeding, it is important to establish high-density planting requirements of a given genotype through preliminary trials to optimize the induction of early flowering. High planting density is one of the low-cost speed breeding strategies suitable for rapid advancement of generations, while maintaining the large population size required for advanced selections.

Carbon Dioxide Levels Modifications

In some plants, increased levels of carbon dioxide (CO2) may enhance rapid plant growth and the speed of the transition from the vegetative to the reproductive stage (Jagadish et al., 2016). However, different crop species and genotypes within a species have varying responses to increased CO2. For example, increased levels of CO2 of 400/700, 350/700, and 350/650/100 ppm reduced days to flowering in soybean, rice, and cowpea by 2, 7, and 12 days, respectively (Springer and Ward, 2007). In contrast, CO2 maintained at 20 μmol/mol2 delayed flowering in soybean by 11 days (Bunce, 2015). In pigeonpea, when CO2 level was increased to 550 μmol/mol2, it delayed flowering by nine days in a short-duration cultivar ICPL 15,011 (Sreeharsha et al., 2015). Nagatoshi and Fujita (2019) reported that a combination of 14 hr light (30°C)/10 hr dark (25°C) cycle and CO2 supplementation > 400 ppm in growth chambers reduced the crop cycle from 102–132 to 70 days in soybean (cv 'Enrei'). They could grow up to 5 generations per year compared to 1–2 generations under field and/or greenhouse conditions. Further, increasing the CO2 level had little impact on flowering time but increased the number of flowers, which was an advantage for developing many crosses. In rice cv. Nipponbare and Yamadawa, increased CO2 (600 ppm) reduced the days to heading from 51 to 52 to 48-49 days, and from 80–88 to 70–74 days, under growth chamber conditions, respectively (Tanaka et al., 2016). Speed breeding involving modification of CO2 levels requires appropriate facilities such as growth chambers, CO2 cylinders and regulators, and operational costs. Also, it is important to adhere to health protocols and safety guidelines while handling and using CO2 cylinders and valves.

Application Of Plant Nutrition And Hormones

In many plants, application of plant nutrition and hormones has accelerated growth and induced flowering and seed set, and germination of immature seed in vitro (Bermejo et al., 2016). Varied responses to plant growth regulators (PGRs) are achieved when used in controlled environments such as greenhouses and growth chambers in which the photoperiod and temperatures can be monitored and controlled. For example, combining auxin and cytokinin hormones, in the form of flurprimidol (0.3 μM), indole-3-acetic acid (5.7 μM) and zeatin (2.3 μM), promoted in vitro flowering at 100% and seed set (90%) in faba bean (Mobini et al., 2015). Increased seed set was achieved by exogenous application of 6-benzylaminopurine (10–5 M BAP) four days after flowering in faba bean (Mobini et al., 2020). Mobini and Warkentin (2016) reported that the application of a combination of flurprimidol (0.9 μM) and 4-chloroindole-3-acetic acid (0.05 μM), resulted in 90 and 80% flowering and seed set in lentils. They also used integration of adjustments to the photoperiod (18/6 hr light/dark), temperatures (22/18°C light/dark), the application of plant growth regulators (i.e., flurprimidol, cytokinin, and auxins), and embryo rescue, to reduce the generation cycle in faba bean and lentil from 102 and 107 days to 54 and 45 days, respectively. With this approach, they could get up to 8 generations per year. They also used an in vivo protocol involving embryonic seed culture grown under 20 hr light (21°C)/4 hr (16°C) dark in a hydroponic system with a vermiculite substrate, scheduled fertilizer applications, and 500 μM/ m2 s−1 light intensity (fluorescent light bulbs). When treated with a plant growth hormone, flurprimidol (0.6 μM), treated and untreated pea plants had 100 and 98% flowering, and seed set in 33 and 68 days, respectively.

Application Of Embryo Rescue Technique

The embryo rescue technique has been successfully applied to reduce the generation time for several crops. The use of immature embryos grown on Murashige and Skoog (1962) culture medium supplemented with 0.175 mg/L indole-3-acetic acid (IAA) and 0 mg/L 6-benzylaminopurine (BAP) was found to be suitable for embryo culture of lentil (Bermejo et al., 2016). Embryo culture of wheat on a half-strength MS supplemented with ten times the normal levels of potassium dihydrogen phosphate (KH2PO4) and 4% sucrose induced 100 and 92% flowering rate and seed set in two wheat cv. 'Emu Rock' and 'Zippy' (Yao et al., 2016).

In cheakpea, the embryos rescued from the flower of 15-20 days old were cultured on Murashige and Skoog Medium modified (½ MS) and incubated at 20-25°C, 40-70% humidity, and 16:8 hour light/dark. After 10 days, the green plants were transferred to MS for regeneration. Ten days later, the green plants were obtained, and they were transferred to peat moss bots (15 Cm diameter) for normal growth. The newly generated plants were flowered after 48 days and until the next embryo rescue was collected again. With this protocol, one generation of chickpea will need about 88 days, and therefore it will be possible to harvest 4 generations per year under controlled-environment growth chambers (Tawkaz et al. 2018).

Application Of Drying And Chilling Methods

Drying and chilling methods can be used to break seed dormancy in the immature seeds of various crops (Ghosh et al., 2018). For example, drying of immature seeds of wheat and barley at 28–35⍰ for 3–5 days in an oven or dehydrator followed by chilling (4–5⍰) for 3 days resulted in germination rates of 80 and 100% for seeds harvested at 2 and 4 weeks after flowering, respectively (Watson et al., 2018). This procedure was subsequently used in combination with a long photoperiod (22/2 hr light/dark) and temperature regulation (22/17⍰ light/ dark) to achieve 4–6 generation cycles per year for wheat, barley, chickpea, pea, and canola. Saxena et al. (2017) reported that immature seeds of rapidly germinating pigeonpea varieties harvested 21 days after flowering achieved germination rates of between 73% and 100% without pre-treatment. Therefore, the use of immature seed and optimal germination conditions are applicable under both field and indoor conditions for rapid generation advancement.

Amenability With Selection Methods

Speed breeding is routinely used for the advancement of generations without phenotypic selection. However, modern technologies (high-throughput genotyping methods, marker-assisted selection, etc.) can be successfully integrated for target trait selection. The combination of speed breeding and effective selection methods can help in the maintenance of a good breeding population and genetic diversity, in such environments that restrict plant growth, and for maximizing yield production (Johnston et al., 2019). Conventional selection methods such as bulk, mass, recurrent, pedigree, and pure line selection require a genetically stable plant population for the selection of optimally yielding genotypes. These methods are not ideal for speed breeding due to the long inbreeding and selection cycles requirement. The most appropriate selection methods amenable to speed breeding are: single seed descent (SSD), single pod descent (SPD), and single plant selection (SPS) methods. These methods are briefly described below.

Single Seed Descent Method

Single seed descent (SSD) is geared towards achieving homozygous populations through continuous inbreeding of segregating populations by retaining one seed from each F2 plant and advancing these individuals to the next generation. Each inbred line developed is traced back to an F2 plant (Fehr, 1991). The time taken to achieve inbred lines with SSD is comparable to that of the doubled haploid (DH) method (Yan et al., 2017). The advantages of the SSD selection method include less growing area and labour requirement during the handling of early generations. It allows for the advancement of progeny under high-density plantings in small nurseries, growth chambers or greenhouses (Funada et al., 2013; Arbelaez et al., 2019). The negative aspect is that SSD is that it carries forward more inferior progenies than pure line, pedigree, and recurrent selection methods. Urrea and Singh (1994) found that the overall mean of seed yields of inbred lines developed through the SSD method was lower than those developed through pedigree and mass selection in common bean. In the case of SSD selection method, a sizeable number of F1 plants (50–100) are required to generate between 2000 and 3000 F2 plants to advance to the F3–F4 generations (Priyadarshan, 2019). At the F5 generation, plants are grown under ideal field conditions at an optimum spacing to attain maximum yield, allowing for the selection of superior F6 genotypes and advancing these using a head-to-row method. Superior lines/rows are selected for preliminary yield trials (F7) and yield trials (F8 – F10). Subsequently, superior lines (F11 – F12) are released as new varieties. Pignone et al. (2015) used SSD to rapidly generate 450 new inbred lines from 500 durum wheat varieties sourced from gene banks before evaluation for agronomic performance under field conditions. SSD was used in a rapid generation technology (RGT) programme using growth chambers, and generated the popular cv. 'Nipponbare' of rice (Tanaka et al., 2016). Bordes et al. (2007) found non-significant differences in the grain yield of inbred lines developed from the same parental genotypes using the doubled-haploid (DH) and SSD methods in maize. Ma et al. (1999) reported that the mean grain yield and kernel weight of SSD lines was higher than those of anther culture-derived lines in maize and spring wheat. Overall, SSD is the best selection method for speed breeding and can be carried out under both field and indoor conditions.

Single Pod Descent Method

The single pod descent (SPD) method involves the selection of one pod per plant from each F2 – F4 plant instead of a single seed. Since each pod contains more than one seed in most legume crops, SPD has a higher chance of maintaining each F2 plant in the advanced generations than SSD selection. Funada et al. (2013) reported that progenies developed from crosses between soybean cv. 'OAC Atwood' and 'RG600RR', with a mean of 2.4 seeds per pod, increased the population from 200 in the F2 to 300 plants in the F3 generation. They found non-significant differences in selection efficiency for lines developed using the SSD, SPD, and bulk methods.

Another advantage of SPD is that it allows for the early selection of pods, so a smaller population can be advanced. In soybean, SPD resulted in progeny with mean yields of 7.96 g plant-1 compared with the mean yield of 6.42 g plant−1 of test lines developed through SSD (Destro et al., 2003). Khosla et al. (2019) reported better yield increases in soybean lines selected using the bulk selection method of 72.6%, followed by pedigree (33%) and SPD (10.4%). It is essential to conduct preliminary trials to determine the efficiency of SPD for the crop and trait being selected for speed breeding.

Single Plant Selection Method

The single plant selection (SPS) method advances each F2 plant by harvesting all the seeds of each selected plant. Accordingly, the next generation will be advanced as plant-to-row. The SPS method has been used in a modified backcross strategy to develop introgression lines (ILs) within two years in barley (Hickey et al., 2017). European barley, cv. 'Scarlett', was crossed with other donor parents to develop lines that were resistant to leaf rust, net blotch, and spot blotch. Hickey et al. (2010) used a speed breeding procedure of continuous light and temperature maintained at 22°C, with 87 of the BC1F3:4 generation being selected from 5,000 BC1F2 plants. They reported that the yield of 12 Scarlett ILs selected for superior multiple disease resistance and agronomic traits was significantly higher than cv. 'Scarlett'. In bread wheat, Alahmad et al. (2018) used the SPS selection method to enhance foliar disease resistance, grain dormancy, seminal root angle, seminal root number, tolerance to crown rot, resistance to leaf rust and plant height in bread wheat. They further reported a better selection response when using SPS in comparison with unselected F3 plants for crown rot and leaf rust resistance, and for better root angle and number. The SPS involves the early selection of plants based on a smaller population than SSD and SPD.

Sb Systems for Fast-Tracking Applied and Basic Research

Early SB activities relied on in vivo–in vitro cycling or full in vitro lifecycle turnover (Croser et al. 2016, Varshney et al. 2021). However, it is the fully in vivo systems that have been most widely applied in improvement programs. Watson et al. (2018) presented three different SB facilities, customizable according to resource availability. SB I consisted of CE plant growth chambers with a photoperiod of 22 h provided by white LED bulbs, far-red bulbs, and ceramic metal hydrargyrum quartz iodide bulbs and 22 °C day/17 °C night temperature. When grown under these conditions, wheat (*Triticum aestivum, T. durum*), barley (*Hordeum vulgare*), and purple false brome (*Brachypodium distachyon*) flowered in half the time of controls grown in unregulated glasshouse conditions during spring and early summer. Germination rates and seed viability remained unaffected by the accelerated growth conditions, validating the potential of this technology for rapid crop improvement. A slightly modified setup, referred to as SB II, employed the same temperature conditions as SB I with a 22 h photoperiod provided by high-pressure sodium vapor lamps. Harvesting of immature seeds and cold treatment were additionally performed under SB II to further reduce generation time. Results from the growth stage for wheat, barley, canola, and chickpea plants demonstrated accelerated plant development and uniformity in time to anthesis under SB conditions compared to the control genotypes grown under identical glasshouse settings without supplemental lights. Under SB II conditions, wheat plants produced significantly more spikes, maintained grain number, and had early maturity within 14 days after flowering. An alternative, cost-effective SB III system included a 3 m3 insulated room, seven LB-8 LED light crates, and a 1.5 HP inverter split system domestic air conditioner. The lighting was adjusted to a 12 h photoperiod for four weeks followed by 18 h. The temperatures were maintained at 18 °C during dark and 21 °C during light. Temperature control systems should be carefully considered in SB systems to influence the rate of plant development. Several generations can be achieved in a plant breeding program in a short time by elevating the temperature, and SB techniques, in general, adopt late-spring-type temperatures. Likewise, 60–70% humidity is recommended for optimum plant growth and accelerated breeding. The combination of photoperiod, temperature, and humidity in a greenhouse chamber increases the rate of plant development in comparison to field conditions or conventional glasshouse conditions (Christopher et al. 2015, Rabilta et al. 2017).

Sb Applications in Research and Breeding

Applications of SB include the development of biparental and more complex mapping populations, pyramiding traits, hastening backcrosses, phenotyping adult plant traits, mutant studies, and genetic transformation experiments (Watson et al. 2018, Jahne et al. 2020). Recent research has shown the power of combining emerging techniques, such as gene editing, high-throughput phenotyping and genotyping, genomic selection (GS), and MAS, with SB for accelerating crop improvement (Atieno et al. 2017, Zaman et al. 2019, Gosal et al. 2020, Croser et al. 2021, Dadu et al. 2021, Khoo et al. 2021, Taylor et al. 2021). Furthermore, the cost and space requirements for producing many inbred lines can be minimized by planting them at high plant densities (Yao et al. 2017). SB helps overcome challenges associated with double haploid (DH) technology such as low germination rate, poor vigour, and sometimes distorted growth (Ferrie 2007). For genetic mapping purposes, recombinant inbred lines (RILs) developed after multiple generations of self-fertilization can be advantageous over DH due to the multiple meiotic events that occur during repeated fertilization and the resulting higher recombination frequency. Similarly, the advancement and evaluation of segregating generations can be done with SSD in a short period under SB conditions (Sinha et al. 2021), which is time-saving and cost-efficient compared to the conventional pedigree breeding method (Johne et al. 2020). This technique was effective in shortening the generation period, resulting in a three times higher generation turnover than shuttle breeding (Ortiz et al. 2007).

SB in Model Species

In *Arabidopsis thaliana*, seed germination on medium incorporating phytohormone benzylaminopurine and plant auxin-mimic picloram led to seed set approximately 40 to 45 days after sowing. The second-generation seeds were sown on half-strength, hormone-free MS medium resulting in accelerated time to flowering and fruiting compared to the first generation. The length of the second and subsequent cycles could, thus, be cut in half compared to the first generation, allowing for up to 13 generations per year. Ribalta et al. (2017) standardized a protocol to achieve up to six generations per year and identified markers for ascertaining the physiological maturity of embryos in peas. Immature seeds, in which embryos had reached physiological maturity, could be reliably harvested, and germinated without in vitro intervention. Further study identified a shift in the expression pattern of hormones related to embryo development, with compression of the development when grown under accelerated SSD, CE conditions. In woody, perennial plants, environmental conditions affect the length of the juvenile phase. In apple, field-grown seedlings typically require 5 years to flower, but growth can be accelerated to the adult reproductive phase in a comparatively short time such as 10 months (Aldwinckle 1975). However, such approaches face challenges at later stages when managing big plants under CE becomes very difficult. The dormancy of temperate-zone trees can also be overcome by maintaining low temperatures under high moisture conditions (Van Nocker 2014), providing a way to hasten the breeding cycle.

SB in Cereals

Several new approaches have been applied to reduce the time required to obtain homozygous lines after hybridization to expeditiously breed cereal varieties. For example, four to six generations of wheat were obtained following the harvesting of immature seeds after 15–20 days of anthesis and the treatment of the seeds with H2O2 at a low temperature (Mukade 1974). A later study by

De Pauw and Clarke (1975) improved the germination response of wheat seeds by extending the duration of H2O2 treatment at a low temperature (11°C) and, depending upon the cultivar, the generation time was reduced by 12–23 days. Similarly, Robertson and Curtis 1967) also observed more than 90% germination for air-dried seeds harvested after 21 days of anthesis. In rice, Japanese researchers used generation advancement methods to breed 'Nipponbare' and several other cultivars. A breakthrough in the generation acceleration procedure of rice was the development of the biotron breeding system (BBS) that accelerates the breeding cycle by regulating temperature, photoperiod, and CO2 level, in combination with embryo rescue and tiller removal (Ohnishi et al. 2011). The utility of the BBS was evidenced by the marked reduction in the generation intervals (two months) of 'Nipponbare'. Tanaka et al. (2016) reduced the generation intervals of 'Nipponbare' by three months using the BBS without embryo rescue and tiller removal, thus, enhancing the feasibility of the BBS adoption for rapid generation advancement. More recently, researchers obtained four to five generations of rice per year (Ghosh et al. 2018).

The SB technique has been used efficiently in wheat for the rapid screening of multiple traits related to diseases, such as leaf rust, and root architecture and for evaluating plant height and flowering time (Ghosh et al. 2018, Alahmed et al. 2018). SB has been implicated for screening drought-tolerance traits in barley (Ghosh et al. 2018, Zhing et al. 2015). A modified, backcrossing methodology, in combination with SB, was used for two years in the development of resistant lines of barley that were otherwise susceptible to different diseases including rust and spot blotch (Hicky et al. 2017). Similarly, the embryo rescue method and direct germination of immature seeds can be applied in sorghum to significantly reduce the time required for the breeding cycle (Rizal et al. 2014). Increasing the photoperiod and a foliar mineral supplement are also shown to reduce time to anthesis for a higher generation turnover in oats (Tanaka et al. 2016).

SB in Oilseeds

The possibility of viable seed production through precocious germination was shown in soybean (Burris 1973). Later, Roumet and Morin (1997) demonstrated a growth cycle truncated from 130–140 to 65–70 days using precocious germination of immature, pre-treated pods. Nagatoshi and Fujita (2019) developed a standardized rapid generation advancement protocol for the high-quality, Japanese, soybean cultivar Enrej, which reduced crop duration from 102–132 days to 70 days. The availability of such methods enables five generations per year instead of one to two generations in a year. In the same way, Watson et al. (2018) optimized an SB protocol in canola to enhance the generation turnover and facilitate the phenotyping of the pod-shattering trait. For this, five canola cultivars susceptible to pod shattering were grown in environment-controlled growth chambers. Using the embryo rescue technique, Dagustu et al. (2012) established a short breeding period protocol for sunflower that can be used to shorten the generation time in a breeding program. For this, seed embryos were cultured in MS media with 2%

sucrose and 0.8% agar at pH 5.6–5.7 after 10–12 days of pollination, as previously used in tobacco (Murashige and Skoog 1962).

SB in Legumes

An in vitro-assisted SSD technique in clover (*Trifolium subterraneum* L.) accelerates the generation cycle by minimizing time to flowering by growth under regulated temperature, along with an expanded photoperiod, truncated seed-filling period, and embryo rescue. Growing immature seeds helps overcome the problem of seed dormancy. This technique enables 2.7–6.1 generations per year over a wide range of clover genotypes (Pazos-Navarro et al. 2017).

Generation acceleration protocols have been optimized in many legume species, especially temperate pulses that positively respond to photoperiod extension owing to their facultative, long-day nature (Croser et al. 2016). For example, continuous light in conjunction with optimal temperature and humidity in a greenhouse facility led to an increased rate of plant growth in peanuts (Jahne et al. 2020). Compared to greenhouse conditions, the in vitro protocols, in combination with in vivo manipulation, worked better for shortening generation cycles in pea and Bambara groundnut, whereas an in vivo-only strategy showed promising results in the case of peas and grass peas (Varshney et al. 2021). Similarly, Espósito et al. (2012) devised an in vitro-only strategy to generate an adequate number of F2 plants in pea breeding programs. Lentil can be grown in vitro using a tissue culture method with agar as the substrate and MS as the medium or a hydroponic system with perlite as the substrate and HS as the medium or agar as the substrate and Hestrin–Schramm (HS) as the medium, which enables eight generations of lentil within one year (Mobini et al. 2014). This study expanded the in vitro rapid generation technology of lentil to faba bean and evaluated the effect of the hydroponic method, tissue culture, and intermediate method on accelerating anthesis time and seed set rates. The generation time was 54 days, including 18 days for immature seeds to be ready for embryo rescue, resulting in 6.8 generations each year as opposed to one in the field and three in the greenhouse. A more recent study in chickpea (*Cicer arietinum* L.) reduced seed-to-seed cycle time based on induction of early flowering and germination of immature seeds (Samineni et al. 2020). In pigeonpea, a rapid generation advancement strategy demonstrating 100% germination from immature seeds harvested from 35-day-old plants provided novel avenues for growing three to four generations in a year (Saxena et al. 2017). Considerations of light quality allowed researchers to extend these SB protocols to soybean, a short-day legume crop. A 10 h photoperiod using a blue-light-enriched, far-red-deprived light spectrum enabled plants to mature within 77 days after sowing, thus, allowing five generations of soybean in a year (Jahene et al. 2020). A recent study in groundnut combined the single-seed chipping (SSC) process with high-throughput genotyping (HTPG) and SB. A small portion of cotyledon from the posterior end of the seed was used for DNA extraction. A germination rate of 95–99% was observed from the chipped seeds. The study led to a time saving of

6–8 months following the implementation of this integrated approach in groundnut research and breeding (Parmer et al. 2021).

The growing numbers of examples of SB in long-day and more recently in short-day plants attest to its broad utility in breeding programs, allowing faster homozygosity, the creation of mapping populations, and a significant reduction in time, space, and resources for cultivar development.

SB in Fruit Crops

Many fruit crops undergo a long juvenile phase before flowering, in some cases, taking >20 years ((Van Nocker 2014). SB techniques have led to vigorous vegetative growth and early flowering in apple (ten months instead of five years) and chestnut (two years instead of seven years) (Van Nocker 2014, Baier et al. 2012). The development of a new cultivar with desirable traits was achieved in apple using SB technology, which is based on transgenic, early-flowering plants and MAS (Flachowsky et al. 2011). Several of the clonally propagated crops, such as banana, roots, and tubers (not fruit crops), have begun to utilize SB in order to reduce flowering time and increase flowering rate, as well as the predictability of flowering, for the introduction of disease-resistance traits, as exemplified by bacterial wilt in banana (Vira et al. 2015, Saiza et al. 2018).

SB in Vegetable Crops

Extending the photoperiod has shortened generation intervals in vegetables, such as pepper, tomato, and amaranth, which respond effectively to increased daylight (Demers et al. 1998, Steller et al. 2016). In tomato, germination of immature seeds from different maturity levels provided new possibilities to achieve five generations instead of the conventionally grown three (Bhattaraj et al. 2009). Similarly, in pepper and tomato, in vitro germination of immature embryos enabled authors to obtain one more generation compared to conventional breeding practice (Manzur et al. 2014, Gebologlu et al. 2011). In grain amaranth, photoperiod manipulation was reported to be helpful in flowering synchronization in different germplasm lines, which, in combination with DNA marker technology, led to the development and identification of true hybrids, thus, accelerating the breeding program (Steller et al. 2016). Other methodologies that can improve generation turnover in vegetables by promoting early flowering involve higher expression of flowering genes such as the CaFT-LIKE gene in pepper (Borovsky et sal. 2020). Similarly, as demonstrated by Velez-Ramirez et al. (2014), in tomato, introgression of the gene CAB-13 can impart tolerance to continuous light, thus, adapting plants to extended photoperiods.

Combining Sb with Modern Breeding and Phenotyping Tools

In the 21st century, crop improvement was revolutionized by DNA marker technology and genomics-assisted breeding. More recently, genome editing techniques, based on site-specific nucleases, are being applied to improve agricultural procedures by developing superior plant varieties. The current genome editing protocols can be enhanced by integration with SB as the edited plants can be grown under SB conditions to rapidly attain edited seeds, thereby accelerating homozygosity and the potential rate of genetic gain. Given this, genome editing using a CRISPR/Cas9 system and SB will likely become popular as the technology is adapted to new species. Desirable lines produced from genome editing can be preselected at the T1 generation with strict evaluation conducted at the T2 generation for the elimination of off-target genotypes (Figure 6.5). The application of SB in combination with genome editing has been demonstrated in *Brassica napus, B. oleracea*, and soybean (Yang et al. 2017, Murovec et al. 2018, Bao et al. 2020).

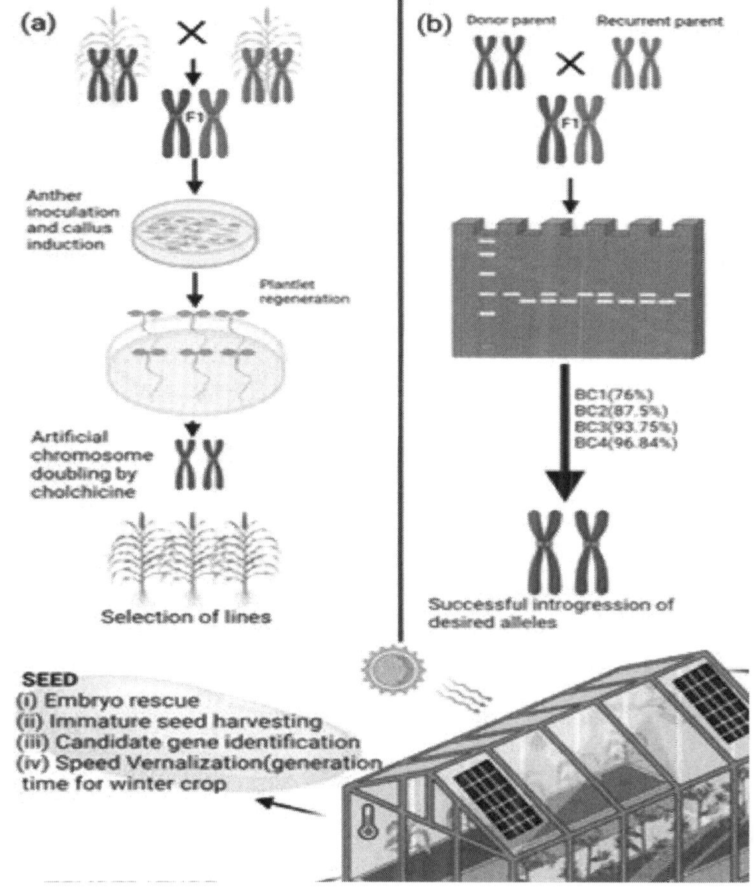

Figure 6.5. Integration of SB with other breeding techniques accelerates the rate of progress. The homozygous lines for research and breeding purposes can be obtained expeditiously through hastening procedures aimed at (a) DH production, (b) marker-assisted selection, and (c) gene editing. (Reproduced from Samantara et al., 2022).

Integration of SB into MAS/marker-assisted backcrossing (MABC) can also serve as a platform for the introgression of beneficial alleles in various crops. MAS/MABC is an established method for improving yield, biotic stress, and abiotic stress among major crops (Varshney et al. 2021). For example, pi21 is known to confer quantitative resistance to rice blast. This gene has been successfully introgressed in select rice cultivars (Angeles-Shim et al. 2020. The DNA-marker-assisted approaches help minimize the problem of linkage drag, i.e., unintended introgression of unfavorable alleles with target loci. Backcrossing procedures demand considerable time for the improvement of the recipient genotypes. Through the integration of SB, the progress of backcrossing or MABC can further be accelerated for quick transfer of the target trait(s).

SB has been applied for evaluating various stages in plant breeding programs. Genomic selection (GS) was combined with SB in spring wheat to increase genetic gain vis-à-vis complex traits (Voss-Fels et al. 2018). SB was used in the phenotyping of specific traits in the training population of wheat, selection candidate development, and selection candidate phenotyping steps. Concerning the indirect selection in targeted-population SB environments, it was concluded that plant height and flowering time can be predicted with accuracy comparable to direct field selection. Speed breeding also facilitates a higher rate of genetic gain compared to direct field phenotyping (Watson et al. 2019). Multi-trait phenotyping protocols have been optimized for evaluating crown root resistance and leaf rust tolerance in wheat under a SB system. The effectiveness of the early-generation selection of F2 population for multiple traits was tested in order to estimate phenotypic response. The method illustrates efficient exploitation of resources by analyzing multiple traits for the same group of plants. Selection in the early generation under SB improves genetic gain in breeding programs, as well as curtailing the time required to incorporate desirable traits in breeding populations (Croser et al. 2021, Ortiz et al. 2007). Phenotyping of wheat or other important crops under SB conditions can be further improved through MAS.

High-throughput phenotyping is among the key breakthroughs of 21st-century agricultural research that have substantially overcome long-standing obstacles in the progress of plant breeding. Conducting high-throughput phenotyping under SB conditions creates novel avenues for the discovery and incorporation of beneficial traits in a resource-efficient manner. For example, targeting proxy traits, such as seminal root number and angle of wheat seedlings, under SB conditions (higher planting density, temperature control, and extended photoperiod) facilitated rapid selection for improved root architecture of mature plants. Similarly, imaging technology permitted the collection of field plot images at 7400 plots/h based on colour traits in wheat (Walter et al. 2019). The technology using non-human-operated aerial vehicles showed a strong correlation with increased grain yield compared to terrestrial sensing.

For the breeding of salt-tolerant rice, Rana et al. (2019) applied SB to rapidly derive a BC3F3 population from highly tolerant and high-yield parents. The population was assayed with SNP markers and provided promising salt-tolerant variety candidates. Plants grown under SB demonstrated significant differences from the control plants in terms of various yield-predicting component traits, viz., panicle length, tiller number, spikelet number per panicle, and normalized difference vegetation index during panicle initiation until mid-grain panicle filling (Phyu et al. 2020). Hence, these studies can be considered to gain more extensive knowledge on plant growth and developmental processes for further crop trait improvement studies.

The single-seed descent (SSD) method under SB conditions enables the generation of near-homozygous lines in a single year and provides greater scope for preserving genetic diversity in a crop breeding program. Likewise, rapid development of recombinant inbred lines (RILs) using SB technology has given impetus to studies aiming at identification of gene-trait associations for breeding applications (Ribalta et al. 2017, Zaman et al. 2019, Dadu et al. 2021, Khoo et al. 2021, Taylor et al. 2021). Again, DNA marker technology in combination with SB helps design new strategies for basic and applied research; for instance, SB can allow the quick development of new generation mapping populations derived from multi-parental populations such as MAGIC and NAM (Kitony et al. 2021, Samantara et al. 2021).

Challenges and Limitations

The use of SB techniques is a valuable approach to accelerate conventional breeding programmes. However, the technology requires expertise, effective and complementary plant phenomics facilities, appropriate infrastructure, and continuous financial support for research and development (Shimelis et al., 2019). Speed breeding methods could accelerate the release of both conventional and genetically modified crop cultivars in SSA. However, the most common challenges hampering the use of speed breeding include: (a) access to suitable facilities, (b) staff trained in the protocol, (c) adopting major changes to breeding programme design and operations, and (d) the need for long-term funding.

A key limitation is access to CE conditions suited to the rapid cycling of the target species. SB settings become expensive if sophisticated CE facilities are not readily available and combining SB with other techniques, such as embryo rescue and MAS, requires additional resources and expertise. Other challenges include a continuous supply of electricity and temperature maintenance.

Once SB is established, species can exhibit genotypic differences in response to intensive growth conditions (Tanaka et al. 2016, Phyu et al. 2020). The intensive growth conditions often result in limited seed yield, which can constrain subsequent field evaluations (Barrios et al. 2020). The use of next-generation sparse phenotyping field trial designs can assist in overcoming low seed numbers (Watson et al. 2018). An excessive photoperiod can limit plant growth and may be correlated to photo-oxidation, high starch production, and elevated levels of stress hormones (Sharma and Jones 2019, Cazzola et al. 2020). Similarly, the harvesting of immature seeds may interfere with the phenotyping of seed traits. Importantly, in pushing for speed, plants are grown at the edge of their physiological capability, and conditions conducive to fast cycle turnover are often detrimental to the plant's ability to defend itself and, without careful management, can lead to catastrophic losses of valuable breeding material. Mitigation comes from the adaptation of conditions to determine photoperiod saturation and temperature limits for each species and, in some cases, for genotypes within species. Maintaining backup seeds from each individual through each generation also provides insurance against genetic loss in the event of one generation being affected by disease, pests, or power loss in CE or similar.

Phenotyping for agronomic traits can be undertaken in conjunction with SB; however, care is required as the phenotypic expression can be biased under CE conditions. As a result, the phenotyping of field crops under SB should be validated in the field to certify that trait expression is associated with the field environment. For example, boron tolerance was reliability discriminated in peas grown hydroponically to integrate with SB conditions, and, in wheat, SB techniques were combined with phenotype screening such as disease resistance (Bennett et al. 2017, Watson et al. 2019). Several agronomic characters, namely flowering time, and plant height, when recorded in SB conditions, are related to field-based determination and production, thus, can facilitate indirect selection (Watson et al. 2018). Nevertheless, several characters may not be consistently phenotyped due to the cross-over interaction between genotypes and growth systems, as observed in the case of plant height and flowering time in oat (Barrios et al. 2020). Other major challenges while growing plants in SB conditions include pest and disease management under such intensive conditions and the need for tracking individuals when developing mapping populations for gene-discovery purposes.

Conclusions

The plant breeders are yet to achieve the scale and speed of plant improvement required to effectively feed a burgeoning world population in the face of a changing climate. The use of speed breeding (SB) can accelerate the development of high-performing cultivars with market-preferred traits by reducing the amount of time, space, and resources invested in the selection and genetic advancement of superior crop varieties. Since its introduction, SB has accelerated the breeding programs of many economically important species. Relying on light and temperature manipulation, along with physical containment, SB impacts different phases of plant breeding by hastening the breeding cycle. SB enables rapid progression to homozygosity and evaluation of already developed or transformed lines, viz., gene-edited crops and transgenic crops. The SB techniques also facilitate resource-efficient genotyping and phenotyping; however, further research is required to assess and mitigate the negative effects of SB conditions on plant growth and development.

The technique allows plant breeders to deliver improved crop varieties more rapidly. Streamlined operations that reduce labour and low-cost facilities are key for the effective integration of speed breeding into a crop improvement programme. Integration of speed breeding with conventional, MAS, and GE breeding approaches can enhance the effective selection of elite genotypes and lines with novel traits, such as higher yield and better nutritional qualities, together with biotic and abiotic stress tolerance. The most appropriate selection methods amenable to speed breeding include SSD, SPD, and SPS methods. The SB protocols are now available for small- or broad-scale adoption and further modifications based on local needs/ innovations. Collaborative international partnerships involving multi-disciplinary teams are needed to encourage the integration of SB systems in basic and applied research, particularly in developing countries.

References

Agbicodo, E.M., Fatokun, C.A., Muranaka, S., Visser, R.G.F., & Van Der Linden, C.G., (2009). Breeding drought tolerant cowpea: Constraints, accomplishments, and future prospects. Euphytica. 167: 353–370. https://doi.org/10.1007/s10681-009-9893-8

Ahmar, S., Gill, R.A., Jung, K.H., Faheem, A., Qasim, M.U., Mubeen, M., & Zhou, W. (2020). Conventional and molecular techniques from simple breeding to speed breeding in crop plants: recent advances and future outlook. Intel. J. Mol. Sci. 21: 2590. https://doi.org/10.3390/ijms21072590

Al Tamimi, N., Brien, C., Oakey, H., Berger, B., ... Saade, S. (2016). Salinity tolerance loci revealed in rice using high-throughput non-invasive phenotyping. Nat. Commun. 7: 13342.

Alahmad, S., Dinglasan, E., Ming Leung, K., Riaz, A., ... Derbal, N. (2018). Speed breeding for multiple quantitative traits in durum wheat. Plant Methods. 14: 36. https://doi.org/10.1186/s13007-018-0302-y

Aldwinckle, H.S. (1975). Flowering of apple seedlings 16–20 months after germination. Hort. Sci. 10: 124–126.

Angeles-Shim, R.B., Reyes, V.P., del Valle, M.M., ... Lapis, R.S. (2020). Marker-assisted introgression of quantitative resistance gene pi21 confers broad spectrum resistance to rice blast. Rice Sci. 27: 113–123.

Anjum, S.A., Ashraf, U., Zohaib, A., Tanveer, M., ... Naeem, M. (2017). Growth and developmental responses of crop plants under drought stress: a review. Zemdirbyste. 104: 267–276. https://doi.org/10.13080/z-a.2017.104.034

Arbelaez, J.D., Tandayu, E., Reveche, M.Y., ... Jarana, A. (2019). Methodology: Ssb-MASS: a single seed-based sampling strategy for marker-assisted selection in rice. Plant Methods. 15: 78. https://doi.org/10.1186/s13007-019-0464-2

Baier, K., Maynard, C., & Powell, W.A., (2012). Early flowering in chestnut species induced under high intensity, high dose light in growth chambers. J. Am. Chestnut Found. 26: 8–10.

Ballare, C.L., Scopel, A.L., Stapleton, A.E., & Yanovsky, M.J. (1996). Solar ultraviolet-B radiation affects seedling emergence, DNA integrity, pant morphology, growth gate, and attractiveness to herbivore Insects in Datura ferox. Plant Physiol. 112: 161–170.

Bao, A., Zhang, C., Huang, Y., Chen, H., Zhou, X., & Cao, D. (2020). Genome editing technology and application in soybean improvement. Oil Crop Sci. 5: 31–40.

Barrios, P.M.G., Bhatta, M., Halley, M., Sandro, P., & Gutiérrez, L. (2020). Speed breeding and early panicle harvest accelerates oat (Avena sativa L.) breeding cycles. Crop Sci. 61: 320–330.

Bennett, R., Ribalta, F.M., Pazos-Navarro, M., Leonforte, A., & Croser, J.S. (2017). Discrimination of boron tolerance in Pisum sativum L. genotypes using a rapid, high-throughput hydroponic screen and precociously germinated seed grown under far-red enriched light. Plant Methods. 13: 70.

Bermejo, C., Gatti, I., & Cointry, E. (2016). In vitro embryo culture to shorten the breeding cycle in lentil (Lens culinaris Medik). Plant Cell, Tissue Organ Culture. 127: 585–590. https://doi.org/10.1007/ s11240-016-1065-7

Bhatta, M., Sandro, P., Smith, M.R., Delaney, O., Voss-Fels, K.P., Gutierrez, L., & Hickey, L.T. (2021). Need for speed: manipulating plant growth to accelerate breeding cycles. Current Opinion Plant Biol. 60: 101986. https://doi.org/10.1016/j.pbi.2020.101986

Bhattaraj, S.P., de la Pena, R.C., Midmore, D.J., & Palchamy, K. (2009). In vitro culture of immature seed for rapid generation advancement in tomato. Euphytica. 167: 23–30.

Bohra, A., Jha, U.C., Godwin, I.D., & Varshney, R.K. (2020). Genomic interventions for sustainable agriculture. Plant Biotechnol. J. 18: 2388–2405.

Bordes, J., Charmet, G., De Vaulx, RD., Lapierre, A., Pollacsek, M., Beckert, M., & Gallais, A. (2007). Doubled-haploid versus single-seed descent and S1-family variation for testcross performance in a maize population. Euphytica. 154: 41–51. https://doi.org/10.1007/s1068 1-006-9266-5

Borlaug, N. (1968). Wheat breeding and Its impact on world food supply. In: Finlay, K.W., Shephard, K.W., (Eds.). Australian Academy of Sciences: Canberra, Australia, p. 1–36.

Borovsky, Y., Mohan, V., Shabtai, S., & Paran, I. (2020). CaFT-LIKE is a flowering promoter in pepper and functions as florigen in tomato. Plant Sci. 301: 110678.

Breseghello, F., & Coelho, A.S.G. (2013). Traditional and modern plant breeding methods with examples in rice (Oryza sativa L.). J. Agril. Food Chem. 61: 8277–8286. https://doi. org/10.1021/jf305531j

Brim, C.A. (1966). A modified pedigree method of selection in soybeans. 19066. Crop Sci. 6: 220.

Bunce, J.A. (2015). Elevated carbon dioxide effects on reproductive phenology and seed yield among soybean cultivars. Crop Sci. 55: 339–343. https://doi.org/10.2135/cropsci2014.04.0273

Burris, J.S. (1973). Effect of seed maturation and plant population on soybean seed quality. Agron. J. 65: 440–441.

Cazzola, F., Bermejo, C.J., Guindon, M.F., & Cointry, E. (2020). Speed breeding in pea (Pisum sativum L.), an efficient and simple system to accelerate breeding programs. Euphytica. 216: 178.

Chiurugwi, T., Kemp, S., Powell, W., & Hickey, L.T. (2018). Speed breeding orphan crops. Theo. Appl. Genetics. 132: 607–616. https://doi.crg/10.1007/s00122-018-3202-7

Christopher, J., Richard, C., Chenu, K., Christopher, M., Borrell, A., & Hickey, L. (2015). Integrating rapid phenotyping and speed breeding to improve stay-green and root adaptation of wheat in changing, water-limited, Australian environments. Procedia Environ. Sci. 29: 175–176.

Cobb, J.N., Juma, RU., Biswas, P.S., Arbelaez, J.D., ... Rutkoski, J. (2019). Enhancing the rate of genetic gain in public-sector plant breeding programs: Lessons from the breeder's equation. Theor. Appl. Genet . 132: 627–645.

Collard, B.C.Y., Beredo, J.C., Lenaerts, B., Mendoza, R., ... Santelices, R. (2017). Revisiting rice breeding methods–evaluating the use of rapid generation advance (RGA) for routine rice breeding. Plant Prod. Sci. 20: 337–352. https://doi.org/10.1080/13439 43X.2017.1391705

Croser, J., Mao, D., Dron, N., Michelmore, S., ... McMurray, L. (2021). Evidence for the application of emerging technologies to accelerate crop improvement: a collaborative pipeline to introgress herbicide tolerance into chickpea. Front. Plant Sci. 12: 779122.

Croser, J.S., Pazos-Navarro, M., Bennett, RG., ... Tschirren, S. (2016). Time to flowering of temperate pulses in vivo and generation turnover in vivo–in vitro of narrow-leaf lupin accelerated by low red to far-red ratio and high intensity in the far-red region. Plant Cell Tissue Organ Cult. 127: 591–599.

Dadu, R.H.R., Bar, I., Ford, R., Sambasivam, P., ... Croser, J. (2021). Lens orientalis contributes quantitative trait loci and candidate genes associated with ascochyta blight resistance in lentil. Front. Plant Sci. 12: 703283.

Dagustu, N., Bayram, G., Sincik, M., & Bayraktaroglu, M. (2012). The short breeding cycle protocol effective on diverse genotypes of sunflower (Helianthus annuus L.). Turkish J. Field Crop. 17: 124–128.

De La Fuente, G.N., Frei, U.K., & Lübberstedt, T. (2013). Accelerating plant breeding. Trends Plant Sci. 18: 667–672. https://doi. org/10.1016/j.tplants.2013.09.001

De Pauw, R.M., & Clarke, J.M. (1976). Acceleration of generation advancement in spring wheat. Euphytica. 25: 415–418.

De Rouw, A., & Winkel, T. (1998). Drought avoidance by asynchronous flowering in pearl millet stands cultivated on-farm and on-station in Niger. Exptl. Agril. 34: 19–39. https://doi. org/10.1017/S0014479798001057

Demers, D.A., Dorais, M., Wien, C.H., & Gosselin, A. (1998). Effects of supplemental light duration on greenhouse tomato (Lycopersicon esculentum Mill.) plants and fruit yields. Sci. Hort. 74: 295–306.

Destro, D., Bizeti, H.S., Garcia, L.A., ... De Batista Fonseca, .I.C. (2003). Comparison between the SPD and the SPDS methods for segregating generation advancement in soybean. Brazilian Archives Biol. Tech. 46: 545–551. https://doi.org/10.1590/S1516-89132003000400008

Dubcovsky, J., Loukoianov, A., Fu, D., Valarik, M., Sanchez, A., & Yan, L. (2006). Effect of photoperiod on the regulation of wheat vernalization genes VRN1 and VRN2. Plant Mol. Biol. 60: 469–480. https://doi.org/10.1007/s11103-005-4814-2

Dwivedi, S.L., Britt, A.B., Tripathi, L., Sharma, S., Upadhyaya, H.D., & Ortiz, R. (2015). Haploids: Constraints and opportunities in plant breeding. Biotech. Adv.. 33: 812–829. https://doi. org/10.1016/j.biotechadv.2015.07.001

El-Hashash, E.F., & El-Absy, K.M. (2019). Barley (Hordeum vulgare L.) breeding. In: Al-Khayri, J., Jain, S., & Johnson, D.V. (Eds.), Advances in Plant Breeding Strategies: Cereals (pp. 1–45). Springer International Publishing. https://doi. org/10.1007/978-3-030-23108-8_1

Espósito, M.A., Almirón, P., Gatti, I., ... Cravero, V.P. (2012). A rapid method to increase the number of F1 plants in pea (Pisum sativum) breeding programs. Genet. Mol. Res. 11: 2729–2732.

Fehr, W. (1991). Principles of cultivar development: theory and techniques. Macmillian Publishing Company. Agronomy Books 1. https://lib.dr.iastate. edu/agron_books/1

Ferrie, A.M.R. (2007). Doubled haploid production in nutraceutical species: a review. Euphytica. 158: 347–357.

Flachowsky, H., Le Roux, P.M., Peil, A., & Patocchi, A. (2011). Application of a high-speed breeding technology to apple (Malus × domestica) based on transgenic early flowering plants and marker-assisted selection. New Phytol. 192: 364–377.

Forster, B.P. (2014). Accelerated plant breeding. CAB Rev. 9: 1–16.

Fukushima, A., Shiratsuchi, H., Yamaguchi, H., & Fukuda, A. (2011). Effects of nitrogen application and planting density on morphological traits, dry matter production and yield of large grain type rice variety bekoaoba and strategies for super high-yielding rice in the tohoku region of Japan. Plant Prod. Sci. 14: 56–63. https:// doi.org/10.1626/pps.14.56

Funada, M., Helms, T.C., Hammond, J.J., Hossain, K., & Doetkott, C. (2013). Single-seed descent, single-pod, and bulk sampling methods for soybean. Euphytica. 192: 217–226. https://doi.org/10.1007/ s10681-012-0837-3

Geboloğlu, N., Bozmaz, S., Aydin, M., & Cakmak, P. (2011). The role of growth regulators, embryo age and genotypes on immature embryo germination and rapid generation advancement in tomato (Lycopersicon esculentum Mill.). Afr. J. Biotechnol. 10: 4895–4900.

Ghosh, S., Watson, A., Gonzalez-Navarro, O.E., ... Ramirez-Gonzalez, R.H. (2018). Speed breeding in growth chambers and glasshouses for crop breeding and model plant research. Nature Protocols. 13: 2944–2963. https:// doi.org/10.1038/s41596-018-0072-z

Gosal, S.S., Pathak, D., Wani, S.H., Vij, S., & Pathak, M. (2020). Accelerated breeding of plants: methods and applications. In: Accelerated Plant Breeding, Volume 1; Gosal, S.S., & Wani, S.H., (Eds). Springer International Publishing: Cham, Switzerland. p. 1–29. ISBN 978-3-030-41865-6

Goufo, P., Moutinho-Pereira, J.M., Jorge, T.F., ... Correia, C.M. (2017). Cowpea (Vigna unguiculata L. Walp.) metabolomics: Osmoprotection as a physiological strategy for drought stress resistance and improved yield. Frontiers. Plant Sci. 8: 586. https://doi.org/10.3389/ fpls.2017.00586

Hatfield, J.L., & Prueger, J.H. (2015). Temperature extremes: effect on plant growth and development. Weather Climate Extremes. 10: 4–10. https://doi. org/10.1016/j.wace.2015.08.001

Hayashi, S., Kamoshita, A., & Yamagishi, J. (2006). Effect of planting density on grain yield and water productivity of rice (Oryza sativa L.) grown in flooded and non-flooded fields in Japan. Plant Prod. Sci. 9: 298–311. https://doi. org/10.1626/ pps.9.298

Hickey, L.T., Germán, S.E., Pereyra, S.A., Diaz, J.E., ... Ziems, L.A. (2017). Speed breeding for multiple disease resistance in barley. Euphytica. 213: 64.

Hickey, L.T.N. Hafeez, A., Robinson, H., ... Jackson, S.A. (2019). Breeding crops to feed 10 billion. Nature Biotech. 37: 744–754. https://doi.org/10.1038/ s41587-019-0152-9

Hickey, L.T., Dieters, M.J., DeLacy, I.H., Christopher, M.J., Kravchuk, O.Y., & Banks, P.M. (2010). Screening for grain dormancy in segregating generations of dormant × non-dormant crosses in white-grained wheat (Triticum aestivum L.). Euphytica. 172: 183–195. https://doi. org/10.1007/s10681-009-0028-z

Hussain, H.A., Hussain, S., Khaliq, A., Ashraf, U., Anjum, S.A., Men, S., & Wang, L. (2018). Chilling and drought stresses in crop plants: Implications, cross talk, and potential management opportunities. Frontiers Plant Sci. 9: 393. https:// doi.org/10.3389/ fpls.2018.00393

Jagadish, S.V.K., Bahuguna, R.N., Djanaguiraman, M., ... Gamuyao, R. (2016). Implications of high temperature and elevated CO2 on flowering time in plants. Frontiers Plant Sci. 7: 913. https://doi.org/10.3389/fpls.2016.00913

Jähne, F., Hahn, V., Würschum, T., & Leiser, W.L. (2020). Speed breeding short-day crops by LED-controlled light schemes. Theor. Appl. Genet. 133: 2335–2342.

Johnston, H.R., Keats, B.J.B., & Sherman, S.L. (2019). Population genetics. In: Pyeritz, R.E., Korf, B.R., & Grody, W.W. (Eds.), Emery and Rimoin's Principles and Practice of Medical Genetics and Genomics. Foundations (p. 359–373). Academic Press. https://doi.org/10.1016/ B978-0-12-812537-3.00012-3

Jones, O.R., & Johnson, G.L. (1991). Row width and plant density effects on Texas high plains sorghum. J. Prod. Agri. 4: 613–621. https://doi.org/10.2134/ jpa1991.0613

Khan, A., Najeeb, U., Wang, L., Tan, D.K.Y., Yang, G., Munsif, F., Ali, S., & Hafeez, A. (2017). Planting density and sowing date strongly influence growth and lint yield of cotton crops. Field Crops Res. 209: 129–135. https://doi.org/10.1016/j. fcr.2017.04.019

Khan, N., Han, Y., Xing, F., Feng, L., Wang, Z., Wang, G., ... Yang, B. (2019). Plant density influences reproductive growth, lint yield and boll spatial distribution of cotton. Agronomy. 10: 14. https://doi.org/10.3390/agronomy10010014

Khoo, K.H.P., Sheedy, J.G., Taylor, J.D., Croser, J.S., ... Hayes, J.E. (2021). A QTL on the Ca7 chromosome of chickpea affects resistance to the root-lesion nematode Pratylenchus thornei. Mol. Breed. 41: 78.

Khosla, G., Gill, B.S., & Sharma, P. (2019). Comparison of different breeding methods for developing superior genotypes in soybean. Agri. Res. J. 56: 628. https:// doi. org/10.5958/2395-146x.2019.00098.x

Kitony, J.K., Sunohara, H., Tasaki, M., Mori, J.I., Shimazu, A., ... Reyes, V. (2021). Development of an aus-derived nested association mapping (Aus-NAM) population in rice. Plants. 10: 1255.

Kouressy, M., Dingkuhn, M., Vaksmann, M., & Heinemann, A.B. (2008). Adaptation to diverse semi-arid environments of sorghum genotypes having different plant type and sensitivity to photoperiod. Agri. Forest Meteorology. 148: 357–371. https://doi. org/10.1016/j.agrformet.2007.09.009

Laux, P., Jäckel, G., Tingem, R.M., & Kunstmann, H. (2010). Impact of climate change on agricultural productivity under rainfed conditions in Cameroon-A method to improve attainable crop yields by planting date adaptations. Agri. Forest Meteorology. 150: 1258– 1271. https://doi.org/10.1016/j. agrformet.2010.05.008

Lin, Z., Cogan, N.O.I., Pembleton, L.W., ... Spangenberg, G.C. (2016). Genetic gain and inbreeding from genomic selection in a simulated commercial breeding program for perennial ryegrass. Plant Genome. 9: 1–12.

Liu, H., Zwer, P., Wang, H., Liu, C., Lu, Z., Wang, Y., & Yan, G. (2016). A fast generation cycling system for oat and triticale breeding. Plant Breed. 135: 574–579.

Lulsdorf, M.M., & Banniza, S. (2018). Rapid generation cycling of an F2 population derived from a cross between Lens culinaris Medik. and Lens ervoides (Brign.) Grande after aphanomyces root rot selection. Plant Breed. 137: 486–491.

Ma, H., Busch, RH., Riera-Lizarazu, O., Rines, HW., & Dill-Macky, R. (1999). Agronomic performance of lines derived from anther culture, maize pollination and single-seed descent in a spring wheat cross. Theo. Appl. Genetics. 99: 432–436. https://doi. org/10.1007/s001220051254

Manzur, J., Oliva-Alarcón, M., & Rodríguez-Burruezo, A. (2014). In vitro germination of immature embryos for accelerating generation advancement in peppers (Capsicum annum L.). Sci. Hort. 170: 203–210.

McClung, C.R., Lou, P., Hermand, V., & Kim, J.A. (2016). The importance of ambient temperature to growth and the induction of flowering. Front. Plant Sci. 7: 1266. https://doi.org/10.3389/ fpls.2016.01266

Mobini, S., Khazaei, H., Warkentin, T.D., & Vandenberg, A. (2020). Shortening the generation cycle in faba bean (Vicia faba) by application of cytokinin and cold stress to assist speed breeding. Plant Breeding. 139: 1181–1189. https://doi. org/10.1111/pbr.12868

Mobini, S.H., Lulsdorf, M., Warkentin, T., & Vandenberg, A. (2014). Plant growth regulators improve in vitro flowering and rapid generation advancement in lentil and faba bean. Vitr. Cell. Dev. Biol. Plant. 51: 71–79.

Mobini, SH., Lulsdorf, M., Warkentin, T.D., & Vandenberg, A. (2016). Low red: Far-red light ratio causes faster in vitro flowering in lentil. Can. J. Plant Sci. 96: 908–918.

Mobini, S.H., & Warkentin, T.D. (2016). A simple and efficient method of in vivo rapid generation technology in pea (Pisum sativum L.). Vitro Cellular Devel. Biology - Plant, 52: 530–536. https:// doi.org/10.1007/s11627-016-9772-7

Moose, S.P., & Mumm, R.H. (2008). Molecular plant breeding as the foundation for 21st century crop improvement. Plant Physiol. 147: 969–977.

Morris, M.L., & Bellon, M.R. (2004). Participatory plant breeding research: Opportunities and challenges for the international crop improvement system. Euphytica. 136: 21–35. https://doi. org/10.1023/B:EUPH.0000019509.37769. b1

Mukade, K. (1974). New procedures for accelerating generation advancement in wheat breeding. JARQ. 8: 1–5.

Munamava, M., & Riddoch, I. (2001). Response of three sorghum (Sorghum bicolor L. Moench) varieties to soil moisture stress at different developmental stages. South African J. Plant Soil. 18: 75–79. https://doi.org/10.1080/02571862.2 001.10634407

Mundia, M. (2020). Speed breeding, a promising approach to crop breeding. News Letter Intel. Livestock Research Institute, Nairobi, Kenya.

Murashige, T., & Skoog, F. (1962). A revised medium for rapid growth and bio assays with tobacco tissue cultures. Physiologia Plantarum. 15: 473–497. https://doi.org/10.1111/j.1399-3054.1962.tb08052.x

Murovec, J., Guček, K., Bohanec, B., Avbelj, M., & Jerala, R. (2019). DNA-free genome editing of Brassica oleracea and B. rapa protoplasts using CRISPR-Cas9 ribonucleoprotein complexes. Front. Plant Sci. 9: 1594.

Nagatoshi, Y., & Fujita, Y. (2019). Accelerating soybean breeding in a co2-supplemented growth chamber. Plant Cell Physiol. 60: 77– 84. https://doi.org/10.1093/pcp/pcy189

Ochatt, S.J., & Sangwan, R.S. (2008). In vitro shortening of generation time in Arabidopsis thaliana. Plant Cell, Tissue Organ Culture. 93: 133–137. https://doi.org/10.1007/s11240-008-9351-7

Ochatt, S.J., Sangwan, R.S., Marget, P., Assoumou Ndong, Y., Rancillac, M., & Perney, P. (2002). New approaches towards the shortening of generation cycles for faster breeding of protein legumes. Plant Breed. 121: 436–440. https://doi.org/10.1046/j.1439-0523.2002.746803.x

O'Connor, D.J., Wright, G.C., Dieters, M.J., ... George, D.L. (2013). Development and application of speed breeding technologies in a commercial peanut breeding program. Peanut Sci. 40: 107–114.

Ohnishi, T., Yoshino, M., Yamakawa, H., & Kinoshita, T. (2011). The biotron breeding system: a rapid and reliable procedure for Genetic studies and breeding in rice. Plant Cell Physiol. 52: 1249–1257.

Ortiz, R., Trethowan, R., Ferrara, G.O., Iwanaga, M., ... Dodds, J.H. (2007). High yield potential, shuttle breeding, genetic diversity, and a new international wheat improvement strategy. Euphytica. 157: 365–384.

Parmar, S., Deshmukh, D.B., Kumar, R., Manohar, S.S., ... Joshi, P. (2021). Single Seed-based high-throughput genotyping and rapid generation advancement for accelerated groundnut genetics and breeding research. Agronomy. 11: 1226.

Pazos-Navarro, M., Castello, M., Bennett, R.G., Nichols, P., & Croser, J. (2017). In vitro-assisted single-seed descent for breeding-cycle compression in subterranean clover (Trifolium subterraneum L.). Crop Pasture Sci. 68: 958.

Pfeiffer, N.E. (1926). Microchemical and morphological studies of effect of light on plants. Bot. Gaz. 81: 173–195.

Phyu, P., Islam, M.R., Cruz, P.C.S., Collard, B.C.Y., & Kato, Y. (2020). Use of NDVI for indirect selection of high yield in tropical rice breeding. Euphytica. 216: 74.

Pignone, D., De Paola, D., Rapanà, N., & Janni, M. (2015). Single seed descent: A tool to exploit durum wheat (Triticum durum Desf.) genetic resources. Genetic Resources Crop Evolution. 62: 1029–1035. https://doi.org/10.1007/s10722-014-0206-2

Priyadarshan, P.M. (2019). Breeding self-pollinated crops. In: Plant breeding: classical to modern (p. 223–241). Springer. https://doi. org/10.1007/978-981-13-7095-3_11

Rahman, M.A., Quddus, M.R., Jahan, N., Rahman, A., Hossain, M.R.A.S.H., & Iftekharuddaula, K M. (2019). Field rapid generation advance: An effective technique for industrial scale rice breeding program. The Experiment. 47: 2659–2670

Rana, M.M., Takamatsu, T., Baslam, M., Kaneko, K., ... Itoh, K. (2019). Salt tolerance improvement in rice through efficient SNP marker-assisted selection coupled with speed-breeding. Int. J. Mol. Sci. 20: 2585.

Ribalta, F.M., Pazos-Navarro, M., Nelson, K., ... Edwards, K. (2017). Precocious floral initiation and identification of exact timing of embryo physiological maturity facilitate germination of immature seeds to truncate the lifecycle of pea. Plant Growth Regul. 81: 345–353.

Richard, C., Hickey, L., Fletcher, S., ... Chenu, K. (2015). High-throughput phenotyping of wheat seminal root traits in a breeding context. Proc. Environ. Sci. 29: 102–103.

Rizal, G., Karki, S., Alcasid, M., Montecillo, F., ... Acebron, K. (2014). Shortening the breeding cycle of sorghum, a model crop for research. Crop Sci. 54: 520–529.

Robertson, L.D., & Curtis, B.C. (1967). Germination of immature kernels of winter wheat. Crop Sci. 7: 269–270.

Roumet, P., & Morin, F. (1997). Germination of Immature soybean seeds to shorten reproductive cycle duration. Crop Sci. 37: 521–525.

Saito, H., Yuan, Q., Okumoto, Y., Doi, K., ... Yoshimura, A. (2009). Multiple alleles at early flowering 1 locus making variation in the basic vegetative growth period in rice (Oryza sativa L.). Theo. Appl. Genet. 119: 315– 323. https://doi. org/10.1007/s00122-009-1040-3

Samantara, K., Bohra, A., Mohapatra, S.R., Prihatini, R., ... Asibe F. (2022). Breeding more crops in less time: a perspective on speed breeding. Biology. 11: 275. https://doi.org/10.3390/biology11020275

Samantara, K., Reyes, V.P., Agrawal, N., Mohapatra, S.R., & Jena, K.K. (2021). Advances and trends on the utilization of multi-parent advanced generation intercross (MAGIC) for crop improvement. Euphytica. 217: 189.

Samineni, S., Sen, M., Sajja, S.B., & Gaur, P.M. (2019). Rapid generation advance (RGA) in chickpea to produce up to seven generations per year and enable speed breeding. Crop J. 8: 164–169. https:// doi.org/10.1016/j.cj.2019.08.003

Saxena, K., Saxena, R.K., & Varshney, R.K. (2017). Use of immature seed germination and single seed descent for rapid genetic gains in pigeonpea. Plant Breeding, 136: 954–957. https://doi.org/10.1111/ pbr.12538

Saxena, K.B., Saxena, R.K., Hickey, L.T., & Varshney, R.K. (2019). Can a speed breeding approach accelerate genetic gain in pigeonpea? Euphytica. 215: 1–7. https://doi.org/10.1007/s10681-019-2520-4

Sharma, A., Jones, J.B., & White, F.F. (2019). Recent advances in developing disease resistance in plants. F1000Research. 8: 1934.

Shavrukov, Y., Kurishbayev, A., Jatayev, S., ... Shvidchenko, V. (2017). Early flowering as a drought escape mechanism in plants: How can it aid wheat production? Frontiers Plant Sci. 8: 1950. https://doi. org/10.3389/fpls.2017.01950

Singh, D., Basu, C., Meinhardt-Wollweber, M., & Roth, B. (2015). LEDs for energy efficient greenhouse lighting. Renew. Sustain. Energy Rev. 49: 139–147.

Singh, V., Nguyen, C.T., van Oosterom, E.J., Chapman, S.C., Jordan, D.R., & Hammer, G.L. (2015). Sorghum genotypes differ in high temperature responses for seed set. Field Crops Res. 171: 32–40. https://doi.org/10.1016/j.fcr.2014.11.003

Sinha, P., Singh, V.K., Bohra, A., Kumar, A., Reif, JC., & Varshney, R.K. (2021). Genomics and breeding innovations for enhancing genetic gain for climate resilience and nutrition traits. Theo. Appl. Genet. 134: 1829–1843.

Souza, L.S., Diniz, R.P., Neves, R., Alves, A.A.C., & de Oliveira, E.J. (2018). Grafting as a strategy to increase flowering of cassava. Sci. Hort. 240: 544–551.

Springer, C.J., & Ward, J.K. (2007). Flowering time and elevated atmospheric CO2. New Phytologist. 176: 243–255. https://doi. org/10.1111/j.1469-8137.2007.02196.x

Sreeharsha, R.V., Sekhar, K.M., & Reddy, A.R. (2015). Delayed flowering is associated with lack of photosynthetic acclimation in pigeonpea (Cajanus cajan L.) grown under elevated CO2. Plant Sci. 231: 82– 93. https://doi.org/10.1016/j.plantsci.2014.11.012

Stetter, M.G., Zeitler, L., Steinhaus, A., Kroener, K., Biljecki, M., & Schmid, K.J. (2016). Crossing methods and cultivation conditions for rapid production of segregating populations in three grain amaranth species. Frontiers Plant Sci. 7: 816. https://doi.org/10.3389/ fpls.2016.00816

Sysoeva, M.I., Markovskaya, E.F., & Shibaeva, T.G. (2010). Plants under continuous light: a review. Plant Stress. 4: 5–17.

Tanaka, J., Hayashi, T., & Iwata, H. (2016). A practical, rapid generation advancement system for rice breeding using simplified biotron breeding system. Breeding Sci. 66: 542–551. https://doi. org/10.1270/jsbbs.15038

Tawkaz, S., Istanbuli, T., & Hamweh, A. (2018). Embryo rescue for accelerating the breeding cycle in cheakpea. 7th. Int. Conf. Food Legume Res. May 2018, Marrakesh, Morocco.

Taylor, C.M., Garg, G., Berger, J.D., Ribalta, F.M., ... Croser, J.S. (2021). A trimethyl guanosine synthase1-like (TGS1) homologue is implicated in vernalisation and flowering time control. Theo. Appl. Genet. 134: 3411–3426.

Urrea, C.A., & Singh, S.P. (1994). Comparison of mass, F2-derived family, and single-seed-descent selection methods in an interracial population of common bean. Can. J. Plant Science. 74: 461–464. https://doi.org/10.4141/cjps94-085

Vadez, V., Hash, T., Bidinger, F.R., & Kholova, J. (2012). Phenotyping pearl millet for adaptation to drought. Frontiers Physiology. 3: 386. https://doi.org/10.3389/fphys.2012.00386

Van Nocker, S., & Gardiner, S.E. (2014). Breeding better cultivars, faster: applications of new technologies for the rapid deployment of superior horticultural tree crops. Hort. Res. 1: 14022.

Varshney, R.K., Bohra, A., Roorkiwal, M., ... Barmukh, R. (2021). Fast-forward breeding for a food-secure world. Trends Genet. 37: 1124–1136.

Varshney, R.K., Bohra, A., Yu, J., Graner, A., Zhang, Q., & Sorrells, M.E. (2021). Designing future crops: genomics-assisted breeding comes of age. Trends Plant Sci. 26: 631–649.

Velez-Ramirez, A.I., Van Ieperen, W., ... Vreugdenhil, D. (2014). A single locus confers tolerance to continuous light and allows substantial yield increase in tomato. Nat. Commun. 5: 4549.

Villar, J.L., Maranville, J.W., & Gardner, J.C. (1989). High density sorghum production for late planting in the Central Great Plains. J. Prod. Agri. 2: 333–338. https://doi.org/10.2134/ jpa1989.0333

Vince-Prue, D. (1994). The duration of light and photoperiodic responses. In: Kendrick, R.E., & Kronenberg, G.H.M. (Eds.), Photomorphogenesis in plants. Dordrecht: Springer. https://doi. org/10.1007/978-94-011-1884-2_17

Vira, B., Wildburger, C., & Mansourian, S. (2015). International Union of Forestry Research Organizations (Eds.) Forests, Trees and Landscapes for Food Security and Nutrition: A Global Assessment Report; IUFRO World Series; IUFRO: Vienna, Austria.

Voss-Fels, K.P., Herzog, E., Dreisigacker, S., ... Sukurmaran, S. (2018). Speed GS to accelerate genetic gain in spring wheat. In: Applications of genetic and genomic research in cereals, 1st ed. In: Miedaner, T., Korzun, V., (Eds.); Woodhead Publishing: Cambridge, MA, USA.

Walter, J., Edwards, J., Cai, J. McDonald, G., Miklavcic, S., & Kuchel, H. (2019). High-throughput field imaging and basic image analysis in a wheat breeding programme. Front. Plant Sci. 10: 449.

Wanga, M.A., Shimelis, H., Mashilo, J., & Laing, M.D. (2021). Opportunities and challenges of speed breeding: a review. Plant Breed. 140:185–194. https://doi.org/10.1111/pbr.12909

Warnasooriya, S.N., & Brutnell, T.P. (2014). Enhancing the productivity of grasses under high-density planting by engineering light responses: From model systems to feedstocks. J. Exptl. Bot. 65: 2825–2834. https://doi.org/10.1093/jxb/eru221

Watson, A., Ghosh, S., Williams, M.J., Cuddy, W.S., Simmonds, J., ... Rey, M.D. (2018). Speed breeding is a powerful tool to accelerate crop research and breeding. Nature Plants. 4: 23– 29. https://doi.org/10.1038/s41477-017-0083-8

Watson, A., Hickey, L.T., Christopher, J., Rutkoski, J., Poland, J., & Hayes, B.J. (2019). Multivariate genomic selection and potential of rapid indirect selection with speed breeding in spring wheat. Crop Sci. 59: 1945–1959.

Wheeler, R.M. (2008). A historical background of plant lighting: an introduction to the workshop. Hort. Sci. 43: 1942–1943.

Wiebbecke, C.E., Graham, M.A., Cianzio, S.R., & Palmer, R.G. (2012). Day temperature influences the male-sterile locus ms9 in soybean. Crop Sci. 52: 1503–1510. https://doi.org/10.2135/crops ci2011.08.0410

Yan, G., Liu, H., Wang, H., Lu, Z., Wang, Y., Mullan, D., Hamblin, J., & Liu, C. (2017). Accelerated generation of selfed pure line plants for gene identification and crop breeding. Frontiers Plant Sci. 8: 1786. https://doi.org/10.3389/fpls.2017.01786

Yan, L., Loukoianov, A., Blechl, A., Tranquilli, G., Ramakrishna, W., ... SanMiguel, P. (2004). The wheat VRN2 gene is a flowering repressor downregulated by vernalization. Science. 303: 1640–1644. https:// doi.org/10.1126/science.1094305

Yang, H., Wu, J.J., Tang, T., Liu, K.D., & Dai, C. (2017). CRISPR/Cas9-mediated genome editing efficiently creates specific mutations at multiple loci using one sgRNA in Brassica napus. Sci. Rep. 7: 7489.

Yang, L., Wang, D., Xu, Y., Zhao, H., Wang, L., Cao, X., Chen, Y., & Chen, Q. (2017). A new resistance gene against potato late blight originating from Solanum pinnatisectum located on potato chromosome 7. Frontiers Plant Sci. 8: 1729. https://doi.org/10.3389/ fpls.2017.01729

Yang, S., Weers, B.D., Morishige, D.T., & Mullet, J.E. (2014). CONSTANS is a photoperiod regulated activator of flowering in sorghum. BMC Plant Biol. 14: 148. https://doi.org/10.1186/1471-2229-14-148

Yao, Y., Zhang, P., Liu, H., Lu, Z., & Yan, G. (2016). A fully in vitro protocol towards large scale production of recombinant inbred lines in wheat (Triticum aestivum L.). Plant Cell, Tissue, Organ Culture. 128: 655– 661. https://doi.org/10.1007/s11240-016-1145-8

Zaman, S.U., Malik, A.I., Kaur, P., Ribalta, F.M., & Erskine, W. (2019). Waterlogging tolerance at germination in field pea: variability, genetic control, and indirect selection. Front. Plant Sci. 10: 95.

Zhang, Z., Wei, W., Zhu, H., Challa, G.S., Bi, C., Trick, H.N., & Li, W. (2015). W3 Is a new wax locus that Is essential for biosynthesis of β-diketone, development of glaucousness, and reduction of cuticle permeability in common wheat. PLoS One. 10: e0140524.

Zheng, Z., Wang, H.B., Chen, G.D., Yan, G.J., & Liu, C.J. (2013). A procedure allowing up to eight generations of wheat and nine generations of barley per annum. Euphytica. 191: 311–316. https://doi. org/10.1007/s10681-013-0909-z

Chapter-7

Sensors for Crop Breeding

Abstract

Due to the rapid rise in the world's population, farming activities have become increasingly complex, competitive, vast, and optimized. The use of technology has led farming operations to be more productive than before thereby increasing what farmers harvest and the quality of products. Sensors have played critical roles in this technological advancement in precision farming. This method of farming relies on data and analytics extracted from IoT sensors and devices that are placed on farm equipment and surveillance devices used in the process. There are different types of sensors used in agriculture and the most salient ones are air and soil sensors. The other forms of sensors that one often comes across in precision farming and smart agriculture are GPS-based sensors, GIS-based sensors, electrochemical sensors, mechanical sensors, soil moisture sensors, airflow sensors, yield monitoring sensors, VRT sensors, etc. Agriculture sensors such as air temperature and humidity, soil moisture, soil pH, light intensity, and carbon dioxide are often used to collect data in all aspects of crop growth such as nursery, growth, and harvest. Agricultural conductivity and agricultural pH sensors are used to monitor water and fertilizer. Modern agriculture uses the Internet of Things (IoT) and sensor technology to provide accurate and timely crop growth data for formulating scientific planting programs, saving labour, optimizing crop varieties, and improving crop quality and yield. Different types of sensors that are currently in use in precision farming and their usefulness are discussed.

Keywords: *Sensors, GPS-based sensors, GIS-based sensors, electrochemical sensors, mechanical sensors, soil moisture sensors, airflow sensors*

Introduction

As a result of the speedy growth of the population, climate change, reduction in the amount of rainfall, and the increasing pressure for nourishment globally, there has been a constant demand for increasing the productivity of the farmland. Thus, agriculture needs to be carried out more intelligently, using new and state-of-the-art technology. It is the only way to provide a solution and meet the endless and growing needs of the world's population.

The evolution of precision agriculture (PA) has the potential to overcome the increasing demand for agricultural produce. Precision agriculture (PA) is the application of specific inputs at various rates to optimize economic efficiency and reduce wastage. The applications of various sensors have made precision agriculture successful. Sensors aid farmers to make a smooth move from the old ways of carrying out farming activities. Sensors are very efficient in precision agriculture because they transmit data that helps farmers not only to monitor but also to improve their products and keep abreast of changes in the field and ecosystem. Intelligent agricultural sensors help to easily identify animals, detect heat, and monitor their health, thus facilitating the isolation and healing of sick cows by identifying, detecting, and following herds. Using smart sensors in agriculture, farmers can now record their crops and keep an eye on their effectiveness remotely, address crop pests, and take swift action to protect their crops from any risk to the environment.

Agricultural Sensors

A sensor is a gadget that perceives and responds to certain inputs which could be illumination, locomotion, pressure, heat, or moisture, and transforms it into a representation or signals that can be read by humans for further reading and processing. There are various types of sensors used in agriculture that assist various smart agricultural operations. Sensors can be classified according to their mechanism of operation, and also by their various applications.

Classification of Agricultural Sensors Based on the Mechanism of Operation

Agricultural sensors classified based on their mechanism of operation are described in the following section.

Optical Sensor

Optical Sensors use light to measure the properties of the soil (Figure 7.1). These sensors can be installed on automobiles, satellites, drones, or robots, for collecting and processing reflected light from the soil. Optical sensors also can collect data on the condition of the clay, organic matter, and moisture content of the soil.

Optical sensors use reflectance spectroscopy to identify the magnitude of reflected and absorbed energy by soil nutrient ions and are mostly reported for the three major nutrients, N, P, and K in the soil. Bogrekci and Lee (2005) invented a portable Raman sensor for soil P detection in dry as well as wet soil. A 600 mW laser light source at 785 nm with a full width at half maximum of about 0.2 nm and a spectrometer that covers 340 cm−1 and 3640 cm−1 was used. The program was written in Visual C++ and partial least squares analysis was used to calibrate the model. Rogovska et. al. (2019) utilized the potential of Diamond-Attenuated Total internal Reflectance (D-ATR) Fourier Transform Infrared (FTIR) spectroscopy as a soil NO3 sensor for rapid field-mobile determination of soil NO3 concentrations. Zhou et. al. (2019) developed an optical sensor for in-situ soil total nitrogen and soil moisture detection. The unit is composed of eight single-band Near-Infrared (NIR) laser sources at 1260, 1330, 1360, 1430, 1530, 1580, 1660, and 1450 nm. The software part of the program was written in the JN5139 microcontroller unit. TD et al. also developed a multichannel, optoelectronic measurement system with automatically switching light sources that consist of peak emission wavelengths of 405 nm, 660 nm, and 515 nm, a photodiode array, and a circuit board with a MCU (Viscarra Rossel and Bouma 2016). They used it for the detection of N, P, and K content in six soil samples. Laskar and Mukherjee (2016) reported the design of a portable soil nutrient detector for N, P, and K based on optical sensors and a microcontroller (Chandel et. al. 2016). Based on experimentation and analysis, the researchers reported 850 nm for nitrogen, 620–630 nm for phosphorous, and 460–470 nm for potassium analysis for the design. Most of these sensor devices are equipped with farm vehicles to map the variability of soil parameters while traversing the farm terrain (Sinfield et. al. 2010). Although optical sensors using reflectance spectroscopy change with the nutrient content in the soil, the reflectance signatures vary with the soil type making it challenging to calibrate the model for quantification. Therefore, it is challenging to obtain consistently good estimates across a range of soils (Ozhikandathil et. al. 2018).

Figure 7.1: Optical Sensor for Agriculture.

Types of Optical Sensors

Various types of optical sensors applicable in agriculture are described in the following section.

Vis-IR spectroscopy

Infrared spectroscopy (IRS) is a technique used to determine the structure of compounds, both inorganic and organic. This has been accomplished by using an ultraviolet, visible, and infrared polychromatic radiation source to quantify light energy absorbed and, or diffusely reflected by soil nutrients. IR spectroscopy is a technique that analyzes molecular vibrations, and the principle of IR spectroscopy is based on the theory of simple harmonic oscillation. An example of a simple harmonic oscillator system would be the shared bonds between two atoms. These spring-like bonds display resonant characteristics dependent upon the "spring" constant which describes the force between them, and the atomic weight at the end of each bond. Thus, the oscillation of a particular chemical bond coincides with a unique frequency and intensity of energy. Further, these unique frequencies of oscillation between bonds are complementary to particular wave numbers cm−1, sharing a positive correlation. Incident radiation is projected upon a sample at similar vibrational frequencies shared between the bonds (Figure 7.2). This radiation becomes adsorbed while others are intently reflected. A spectrophotometer records the reflectance, forming reflectance spectra that reveal the magnitude of energy captured as a function of wavelength. Therefore, a molecule can be identified by comparing its absorption peak to a database of spectrums. Some of these sensors have been designed to be highly portable and can offer a full spectrum in less than 1 second.15 These sensors have also been integrated into microfluidic devices such as the one presented in.16. A lab on a chip integrated with a light-emitting diode and photoresistor to measure the changes in absorption and detect levels of ammonia and amino acids has been developed. Many soil properties such as moisture, organic matter, nutrients, and pH, can be identified rapidly and cost-efficiently using diffuse reflectance spectroscopy (Shibusawa et. al. 2000, Birrell and Hummel 2001, Soriano-Dista et. al. 2014, Nocita et. al. 2015)

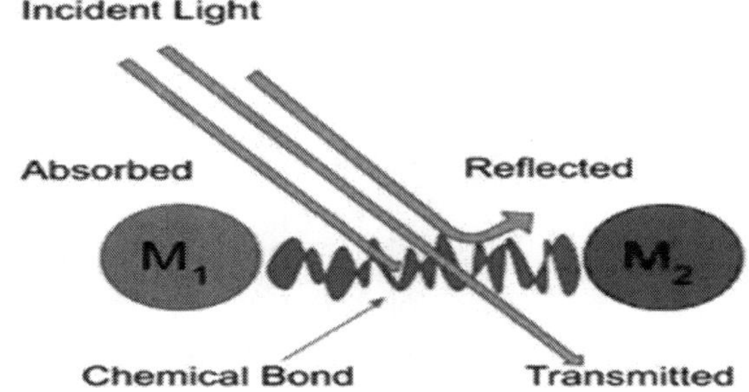

Figure 7.2. In IR spectroscopy an incident light is directed to molecular bonds between two atomic masses, and the energy is either absorbed, transmitted, or reflected (Reproduced from Burton et. al. 2020)

Attenuated Total Reflectance Spectroscopy (ATR)

Attenuated total reflectance (ATR) spectroscopy operates using the same principles of infrared energy absorption by particles and molecular bonds. However, instead of targeting samples with infrared spectrums and acquiring the diffuse reflectance, a crystal directly adjoining a target receives the incident IR energy (Figure 7.3). The crystal is usually made of diamond, zinc selenide, geranium, and, or thallium iodide, however, certain criteria should be considered before selecting the proper ATR crystal, such as the refractive index, spectral range, and physical and chemical properties of the sample. When incident radiation is directed upon the crystal, an evanescent field is created between the crystal and sample due to the incident reflection. The energy then departs the ATR crystal and travels toward a spectrometer. Here, a spectrum will be produced for the sample. ATRS instrumentation is expensive and delicate, therefore it is not practical for use in field settings.

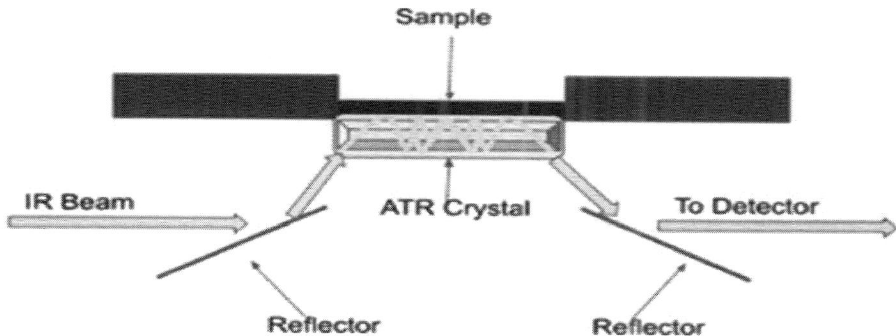

Figure 7.3. An IR beam is directed into the ATR device which is reflected from a series of reflectors that direct the incident light toward the ATR crystal. The crystal, in contact with a soil sample, reflects the light, creating an evanescence. The reflected energy exits the ATR crystal which is directed toward a spectrometer to produce a reflectance spectrum for the sample (Reproduced from Burton et. al. 2020).

Raman Spectroscopy

Raman spectroscopy is a technique that involves measuring the changes in wavelength and intensity of scattered light upon interaction with a sample. Incident light energy is absorbed and reemitted from the sample at different frequencies known as the Raman scatter or Raman effect. The Raman spectrum from the observed sample is a unique fingerprint of a sample that offers information about the chemical structure and identity of a sample, the polymorphism, intrinsic strain, and sample contamination. Lee and Bogrekci (2007) have explored and developed a portable Raman spectroscopy device and method that provided good results for the determination of phosphorus and other nutrients in soils and vegetation. The device is also capable of detecting nutrients in both wet and dry soils and contains compartments for drying, sieving, and grinding soil samples. This eliminates further laboratory needs and provides in situ determination. The device also features an onboard microprocessor which allows the data acquired by the spectrometer to calculate the concentration of phosphorus within the sample.

The advantages of optical sensors are that they are a highly flexible form of analysis, capable of measuring irregular surfaces non-destructively, and require little to no sample preparation. (Figure 7.4). This method of measurement eliminates the need for additional sample solutions and preparations such as extractants, sample grinding, and drying. It has also been noticed that these reproducible measurements can be produced at an extremely low cost (Viscarra Rossel and Bouma 2016). Since measurements are taken in a non-destructive manner (Kim et. al 2009) and they can be deployed subsoil for rapid in situ real-time analysis, this makes them perfect for in situ soil measurements. Collecting proper soil samples is a critical aspect of soil testing which must consider the spatial distribution, depth, and time of day. Therefore, large-scale deployment of these devices has the potential to offer a higher resolution of nutrient-sensing within the terrain.

Although these sensors are becoming increasingly popular and less expensive to assemble, they are still affected by many soil physical and chemical properties to different degrees (Baumgardner et. al. 1986). Ehsani et. al. (2000) while exploring MIR for soil nutrient sensing, discovered that measurements in desiccated soils were extremely difficult. Later, it was recommended that the soils be prepared moist and paste-like to facilitate adequate soil-crystal contact for improving results (Jahn et. al. 2006, Shi et. al. 2015). A recent study by Viscarra et. al. (2006) revealed the effects of moisture content and the environment when deriving proximal sensed Vis-IR data. Researchers also state that to successfully implement in situ NIRS, calibration is required for each site, which can potentially increase sample preparation time and cost.

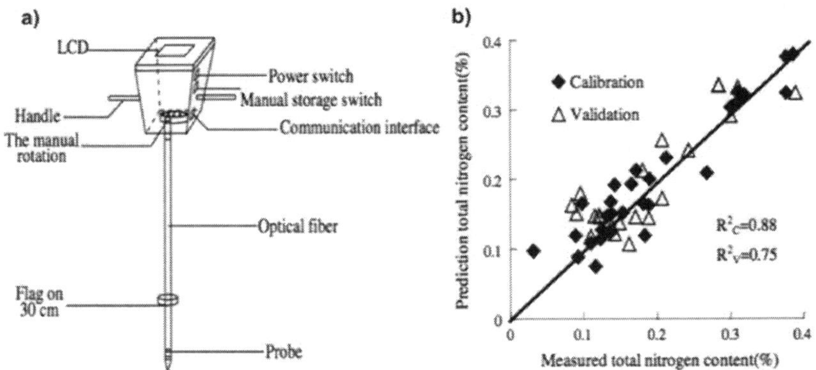

Figure 7.4. (a) Configuration of a TN prototype for detecting total N, (b) comparison between predicted and measured values of soil TN content (Reproduced from Burton et. al. 2020).

Electrochemical Sensor

Plant growth is influenced by complex and mutual relationships, including adequate nutrition, various environmental conditions, photosynthesis, respiration, pests, and diseases (Schans and Arntzen, 1991). To ensure significant agricultural production, it is essential to monitor the factors affecting plant growth in real-time. Electrochemical sensors have the potential to systematically monitor plant health and provide early diagnoses of disease and stress. Thus, these can replace conventional approaches (such as image-sensing technology) for monitoring plant health. Electrochemical sensors used for monitoring plant signals are presented in Table 7.1.

The electrochemical sensors aid in the collection, processing, and mapping of the chemical data of the soil (Figure 7.5). This includes the nutrient levels of the soil and pH. Error-free measurements are carried out with the use of an ion-selective electrode. These electrodes notice the pursuit of specified ions, such as hydrogen, nitrate, and potassium. They are usually mounted on a specially designed sledge.

Table 7.1. Electrochemical system-based plant signal sensors.

Target material	Sensing material	Detection signal	Applications	References
Humidity sensor	GO	Capacitance changes	Plant leaf water detection	Lan et. al. 2020
Humidity sensor	Polyamide	Capacitance changes	Plant leaf water detection	Im et. al. 2018
NO2 sensor	Silver/rGO	Resistance change	Toxic gas detection	Li et. al 2021
Nutrient sensor	rGO aerogel/ ISE	Potential change	Plant sap monitoring	Kim et. al. 2021
Nutrient sensor	LIG/ ISE	Potential changes	Nitrogen sensing in soil	Garland et. al. 2018
03 sensor	PEDOT-CT	Mpdance and phase changes	03 damage detection	Kim et. al. 2020
Pesticide sensor	LIG-OHP/ AnNPS	Current change	In situ analysis of pesticide	Zhao et. al. 2020
VOC sensor	CNT/ Graphite	Resistance change	Gas detection from plants	Li et. al. 2021
VOC sensor	rGO/ AnNPS	Resistance change	Plant health monitoring	Calisgen et. al. 2020

Note: *GO, graphene oxide; rGO, reduced graphene oxide; CNT, carbon nanotube; AuNPS, gold nanoparticle; PEDOT, poly (3,4-ethylenedioxythiophene); LIG, Lase-induced graphene; ISE, ion selective electrode; VOC, volatile organic compounds.*

Figure 7.5. Electrochemical sensors.

Types of Electrochemical Sensors

Electrochemical sensors are a common approach to understanding the chemical content of a solution and are becoming increasingly popular due to their miniature size and ease of integration into autonomous systems. The electrochemical device functions by coupling a chemically selective layer, known as a recognition element, to an electrochemical transducer. When the recognition layer is in contact with the ion of interest, the transducer converts the chemical energy of the selective membrane interaction into an electrical signal. The electrical techniques used for the transduction of the target analytes allow electrochemical sensors to be further organized into subcategories. These categories include potentiometric, which measures the changes in membrane potential, conductometric for measuring changes in conductance, impedimetric sensors which measure changes in impedance, and amperometric sensors for measuring changes of current at the sensing membrane. The two most employed electrochemical sensors for determining soil nutrients are the ion-selective field-effect transistors (ISFET) and ion-selective electrodes (ISE).

Ion Selective Electrode (ISE)

Ion-selective electrodes are conductive pieces of matter with a specialized membrane (glass/polymer) whose potential reflects the concentration of the selective target being measured. These ion concentrations are determined by calculating the voltage drop across the membrane using the Nernst equation. The Nernst equation is by far the most fundamental equation to the understanding of electrochemical cells. Formulated by German scientist Walther Nernst, this equation considers standard potential, temperature, and concentrations in an electrochemical reaction to relate with the potential of the working electrode.

The Nernst equation is depicted as:

$$\phi = \phi^0 + RT/z_1F(\ln a_1^{solution})$$

where ϕ is the electrode potential, ϕ^0 is the standard value of the potential, a_1 is concentration and z_1 is the target ion charge number, R is the gas constant, T is the absolute temperature, and F is the Faraday constant. Usually, in a galvanic cell, the difference between two electrodes submerged in a solution is measured as the electromotive force (EMF). EMF is described as a source of energy that enables electrons to move around an electric circuit (Smith and Holroyd 1969) and is determined by the difference between oxidizer and reducer potentials (Yefimov and Naboychenko 2009). in a reduction-oxidation reaction. Under certain conditions, when the reference potential remains constant in a galvanic system and the active electrode follows the Nernst equation, the EMF equation can be described as:

$$E = E^0 + S\log a_1$$

where E is the measured EMF, E^0 is the standard EMF value at al = 1, and S is the response slope. The slope is usually $S_{25} = 59.18/z_1$ mV at 25 C. However, according to the literature, the slope can differ with a freedom of about 0.2–2 mV (Mikhelson (2013). The slope is derived by obtaining a calibration curve using at least 3 standards. Once this is done the ISE can be used to measure concentrations of target analytes in solutions. The sensitivity can be calculated by dividing the linear part of the calibration slope by the surface area of the working electrode.

ISEs are proven to have a wide dynamic range, appearing to distinguish between variations in concentrations of residual nitrogen (0.1–10,000 ppm N) and nutrients. ISEs are traditionally employed in laboratory facilities to obtain nutrient measurements. A common approach for using ISEs is integrating them into microfluidic structures. Microfluidics allows the controlling of liquid samples at micro and Pico levels while the integrated ISEs allow rapid analysis of nutrients. Chen et al. (2018) presented a flow-loop microfluidic device integrated with an ISE sensor for in situ analysis of phosphate.

Electrochemical sensors can be employed in a variety of environments. The majority of electrochemical sensors are used to detect nutrients, toxins, and pollutants in waters and aqueous solutions (Li et. al. 2016, Kumar and Sundramoorthy 2018, Van Staden et. al. 2018, Esmaeilli et. al. 2019, Tyszczuk-Rotko and Jedruchniewicz 2019). However, ISE's have also been known to gather rapid measurements in slurries, unfiltered soil extract, and naturally moist oils (Adamchuk 2005, Lobsey et. al. 2010a, Lobsey et. al 2010b). Wang et. al (2018) developed a portable multi-ion measurement system developed allowed site-specific nutrient analysis and characterization of the spatial variability of surface soil nutrients. Adamchuk et. al (2005) developed a system for on-the-go mapping of nutrients using direct soil measurement methods that allowed nutrient electrodes to distinguish between very high and very low concentrations. In addition to aqueous solutions, moist soils, and slurries, electrochemical sensors have also been used to determine ionic concentrations in gaseous forms.

Though many recognition elements for nutrient detection exists, the most popular membranes are based on PVC-ligand membranes and molecular imprinted polymers. Glass membrane-based ISE's are popularly used for determining pH, but are not commonly reported for detecting nutrients. Some of the most popular membranes for detecting nutrients include the use of doped polymer membranes such as Nitrogen-doped polypyrrole (N-doped PPy) for the detection of soil nitrogen, tetradodecylammonium o-nitrophenyl octyl ether (TDDA-NPOE) based electrodes for detecting nitrate, valinomycin-bis (2- ethylhexyl) sebacate (DOS) based electrodes for detecting potassium, and cobalt rod-based electrodes for measuring phosphate. MIP-based membranes for detecting soil nitrogen have been explored in the electrochemical doping was used to fabricate a nitrate sensor. The researchers further increased the sensitivity of their sensor by doping polypyrrole nanowires. Optimal membrane compositions and preparation conditions have been determined for polypyrrole (PPy) for determining soil nitrate concentrations. The researchers reported that the lifetime of the membranes can also be extended by adding appreciable amounts of plasticizers to the membranes. Overall, due to the ease of preparation, MIPs and PVC are suitable for the fabrication of selective ISE's.

Ion Selective Field Effect Transistor (ISFET)

Ion-selective field-effect transistors (ISFET's) are simply ISE's joined with a field-effect transistor (FET) and can also be referred to as chemically modified field-effect transistors (CHEMFET's). The function and structure of a FET are quite interesting, as it uses an electric field within a region of the device to control the flow of current and is comprised of three terminals; the source, drain, and gate. There are two types of FET devices, p-type and n-type. The device type dictates which carrier conducts the flow of current. There are two carrier types, electrons, and holes. In an n-type FET, a gate receiving a positive voltage causes a carrier depletion region due to the repelling of holes in the region beneath the gate. As the holes are repelled, electrons are pulled from the substrate into the channel region, assembling a bridge between the source and drain. Now, as a voltage is applied between the drain and the source, electrons can flow freely through this channel. Quite the opposite effect occurs within a p-type device, as a positive voltage will deplete the carriers and reduce the conductance. By applying a voltage to the gate, an electric field can be generated to control the amount of charge in the channel and therefore influence the conductivity of the channel. This applied voltage is referred to as the threshold voltage, as it represents the number of volts required to sufficiently accumulate a number of electrons for a conductive channel. In the case of a CHEMFET or ISFET, the ISE membrane is appended to the insulating layer of the FET, thereby chemically modulating the threshold voltage as a function of the concentration within the solution.

The use of ISFET's with flow injection analysis (FIA) has been discussed in.6 Researchers have investigated technology to successfully measure soil nutrients from prepared extracts. The system based on the multi-ISFET technology featured a rapid extraction ISFET system for real-time in-field measurements of soil nitrogen in less than 5 seconds. Literature has also discussed ISFET devices that have demonstrated the ability to analyze various soil nutrients,7 however, like many previous devices the sensor relies on an extraction system. It is reported that the flow injection analysis system helps to mitigate ISFET sensor drift and increase the efficiency and performance of the sensor system. Compared to ISE's, ISFET's are relatively smaller in dimensions, have a higher signal-to-noise ratio, and rapid response time.

Advantages and Disadvantages of ISFET

The ion-selective electrodes can make an ideal platform for in situ soil nutrient measurements. These soil sensor measurements can be acquired in a rapid manner with lab-grade accuracy (Sibley et. al. 2010). The measurements can also be performed in slurry solutions, unfiltered extractants, and naturally moist soils. Additionally, electrochemical sensors can be fabricated using low-cost materials, such as the ammonia sensor based on paper substrates (Sekhar and Kyser 2017). Printed sensor platforms can be mass-produced using common industrial printing techniques, then interfaced with a single microcontroller for real-time soil nutrient analysis (Burton et. al 2018).

Disadvantages of the ion-selective electrodes are that it requires frequent calibration, may require additional extraction solutions, and relies on soil moisture for accurate nutrient readings (Antonacci et. al. 2018). Another disadvantage is that nutrient sensor response is affected by various properties of soil such as the textural class, particle size, volumetric water, organic matter, and interfering ions. In addition, the sensor requires periodic rinsing.

Mechanical Sensor

These sensors are used to measure soil compression or mechanical resistance (Figure 7.6). This sensor uses an application that passes through the soil. This sensor then records the force calculated by pressure scales or load cells. When a sensor passes through the soil, it documents the holding forces that result from the cutting, smashing, and displacing of soil. Soil mechanical resistance is recorded in a unit of pressure and points out the ratio of the force necessary to go into the soil channel to the frontal area of the tool engaged with the soil.

Figure 7.6. Mechanical soil sensor.

Dielectric Soil Moisture Sensor

This sensor calculates the moisture levels in the soil with the assistance of a dielectric constant. This is an electrical property that substitutes depending on the moisture content in the soil. The moisture sensors are used in association with precipitation check locations placed around the farm area (Figure 7.7).

Figure 7.7. Dielectric soil moisture sensor.

GPS or Location Sensor

They are also known as agricultural weather stations. They are positioned at different places throughout the fields (Figure 7.8). These precision agriculture sensors are used to determine the variety, distance, and height of the plants at any position within the required area. They take the help of GPS satellites for this purpose.

GPS sensors are also used for other farming activities such as: tilling the field, plant harvesting, and other related activities. Making use of auto-guided systems can enhance field routing, reduce overlapping processes, and eventually reduce the quantity of time required to complete a task.

Figure 7.8. GPS or location sensors for agriculture.

Airflow Sensors

These sensors are used to measure air permeability or soil air penetration. They can be used in a fixed position or dynamically while in motion. The expected result is the pressure needed to push a decided amount of air into the ground at a prescribed depth. Various types of soil properties, including moisture levels, soil type compaction, and structure, produce a different identifying signature.

Temperature Sensor

Temperature sensors are crucial in two key categories of agriculture (Figure 7.9). These categories are ambient condition monitoring and mechanical asset monitoring. For instance, ice wine harvesting usually occurs within a narrow temperature window when temperatures first reach between -10°C and -12°C during a harvesting season. The ice wine industry requires a highly accurate temperature and humidity sensor to give a precise prediction of the temperature forecast. Temperature sensors do not only play an important role in monitoring the ambient conditions of physical space, but they play a crucial role in virtually all smart agriculture asset monitoring applications

Temperature sensors are also used to evaluate the performance of the equipment used in various agricultural operations. Whenever an equipment system is in need of minor maintenance, is underperforming, or is critically failing, the temperature sensor dishes out an alert. They are highly effective in virtually every activity that relates to the predictive and reactive maintenance system. This in turn protects against overheating and detrimental failure of equipment.

Figure 7.9. Temperature Sensors.

Accelerometer Sensor

This is quite similar to the use of temperature sensors in maintenance prediction. Accelerometers are vastly made use of across the smart agriculture industry to predict and assist with required maintenance. They are mostly used on moving components and motors. Their major aim is to detect slight variations in movement and vibration inconsistencies and foretell when standard maintenance is required or a compromised component needs to be replaced. However, this sensor is usually associated with farming and other agriculture, accelerometers play an indispensable role in the maintenance of vital smart agriculture equipment. Accelerometers can also be used in various automated systems and tracking methods. For instance, a low-power accelerometer makes it easier and faster to monitor the status of an adjustable spray nozzle on the end of a fertilization beam. With adverse technology, the use of autonomous drones in smart agriculture depends critically on accelerometers and IMU (inertial measurement units) to track motion, speed, crash events, and even position in space.

Smart Cameras in Agriculture

Smart cameras have been increasingly adopted for a variety of smart agriculture applications. Smart camera technology is used to detect weeds and other plant locations. Following this, an automatic and accurate dispensation of herbicides and fertilizer can be carried out. This helps in the judicious application of chemicals and increases overall productivity.

One of the most significant challenges in agriculture is the issue of pest control. With the use of smart cameras, farmers can now detect pests in real-time and effectively monitor actions against pests without necessarily harming helpful agricultural insects. Smart cameras can also take the place of semi-legacy sensing devices such as ambient light monitoring, thereby enabling a simplified system and a reduction in the component count.

Common Sensors and Detection Parameters in Agriculture

Some of the common sensors and detection parameters including soil temperature, soil moisture, conductivity, salinity level of the soils, plant wetness, plant moisture, photosynthesis, and CO^2 level used in agriculture are presented in Table 7.2.

Table 7.2. Common sensors and detection parameters in agriculture (Abbasi et al., 2014).

Name of the Sensors	Parameters monitored
ECH20 Soil moisture sensor	Soil temperature, soil moisture, conductivity
Hydra probe II soil sensor	Soil temperature, salinity level, soil moisture, conductivity
MP 406 Soil moisture sensor	Soil moisture, soil temperature
EC sensor (EC250)	Soil temperature, salinity level, soil moisture, conductivity
Pogo portable soil sensor	Soil temperature, soil moisture
237 Leaf wetness sensor	Plant moisture, plant wetness, plant temperature
YSI 6025 Chlorophyll sensor	Photosynthesis
TT4 Multi-sensor thermocouple	Plant moisture, plant temperature
TPS-2 Portable photosynthesis	Photosynthesis, plant moisture, CO^2
CM-100 Compact weather sensor	Air temperature, air humidity, wind speed, air pressure
XFAM-115KPASR	Air temperature, air pressure, air humidity

Classification of Agricultural Sensors Based on Their Applications

Agricultural sensors classified according to their applications are presented in the following section.

Plant Nutrient Sensor

Several Inorganic nutrients are essential for plant growth. Among them, sixteen are known to be essential elements that are required by crops (Muñoz-Huerta et al., 2013) The essential nutrients necessary for plant metabolism include carbon, hydrogen, and oxygen (which can be absorbed from water and carbon dioxide), macronutrients (nitrogen, potassium, calcium, phosphoric acid, sulfur, and magnesium), and micronutrients (chlorine, boron, iron, manganese, zinc, copper, nickel, and molybdenum. (Kashyap and Kumar, 2021). The demand for these ions differs between plant types and growth stages, and the ions interact intricately through various absorption mechanisms in plant roots (Gieling et al., 2005; Guignard et al., 2017). Therefore, accurately monitoring these nutrients can reduce the waste of fertilizers and protect the environment (Kashyap and Kumar, 2021; Yin et al., 2021). A representative method for analyzing nutrients in crops *in situ* is to use an ion-selective electrode (ISE) (Cheong et al., 2021).

The ISE measures the potential difference generated by the binding of a target ion to the selective membrane using a membrane that selectively reacts with a specific analyte. Unlike other analytical methods, ISE measurements can be conducted on-site without the need for pre-treatment. Therefore, it is a simple and rapid method to analyze ions in the environmental, water, food, pharmaceutical, and biological samples (Shao et al., 2020). An ISE can be created by attaching an electrode polymer membrane—comprising an ion-sensitive material (ionophore) that reacts with the ionic material—to the electrode surface. Ionophore-doped polymeric membrane-based ISEs exhibit excellent processability, and enable the detection of various analytes (Chen et al., 2018). Conventional membrane-based ISEs are liquid contact-type ISEs with an internal filling solution on an electrode, designed as a glass membrane. Liquid-contact ISEs are relatively stable in a research environment, and are already commercially available; however, they are difficult to maintain, and miniaturization is disadvantageous because of the presence of an internal liquid. Compared to liquid-contact ISEs, solid-contact ISEs are stable, simple to manufacture, and can be miniaturized (due to the absence of an internal liquid). Recently, solid-contact ISEs have shown excellent performance improvements when various solid-contact functional materials were applied as ion-to-electron converters (Hu et al., 2016). These instruments can be miniaturized and require simple assembly. Moreover, the use of disposable screen-printed electrode (SPE) substrates makes it economical.

Recent studies have reported the detection of nutrients in plant sap, soil, and nutrient solutions using a potentiometric ion sensor based on a solid contact-ion selective electrode (SC-ISE). Kim et al. (2021) reported an increase in the stability of the sensor by using a disposable SPE-based reduced graphene oxide aerogel (rGOA) as an electron-ion converter to measure Ca^{2+} and NO_3^- in plant sap (Figure 7.10). Using rGOA with high capacitance as an electron-ion converter, the sensor showed high stability. In addition, the sensor measured Ca^{2+} and NO_3^- in the sap of perilla leaves, and the results displayed high performance similar to those of the ICP and IC methods. Artigas et al. (2003) developed a sensor probe incorporating a screen-printed electrode-based nitrate-selective electrochemical sensor equipped with a copper plate reference electrode. This sensor has shown promising results in monitoring the ionic state of inorganic nitrogen in soil using the potentiometric method. Using laser-induced graphene (LIG) as the solid electrode layer of SC-ISE also reduces toxicity, and requires a simple manufacturing procedure (Garland et al., 2018). In this method, a flexible LIG electrode is attached to polyimide (PI) film through a manufacturing procedure free of toxic chemicals. The LIG-based SC-ISE has been used to measure NO_3^- and NH_4^+ levels in soil slurry, and showed excellent detection abilities. Jiang et al. (2017) developed an ion sensor including an ion-selective electrode (that uses the potentiometric method) to measure the concentrations of $H_2PO_4^-$ and NO_3^- in the growth medium of rice plants. In this device, an ion-selective electrode is attached to the bottom of a poly (methyl methacrylate) (PMMA) chamber filled with growth medium, and a plastic mesh is mounted on the top of the sensor to prevent contact with plant roots. This potentiometric sensor successfully quantified $H_2PO_4^-$ and NO_3^- levels in plants for 2 weeks, and demonstrated the possibility of monitoring the nutrient utilization efficiency of plants in real-time. Vardar et al. (2015) used SC-ISE to detect ions in a hydroponic nutrient solution. NO_3^-, K^+, and Ca^{2+} sensors were prepared using the previously described ion-selective membrane, and these sensors could detect changes in the concentration of ions in hydroponic nutrient solutions according to the growth stage of crops. Monitoring the levels of NO_3^-, K^+, and Ca^{2+} in cucumber and tomato crops revealed that tomatoes consumed less nitrate than cucumbers during the first 2 months of growth, and that potassium intake was higher than that of other nutrients. The results of these studies indicate that the types and amounts of nutrients absorbed by plants differ between crop types, growth stages, and seasons. Kim et al. (2013) reported the direct measurement of NO_3-N, K, and Ca concentrations using ISE in a paprika hydroponic nutrient solution. An ISE array coupled with a computer-controlled measurement system has been shown to monitor the ion concentrations of nutrient solutions over time. Therefore, systematically generating a nutrient absorption mechanism and prescription map for each crop and growth stage and help improve the possibility and performance of PA.

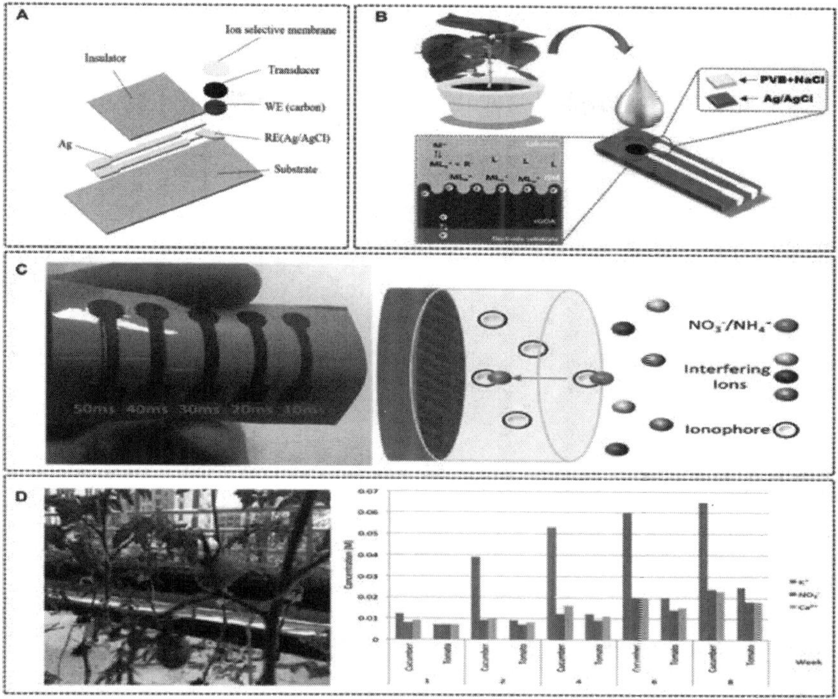

Figure 7.10. (A) Schematic representation of the SC-ISE-based SPE (Kim et. al. 2021). (B) Reduced graphene oxide aerogel-based SC-ISE for Ca2+ and NO3 – detection in the plant sap (Kim et. al. 2021), (C) Laser-induced graphene-based SC-ISE for the determination of NO3 – and NH4 + levels in the soil sample (Garland et. al. 2018). (D) The hydroponic greenhouse control systems (left) and mineral uptake by cucumber and tomato plants in nutrition reservoirs (right) (Reproduced from Vardar et. al. 2015).

Soil Nutrient Sensor

Advanced technology and communications are allowing faster and more stable transmission of sensor data worldwide, and are becoming more affordable due to miniaturization and lower material costs. Various types of sensors, both commercially available and under development, to measure soil chemical properties are presented in Table 7.3 (Bali et. al. 1999, Laskar and Mukherjee 2016, Viscarra Roissel and Bouma 2016). These can be employed in either map-based (Chandel et. al. 2016) or real-time systems. Among all the available techniques, the majority of in situ soil sensing involves optical and, or, electrochemical methods (Figure 7.11). Optical sensors use spectroscopy based on reflectance spectroscopy to identify the magnitude of reflected and absorbed energy by soil nutrient ions. Electrochemical sensors for soil nutrient determination function by using ion-selective electrodes to initiate a current or voltage output that reflects the concentration of target ions.

Table 7.3. Types of soil nutrient sensors.

Sensor concept	Development Status	Current Results	References
Vis-NIR	Lab/Field	Soil pH and Nutrients	Viscarra Rossel et. al. 2006.
Vis-MIR	Lab	Soil Mineral Nitrogen	Ehsani et. al. 2000.
ATR Spec	Lab/Field	Soil Nutrient	Christy et. al. 2003.
Ramen	Lab/Field	Soil Nutrients	Jahn et. al. 2006.
ISE	Lab/Field	Soil pH and Nutrients	Van Staden et. al. 2018.
ISFET	Lab/Field	Soil pH and Nutrients	Soriano-Disla et. al. 2014.

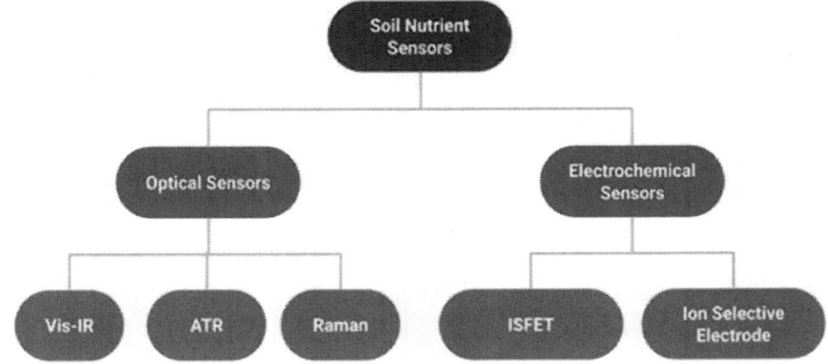

Figure 7.11. Flow chart for real-time soil nutrient sensors used in precision agriculture (Reproduced from Burton et. al, 2020).

Plant Hormone Sensor

The possibility of measuring phytohormones *in situ* in agricultural fields is a remarkable achievement, as phytohormones play a crucial role in regulating physiological processes and defensive signals in plants (Giraldo et al., 2019). Plant hormones are important molecular aggregates that regulate plant growth, development, and other physiological processes (Santner and Estelle, 2009). Because these plant hormones are directly related to crop yields, a precise monitoring technology is required to measure phytohormones (Giraldo et al., 2019). Plants exhibit hormonal changes due to various environmental stresses, and these changes are associated with different electrochemical patterns (Kinpara et al., 2001). Therefore, the technology to monitor plant hormones can help predict various diseases and stress responses, such as responses to drought, salinity, and pest stress (Suzuki et al., 2014).

Auxins play a central role in the regulation of plant growth and development. Auxins increase the permeability of cell membranes and promote the growth of individual cells by promoting cell division, irregular root formation, and flower bud formation (Sun et al., 2017). Auxins are produced in new leaves and are transferred to the roots (in the direction of gravity) to promote root development. Recently, an electrode using a stainless steel (SS) wire was developed for the detection of IAA (a representative plant growth hormone) (Li et al., 2019). The proposed electrochemical microsensor showed a low detection limit of 43 pg ml−1, and successfully detected IAA in vivo in the stems of soybean seedlings under saline stress conditions (Figure 7.12). Coppedè et al. (2017) proposed a biomimetic fiber-based electrochemical biosensor that could be directly inserted into plant tissues. This sensor can monitor changes in the solute content of plant sap in vivo in real time. These changes can indicate signs of stress, monitoring which can help improve plant growth. Salicylic acid (SA) is another type of plant hormone that regulates central signals for the physiological behaviours of plants (Sun et al., 2014). A disposable working electrode was developed by coating a carbon tape with a mixture of multi-walled carbon nanotubes and Nafion, to detect salicylic acid released from tomato leaves. The electrochemical reaction of salicylic acid was greatly enhanced by treating the modified carbon tape electrode with oxygen plasma. The results showed that the amount of salicylic acid released from tomato leaves was statistically different between normal and diseased tomato leaves infected with *Botrytis cinerea.* Therefore, this electrode could potentially detect plant conditions under biotic and environmental stresses by quantifying salicylic acid at nanogram levels in real time.

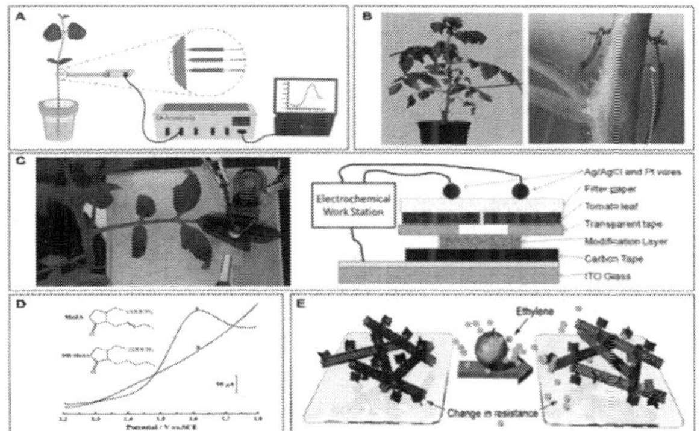

Figure 7.12. Electrochemical sensor for monitoring plant's hormones. (A) A schematic illustration of in vivo determination of auxins in soybean seedlings (Li et al. 2019), (B) The textile sensor device is implanted in a tomato plant stem for monitoring variations in the solute content of the plant sap (Coppedè et al. 2017), (C) A paper-based electroanalytical device (left) and its cross-section view (right) (Sun et al. 2014), (D) Square-wave voltammograms of methyl jasmonate (MeJA) and DH-MeJA (Gan et al. 2010), (E) Schematic image of ethylene sensing using a chemo-resistive sensor (Reproduced from Esser et al. 2012).

Jasmonic acid and ethylene are two representative phytohormones that respond to environmental changes (Xu et al., 2020). Gan et al. (2010) reported the direct electrochemical determination of methyl jasmonate (MeJA) using a nano-montmorillonite-modified glassy carbon electrode (nano-MMT/GCE). This electrode exhibited a sensitive electrochemical signal for the direct oxidation of MeJA, and this method was applied to the detection of MeJA released from wheat spikelet samples. Ethylene, the smallest phytohormone, plays a role in initiating the ripening of fruits, is involved in several plant developmental processes, and promotes seed germination, flowering, and senescence of leaves and flowers. Ethylene is difficult to detect selectively because of its small size and nonpolar chemical properties. However, ethylene was selectively detected using carbon nanotubes and a copper (I) complex one based on a fluorinated tris(pyrazolyl) borate ligand (Esser et al., 2012). The copper (I) complex one combines with ethylene to form complex 2, which shows reduced interaction with the surface of the single-walled carbon nanotube (SWCNT). This change causes a corresponding change in the resistance of the SWNT network. These changes were measured as electrochemical signals and used for the detection of ethylene.

Airflow Sensor for Indoor Farming

As cities become more and more pressed for space, the need for alternative solutions to field farming is becoming more and more essential to supply food for ever-growing populations. Forward-thinking agriculturalists have moved toward indoor farming, which can be done on both horizontal and vertical planes, which both have their fair share of strengths and weaknesses. Of course, some crops grow much better in an indoor environment than others, like carrots, beans, tomatoes, avocados, and even marijuana, while some crops are more suited for outdoor farming, like rhubarb or squash and other plants that require plenty of room for sprawling vines or bushy leaves. Nevertheless, indoor farming is a huge leap forward for space-saving food production, and it can be done effectively right in the middle of a metropolitan area.

Challenges in Indoor Farming

The control parameters for indoor farming, include lighting, water/moisture, CO2 levels and photosynthesis rates. Recently, the airflow of these indoor farming facilities has become a topic of much discussion.

Airflow is crucial in indoor farming for several reasons. One, it forces air through the filtration systems that limit the number of microbes that enter the facility in the first place. Many believe that indoor facilities are immune to a lot of the challenges that face outdoor farms in terms of bacteria, pests, and mold or rot, but indoor facilities are just as much at risk for these types of crop-killers as field-growing operations.

Airflow circulates CO2 through these indoor farming facilities, which both supply the plants with the crucial gas for growth and reduce the level of humidity. Too much moisture gathering on the cops can create a quite hospitable environment for mold and mildew. The circulated air also reduces the build-up of heat, which can create the perfect environment for pests, pathogens, diseases, and bacteria. In fact, without constant and substantial airflow through these indoor farms, significant growth would be downright impossible.

Solutions to airflow challenges in indoor farming

It is evident that too much of airflow can be as detrimental to crop growth as too little. Otherwise, farmers can face devastating losses and costly repairs. Thus, to have optimal airflow, it must be monitored precisely. Therefore, it's necessary to have multiple, sometimes many, test points that work on one system. Ideally, those test points should be designed for fixed installation so they can monitor the air inflow from all the sources (fans, pumps, vents, etc.), continuously.

Airflow transducers have now been developed as the perfect solution to these airflow monitoring challenges. They can be installed near the enter/exit points of the facility and they feature an analog output, which allows them to be fed information to data loggers, alarms, or controllers. The airflow transducers can measure 24/7 and all setting adjustments and measurement collections can be done from one terminal. Indoor farm hands will know immediately if the airflow is either insufficient or excessive immediately so adjustments can be made quickly, avoiding waste or loss of crops.

The instruments measure air velocity with hot-wire technology, meaning that the speed of the moving air is translated by the time it takes for the inner mechanism to cool, resulting in extremely accurate results. This, paired with the wide selection of probes available for the transducers promises the most precise measurements possible for 24/7, fixed installation monitoring. The probes can be selected specifically for individual tasks, which makes these instruments very versatile.

Biological Stress Sensor

A variety of other parameters can be used to monitor plant growth status and physiological conditions. Plants have evolved signalling networks that trigger appropriate physiological processes as a response to changing environmental conditions (Lee et al., 2021). Detecting plant stress markers based on chemical properties (such as plant signalling molecules) may help diagnose specific environmental or biological stresses in plants. Electrochemical biosensors using organic electrochemical transistors (OECTs) are suitable systems for real-time monitoring of biological processes. The direct bonding between the OECT channel and the electrolyte allows for signal amplification and a prompt sensor response, while maintaining a close interface with biological systems.

Glucose Monitoring Sensor

An OECT sensor that uses glucose oxidase cross-linked with a chitosan matrix as a sensing mechanism has been reported to detect glucose levels from the chloroplast (a plant organelle that synthesizes glucose) (Diacci et al., 2020). Glucose synthesized from chloroplasts was monitored in real-time at two distinct plant metabolic stages (Figure 7.13), and these results showed improvements over existing methods. Electrochemical sensors for glucose detection have been developed using various materials (Naikoo et al., 2021). These monitoring techniques are important because the production of glucose in chloroplasts is directly related to the yield and sugar content of fruits (Naikoo et al., 2021).

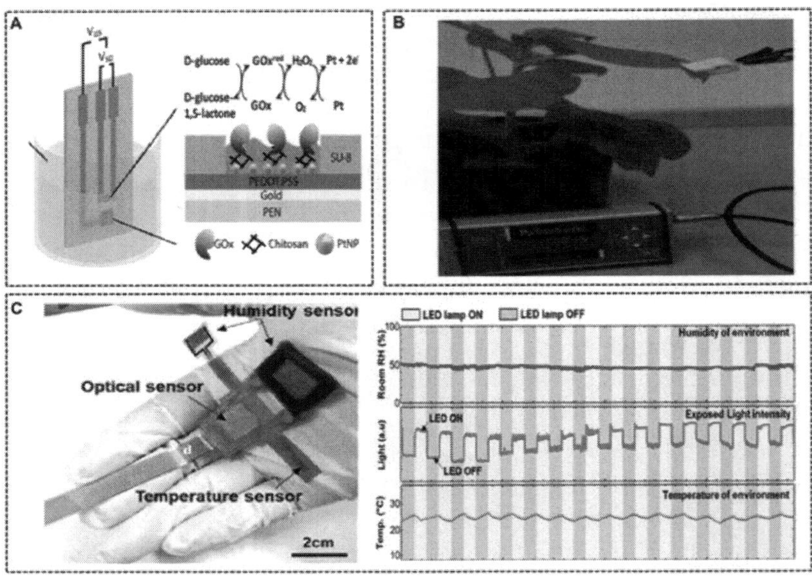

Figure 7.13. Electrochemical plant sensors. (A) Overview of the experimental design for the detection of glucose in isolated chloroplasts (Diacci et. al. 2020). (B) A 3D-printed chip holder containing a hand-held potentiostat and sensor chip (Pandey et al. 2018). (C) A multimodal flexible plant sensor integrated with humidity, optical, and temperature sensors (left) and the results of transpiration monitoring using a multimodal flexible sensor device for over 350 h (right) (Reproduced from Lu et. al. 2020).

Gus Monitoring Sensor

A technique has been developed for the real-time monitoring of β-glucuronidase (GUS) expression in transgenic tobacco plants using biomarker activity (Pandey et. al. 2018). Microchip electrodes were designed with a three-electrode system using the chronoamperometric technique, and were used to convert the expression of GUS enzymes in Msk8 tomato cells and BY2 tobacco cells into measurable current signals. The detection sensitivity of the GUS expressed in tomato cells was 0.076 mA/mM cm2, and the detection limit was calculated to be 0.1 mM.

Plant-Wearable Sensor

Continuously monitoring the growth requirements of crop plants is essential for increasing yields, especially in harsh environments. Recently, plant-wearable sensors have emerged in the field of plant health monitoring. These devices can monitor plant health conditions in real-time, and can help detect diseases or abnormalities in plants at an early stage (Qu et. al. 2021). For instance, plant leaves release volatile organic compounds (VOCs) when infected with *Phytophthora infestans*, and the presence of this disease can be monitored by detecting the released VOCs (Li et. al. 2021). A sensor was designed by integrating a graphene-based sensing material and a silver nanowire electrode on a stretchable substrate, and could detect diseases in tomato plants with >97% accuracy (Figure 7.14). Zhao et. al. (2020) used LIG technology to develop a field analysis device for organophosphorus pesticides based on a 3-electrode system. A stretchable and flexible LIG-based electrode was prepared with polydimethylsiloxane (PDMS). After being modified with an organophosphorus hydrolase (OPH), it could selectively detect the methyl parathion remaining on the surface of crops. OPHs hydrolyze methyl parathion to release p-nitrophenol, which electrochemically controls the catalytic reaction. The information was transmitted to a smartphone, which allowed real-time monitoring of the data. Nassar et. al. (2018) developed an ultra-lightweight plant-wearable sensor with ultra-low power requirement. The 3D-based stretchable plant-wearable sensor continuously monitors the local humidity and temperature on the leaf surface. Moreover, an innovative plant-wearable device with a 3D-printed origami-based plant copter design was used to improve plant-wearable sensors. Li et. al. (2018) developed a device that directly prepared a target object by introducing an atomized gas sensor array using a silver nanoparticle (AgNP)–all-carbon hybrid nanostructure. Reduced graphene oxide (rGO) decorated with AgNPs was used as a sensing layer on a metal single-star carbon nanotube as a conductive electrode base, and this sensor detected a low NO2 concentration (0.2 ppm) at room temperature. Lan et. al. (2020) designed a flexible and wearable humidity sensor that uses laser-induced graphene interdigital electrodes. The highly flexible graphene oxide (GO)-based humidity sensor could be attached to the surface of a plant leaf without physical damage, and using GO to detect humidity ensured high sensitivity (3,215.25 pF/% RH) and long-term stability (variation less than ±1%). As mentioned above, plant-wearable sensors with good flexibility can be attached to leaves or stems without causing physical damage, thus enabling real-time and long-term monitoring.

Multifunctional plant-wearable sensing devices have been developed using advanced nano-electronic technology to monitor potential health issues of plants. A multimodal flexible sensor system for monitoring changes in plant growth status has been proposed using stacked ZnIn2S4 (ZIS) nanosheets as the sensing medium (Lu et. al., 2020). The ZIS-based flexible sensor detected the illuminance and monitored the humidity in the vicinity of the leaves (i.e., where plant transpiration occurs). Monitoring humidity levels according to the opening and closing of plant stomata can also help monitor the dehydration state of plants.

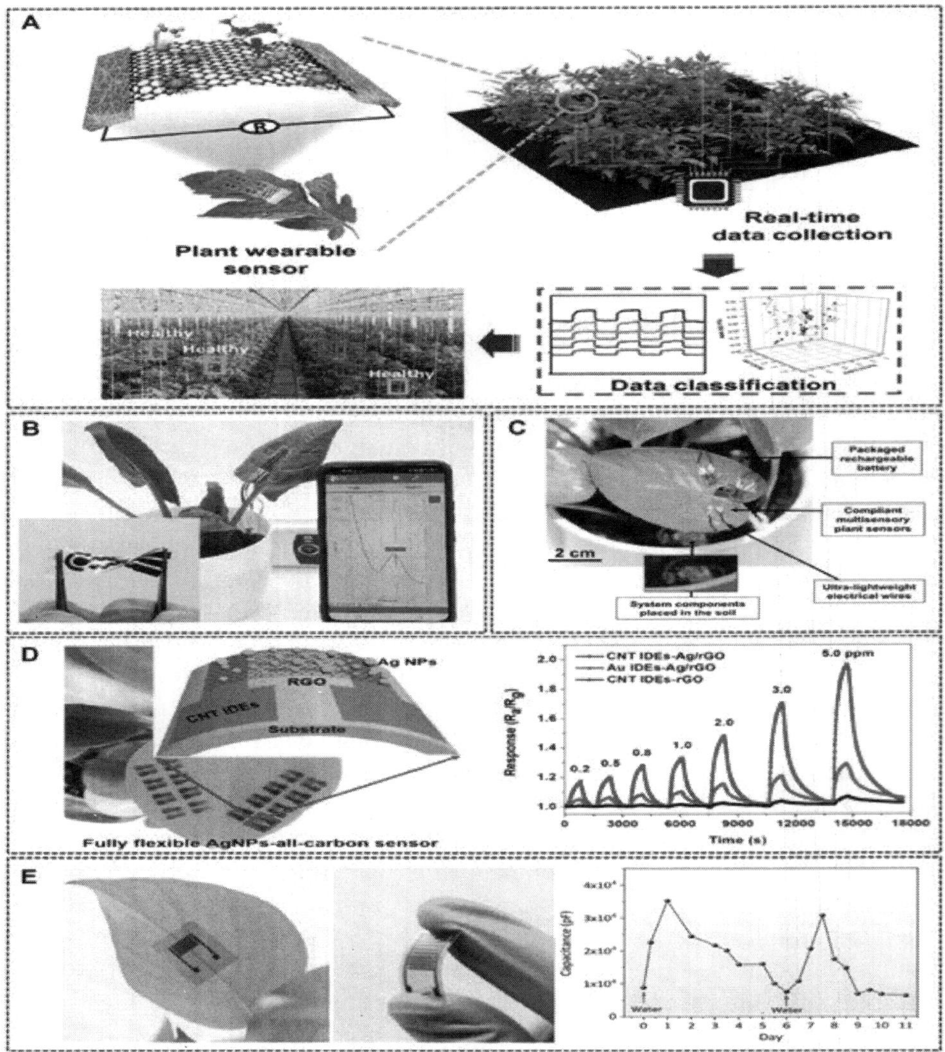

Figure 7.14. Plant-wearable sensors for determining plant physiological signals. (A) Late blight diagnosis in tomatoes through an abiotic stress monitoring system for sensing volatile organic compounds using a leaf-attached chemical sensor (Li et al. 2021), (B) In-situ organophosphorus pesticide analysis on the surfaces of agricultural products (inset: the fabricated PDMS/LIG electrode) (Zhao et al. 2020). (C) An environmental plant sensor platform on the surface of a plant leaf (Nassar et al. 2018). (D) A gas sensor array is attached to a real plant leaf, and a schematic illustration of the flexible sensing device (left) and real-time response curves (right) (Li et al. 2018). (E) Plant humidity sensor for real-time monitoring; a humidity sensor attached to the back surface of a leaf (left) and the graphene oxide-based sensor (middle), and the capacitance response to drought stress (right) (Reproduced from Lan et al., 2020).

Multi-Measurement Sensor

Multi-measurement sensors can simultaneously monitor multiple external signals and indicate the health conditions of plants (Figure 7.15). However, the existing multi-measurement sensors only detect microclimatic factors such as temperature and humidity in the vicinity of plants, and systematic monitoring using multi-measurement sensors has limitations (Table 7.4). Therefore, it is essential to focus on the development of multi-measurement sensors that monitor plant biological signals and correlate them with environmental signals to elucidate plant physiological conditions. Various parameters can be used for monitoring the ov7.erall health of plants. However, these should be integrated with each other, and future plant sensors should include sensor arrays capable of multiple measurements. It is essential to improve the selectivity of biomarkers or catalysts that selectively react with active target molecules (Segev-Bar et. al. 2017). It should also be possible to simultaneously monitor plant responses to abiotic and biological stresses by combining existing sensors—such as temperature, humidity, and pH sensors—with an electrochemical measurement system. This would make it possible to predict and prevent a decrease in crop yield due to plant diseases and stress.

Table 7.4. Multi-detection plant sensors.

System	Components	Sensing materials	Detection signal	Applications	References
Humidity/ temperature/ light sensor	ZIS	CNT resistance	Photoresistor, Thermistor	Plant health monitoring	Lu et. al. 2020
Hydration/ strain/ light/ temperature sensor	PI,Cu, Silicon- based phototransistor	Impedance, thermistor, piezoresistance	Photoresistor	Plant physiology and microclimate	Zhao et. al. 2019
Humidity/ UV sensor	p-doped pproDOT	Capacitance	Impedance	Plant tissue damage monitoring	Kim et. al. 2019
Humidity/ temperature sensor	Titanium/ gold	Capacitance	Thermistor	Microclimate monitoring around the plant	Nasser et. al. 2018
Note: *ZIS: ZnLn2S4; PI: polyimide; p-doped pproDOT: p-doped poly (3,4-propylenedioxy -thiophene)*					

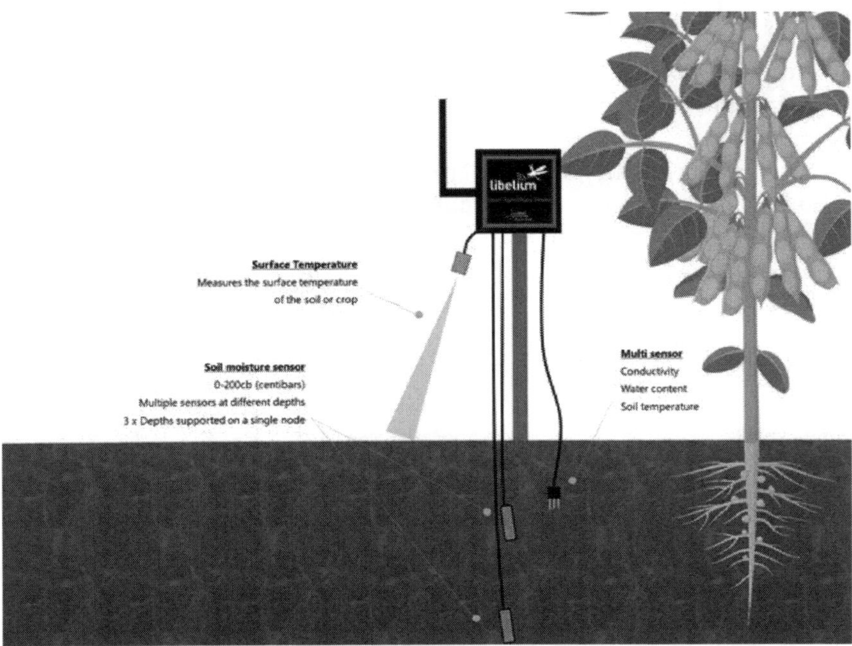

Figure 7.15. Smart Agriculture Node – monitoring moisture, conductivity, surface temperature, and soil temperature.

Multifunctional Stretchable Sensors for Continuous Monitoring of Long-Tetrm Leaf Physiology and Microclimate

Communication with plants to understand their growth mechanisms and interaction with the surrounding environment may improve production yield in agriculture and facilitate the prevention of plant diseases and the negative influence of environmental stress. Typical sensing technologies in plant biology and precision agriculture largely rely on techniques with low spatial and temporal resolutions, and fail to determine localized variation continuously and precisely in leaf physiology and microenvironments. Here, techniques to develop a multifunctional stretchable leaf-mounted sensor have been developed to offer optimized adaptability to plant growth and monitor leaf physiological and environmental conditions in continuous and highly sensitive manners. The multifunctional leaf sensor contains multiple heterogeneous sensing elements made of metal, carbon nanotube matrix, and silicon, leading to temperature, hydration, light illuminance, and strain sensing capabilities on a leaf. Evaluation under a controlled environment indicates excellent precision and accuracy of the sensor compared to conventional devices. Furthermore, indoor and outdoor experiments have demonstrated the multifunctional monitoring ability of the sensor in real situations. The multifunctional stretchable sensor holds the promise to advance monitoring techniques in plant biology and precision agriculture, resulting in improved capability to record slow and subtle physiological changes in plants and plant/environment interaction.

Zhoa et. al. (2019) developed a stretchable multifunctional sensor (Figure 7.16a) with a dimension of 13 000 × 8800 × 30 μm3 and a mass of 17 mg containing multiple heterogeneous sensing elements made of metal, carbon nanotube (CNT), and silicon. The sensor was fabricated by combining CMOS, printable electronics, and transfer printing techniques, leading to hydration, temperature, strain, and light illuminance sensing capabilities on leaves (Figure 7.16b). These elements related to meshed structures, forming an island-bridge configuration that can be engineered to optimize strain distribution within the ultrathin composition materials when being stretched. In addition, the self-similar serpentine interconnects that contain secondary hierarchy structures offer stretchability that allows the entire device to synchronize its growth with the hosting leaf. A perforated silicone membrane with an ultralow modulus (~3.0 kPa) supports and protects the sensor while allowing light, gas, and water vapor to penetrate through. This membrane employs a healthcare-grade material with biocompatibility to enable excellent adhesion between the leaf and the sensor. The adhesion of the sensor may be subjected to the influence of rain and snow, causing the variation of adhesion forces. To quantify the influence, the adhesion forces between the sensors and leaves were characterized using a tensile tester by 180° peel tests. Sensors supported by a commercial fabric (10 × 70 mm2) were attached to leaves and were peeled off mechanically from the leaves at a speed of 50 mm/min−1. Peel forces in situations without and with immersing the sensors on leaves into the water were obtained. The average peel forces are 0.32, 0.29, 0.28, and 0.27 N, respectively, corresponding to immersing times of 0, 10, 20, and 30 min. As the immersing time increased to 30 min, the average peel force only decreased by 0.05 N, indicating that the sensor still maintains good adhesion to the leaf under a wet environment. The sensor measures comprehensive information for a better understanding of the effect of photosynthesis, plant nutrition, environmental stress, and transpiration on plant growth and crop yield. The entire sensor can withstand different deformations (Figure 7.16c) and grow together with the host leaf in a measurement period of 7 days (Figure 7.16d). In addition, the sensor only applies minimized contact pressure (<170 μN) to leaves with almost no physical constraint to leaf growth.

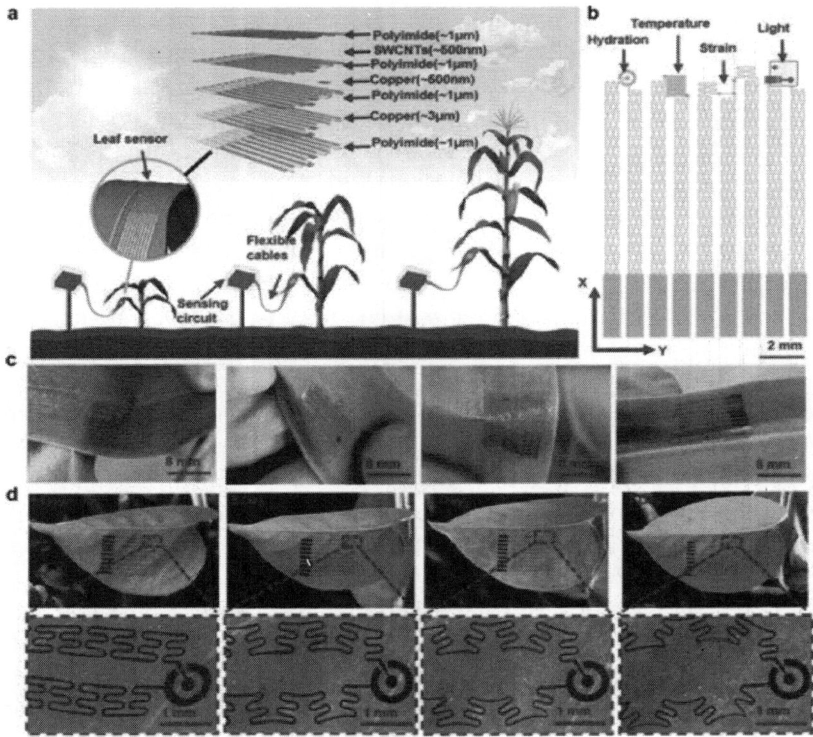

Figure 7.16. Schematics and demonstration of a multifunctional stretchable sensor on a leaf. (a) Schematic diagram and exploded view of a leaf sensor. (b) Top view of the leaf sensor. (c) A sensor can adapt to the morphology of a corn leaf and deform together with the leaf. (d) A sensor attached to a leaf grows together with the leaf for 7 days. (Reproduced from Zhao et. al. 2019)

Critical Issues With The Sensors

Over the past few decades, precision agriculture (PA) has been used to reduce costs, improve crop yields, and protect the environment. However, the adoption of these novel technologies by farmers is still limited due to the following reasons.

Robustness of Sensors

Under field conditions, it is important for sensors to exhibit long-term stability under harsh environmental conditions (such as large fluctuations in temperature and humidity) and resist mechanical aging due to direct sunlight, rain, wind, and exposure to chemicals (such as pesticides). Thus, to improve the durability of sensors, it is necessary to incorporate a variety of materials. Research on the durability of plant sensors is still in its infancy, and a shift in the technologies used in human medicine may provide insights into improving the durability of plant sensors. It is important to develop a next-generation electrochemical sensor that must include a compact and ultra-small structure with greater rigidity, corrosion resistance, and an encapsulation layer with excellent waterproof performance.

Data Accumulation and Management

Big data, should be a key element of the agricultural data management system, which will replace farmers' insights about cultivation strategies. Big data provides precise recommendations to growers, so that even beginners in agriculture can easily harvest crops according to the information provided. However, there is no specific method for obtaining such crucial big data. Generally, big data in agriculture are obtained through interviews with numerous farmers. However, these methods are considerably labour- and time-consuming. Therefore, there is an urgent need to develop the technology to collect agronomic data in a prompt and accurate manner using sensors and information communication technologies.

Typically, big data for PA have been obtained using physical sensors to measure characteristics such as meteorological phenomena and soil moisture, pH, temperature, and electrical conductivity. Recently, data from image sensing by drones or satellites have been used to judge whether crop growth or yield can be predicted, and to decide whether a crop is infected with pests or diseases (Mogili and Deepak, 2018). High-resolution images of crops are obtained from the drone and then further processed to extract information, which is then used to make decisions and predictions. However, image sensing is insufficient for precise prescriptions, as it relies on a consequential point of view and judges the degree of growth of crops as a result of observing the outward appearance of crops.

PA involves data collection and processing for several parameters related to crop health, such as moisture, temperature, and nutrients. Thus, PA allows farmers to know exactly what parameters are needed to grow healthy crops, what are the requirements at any given point in time, and which parameters require attention. This implies gathering vast amounts of information from a variety of sources, such as nutrients in the soil and sap, the presence of pests and diseases, the chlorophyll content of plant leaves, and surrounding weather conditions. The collected information should be analyzed so as to generate agronomic recommendations by specialists. For instance, when considering the developmental stages of a plant, the size and colour of a plant's leaves can reveal deficiencies or excesses of nutrients. However, the phenotype of a plant cannot accurately determine which element is deficient, and visual examination only provides a rough estimate. Therefore, a technique is required that can use the plant phenotype to identify the degree of deficiency or excess of a specific nutrient. The technology to achieve this includes an ion sensor that detects ions (nutrients) in a crop sap or hydroponic culture medium. These nutrient data are combined with the weather conditions and the characteristics and nutrient distribution of the soil in which the plants are grown. All the combined information is then used to determine exactly when and how much of a specific fertilizer should be applied. Thus, incorporation of the ion sensor and an electrochemical technique as a data collection method for PA will provide the necessary inputs to provide accurate prescriptions for future crop cultivation. Moreover, a multi-measurement system equipped with electrochemical sensors (to measure chemical and physical parameters) would form an integrative approach to providing comprehensive prescriptions.

Hardware Costs

PAs rely primarily on hardware such as sensing devices, wireless nodes, and automated systems, which are used to monitor the overall plant growth status in real-time. These sensors have multiple limitations, including a high level of development, maintenance, and deployment cost. Therefore, it is necessary to develop a low-cost multi-measurement sensing device that can measure several parameters simultaneously.

Self-Powered Sustainable Agriculture System

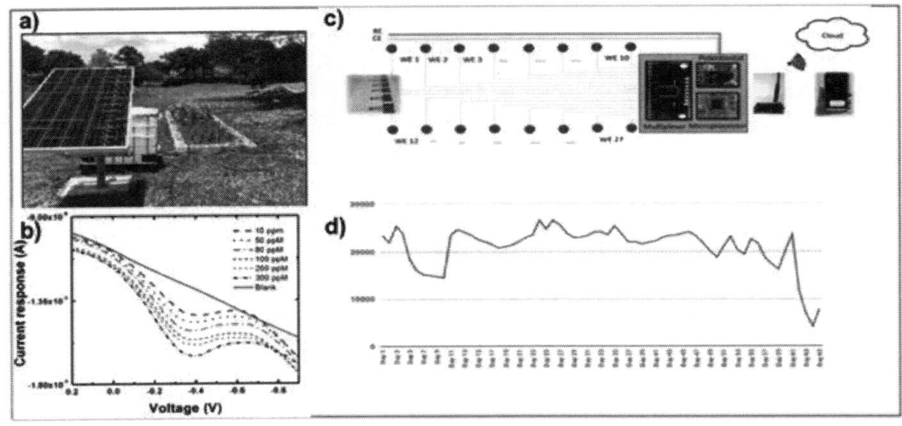

Figure 7.17. (a) Solar-powered IoT sensor garden (Mekonnen et. al. 2020), (b) nitrate calibration response using fabricated in situ nitrate ISE (Alam et. al. 2018), (c) Layout of the IoT-enabled soil sensor sheet system (Burton et al. 2020), (d) Real-time soil moisture content data used to correlate with in situ soil nitrate sensor (Reproduced from Burton et. al 2020).

Advances in various sensing techniques have enabled automated plant growth monitoring. However, developing sustainable power supplies for electronic devices for sensing remains a challenge (Lan et al., 2021). To date, various types of sensing devices have been introduced that are fixed to the soil or directly attached to plant leaves or stems when monitoring environmental conditions or plant health. Additionally, the convergence of the IoT and sensing systems has made improved the automation of systems for monitoring and sensing information. On average, vegetables are harvested every 3–4 months, and the sensing device (including the battery) can operate for at least one cycle. However, a sustainable power supply for these electronic devices remains a fundamental problem. Currently, traditional batteries are used as temporary power sources, and must be replaced or recharged periodically, which complicates their maintenance and causes environmental problems. To overcome these issues, researchers have proposed the idea of a smart farm that uses renewable energy and utilizes surplus renewable energy for the agricultural field (Lin et al., 2021). An efficient energy consumption system (reactive energy utilization technology) was used to predict the optimal scenario that minimizes the total power loss with the lowest voltage in each scenario. Thus, a system that introduces renewable energy into the agricultural sector will provide an opportunity to expand the possibilities of sustainable precision agriculture (Figure 7.17).

Conclusions

Given all the sensors mentioned above, there exists an urgent need for incorporating them with IoT and Artificial Intelligence. This would allow real-time monitoring and control of farm systems, thus making it easier for farmers to optimize production and minimize resource utilization. Due to the excessive utilization of fertilizers and pollution effects on the environment, a growing concern about pollution has warranted the need for technologies that can better monitor soil nutrients and their fate. The conventional laboratory methods may offer highly accurate analysis of soil chemistry, but in situ-based soil nutrient sensors that offer real-time feedback are needed in order to truly increase the efficiency of farming and managing the environment. As compared to conventional lab instruments for soil nutrient analysis, in situ-based sensors are more advantageous due to their low-cost, and high-density measurement capability for large-area soil nutrient mapping. Although these sensors are becoming less expensive to manufacture and can provide comparable results to laboratory soil analysis, there is a need to understand the effects of soil heterogeneity on the response of both optical and electrochemical sensors. More durable and accurate sensor systems that consider the effects of soil heterogeneity and offer soil-specific calibrations are needed to move toward commercializing these platforms. When interfaced with long-range low-powered radio modules, these sensor platforms can implement real-time prediction, control, and decision management for large-scale precision agricultural practices.

The crucial focus for the future of precision agriculture (PA) should be on the application of electrochemical multi-sensors to monitor plant health. The continuous collection of plant growth and environmental data in real time requires the development of a multi-measurement sensing platform that can be deployed on a large scale. The excellent performance, miniaturization, low cost, and environment-friendly characteristics of electrochemical plant sensors are expected to help upgrade the detection systems for PA.

References

Adamchuk, V.I., Lund, E.D., Sethuramasamyraja, B., Morgan, M.T., Dobermann, A., & Marx, D.B. (2005). Direct measurement of soil chemical properties on-the-go using ion-selective electrodes Comput. Electron. Agric. 48: 272.

Adamowski, M., & Friml, J. (2015). PIN-dependent Auxin Transport: Action, Regulation, and Evolution. Plant Cell. 27: 20–32. doi:10.1105/tpc.114.134874

Alexandratos, N., & Bruinsma, J. (2012). World Agriculture towards 2030/2050: The 2012 Revision. Rome: FAO. doi:10.22004/ag.econ.288998

Antonacci, A., Arduini, F., Moscone, D., Palleschi, G., & Scognamiglio, V. (2018) Nanostructured (bio) sensors for smart agricul-ture TrAC. Trends Analytic. Chem. 98: 95-98.

Aqeel-ur-Rehman, A.Z., Abbasi, A.Z., Islam, N., & Shaikh, Z.A. (2014). A Review of Wireless Sensors and Networks' Applications in Agriculture. Comp. Stand. Inter. 36: 263–270. doi:10.1016/j.csi.2011.03.004

Aravamudhan, S., & Bhansali, S. (2008). Development of micro-fluidic nitrate-selective sensor based on doped-polypyrrole nanowires. Sens. Actuators B. 132: 623-628.

Artigas, J., Jimenez, C., Lemos, S.G., Nogueira, A.R.A., Torre-Neto, A., & Alonso, J. (2003). Development of a screen-printed thick-film nitrate sensor based on a graphite-epoxy composite for agricultural applications. Sensors Actuators B: Chem. 88: 337–344. doi:10.1016/S0925-4005(02)00399-4

Bailey-Serres, J., Parker, J.E., Ainsworth, E.A., Oldroyd, G.E.D., & Schroeder, J.I. (2019). Genetic strategies for improving crop yields. Nature. 575 (7781): 109–118. doi:10.1038/s41586-019-1679-0

Ball, A.S., Wilson, W.S., & Hinton, R. (1999). Managing risks of nitrates to humans and the environment. Cambridge, UK: The Royal Society of Chemistry. P. 237.

Baumgardner, M.F., Silva, L.F., Biehl, L.L., & Stoner, E.R. (1986). Reflectance properties of soils. Adv. Agronomy. 38: 112-118.

Birrell, S.J., & Hummel, J.W. (2001). Real-time multi isfet/fia soil analysis system with automatic sample extraction Comp. Electron. Agr. 32: 45-49.

Bogrekci, J., & Lee, W.S. (2005). Design of a portable Raman sensor for phosphorous sensing in soils. ASABE Annual Meeting. Paper No. 051040. Doi:10.13031/2013.15769/@2005

Burton, L, Jayachandran, K., & Bhansali, S. (2020). The real-time revolution for in situ soil nutrient sensing. J. Electrochemical Soc. 167: 569-576.

Burton, L., Dave N., Fernandez R.E., Jayachandran K. & Bhansali S. (2018) Smart gardening iot soil sheets for real-time nutrient analysis. J. Electrochem. Soc. 165: B3157.

Calisgan, S.D., Rajaram, V., Kang, S., Risso, A., Qian, Z., & Rinaldi, M. (2020). Micromechanical switch-based zero-power chemical detectors for plant health monitoring. J. Microelectromech. Syst. 29: 755–761. doi:10.1109/JMEMS.2020.3007309

Chandel, N.S., Mehta, C.R., Tewari, V.K., & Nare, B. (2016). Digital map-based site-specific granular fertilizer application system. Curr. Sci. 111: 1208-1212.

Chen, Y., Guo, X.L., Yan, J.C., Zhao, Y.F., Pang, Y., Jian, J.M., Morikado, M., Wu, X.M., Yang, Y., & Ren, T.L. (2018a). Toward an in situ phosphate sensor in natural waters using a microfluidic flow loop analyzer. J. Electrochem. Soc. 165: B737.

Chen, X., Zhou, G., Mao, S., & Chen, J. (2018b). Rapid detection of nutrients with electronic sensors: a review. Environ. Sci. Nano. 5: 837–862. doi:10.1039/C7EN01160A

Cheong, Y.H., Ge, L., & Lisak, G. (2021). Highly reproducible solid contact ion selective electrodes: emerging opportunities for potentiometry: a review. Analytica Chim. Acta. 1162: 3383-3394. doi:10.1016/j.aca.2021.338304

Christy, C.D., Drummond, P., & Liard, D. (2003). An on-the-go spectral reflectance sensor for soil. ASAE Annual Meeting 2003. Am. Soc. Agril. Biol. Engg., p.1.

Coppedè, N., Janni, M., Bettelli, M., Maida, CL., Gentile, F., ... Villani, M. (2017). An in vivo biosensing, biomimetic electrochemical transistor with applications in plant science and precision farming. Sci. Rep. 7: 1–9. doi: 10.1038/s41598-017-16217-4

de Araujo Zanella, A.R., da Silva, E., & Pessoa Albini, L.C. (2020). Security challenges to smart agriculture: current state, key issues, and future directions. Array 8: 648-652. doi:10.1016/j.array.2020.100048

Diacci, C., Lee, J.W., Janson, P., Dufil, G., Méhes, G., ... Berggren, M. (2020). Real-time monitoring of glucose export from isolated chloroplasts using an organic electrochemical transistor. Adv. Mater. Technol. 5: 1900-1911. doi: 10.1002/admt.201900262.

Duhan, J.S., Kumar, R., Kumar, N., Kaur, P., Nehra, K., & Duhan, S. (2017). Nanotechnology: the new perspective in precision agriculture. Biotechnol. Rep. 15: 11–23. doi:10.1016/j.btre.2017.03.002

Ehsani, M.R., Keff, C.K., Beatle, J., ... Jaffer, U. (2000). Quantitative measurement of soil nitrate content using mid-infrared diffuse reflectance spectroscopy. Milwaukee, Wisconsin, 2000, American Society of Agricultural Engineers (ASAE). p. 1

Esmaeili, C., Norouzi, P., Zar, M.S., Eskandari, M., Faridbod, F., & Ganjali, M.R. (2019). A fft square wave voltammetry sensing method for highly sensitive detection of phytic acid using a cerium oxide nanoparticles decorated graphene oxide J. Electrochem. Soc.166: B1630.

Esser, B., Schnorr, J.M., & Swager, T.M. (2012). Selective detection of ethylene gas using carbon nanotube-based devices: utility in determination of fruit ripeness. Angew. Chem. Int. Ed. 51: 5752–5756. doi:10.1002/anie.201201042

Gan, T., Hu, C., Chen, Z., & Hu, S. (2010). Direct electrochemical determination of methyl jasmonate in wheat spikelet at a nano-montmorillonite film modified electrode by derivative square wave voltammetry. J. Agric. Food Chem. 58: 8942–8947. doi:10.1021/jf101531c

Garland, N.T., McLamore, E.S., Cavallaro, N.D., Mendivelso-Perez, D., Smith, E.A., ... Jing, D. (2018). Flexible laser-induced graphene for nitrogen sensing in soil. ACS Appl. Mater. Inter. 10: 39124–39133. doi:10.1021/acsami.8b10991

Gieling, T., Vanstraten, G., Janssen, H., & Wouters, H. (2005). ISE and chemfet sensors in greenhouse cultivation. Sensors Actuators B: Chem. 105: 74–80. doi:10.1016/j.snb.2004.02.04510.1016/s0925-4005(04)00113-3

Giraldo, J.P., Wu, H., Newkirk, G.M., & Kruss, S. (2019). Nanobiotechnology approaches for engineering smart plant sensors. Nat. Nanotechnol. 14: 541–553. doi:10.1038/s41565-019-0470-6

Guignard, M.S., Leitch, A.R., Acquisti, C., Eizaguirre, C., Elser, J.J., ... Hessen, D.O. (2017). Impacts of nitrogen and phosphorus: from genomes to natural ecosystems and agriculture. Front. Ecol. Evol. 5: 70-78. doi:10.3389/fevo.2017.00070

Hofmeister, F. (1888). On the theory of the effects of salts arch. Exp. Pathol. Pharmacol. 24: 247.

Hosseinzadeh, S., Verheust, Y., Bonarrigo, G., & Van Hulle, S. (2017). Closed hydroponic systems: operational parameters, root exudates occurrence and related water treatment. Rev. Environ. Sci. Biotechnol. 16: 59–79. doi:10.1007/s11157-016-9418-6

Hu, J., Stein, A., & Bühlmann, P. (2016). Rational design of all-solid-state Ion-selective electrodes and reference electrodes. Trac. Trends Anal. Chem. 76: 102–114. doi:10.1016/j.trac.2015.11.004

Im, H., Lee, S., Naqi, M., Lee, C., & Kim, S. (2018). Flexible PI-based plant drought stress sensor for real-time monitoring system in smart farm. Electronics. 7: 114-120. doi:10.3390/electronics7070114

Jahn, B.R., Linker, R., Upadhyaya, S.K., Shaviv, A., Slaughter, D.C., & Shmulevich I. (2006). Mid-infrared spectroscopic determination of soil nitrate content. Biosyst. Eng. 94: 505-512.

Jawad, H., Nordin, R., Gharghan, S., Jawad, A., & Ismail, M. (2017). Energy-efficient wireless sensor networks for precision agriculture: a review. Sensors. 17: 1781-1788. doi:10.3390/s17081781

Jiang, H., Ali, MA., Jiao, Y., Yang, B., & Dong, L. (2017). In-situ, real-time monitoring of nutrient uptake on plant chip integrated with nutrient sensor. 19th Intel. Conf. Solid-State Sensors, Actuators and Microsystems (TRANSDUCERS), Kaohsiung, Taiwan: IEEE), 289–292. doi:10.1109/transducers.7994045

Kashyap, B., & Kumar, R. (2021). Sensing methodologies in agriculture for soil moisture and nutrient monitoring. IEEE Access 9: 14095–14121. doi:10.1109/ACCESS.2021.3052478

Khanna, A., & Kaur, S. (2019). Evolution of internet of things (IoT) and its significant impact in the field of precision agriculture. Comput. Elect. Agric. 157: 218–231. doi:10.1016/j.compag.2018.12.039

Kim, H.J. (2006). Ion-selective electrodes for simultaneous real-time analysis for soil macronutrients. Ph.D. Thesis University of Missouri-Columbia,USA.

Kim, H.J., Sudduth, K.A., & Hummel, J.W. (2009). Soil macronutrient sensing for precision agriculture. J. Environ. Monitor. 11: 1810-1817.

Kim, M.Y., & Lee, K.H. (2022). Electrochemical sensors for sustainable precision agriculture: a review. Front. Chem. 10: 848-852. doi: 10.3389/fchem.2022.848320

Kim, H.J., Kim, W.K., Roh, M.Y., Kang, C.I., Park, J.M., & Sudduth, K.A. (2013). Automated sensing of hydroponic macronutrients using a computer-controlled system with an array of ion-selective electrodes. Comput. Elect. Agric. 93, 46–54. doi:10.1016/j.compag.2013.01.011

Kim, J.J., Allison, L.K., & Andrew, T.L. (2019). Vapor-printed polymer electrodes for long-term, on-demand health monitoring. Sci. Adv. 5: 463-467. doi: 10.1126/sciadv.aaw0463

Kim, J.J., Fan, R., Allison, L.K., & Andrew, T.L. (2020). On-site identification of ozone damage in fruiting plants using vapor-deposited conducting polymer tattoos. Sci. Adv. 6: 329-335. doi:10.1126/sciadv.abc3296

Kim, M.Y., Lee, J.W., Park, D.J., Lee, J.Y., Myung, N.V., ... Kwon, S.H. (2021). Highly stable potentiometric sensor with reduced graphene oxide aerogel as a solid contact for detection of nitrate and calcium ions. J. Electroanalytical Chem. 897: 1155-1163. doi:10.1016/j.jelechem.2021.115553

Kinpara, T., Murakami, Y., Yokoyama, K., & Tamiya, E. (2001). Electrochemical monitoring of plant stress responses. Electroanalysis. 13: 451–456. doi:10.1002/1521-4109(200104)13:6<451::AID-ELAN451>3.0.CO;2-9

Kumar, T.H.V., & Sundramoorthy, A.K. (2018). Non-enzymatic electrochemical detection of urea on silver nanoparticles anchored nitrogen-doped single-walled carbon nanotube modified electrode. J. Electrochem Soc. 165: B3006.

Lan L., Le X., Dong H., Xie J., Ying Y., & Ping J. (2020). One-step and large-scale fabrication of flexible and wearable humidity sensor based on laser-induced graphene for real-time tracking of plant transpiration at bio-interface. Biosens. Bioelectron. 165: 112-119. doi:10.1016/j.bios.2020.112360

Lan, L., Xiong, J., Gao, D., Li, Y., Chen, J., ... Lv, J. (2021). Breathable nanogenerators for an on-plant self-powered sustainable agriculture system. ACS Nano. 15: 5307–5315. doi:10.1021/acsnano.0c10817

Laskar, S., & Mukahajee, S. (2016). Optical sensing methods for assessment of soil micro-nutrients and other properties for application in precision agriculture: a review. ADBU-J. Engg. Tech. 4: 206-210.

Lee, W., & Bogrekci I. (2007). Portable Raman sensor for soil nutrient detection. US Pat. App. 11: 475,501.

Lee, G., Wei, Q., & Zhu, Y. (2021). Emerging wearable sensors for plant health monitoring. Adv. Funct. Mater. 31, 2106475. doi:10.1002/adfm.202106475

Li, Y., Jiang, T., Yu X., & Yang, H. (2016). Phosphate sensor using molybdenum. J. Electrochem. Soc. 163: B479.

Li, H., Wang, C., Wang, X., Hou, P., Luo, B., ... Song, P. (2019). Disposable stainless steel-based electrochemical microsensor for in vivo determination of indole-3-acetic acid in soybean seedlings. Biosens. Bioelectron. 126: 193–199. doi:10.1016/j.bios.2018.10.041

Li, W., Teng, C., Sun, Y., Cai, L., Xu, JL., ... Sun, M. (2018). Sprayed, scalable, wearable, and portable NO2 sensor array using fully flexible AgNPs-all-carbon nanostructures. ACS Appl. Mater. Inter. 10: 34485–34493. doi:10.1021/acsami.8b11254

Li, Z., Liu, Y., Hossain, O., Paul, R., Yao, S., ... Wu, S. (2021). Real-time monitoring of plant stresses via chemo resistive profiling of leaf volatiles by a wearable sensor. Matter 4: 2553–2570. doi:10.1016/j.matt.2021.06.009

Lin, X., Sun, X., Manogaran, G., & Rawal, B.S. (2021). Advanced energy consumption system for smart farm based on reactive energy utilization technologies. Environ. Impact Assess. Rev. 86: 106496. doi:10.1016/j.eiar.2020.106496

Liu, Y.W., & Huang, C.K. (2019). Effects of the circulation pump type and ultraviolet sterilization on nutrient solutions and plant growth in plant factories. Hort Technology. 29: 189–198. doi:10.21273/HORTTECH04244-18

Lobsey, C., Burn, P., Smith, K., ... King, S. (2010). An automated system for rapid in-field soil nutrient testing. 19th World Congress of Soil Science, Brisbane, Australia, August. Soil Solutions for a Changing World. p. 1-6.

Lobsey, C.R., Viscarra Rossel, R.A., & Mcbratney, A.B. (2010). Proximal soil nutrient sensing using electrochemical sensors. Proximal Soil Sensing, Dordrecht: Springer. p. 77

Lu, Y., Xu, K., Zhang, L., Deguchi, M., Shishido, H., ... Arie, T. (2020). Multimodal plant healthcare flexible sensor system. ACS Nano. 14: 10966–10975. doi:10.1021/acsnano.0c03757

Marios, S., & Georgiou, J. (2017). Precision agriculture: challenges in sensors and electronics for real-time soil and plant monitoring. IEEE Biomedical Circuits and Systems Conference (BioCAS), Turin, Ital. IEEE. doi:10.1109/BIOCAS.2017.8325180

Martinelli, F., Scalenghe, R., Davino, S., Panno, S., Scuderi, G., ... Ruisi, P. (2015). Advanced methods of plant disease detection: a review. Agron. Sustain. Dev. 35: 1–25. doi:10.1007/s13593-014-0246-1

Martinière, A., Desbrosses, G., Sentenac, H., & Paris, N. (2013). Development and properties of genetically encoded pH sensors in plants. Front. Plant Sci. 4: 523-530. doi:10.3389/fpls.2013.00523

McBratney, A., Whelan, B., Ancev, T., & Bouma, J. (2005). Future directions of precision agriculture. Precision Agric. 6: 7–23. doi:10.1007/s11119-005-0681-8

Mekonnen, Y., Namuduri, S., Burton, L., & Sarwat., A. (2020). Machine learning techniques in wireless sensor network-based precision farming. J. Electronic Soc. 167: 522-530.

Mekonnen, Y., Burton, L., Sarwat, A., & Bhansali, S. (2018). Iot sensor network approach for smart farming: An application in food, energy and water system. IEEE Global Humanitarian Technology Conference (GHTC).

Mikhelson, K.N. (2013). Ion-selective electrodes. Berlin, Germany: Springer. Vol. 81

Mogili, U.R., & Deepak, B.B. (2018). Review on application of drone systems in precision agriculture. Proc. Comp. Sci. 133: 502–509. doi:10.1016/j.procs.2018.07.063

Moss, B. (2008). Water pollution by agriculture. Phil. Trans. R. Soc. B 363. 1491: 659–666. doi:10.1098/rstb.2007.2176

Muñoz-Huerta, R., Guevara-Gonzalez, R., Contreras-Medina, L., Torres-Pacheco, I., Prado-Olivarez, J., & Ocampo-Velazquez, R. (2013). A review of methods for sensing the nitrogen status in plants: advantages, disadvantages and recent advances. Sensors. 13: 10823–10843. doi:10.3390/s130810823

Naikoo, G.A., Salim, H., Hassan, I.U., Awan, T., Arshad, F., ... Pedram, M.Z. (2021). Recent advances in non-enzymatic glucose sensors based on metal and metal oxide nanostructures for diabetes management: a review. Front. Chem. 9, 748957. doi:10.3389/fchem.2021.748957

Nassar, J.M., Khan, S.M., Villalva, D.R., Nour, M.M., Almuslem, A.S., & Hussain, M.M. (2018). Compliant plant wearables for localized microclimate and plant growth monitoring. Npj. Flex Electron. 2: 1–12. doi:10.1038/s41528-018-0039-8

Nocita, M., Singh, P.K., Kusum, S., ... Barman, D. (2015). Soil spectroscopy: an alternative to wet chemistry for soil monitoring. Advances in Agronomy, Amsterdam, Netherlands: Elsevier. Vol. 132 p. 139.

Ozhikandathil, J., Badilescu, S., & Packirisamy, M. (2018). Polymer composite optically integrated lab on chip for the detection of ammonia. J. Electrochem. Soc. 165: B3078.

Pandey, R., Teig-Sussholz, O., Schuster, S., Avni, A., & Shacham-Diamand, Y. (2018). Integrated electrochemical chip-on-plant functional sensor for monitoring gene expression under stress. Biosens. Bioelectron. 117: 493–500. doi: 10.1016/j.bios.2018.06.045.

Pandikumar, A., & Rameshkumar, P. (2018). Graphene-based electrochemical sensors for biomolecules. Elsevier.

Pawlak, K., & Kołodziejczak, M. (2020). The Role of agriculture in ensuring food security in developing countries: considerations in the context of the problem of sustainable food production. Sustainability 12: 5488. doi:10.3390/su12135488

Pedersen, S.M., & Lind, K.M. (2017). Precision agriculture: technology and economic perspectives. Springer.

Qu, C.C., Sun, X.Y., Sun, W.X., Cao, L.X., Wang, X.Q., & He, Z.Z. (2021). Flexible wearables for plants. Small 17: 2104482. doi:10.1002/smll.202104482

Ratnaparkhi, S., Khan, S., Arya, C., Khapre, S., Singh, P., ... Diwakar, M. (2020). Smart agriculture sensors in IoT: a review. Mater. Today Proc. doi:10.1016/j.matpr.2020.11.138

Rogovaska, N., Laird, D.A., Chiou, C.P., & Bond, L. (2019). Development of field mobile soil nitrate sensor technology to facilitate precision fertilizer management. Precision Agriculture. 20: 263-268.

Sadeghi-Tehran, P., Sabermanesh, K., Virlet, N., & Hawkesford, M.J. (2017). Automated method to determine two critical growth stages of wheat: heading and flowering. Front. Plant Sci. 8: 252. doi:10.3389/fpls.2017.00252

Santner, A., & Estelle, M. (2009). Recent advances and emerging trends in plant hormone signalling. Nature. 459: 1071–1078. doi:10.1038/nature08122

Schans, J., & Arntzen, F.K. (1991). Photosynthesis, transpiration and plant growth characters of different potato cultivars at various densities of Globodera pallida. Neth. J. Plant Pathol. 97: 297–310. doi:10.1007/BF01974225

Segev-Bar, M., Bachar, N., Wolf, Y., Ukrainsky, B., Sarraf, L., & Haick, H. (2017). Multi-parametric sensing platforms based on nanoparticles. Adv. Mater. Technol. 2: 1600206. doi:10.1002/admt.201600206

Sekhar, P.K., & Kysar, J.S. (2017). An electrochemical ammonia sensor on paper substrate. J. Electrochem. Soc. 164: B113.

Shafi, U., Mumtaz, R., García-Nieto, J., Hassan, S. A., Zaidi, S. A. R., & Iqbal, N. (2019). Precision agriculture techniques and practices: from considerations to applications. Sensors 19: 3796. doi:10.3390/s19173796

Shao, Y., Ying, Y., & Ping, J. (2020). Recent advances in solid-contact ion-selective electrodes: functional materials, transduction mechanisms, and development. Trends. Chem. Soc. Rev. 49: 4405–4465. doi:10.1039/C9CS00587K.

Shi, Z., Ji, W., Viscarra Rossel, R.A., Chen, S., & Zhou, Y. (2015). Prediction of soil organic matter using a spatially constrained local partial least squares regression and the chinese vis-nir spectral library. Eur. J. Soil Sci. 66: 679.

Shibusawa, S., Sato, H., Sasao, A., Hirako, S., & Otomo, A. (2000). A revised soil spectrophotometer. IFAC Proc. 33: 231.

Sibley, K.J., Brewster, G.R., Astatkie, T., Adsett, J.F., & Struik, P.C. (2010). In-field measurement of soil nitrate using an ion-selective electrode. Advances in Measurement Systems, United Kingdom: IntechOpen.

Sinfield, J.V., Fagerman, D., & Colic, O. (2010). Evaluation of sensing technologies for on-the-go detection of macro-nutrients in cultivated soils. Comput. Electron. Agric. 70: 1.

Smith, K.M. & Holroyd, P. (1969). Engineering principles for electrical technicians, 1st Ed. Oxford, London: Pergamon Press.

Son, J.E., Kim, H.J., & Ahn, T.I. (2020). Hydroponic Systems. In: Plant Factory. Elsevier, p. 273–283. doi:10.1016/B978-0-12-816691-8.00020-0

Soriano-Disla, J.M., Janik, L.J., Viscarra Rossel, R.A., Macdonald, L.M., McLaughlin, M.J. (2014). The performance of visible, near-, and mid-infrared reflectance spectroscopy for prediction of soil physical, chemical, and biological properties. Appl. Spectroscopy Rev. 49: 139.

Sun, L.J., Feng, Q.M., Yan, Y.F., Pan, Z.Q., Li, X.H., ... Song, F.M. (2014). Paper-based electroanalytical devices for in situ determination of salicylic acid in living tomato leaves. Biosens. Bioelectron. 60: 154–160. doi: 10.1016/j.bios.2014.04.021

Sun, L.J., Xie, Y., Yan, Y.F., Yang, H., Gu, H.Y., & Bao, N. (2017). Paper-based analytical devices for direct electrochemical detection of free IAA and SA in plant samples with the weight of several milligrams. Sensors Actuators B: Chem. 247: 336–342. doi:10.1016/j.snb.2017.03.025

Suzuki, N., Rivero, R.M., Shulaev, V., Blumwald, E., & Mittler, R. (2014). Abiotic and biotic stress combinations. New Phytol. 203: 32–43. doi:10.1111/nph.12797

Tyszczuk-Rotko, K., & Jedruchniewicz, K. (2019). Ultra-sensitive sensor for uranium monitoring in water ecosystems. J. Electrochem. Soc. 166: B837.

van Staden, J.K.F., Nuta, R.G., & Arnold, G.L.T. (2018). Determination of nitrite from water catchment areas using graphite based electrodes J. Electrochem. Soc. 165: B565.

Vardar, G., Altıkatoğlu, M., Ortaç, D., Cemek, M., & Işıldak, İ. (2015). Measuring calcium, potassium, and nitrate in plant nutrient solutions using ion-selective electrodes in hydroponic greenhouse of some vegetables. Biotechnol. Appl. Biochem. 62: 663–668. doi:10.1002/bab.1317

Viscarra Rossel, R.A., & Bouma, J. (2016). Soil sensing: a new paradigm for agriculture. Agric. Syst. 148: 71.

Viscarra Rossel, R.A., Walvoort, D.J.J., McBratney, A.B., Janik, L.J., & Skjemstad, J.O. (2006). Visible, near-infrared, mid infrared or combined diffuse reflectance spectroscopy for simultaneous assessment of various soil properties. Geoderma. 43: 131-159.

Voke, E., Pinals, R.L., Goh, N.S., & Landry, M.P. (2021). In Planta nanosensors: understanding biocorona formation for functional design. ACS Sens. 6: 2802–2814. doi:10.1021/acssensors.1c01159

Wu, S., Huang, F., Lan, X., Wang, X., Wang, J., & Meng, C. (2013). Electrochemically reduced graphene oxide and nafion nanocomposite for ultralow potential detection of organophosphate pesticide. Sensors Actuators B: Chem. 177: 724–729. doi:10.1016/j.snb.2012.11.069

Xu, P., Zhao, P.X., Cai, X.T., Mao, J.L., Miao, Z.Q., & Xiang, C.B. (2020). Integration of jasmonic acid and ethylene into auxin signaling in root development. Front. Plant Sci. 11: 271. doi:10.3389/fpls.2020.00271

Yefimov, N.A., & Naboychenko, S. (2009). Handbook of non-ferrous metal powders: technologies and applications, 1st Ed. Amsterdam, The Netherlands: Elsevier.

Yin, H., Cao, Y., Marelli, B., Zeng, X., Mason, A.J., & Cao, C. (2021). Soil sensors and plant wearables for smart and precision agriculture. Adv. Mater. 33: 2007764. doi:10.1002/adma.202007764

Zhao, F., He, J., Li, X., Bai, Y., Ying, Y., & Ping, J. (2020). Smart plant-wearable biosensor for in-situ pesticide analysis. Biosens. Bioelectron. 170: 112636. doi:10.1016/j.bios.2020.112636.

Zhao, Y., Gao, S., Zhu, J., Li, J., Xu, H., Xu, K., Cheng, H., & Huang, X. (2019). Multifunctional stretchable sensors for continuous monitoring of long-term leaf physiology and microclimate. ACS Omega 4: 9522–9530. doi:10.1021/acsomega.9b01035

Chapter-8

Internet of Things for Crop Breeding

Abstract

The growth of the global population coupled with a decline in natural resources, farmland, and the increase in unpredictable environmental conditions leads to food security has become a major concern worldwide. To overcome these problems agricultural farming is gradually transforming towards smart agriculture with the application of the Internet of Things (IoT) and big data solutions, to improve operational efficiency and productivity. The IoT integrates a series of existing state-of-the-art solutions and technologies, such as wireless sensor networks, cognitive radio ad hoc networks, cloud computing, big data, and end-user applications. The opportunities for the application of IoT in agricultural farming are huge. The vision of IoT-enabled smart agriculture ecosystems has been discussed by evaluating their architecture (IoT devices, communication technologies, big data storage, and processing), and their applications. However, IoT applications for smart agriculture also have several challenges. But it is inevitable for the farmers to adopt IoT solutions to improve the productivity and quality of the agriculture sector as well as facilitate the transition towards a future sustainable environment with an agroecological approach.

Keywords: *Internet of Things (IoT), IoT devices, big data, communication technology, smart agriculture, data analysis and storage, greenhouse production*

Introduction

As per the United Nations (UN 2019) statistics, the world population is estimated to grow to 10 billion by 2050. Consequently, the requirements for agricultural products are continually increasing. However, farmlands are declining, natural resources are increasingly depleted, and the rise of unpredictable natural challenges, such as global warming, salinization, and flooding, make food security the most concerning problem for all nations worldwide.

In order to meet the current global needs of humanity, new solutions, and technologies are being constantly proposed and implemented. This has led to the advent of the Internet of Things (IoT). IoT is defined as the network of all objects that are embedded within devices, sensors, machines, software, and people through the Internet environment to communicate, exchange information and interact in order to provide a comprehensive solution between the real world and the virtual world (Elijah et. al. 2018). The use of IoT sensor devices has enabled farmers to produce high yields efficiently through farm data tracking and analysis of weather conditions, soil moisture, temperature, etc. This is known as smart farming or smart agriculture. In short, the use of IoT in agriculture would enable agricultural enterprises to have better supervision over the internal workings of the system. A series of other IoT applications have been described (Shafique et al. 2020). Diverse applications of IoT in smart agriculture are presented in Figure 8.1.

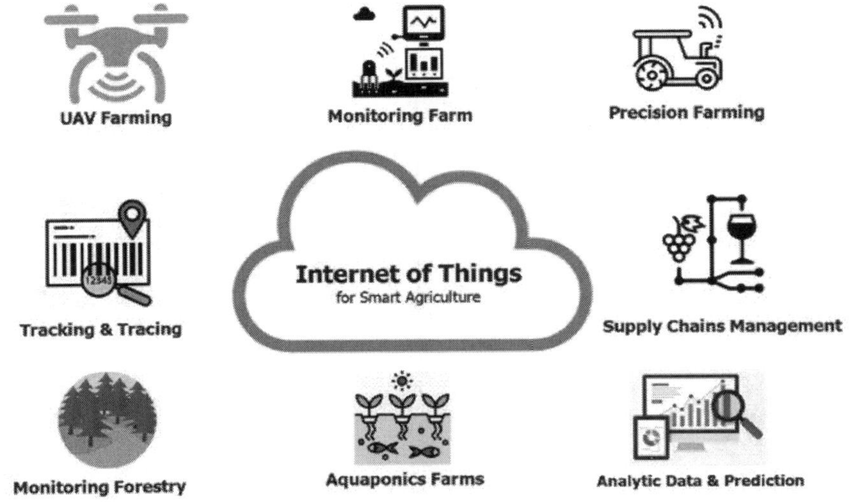

Figure 8.1. An illustration of IoT applications for smart agriculture. (Reproduced from Quy et. al. 2022).

Over the years, wireless sensor networks (WSN) have been strongly applied in the agricultural sector, building the foundation for developing smart agriculture (Saad et. al. 2020). The unique characteristics of WSN, such as the ability to self-organize, self-configure, self-establish, and self-recover, make it suitable for smart agriculture (Tyagi et. al. 2020). The sensor device consists of a radio frequency (RF) transceiver, sensor, microcontroller, and battery power (Li et. al. 2020). The WSN focuses on applications such as environmental monitoring, machine control automation, and traceability (Abdelnour et. al. 2018; Spachos et. al. 2019; Udutalapally et. al. 2020; Gopalakrishnan et. al. 2021).

IoT integrates a series of existing solutions and technologies, such as WSN, cognitive radio, ad hoc networks, cloud computing, and end-user applications (Friha et. al. 2021). In the smart agricultural sector, automation solutions and technologies, mechanical machines, knowledge, decision-making tools, services, and software are integrated seamlessly to help farmers improve productivity, product quality, and profitability (Farooq et. al. 2019).

The Internet of Things (IoT)

The Internet of Things (IoT) network comprises systems and a network of web-connected intelligent devices that employ encoded networks like sensors, processors, and interactive hardware to receive, send, and store data. The network in the IoT uses the stored data. To use the stored data, the IoT device sends out the collected sensor information through a connection with a gateway on the IoT or through other means such as using devices with edge tools where data is either sent for analysis on the local network or in the cloud. Connected devices on the IoT network interact with each other and use the shared data they acquire to solve problems (Ganapathy, 2016b). The interactions within the network are done without human supervision or interference, and most of the job is carried out by the devices. Even though humans may communicate with the devices to set them up, writing scripts for protocols and commands provides access to information.

Different applications for which IoT is used shall determine the protocol and system of connection, communication, and network. Artificial Intelligence (AI) and machine learning in IoT improve data collection and processing by the connected devices in a network system (Neogy and Paruchuri, 2014). Machine learning makes it dynamic. The Internet of Things connects the various things on the Internet, making it easier to live and allowing jobs to be done more smartly. It also gives total control to the users. For business enterprises, IoT gives real-time analysis of their system's performance and provides information regarding the functions of the devices in the supply network and logistics. IoT through artificial intelligence and machine learning reduces labour costs by automating jobs and processes for companies (Vadlamudi, 2015). It reduces waste and makes service delivery more efficient. In the long run, client transactions become more transparent, and the cost of production and delivery reduces remarkably. The advantages of the application of IoT to agricultural business enterprises are as follows:

- Monitors the process of the business
- Makes the customer experience (CX) more efficient
- The automated processes help save time and reduce cost
- The connected network of data helps improve employees' output
- Can connect with various business models
- Data sharing allows agri-businesses to make better decisions
- Save time and cost to help improve revenue

Through machine learning, IoT manages infrastructure within the organization making these industries more digitalized and transformed. Information and data about the weather can be collected by sensors connected to the Internet of Things which detect rainfall, temperature, and the content of the soil, along with other information that may help improve the farming industry. The need to manage the operations of machines, equipment, and infrastructure can also be solved using IoT.

IoT and Agricultural Sensors

With the increase in adoption of the Internet of Things (IoT) the ability to connect various devices have been implemented in virtually every aspect of our life. It only makes sense that automation also finds its own application in agriculture as it will have a great impact on it.

The sensor provides real-time information on what is happening in the field, the information including air temperature, soil temperature at various depths, rainfall, leaf wetness, chlorophyll, wind speed, dew point temperature, wind direction, relative humidity, solar radiation, and atmospheric pressure.

This indicates that farmers know when their crops are due for harvest, the quantity of water being used, the soil health, and the requirement of any additional input. These are measured and recorded at scheduled intervals. Many sensors are used in agriculture IoT systems which leads to solutions for Smart Farming. The use of precision agriculture sensors will transform the agricultural industry by increasing crop production, adopting a pest-free high-yield variety in crops, and keeping up with the increasing demand for food.

IoT Ecosystem Architecture for Smart Breeding

A common framework of an IoT ecosystem for smart breeding is based on three main components, such as: (i) IoT devices, (ii) communication technologies, and (iii) data process and storage solutions (Figure 8.2).

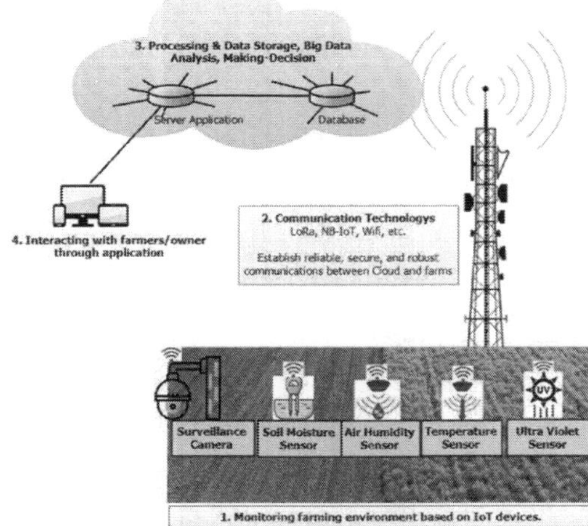

Figure 8.2. An illustration of IoT ecosystems' architecture for smart agriculture. (Reproduced from Quy et. al. 2022).

IoT Devices

The common architecture of an IoT device consists of sensors to collect information from the environment, actuators based on wired or wireless connections, and an embedded system that has a processor, memory, communication modules, input–output interfaces, and battery power (Mishra et. al. 2021). The common architecture of a typical IoT device for smart agriculture is presented in Figure 8.3.

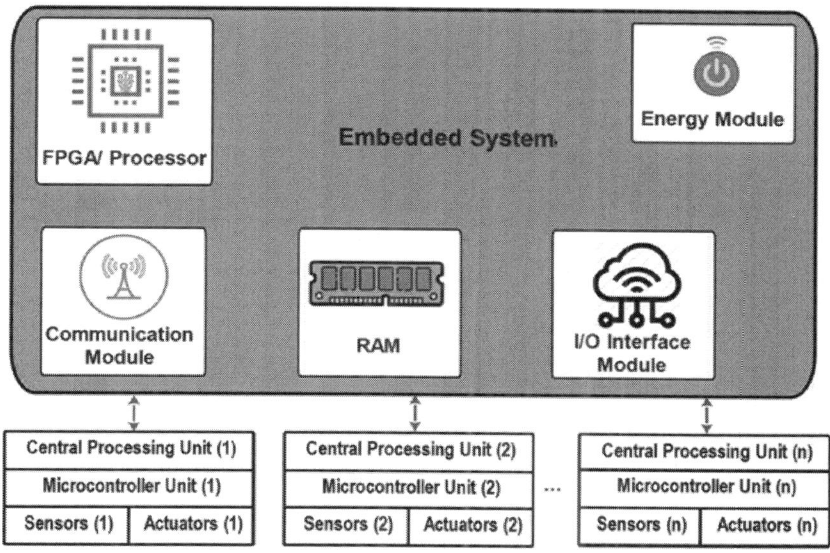

Figure 8.3. An illustration of the common architecture of an IoT device. (Reproduced from Qủy et. al. 2022).

Embedded systems are programmable interactive modules, namely FPGAs (field programmable gate arrays). Sensor devices are specially designed to operate in open environments, in nature, in soil, water, and air to measure and collect environmental parameters that affect production, such as soil nutrients, humidity, temperature, etc. Smart farming solutions are agricultural operations that are often deployed on large farmlands, outdoors, so the devices that support solutions need some unique characteristics, such as the ability to withstand the effects of weather, humidity, and temperature instability throughout their service lifecycle. Some of the main features which make IoT devices suitable for smart agriculture solutions are presented in Figure 8.4 (Wang et. al. 2018; El-Basioni et. al. 2020; Poyen et. al. 2021).

Depending on the required operation, there are several typical sensors applied in the smart agriculture sector. Sensors can be divided into several typical categories, such as

- ❑ Location sensor,

- ❑ Optical sensor,
- ❑ Mechanical sensor,
- ❑ Electrochemical sensor, and
- ❑ Air flow sensor.

These sensors are used to collect information such as air temperature, soil temperature, air humidity, soil moisture, leaf moisture, precipitation, wind speed, wind direction, solar radiation, and barometric pressure (Friha et. al 2021; Song et. al. 2021; Yang et. al 2021)

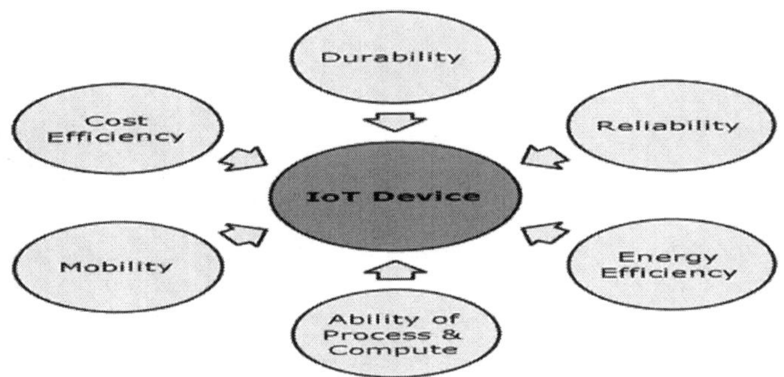

Figure 8.4. The main characteristics of IoT devices. (Reproduced from Quy et. al. 2022).

Communication Technology

To integrate IoT into the smart breeding sector, communication technologies must progressively improve to match the evolution of IoT devices (Alam et. al. 2018; Chetri et. al. 2020). The existing communication solutions can be classified as: protocol, spectrum, and topology.

Protocols: many wireless communication protocols have been proposed for the smart agriculture sector. Based on these protocols, devices in a smart agricultural system can interact, exchange information, and make decisions to monitor and control farming conditions and improve yields and production efficiency. The low-power communication protocol numbers commonly used in smart agriculture can be divided into the following categories:

- ❑ **Short-range**
    - ☆ NFMI (near-field magnetic induction) (Pal and Kant 2019),
    - ☆ Bluetooth (Collotta et. al. 2018),
    - ☆ ZigBee (Bacco et. al. 2018),
    - ☆ Terahertz (Z-Wave) (Gente et. al 2016; Afsharinejad et. al. 2017), and
    - ☆ RFID (Wang et. al. 2019).

- **Long-range:**
  - ☆ LoRa (Ali et. al. 2021),
  - ☆ Sigfox (Joris et. al. 2019; and
  - ☆ NB-IoT (Narrowband IoT) (Popil et. al. 2019).

Some typical communication technologies used in smart agriculture are presented in Table 8.1. It is apparent that short-range communication technologies have a transmission distance of less than 20 (m), high energy efficiency, and low data rate. These protocols are often employed in sensor networks, while long-range communication technologies have transmission distances of up to several kilometers, consume more energy, and are installed for backhaul device-to-device communications. Spectrum: Each radio device uses certain frequency bands for communication. The FCC (Federal Communications Commission) has defined unlicensed spectrum bands for unlicensed operations for scientific, industrial, and medical purposes (Xing and Li 2020). These spectrum bands are often applied for low-power levels and short-range applications. Consequently, a series of common technologies for the smart agriculture sector, from wireless machine control and UAVs to communication technologies such as Wi-Fi and Bluetooth, use unlicensed spectrum bands (Jiang et. al. 2021). However, the use of unlicensed spectra faces several challenges, such as; lack of quality-of-service guarantee, high cost of setting up the initial infrastructure, and interference generated by the huge number of IoT devices (Lagen et. al 2020; Shah et. al. 2020).

Table 8.1. Some typical communication technologies for smart agriculture.

Type	Spectrum	Transmission Distance	Type of Network	Frequency	Data Rate
NFC	Unlicensed	20 cm	D2D	13.56 MHz	424 Kbps
6LoWPAN	Unlicensed	30 m	WHAN	Zwave	250 Kbps
Z-Wave	Unlicensed	100 m	WHAN	900 MHz	100 Kbps
Bluetooth	Unlicensed	100 m	WPAN	2.4 GHz	2-26 Mbps
ZigBee	Unlicensed	1 km	WHAN	2.4 GHz	250 Kbps
802.11a/b/g/n/ac	Unlicensed	100 m	WLAN	2.4-5 GHz	2-700 Mbps
802.11ah	Unlicensed	1000 m	WLAN	Several Sub-GHz	78 Mbps
802.11p	Licensed	1 km	WLAN	5.9 GHz	3-27 Mbps
802.11af	Licensed	1 km	WLAN	54-790	25-550 Mbps
SigFox	Licensed	50 km (rural) 10 km (urban)	LPWA	Zwave	100-600 bps
LoRaWAN	Licensed	20 km	LPWA	Several Sub-GHz	0.3-100 Kbps
NB-IoT	Licensed	35 km	LPWA	Zwave	250 Kbps
LTE-3GPP	Licensed	5 km	WWAN	1.4MHz	200 Kbps
EC-GPRS	Licensed	5 m	WWAN	GSM bands	240 Kbps
WiMAX	Hybrid	50-80 km	WWAN	Several Sub-GHz	70 Mbps

A licensed spectrum usually is allocated to mobile networks. It provides efficient network traffic and reliability, enhanced quality of service (QoS), security, extensive coverage, and involves lower initial infrastructure costs. However, licensed spectrum bands have some limitations, such as high data transmission costs and low energy efficiency of IoT devices (Lu et. al. 2019).

Several studies have demonstrated the efficiency of unlicensed spectrum bands in the mm-wave range. It uses extremely low power but provides large transmission distances and high data rates (Patriciello et. al. 2020; Razavieh et. al. 2021). One limitation of the mm-wave spectrum is that the data rate is strongly affected by weather conditions, especially rain (Mezzavilla et. al. 2018).

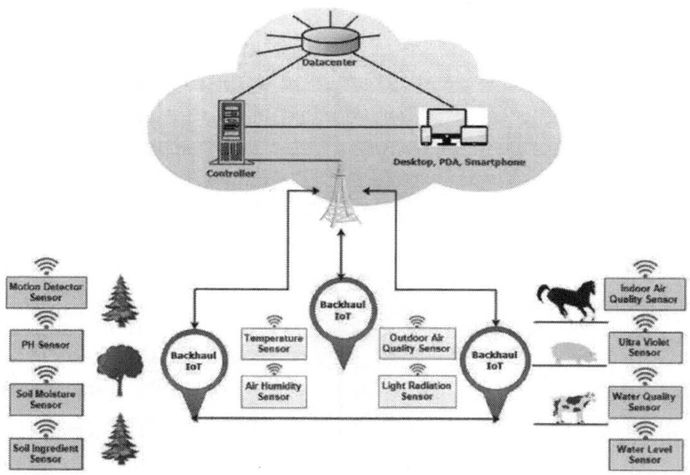

Figure 8.5. An illustration of the common IoT-based smart agriculture topology. (Reproduced from Quy et. al. 2022).

Topology: The establishment of the communication spectrum band and operation protocol of IoT devices depends on the structure that deploys IoT devices for smart agriculture applications. Network structures for smart agriculture usually have two main types of nodes: sensor and backhaul nodes (Kassim 2020). The common characteristics of IoT sensor nodes are short communication distance, low data rate, and high energy efficiency. In contrast, IoT backhaul nodes often require large transmission distances, high throughput, and data rates. Therefore, based on the role of each IoT network node, the sensor node or backhaul node selects and installs appropriate communication technologies (Boursianis et. al. 2020). A typical low-power network topology designed for measuring and monitoring factors in a smart farm is presented in Figure 8.5. The system includes:

- IoT sensor nodes collect information from the farming environment, such as soil moisture, air humidity, temperature, nutrient ingredients of soil, pest images, and water quality, then transmit collected data to IoT backhaul devices.

- Depending on the operation purpose and installation location, IoT sensor nodes can be installed as RFDs (reduced-function devices), which only communicate with FFDs (full-function devices). These nodes cannot communicate with the other RFDs, aiming to save energy and decrease investment costs.
- IoT backhaul nodes, besides having the role of an IoT sensor node, also play a role as intermediate nodes to receive information from other IoT nodes and transmit it to the control center. IoT backhaul nodes are often installed as an FFD device, which can be connected to other FDD and RFD devices.

Data Analytics And Storage Solutions

In the smart breeding domain, besides the main problems of sensing, collecting data, and controlling devices to respond to the real farming environment, data storage, and processing are also important problems and face some challenges (Kour and Arora 2020; Alfred et. al. 2021). The number of collected data is huge, and traditional data storage, organization, and processing solutions are not feasible. Therefore, big data processing solutions need to be researched and applied for smart agriculture (Lopez et. al. 2020; Bhat et. al. 2021).

The complexity of data storage and processing is due to the unique characteristics of the smart agriculture field, including unstructured data and various formats, such as text, images, audio and video, economic figures, and market information. Recent solutions and technologies have introduced the use of cloud platforms for storage and data analytics, which are collected from farms (Firha et. al. 2021; Roopaei et. al. 2017). In addition, cloud-assisted big data analytic solutions, such as edge computing (Zhang et. al 2020) or fog computing (QUy et. al. 2021), are also proposed to reduce latency and costs and support QoS.

Commercialized applications have now been developed to provide solutions and services to farmers to manage their farms and fields, aiming to increase productivity, reduce human labour, and enhance farming efficiency. A few examples are as follows:

OnFarm: It is part of the SWIM family, which specializes in providing solutions and technology for smart agriculture. OnFarm is a technology platform that allows farmers to manage and use data in the simplest way. It is also a comprehensive solution for managing, using, and controlling water on smart farms. (http://www.onfarm.com)

Farmobile: It is a commercialized online platform to manage smart farms that allows farmers, traders, scientists, and insurance companies to operate and communicate centrally on an online platform. (http://.www.farmobile.com)

Silent Herdsman Platform: It is a platform that allows monitoring of the activities of cow colonies and predicting their milk production. ((https://www.thoughtworks.com)

CropX: It is a platform that will enable monitor and control nutrients of farming soil based on a sensor system and big data analytic solutions. (https://www.cropx.com)

FarmX: It is an all-in-one platform for tree crops. FarmX provides a series of diverse farming management solutions, including irrigation, fertilizer, cropland management systems, environmental monitoring, and crop production forecasting. (https://www.farmx.co)

Easyfarm: It is a platform that provides software to help farmers manage and account for figures of farms. Easyfarm provides visual figures, including input and output supplies management, production forecasting, and market connectivity, to help farmers fully manage their farms. (https://www.easyfarm.com)

KAA: It is a cloud-based IoT platform that aims to provide comprehensive, end-to-end solutions for farmers, including: (1) connecting and managing IoT devices in farms or fields, and (2) monitoring and controlling behaviours of devices based on data analysis results. (https://www.kaaproject.org)

Farmlogs: It is a platform that provides tools and solutions for: (i) automating the production cost calculation process, (ii) managing the day-to-day activities of the farm in real-time, and (iii) supporting marketing and increasing sales of products. (https://www.farmlogs.com)

Benefits of Iot Farming

The connected network of IoT allows for access to data. Information may be accessed from anywhere irrespective of time and using any web-enabled device. IoT improves interaction among electronic devices. IoT saves time and resources as data packets are moved through a connected network system. Automated machine function helps develop quality-of-service delivery and reduces the importance of human interference. IoT is no doubt transforming farming and agriculture. The following are ways through which IoT beneficially shapes and improves the use of technologies in agriculture.

Data Gathering

Smart IoT-connected agriculture sensors gather huge amounts of data on crops and animal growth, weather conditions, cattle health, soil quality, and so on. The state and conditions of the agricultural enterprise, workers' performance, and efficient equipment can be monitored using this data.

Advanced Management

The use of IoT in agriculture would enable agricultural enterprises to have better supervision over the internal workings of the business. Advanced management and supervision would reduce the risks involved in production. The collected data can also be used to determine production output, giving farmers the chance to organize for better distribution of products. Getting data on total production output (harvest) gives you the ability to plan for efficient delivery, so your harvest doesn't just lie unsold.

Reduction in Management of Cost and Waste

Following an efficient product management system, it will be possible to increase production efficiency which will directly reduce waste. The ability to detect abnormalities in livestock health and crop growth will provide the opportunity to reduce the possibilities of yield loss.

Increase in Business Efficiency

Automation of business processes could improve business efficiency. Through the use of smart IoT devices, multiple processes can be automated across numerous production processes, such as pest control, irrigation, or fertilizing.

Improved Quality and Volumes of Products

With advanced control over the processes involved in the production, it will be able possible to maintain a greater quality of crops and capacity for growth using automated processes. The combined effects of all the stated factors can ultimately result in greater dividends.

Disadvantages of Iot Farming

Some of the downsides of IoT are described in the following section.

The Risk of Data and Information Theft by Hackers

The growth of IoT means an increase in the connection of devices and the growth of data sharing between connected devices. The impact and likely hood of a hack would also be huge.

Difficulty in Managing Large Volumes of Data

This would make it a more complicated task to manage the large inflow of data in IoT by persons and content management systems. Through the shared network, whenever a bug affects a system, there is the tendency that it may also affect all the devices connected and corrupt them. IoT lacks an international standard for the compatibility of the device. In addition, issues may come up due to the difference in the manufacturing devices to communicate with each other.

Applications of Iot in Agriculture

Some of the applications of IoT in agriculture are described in the following section.

Smart Agriculture

There are several ways by which smart and modern agriculture can be termed. For instance, it can be termed AgricTech, which is generally used in agricultural practices and farming. On the other hand, smart Agriculture is primarily used to refer to the application and use of Internet of things connected devices and solutions in agriculture. Through the use of the Internet of things network devices like sensors to gather particular machines and environmental data, farmers can take informed actions to relatively enhance approximately every area of their work, from crop to livestock farming. For instance, smart agriculture sensors can be used for monitoring the condition of crops. This can enable farmers to accurately access the exact amounts of fertilizers and pesticides to maximize efficiency. This is also what smart farming entails also.

However, consumer-connected devices are more popular than smart industrial and agricultural IoT devices, even though the smart agriculture IoT devices market remains very flexible. The application of IoT frameworks and systems in agriculture grows constantly. Disturbances in the chain of supply and the inadequacy of qualified workers have boosted its compound annual growth rate to 9.9%. According to reported research, the smart farming market stocks could get to about $8 billion by 2022, and by 2025, the world smart agriculture market size will go up by 300% to about $16 billion. Numerous IoT sensors are available for use in agriculture as well as other IoT devices in general, which include the following:

Weather Monitors

Weather stations are presumably the most known and used IoT-connected agriculture devices (Figure 8.6). They combine several IoT farming sensors. They gather numerous data from different positions in the environment and across the field and send them to the cloud. The analysis provided can be used in mapping the conditions of climate, selecting reasonable crops, and implementing the needed strategies to enhance crop capacity. Smart Elements Pycno and allMETEO are examples of some agriculture IoT technology.

Figure 8.6. Weather monitoring in an agricultural field. (Source: easternpeak.com)

Following smart agricultural sectors are applying IoT monitoring solutions:

Crop Farming: In this sector, some vital factors that affect the farming process and production efficiency include air temperature, precipitation, air humidity, soil moisture, salinity, solar radiation, pest status, soil nutrient ingredients, etc. An IoT device called FarmFox allows real-time collection and analysis of the composition of the farming soil and transmits the information to farmers/owners via the Internet (Sengupta et. al. 2021). The results demonstrate the health of the soil is monitored in real-time to provide timely recommendations to farmers aiming to increase productivity and farming efficiency.

A device called Weather Radar can automatically turn on the warning mode using the light signal and send messages to the farmer when the temperature or humidity exceeds a pre-installed threshold.

Monitoring information, such as soil condition, moisture, and temperature, and the prediction of natural factors, such as rainfall and weather, support the control of growing conditions of crops, helping farmers plan and make irrigation decisions to optimize production and reduce labour costs. In addition, the collected data, combined with big data processing technology, can provide recommendations for implementing preventive and remedial solutions against pests and diseases in farming.

Aquaponics: The integration of aquaculture and hydroponics is called aquaponics. It is a farming technique where fish waste becomes a source of nutrients needed by plants. One of the most important issues in such farms is constantly monitoring water quality, water level, temperature, salinity, pH, sunlight, etc. This system is equipped with a control system of water metrics to keep the fish habitat stable and an automatic fish-feeding function to increase the productivity of the fish. Aquaponics farm for households/urban areas based on IoT has also been designed (Lee and Jhang 2019). This system recommends the proper ratio of fish and plants.

Forestry: Forests play a vital role in the carbon cycle and provide a habitat for more than two-thirds of animal species in the world. Forests also have the effect of protecting watersheds, limiting floods, and mitigating climate change. The main factors that need to be monitored in a forest include soil ingredients, air temperature, humidity, and concentration of several different gases, such as oxygen, methane, ammonia, and hydrogen sulfide. A series of forest control systems and solutions based on IoT and big data analytics are available (Marcu et. al 2019, Zou et. al. 2019).

Livestock Farming: It is defined as the process of raising domesticated animals, such as cows, pigs, sheep, goats, and chickens, etc., in an agricultural environment to obtain traction, serve production, and obtain products such as meat, eggs, milk, fur, leather, etc.

The VetLink system can diagnosis, prevention, and treatment of diseases for livestock (Yang et. al. 2020). This system can provide recommendations for animal health for farmers in rural areas where it is difficult to access veterinary doctors immediately. A non-contact temperature measurement system can ensure early detection of diseases and animal health. This system can be used for remote monitoring of animal health and timely anomaly detection.

The monitoring data of water, feed, and animal health for livestock in the farming process helps farmers set up livestock plans, reduce labour costs, and enhance production efficiency. While a series of solutions have been provided for monitoring large-scale farms, their application in small and medium-sized farms is very limited, especially in developing countries. This can be attributed to the high cost and the lack of knowledge needed to set up, manage, and operate IoT systems.

Cattle Management and Monitoring

Like in monitoring and managing crops, several IoT agriculture sensors may be attached to farm livestock to keep track of their health status and record performance. In addition, tracking and monitoring animals on the farm can help with data collection on the well-being, health, and physical location of stocks. For instance, sick animals can be identified by these sensors to enable farmers to isolate them from the rest to prevent the spread of the infection. Real-time cattle tracking can be done by using drones. It will also reduce expenses (Ganapathy, 2015). Collar tags, which are smart Agriculture sensors, are used to collect data on the overall health of the animals. They can also be used to provide specific and collective nutrition data.

Tracking and Tracing

In order to meet the needs of consumers and increase profit value, farms need to demonstrate that products offered to the market are clean and can be tracked and traced conveniently, thereby enhancing the trust of consumers in product safety and health-related issues. In order to solve this problem, a series of tracking- and tracing-based problems for the smart agricultural sector has been developed

SISTABENE is an information system that allows tracking and tracing of agricultural products and foods such as dairy and vegetables (Tradigo et. al. 2019). This system helps suppliers track the production process and errors arising in the supply chain, and helps end-users trace the origin of food.

Using blockchain technology, a food supply chain traceability system was developed (Wang et. al. 2021). It helps to track and trace agrifood supply chains' production processes and trace the origin of agricultural products. Although there are some limitations, the system can successfully support the tracing of food and agricultural products through QR codes, improving product quality and ensuring the clear traceability of products.

Global Positioning System in Farming

The advent of the GPS (global positioning system) has created breakthrough advances in many fields of science and technology. The GPS provides the most important parameters for locating a device, such as location and time. GPS systems have been successfully deployed in many fields, such as smartphones, vehicles, and IoT ecosystems. However, GPS is only good support for outdoor systems and the sky. Meanwhile, the demand for locating and navigating systems in the home and on the streets of smart cities is growing rapidly. Aiming to solve this problem, an advanced global navigation satellite system (GNSS) is being deployed (Kong et. al. 2019). Based on GPS and GNSS systems, suitable farming maps have been established for fields and farms. As a result, agricultural machinery and equipment can be operated autonomously (Alatise et. al. 2020). An illustration of the typical cloud-assisted, IoT-based precision agriculture platform is presented in Figure 8.7.

Huge amounts of data from every aspect of the area of the ecosystem and microclimate can be gathered by farmers. Data on things such as pest infections, lighting, soil condition, CO2 levels, and temperature can be obtained. Farmers can use this data to enhance the measure and quantity of pesticides, fertilizers, and water that their products need to lower costs and increase the level of health of crops. For instance, IoT soil sensors developed by CropX measure temperature, the conductivity of electricity, and soil moisture, allowing farmers to tackle each crop based on their unique needs. This technology combines with geospatial data to help generate accurate soil maps for all the fields. Other IoT applications offer comparable services to farmers like improving yields, reducing waste, and increasing farm sustainability.

Crop management products are another type of IoT device in agriculture with an element of precision farming. Sensors are placed in different parts of the agricultural field to gather data. Data ranging from overall crop health to leaf water potential and precipitation are collected. Crop growth can then be monitored, and abnormalities can be detected efficiently to avoid any harm from infections or diseases.

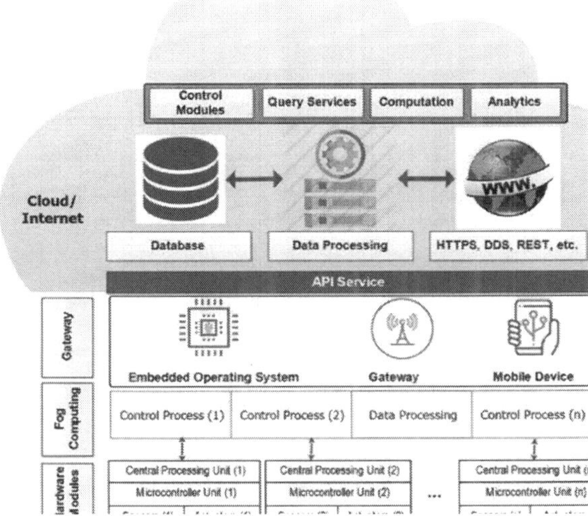

Figure 8.7. Cloud-assisted IoT-based precision agriculture platform. (Reproduced from Quy et. al. 2022).

In smart precision farming, one of the most important applications is the use of drones in monitoring and farming activities. Some common farming tasks using UAVs include spraying pesticides, fertilizing, sowing seeds, evaluating and mapping, and monitoring crop growth.

It has been observed that smart precision agricultural equipment, such as irrigation systems, unmanned aerial vehicles (UAV), and smart agricultural equipment, etc., are configurable in an autonomous-control mode based on certain conditions or can be controlled remotely by the farmer via the Internet (Zhou and Zhou 2020). Smart precision farming helps to improve productivity and production efficiency and is suitable for large-scale farms. Currently, suppliers of precision agricultural equipment have IoT modules built into their machines, allowing machines to operate autonomously and remotely via the Internet (https:www.claasofamerica.com).

Greenhouse Production

A greenhouse consists of walls and a roof, which are usually made from transparent materials, such as plastic or glass. In a greenhouse, plants are grown in a controlled environment, including controlling for moisture, nutrient ingredients of the soil, light, temperature, etc.

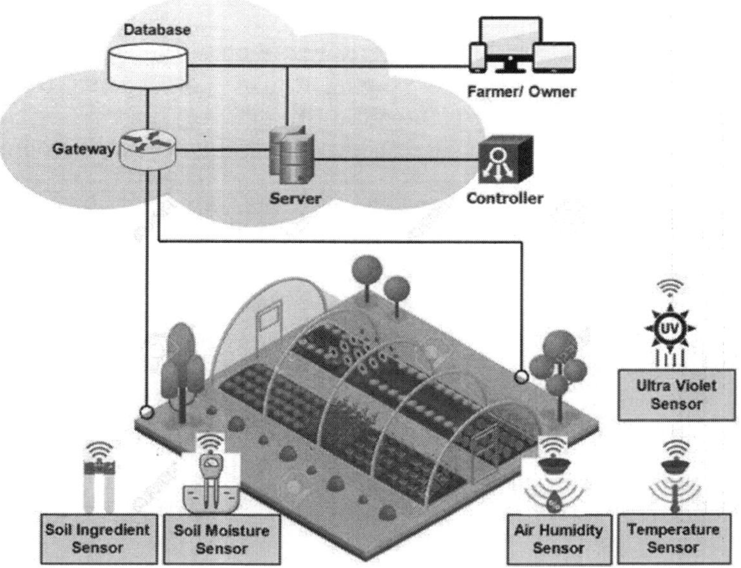

Figure 8.8. An illustration of IoT application for monitoring farming conditions in a greenhouse. (Reproduced from Quy et. al. 2022).

The traditional means of intervention in greenhouse environment management by farmers is manual. With IoT sensors, accurate real-time data on the temperature, humidity, lighting, soil conditions, etc. of the greenhouse can be obtained. Consequently, greenhouse technology makes it possible for humans to grow any plant, at any time, by providing suitable environmental conditions (Tripathy et. al. 2021). A smart agriculture IoT system for monitoring greenhouse farming factors based on IoT ecosystems is presented in Figure 8.8.

An IoT-based greenhouse environmental monitoring system for multipoint monitoring in large greenhouses was developed (Geng et. al. 2019). Instead of using multiple sensors at different locations, this solution involves a drive system that allows the sensor system to move to different locations in the greenhouse. An intelligent IoT system to monitor and control greenhouse temperature for energy efficiency and improve crop productivity was developed (Subhi and Bouazza 2020). The experimental results for the Kingdom of Saudi Arabia, where daytime temperatures can be above 50°C, demonstrate the efficiency of the proposed solution, including saving energy and predicting the rate of plant growth. Recent studies indicated that solutions integrating IoT, big data processing, and artificial intelligence could be applied in greenhouses to reduce labour and energy efficiency (Bai et. al. 2019, Misra et. al. 2020, Munoz et. al. 2020, Ullal et. al. 2020).

Predictive Analytics In Smart Agriculture

To understand IoT smart sensor data, farmers utilize data analytics. It also helps them in essential predictions: the potential infections and diseases, yield volume, crop harvesting time, and so on. Agriculture is majorly reliant on climate conditions; utilizing analytic data devices would make farming more predictable and manageable. For instance, farmers are assisted by Crop Performance systems in assessing the amount and quality of farm outputs beforehand. Farmers may also access data on the risk of adverse events like bad weather conditions, like drought and floods. In addition, predictive analytics allows farmers to enhance nutrients and water supply for all crops. For example, SoilScout is applied in farming and provides farmers with solutions to reduce irrigation water use by up to 50%, reducing fertilizer lost due to excess watering.

Enhanced Farm Management Systems

Farm productivity management systems are more complicated methods of IoT product application. This system consists of installations of several farm IoT sensors and devices on the farm environment along with an effective dashboard with analytical features and preinstalled reporting and accounting system. This enables farmers to remotely monitor and reduce most of the farm tasks drastically. Crop and FarmLogs provide similar services. Aside from farm activity tracking and the other listed IoT farm uses, there are other general applications of these technologies in areas like logistics, storage management, and vehicle tracking, and so on (Paruchuri, 2017).

Soil Nutrient

A system of devices or objects assigned unique identifier tags and exercises the ability to transmit and, or receive signal over a network can be referred to as the internet of things (IoT). an interesting concept behind this system is that its devices can function perfectly without human-to-human, or, human-computer interaction. The unique identifier bridges devices to the cloud, giving the in-situ soil nutrient sensor systems the advantage of real-time data visualization and sharing. The support of communication technologies such as SMS for text messaging, WiFi internet, ZigBee, and LoRa low-powered long-range radio modules have allowed innovative systems for the control, prediction, and monitoring of many soil parameters. This is done in a manner where interaction between the sensor systems and soil can be handled via a web application. A typical IoT sensor system consists of a sensor interfaced with a microcontroller; the microcontroller relies on some battery source to power its function of commands from the microprocessor and is also interfaced with a radio module. The radio module, which holds the unique numeric identification or IP address, transmits and receives data over a network to a cloud, smart device, or personal computer. The personal computer and cloud web application can feature deep learning and algorithms for predicting yield and crop growth parameters. The microcontroller and wireless radio module configuration were also used where researchers deployed a wireless sensor microheater network for smart agriculture use (Popli et. al. 2019). In addition, real-time soil nutrient monitoring techniques have been used to improve crop cultivation and production efficiency in precision agriculture with the use of additional actuator systems for control, such as sensor-dictated fertigation. Real-time, in-field, automated soil monitoring systems have been further explored (Joris et. al. 2019, Sundaram et. al. 2020, Xing et. al. 2020, Ali et. al. 2021).

The interest in real-time in situ sensors and IoT systems has increased in agriculture, especially as farmers demand greater decision management systems for improving yield production and nutrient fertilizer application. Decision support systems incorporate various data elements such as climate, irrigation, crop genetics, energy, land terrain, human activity, and economic resources to provide an analysis of how their interrelations influence productivity. A typical decision management system for nutrient application would include deployable sensor systems for acquiring real-time soil nutrient conditions as presented in Figure 8.9. The sensor systems then forward the data to a gateway node, where the data can be logged or transmitted to a cloud, cellular, or PC network. After this, the data can be further analyzed using machine learning, and neural networks to help predict soil nutrient requirements and present countermeasures for the overloading of nutrient fertilizers. The current state-of-the-art decision management system, SMART Fertilizer Management, is a cloud-based software that offers nutrient requirement data for over 250 different crops. According to SMART, the platform helps farmers maximize yields by an average of 15 percent, save costs, and increase profits.

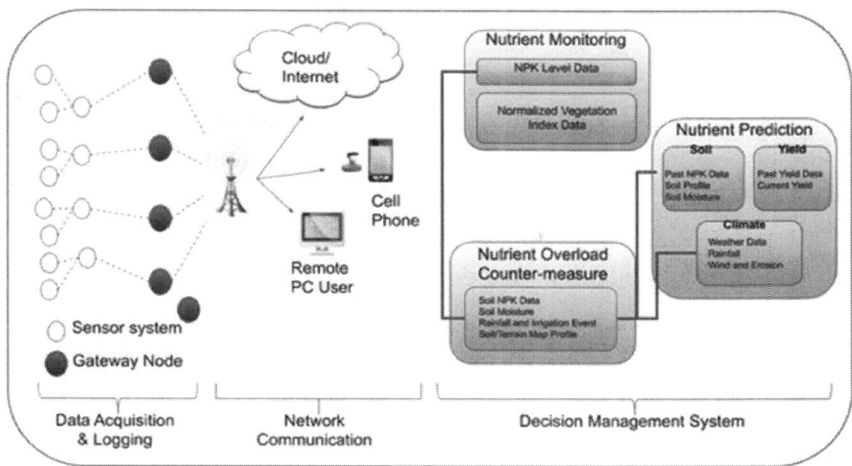

Figure 8.9. Schematic of a decision management system for nutrient fertilizers. (Reproduced from Quy et. al. 2022).

Things to Consider in Designing an Iot Farm Application Tool

Although the application of IoT in agricultural farming is quite numerous, nonetheless, designing IoT systems can prove to be a difficult task. Therefore, before investing in smart farming, it is important to be aware of the following challenges.

Hardware

In building an IoT application for agricultural farming, there is a need to select device sensors or even design a custom one. The type of data you want to gather, and the general purpose of the application will inform your choice. The success of any product deals relatively with its quality which is how reliable and accurate it gathers data.

The Brain

Every IoT agricultural application must be centered on data analytics. The gathered data will be useless to the farmer if it is inaccurate and makes little or no sense. Therefore, there is a 100% need to have high-end data analytics features and connect them to machine learning and predictive algorithms to get usable knowledge based on gathered data.

Maintenance

Hardware maintenance issues are of significant importance for IoT devices in agricultural farming. Sensors and other IoT tools deployed in the field can get damaged easily. This makes it necessary to design your hardware in a way that it can easily be maintained for durability. If this is not done, the alternative is the constant replacement of sensors.

Mobility

IoT devices for smart agriculture farming should be designed for field use. Users (farm admins or owners) must be able to access data and information remotely using mobile phones or computers. It should also be accessible on-site. Also, each IoT device must be independent and possess a sufficient network range to interact with other devices as well as transfer the data to the main server.

Infrastructure

There is a need for a robust infrastructure to make sure that agricultural farming applications perform at optimal levels. It will also enable it to handle huge loads of data. Security of infrastructure is also an important aspect of the system. Systems have to be secure to prevent hacking, data theft, or even getting control of an autonomous device.

Connectivity

Transmission of data to and from numerous agricultural farming establishments is still a huge issue faced in the application of IoT farming. However, the connections must be strong enough to resist harsh weather restrictions so that the operations are not disrupted. There are different connection protocols used by IoT devices presently. However, plans are currently underway to provide a unified network in this area. The development of technologies like 5G and space-based internet perhaps may be a solution

Frequency of data collection: Due to the numerous variations of data in the farming sector, making sure data collection frequency is at an optimal level can be very complicated. There are lots of regulations and restrictions currently placed on information and data gathered from apps, machinery, environmental sensors, equipment, field-based, and even processed analytical data (Vadlamudi, 2017). Sharing and delivering data on time and safely is part of the issues faced by smart farming applications.

Factors Affecting The Mass Application Of Iot In Smart Farming

The Internet of Things plays a critical role in smart agricultural farming and other areas. In the future, as the Mobile network connectivity system develops and farmers can access more loans, IoT applications will also grow. At present, IoT application in farming is still at its take-off stage. Several IoT solutions have been deployed on large-scale farms/fields. However, the widespread deployment of IoT in the agricultural sector still presents some challenges. The following challenges have slowed down the mass application of IoT devices in agriculture and farming.

Economic Efficiency

In agricultural economics, one of the most important characteristics is the low rate of profit of an investment project, which presents many risks from natural conditions. The benefit–cost of new technology-seeking deployment in agriculture should be carefully calculated to ensure a trade-off between the cost of technology implementation and the profit potential. Therefore, it is important to discuss various economic aspects related to IoT implementation in smart agriculture.

There are several types of costs related to the implementation of IoT in agriculture, including the system initialization cost and the system operating cost. The system initialization cost includes hardware purchases (IoT devices, gateways, base station infrastructure). The system operating cost includes service registration cost and the cost of labour to manage IoT devices. Furthermore, additional operating costs include incurred costs from energy consumption, maintenance, and data exchange among IoT devices, gateways, and cloud servers. Turgut and Boloni (2017) emphasised that successful deployment of IoT technologies will only happen if the customer benefits (customers need to know the benefits and potential) that IoT systems provide exceed their physical value and privacy costs.

There is still a gap between service providers and service users (farmers or the owner of the farm), leading to the slow deployment of IoT applications in smart agriculture. In terms of the economic aspect, the analyzed results show that the need for a support policy from regulatory agencies and governments to allow service providers and service users to use IoT-based smart agriculture applications in their infancy can be met. To promote smart agriculture, the European Union has issued supportive economic policies, the so-called European CAP (Common Agricultural Policy), whose annual budget amounts to approximately EUR 59 billion and is paid for by the nations of the EU (Rousi et. al. 2021).

To be able to apply IoT in the field of smart agriculture, service costs (Cpay) and the operating and system initialization cost of IoT (Cuserh) need to constantly be improved and optimized to reduce the cost of the IoT services for farmers. In addition, IoT businesses (service providers) also need to maximize the value of information obtained (Vinfo) to improve the profitability of the service providers.

Service providers may commercially exploit the information received (Vinfo) in the period of providing services for farms, aiming to encourage the deployment of IoT applications in smart agriculture. Nowadays, several IoT platform providers allow free registration and use of services with some limitation conditions regarding services' functionality and ability processing; the number of connected IoT devices; and the number of data stored while premium functions and services charge users a fee.

One of the significant factors slowing down IoT adoption in agriculture is farmers' knowledge and ability to use IoT devices. In developed countries, this issue can be easily solved due to the accessibility of new technologies of farmers. Otherwise, in developing countries, where the majority of farmers in rural areas have very limited access to advanced technologies, this issue is a significant challenge (Ammad Uddin et. al. 2019, Joshi et. al. 2019).

Weak Network

Interference: Deploying a huge number of IoT devices for smart agriculture can cause interference to different network systems, especially some IoT networks using short spectrum bands such as ZigBee, Wi-Fi, Sigfox, and LoRa (Table 1). Interference can degrade system performance as well as reduce the reliability of IoT ecosystems. IoT networks that use cognitive technology to reuse unlicensed spectra increase the cost of the device. the advent of the 6G network (De Lima et. al. 2021) will allow a huge number of devices to connect to the Internet with an extremely high access speed and extremely large bandwidth. Thereby, the interference problem of IoT networks could be solved.

There are usually poor connectivity issues in remote and rural areas around the world. However, satellite internet and mobile network coverage improved substantially towards the end of the 2010s. Nevertheless, network connectivity issues persist, especially in the growth of the Internet of Things in smart Agriculture, mostly in Asia and Africa. Some areas lack the required internet connection power to transmit a huge volume of data from the fields to the mobile applications held by the farm admin or owner.

Security Threats

Data protection is essential in any enterprise. Several marketers sell data transmission and connectivity from the data-gathering stage through IoT sensors, the data processing stage in which the data is displayed and monitored on the farmer's screen. There are several farm data securities that can be breached (Ganapathy and Neogy, 2017). There are many sensitive and secret data even though the agricultural sector relies substantially on the abundance of shared and open data. Some companies have been established to deal with such vulnerabilities by developing Agritech software. With the increased number of security breaches reported yearly, this issue has become a paramount focus for several investors (Vadlamudi, 2016).

Security and Privacy: One of the most important problems of applying IoT in smart agriculture is the security problem, including the protection of data and systems from attacks on the Internet. In regard to system security, IoT devices' limited capacity and ability led to complex encryption algorithms that are impossible to implement on IoT devices. As a result, IoT systems can be attacked using the Internet to gain system control rights; IoT gateways are also attacked via denial of service (Jia et. al 2020, Sandal et. al. 2020, Wang et. al. 2021). In addition, cloud servers can be attacked by data spoofing to perform unauthorized tasks that affect the autonomous farming processes of farms. Cloud infrastructures can also be controlled by attackers (Somani at. Al. 2017, Agrawal and Tapaswi 2019). According to Neshenko et. al. (2019), the IoT data security issue is one of the biggest problems slowing down IoT adoption in smart agriculture.

Regarding data security, the obtained information from IoT systems in farms is collected, processed, and commercially exploited by service providers to varying degrees. Therefore, one of the most important problems of policies regards the validity and legal status of farm data (Chaterji et. al. 2021). In reality, these data are of great value when aggregated and analyzed for large-scale agricultural activities. Consequently, without policies, the data privacy and security of farms can affect the competitive advantage of farmers/farm owners. In our opinion, using cryptography coupled with access keys is a possible solution to solve this problem. Keys could be made available based on a regional user group and to those who contributed to the database. For further complex cases, secure multiparty computation can be used, where the homomorphic encryption method (Jin et. al 2018, Harn et. al 2021), or this method combined with the blockchain (Jia et. al. 2022), can be applied for the purpose of balancing privacy and data utility.

Reliability

Most IoT devices are expected to be deployed outdoors (in fields and farms). Harsh work environments lead to the rapid degradation of IoT devices' quality and can lead to unexpected manufacturer failures. The mechanical safety of IoT devices and systems must be ensured so they can withstand extremes of weather, such as temperature, humidity, rainstorms, and floods (Ballal et. al. 2020). New materials and technologies need to continue to be studied to improve the durability of devices.

Conclusions

The Internet of Things is part of the technological advancements that are changing the mode of human operations worldwide. Farmers can use sensors and other IoT devices to gather data and carry out analyses for accurate estimates. With IoT capabilities, farmers can be able to optimize functions and improve yield at a lower rate. These may also result in reduced costs for consumers.

There are numerous challenges that have limited the full adoption of IoT in the agricultural sector. The affordability of the majority of farmers, weak connectivity, and security risks have limited the full adoption of IoT in farming. In the future, with advancements and the development of IoT, the limitations will be bypassed, and more farmers will be able to access IoT for smart farming. Security technologies need to be continuously improved. However, the application of IoT solutions for smart agriculture is inevitable and will enhance productivity, provide clean and green foods, support food traceability, reduce human labour, and improve production efficiency.

References

Abdelnour, A., Buchin, F., Kaddour, D., & Tedjini, S. (2018). Improved traceability solution based on UHF RFID for cheese production sector. IEEE J. Radio Freq. Identif. 2: 68–72.

Afsharinejad, A., Davy, A., & Naftaly, M. (2017). Variability of terahertz transmission measured in live plant leaves. IEEE Geosci. Remote Sens. Lett. 14: 636–638.

Agrawal, N., & Tapaswi, S. (2019). Defense mechanisms against DDoS attacks in a cloud computing environment: state-of-the-art and research challenges. IEEE Commun. Surv. Tutor. 21: 3769–3795.

Alam, M.M., Malik, H., Khan, M.I., Pardy, T., Kuusik, A., & Le Moullec, Y. (2018). A survey on the roles of communication technologies in IoT-based personalized healthcare applications. IEEE Access. 6: 36611–36631.

Alatise, M.B., & Hancke, G.P. (2020). A review on challenges of autonomous mobile robot and sensor fusion methods. IEEE Access. 8: 39830–39846.

Alfred, R., Obit, J.H., Chin, C.P.Y., Haviluddin, H., & Lim, Y. (2021). Towards paddy rice smart farming: a review on big data, machine learning, and rice production tasks. IEEE Access. 9: 50358–50380.

Ali, S., Glass, T., Parr, B., Potgieter, J., & Alam, F. (2021). Low cost sensor with IoT LoRaWAN connectivity and machine learning-based calibration for air pollution monitoring. IEEE Trans. Instrum. Meas. 70: 5500-5511.

Ammad Uddin, M., Ayaz, M., Aggoune, E.H.M., Mansour, A., & Le Jeune, D. (2019). Affordable broad agile farming system for rural and remote area. IEEE Access. 7: 127098–127116.

Bacco, M., Berton, A., Gotta, A., & Caviglione, L. (2018). IEEE 802.15.4 Air-ground UAV communications in smart farming scenarios. IEEE Commun. Lett. 32: 1910–1913.

Bai, X., Wang, Z., Sheng, L., & Wang, Z. (2019). Reliable data fusion of hierarchical wireless sensor networks with asynchronous measurement for greenhouse monitoring. IEEE Trans. Control Syst. Technol. 27: 1036–1046.

Ballal, K.D., Dittmann, L., Ruepp, S., & Petersen, M.N. (2020). IoT devises reliability study: multi-RAT communication. Proc. IEEE 6th World Forum Internet of Things (WF-IoT), New Orleans, LA, USA; p. 1–2.

Bhat, S.A., & Huang, N.F. (2021). Big data and AI revolution in precision agriculture: survey and challenges. IEEE Access, 9: 110209–110222.

Boursianis, A.D., Papadopoulou, M.S., Gotsis, A., Wan, S., Sarigiannidis, P. Nikolaidis, S., & Goudos, S.K. (2020). Smart irrigation system for precision agriculture: the ARETHOU5A IoT platform. IEEE Sens. J. 21: 17539–17547.

Chaterji, S., DeLay, N., Evans, J., Mosier, N., ... Engel, B. (2021). A vision for machine learning, data engineering, and policy considerations for digital agriculture at scale. IEEE Open J. Comput. Soc. 2: 227–240.

Chettri, L., & Bera, R. (2020). A comprehensive survey on Internet of Things (IoT) toward 5G Wireless Systems. IEEE Internet Things J. 7: 16–32.

Collotta, M., Pau, G., Talty, T., & Tonguz, O.K. (2018). Bluetooth 5: A concrete step forward toward the IoT. IEEE Commun. Mag. 56: 125–131.

De Lima, C., Belot, D., Berkvens, R., Bourdoux, A., ... Dardari, D. (2021). Convergent communication, sensing and localization in 6G systems: an overview of technologies, opportunities and challenges. IEEE Access. 9: 26902–26925.

El-Basioni, B.M.M., & El-Kader, S.M.A. (2020). Laying the foundations for an IoT reference architecture for agricultural application domain. IEEE Access. 8: 190194–190230.

Elijah, O., Rahman, T.A., Orikumhi, I., Leow, C.Y., & Hindia, M.N. (2018). An overview of Internet of Things (IoT) and data analytics in agriculture: benefits and challenges. IEEE Internet Things J. 5: 3758–3773.

Farooq, M.S., Riaz, S., Abid, A., Abid, K., & Naeem, M.A. (2019). A survey on the role of IoT in agriculture for the implementation of smart farming. IEEE Access. 7: 156237–156271.

Friha, O., Ferrag, M.A., Shu, L., Maglaras, L., & Wang, X. (2021). Internet of things for the future of smart agriculture: a comprehensive survey of emerging technologies. IEEE/CAA J. Autom. Sin. 8: 718–752.

Ganapathy, A. (2015). AI fitness checks, maintenance and monitoring on systems managing content and data: a study on CMS world. Malaysian J. Med. Biol. Res. 2: 113-118. https://doi.org/10.18034/mjmbr.v2i2.553

Ganapathy, A. (2016a). Blockchain technology use on transactions of crypto currency with machinery and electronic goods. Amer. J. Trade Policy. 3: 115-120.

Ganapathy, A. (2016b). Speech emotion recognition using deep learning techniques. ABC J. Adv. Res. 5: 113-122. https://doi.org/10.18034/abcjar.v5i2.550

Ganapathy, A. (2017). Friendly URLs in the CMS and power of global ranking with crawlers with added security. Engg. Int. 5: 87-96. https://doi.org/10.18034/ei.v5i2.541

Ganapathy, A., & Neogy, T.K. (2017). Artificial intelligence price emulator: a study on cryptocurrency. Glob. Disc. Econ. business, 6: 115-122. https://doi.org/10.18034/gdeb.v6i2.558

Geng, X., Zhang, Q., Wei, Q., Zhang, T., Cai, Y., Liang, Y., & Sun, X. (2019). A mobile greenhouse environment monitoring system based on the Internet of Things. IEEE Access. 7: 135832–135844.

Gente, R., Busch, S.F., Stubling, E.M., ... Schneider, L.M. (2016). Quality control of sugar beet seeds with THz time-domain spectroscopy. IEEE Trans. Terahertz Sci. Technol. 6: 754–756.

Gopalakrishnan, S., Waimin, J., Raghunathan, N., Bagchi, S., Shakouri, A., & Rahimi, R. (2021). Battery-less wireless chipless sensor tag for subsoil moisture monitoring. IEEE Sens. J. 21: 6071–6082.

Harn, L., Hsu, C.F., Xia, Z., & He, Z. (2021). Lightweight aggregated data encryption for wireless sensor networks (WSNs). IEEE Sens. Lett. 5: 1–4.

Javed, F., Afzal, M.K., Sharif, M., & Kim, B. (2018). Internet of things (IoT) operating systems support, networking technologies, applications, and challenges: a comparative review. IEEE Commun. Surv. Tutor. 20: 2062–2100.

Jia, B., Zhang, X., Liu, J., Zhang, Y., Huang, K., & Liang, Y. (2022). Liang blockchain-enabled federated learning data protection aggregation scheme with differential privacy and homomorphic encryption in IoT. IEEE Trans. Ind. Inform. 18: 4049–4058.

Jia, Y., Zhong, F., Alrawais, A., Gong, B., & Cheng, X. (2020). FlowGuard: An intelligent edge defense mechanism against IoT DDoS attacks. IEEE Internet Things J. 7: 9552–9562.

Jiang, X., Zhang, H., Yi, E.A.B., ... Raghunathan, N., (2021). Hybrid low-power wide-area mesh network for IoT applications. IEEE Internet Things J: 8, 901–915.

Jin, B., Jiang, D., Xiong, J., Chen, L., & Li, Q. (2018). D2D data privacy protection mechanism based on reliability and homomorphic encryption. IEEE Access. 6: 51140–51150.

Joris, L., Dupont, F., Laurent, P., Bellier, P., Stoukatch, S., & Redouté, J. (2019). An autonomous sigfox wireless sensor node for environmental monitoring. IEEE Sens. Lett. 3: 5500604.

Joshi, A., Dandekar, I., Hargude, N., Shrotri, A.P., & Dandekar, A.R. (2019). Application of internet of the things (IoT) for the water conservation and entrepreneurship in the rural area. Proc. IEEE Pune Section Intel. Conf. (PuneCon), Pune, India, Dec.18–20. p. 26-34

Kassim, M.R.M. (2020). IoT applications in smart agriculture: issues and challenges. Proc. IEEE Conf. Open Systems (ICOS), Kota Kinabalu, Malaysia, Nov. 17–19, p. 19–24.

Kong, S., López-Salcedo, J.A., Wu, Y., & Kim, E. (2019). IEEE access special section editorial: GNSS, localization, and navigation technologies. IEEE Access. 7: 131649–131652.

Kour, V.P., & Arora, S. (2020). Recent developments of the internet of things in agriculture: a survey. IEEE Access. 8: 129924–129957.

Lagen, S., Giupponi, L., Goyal, S., Patriciello, N., ... Bojovic, B. (2020). New radio beam-based access to unlicensed spectrum: design challenges and solutions. IEEE Commun. Surv. Tutor. 22: 8–37.

Lee, C., & Jhang, J. (2019). System design for internet of things assisted urban aquaponics farming. Proc. IEEE 8th Global Conference on Consumer Electronics (GCCE), Osaka, Japan, Oct. 15–18; p. 986–987.

Li, X., Pu, T., Li, L., & Ao, J. (2020). Enhanced sensitivity of GaN-based temperature sensor by using the series schottky barrier diode structure. IEEE Electron Device Lett. 41: 601–604.

López, I.D., Figueroa, A., & Corrales, J.C. (2020). Multi-dimensional data preparation: a process to support vulnerability analysis and climate change adaptation. IEEE Access. 8: 87228–87242.

Lu, X., Petrov, V., Moltchanov, D., Andreev, S., Mahmoodi, T., & Dohler, M. (2019). 5G-U: conceptualizing integrated utilization of licensed and unlicensed spectrum for future IoT. IEEE Commun. Mag. 57: 92–98.

Marcu, A.E., Suciu, G., Olteanu, E., Miu, D., Drosu, A., & Marcu, I. (2019). IoT system for forest monitoring. Proc. 42nd Intel. Conf. Telecom. and Signal Processing (TSP), Budapest, Hungary, July 1–3, p. 629–632.

Mezzavilla, M., Polese, M., Zanella, A., ... Dhananjay, A. (2018). Public safety communications above 6 GHz: challenges and opportunities. IEEE Access. 6: 316–329.

Mishra, D., Zema, N.R., & Natalizio, E. (2021). A high-end IoT devices framework to foster beyond-connectivity capabilities in 5G/B5G architecture. IEEE Commun. Mag. 59: 55–61.

Misra, N.N., Dixit, Y., Al-Mallahi, A., Bhullar, M.S., Upadhyay, R., & Martynenko, A. (2020). IoT, big data and artificial intelligence in agriculture and food industry. IEEE Internet Things J. 1: 15-20.

Muñoz, M., Guzmán, J.L., Sánchez, J.A., Rodríguez, F., Torres, M., & Berenguel, M. (2020). A new IoT-based platform for greenhouse crop production. IEEE Internet Things J. 1: 8-11.

Neogy, T.K., & Paruchuri, H. (2014). Machine learning as a new search engine interface: an overview. Engg. International. 2: 103-112. https://doi.org/10.18034/ei.v2i2.539

Neshenko, N., Bou-Harb, E., Crichigno, J., Kaddoum, G., & Ghani, N. (2019). Demystifying IoT security: an exhaustive survey on IoT vulnerabilities and a first empirical look on internet-scale IoT exploitations. IEEE Commun. Surv. Tutor. 21: 2702–2733.

Pal, A., & Kant, K. (2019). NFMI: Connectivity for short-range IoT applications. Computer: 52, 63–67.

Paruchuri, H. (2015). Application of artificial neural network to ANPR: an overview. ABC J. Adv. Research. 4: 143-152. https://doi.org/10.18034/abcjar.v4i2.549

Paruchuri, H. (2017). Credit card fraud detection using machine learning: a systematic literature review. ABC J. Adv. Research, 6: 113-120. https://doi.org/10.18034/abcjar.v6i2.547

Patriciello, N., Lagén, S., Bojović, B., & Giupponi, L. (2020). NR-U and IEEE 802.11 technologies coexistence in unlicensed mmWave spectrum: models and evaluation. IEEE Access. 8: 71254–71271.

Popli, S., Jha, R.K., & Jain, S. (2019). A survey on energy efficient narrowband internet of things (NBIoT): architecture, application and challenges. IEEE Access. 7: 16739–16776.

Poyen, FB., Ghosh, A., Kundu, P., Hazra, S., & Sengupta, N. (2021). Prototype model design of automatic irrigation controller. IEEE Trans. Instrum. Meas. 70: 9502217.

Quy, VK., Hau, N.V., Anh, D.V., & Ngoc, L.A. (2021). Smart healthcare IoT applications based on fog computing: architecture, applications and challenges. Complex Intell. Syst. 21: 52-60.

Quy, V.K., Hau, N.V., Anh, D.V., Quy, N.M., ... Ban, N.T. (2022). ioT-enabled smart agriculture: architecture, applications, and challenges. Appl. Sci. 12: 3396 - 3405. https://doi.org/10.3390/ app12073396

Razavieh, A., Chen, Y., Ethirajan, T., Gu, M., ... Cimino, S. (2021). Extremely-low threshold voltage FinFET for 5G mmWave applications. IEEE J. Electron Devices Soc. 9: 165–169.

Roopaei, M., Rad, P., & Choo, K.R. (2017). Cloud of things in smart agriculture: intelligent irrigation monitoring by thermal imaging. IEEE Cloud Comput. 4: 10–15.

Rousi, M., Sitokonstantinou, V., Meditskos, G., ... Papoutsis, I. (2021). Semantically enriched crop type classification and linked earth observation data to support the common agricultural policy monitoring. IEEE J. Sel. Top. Appl. Earth Obs. Remote Sens. 14: 529–552.

Saad, A., Benyamina, A.E.H., & Gamatié, A. (2020). Water management in agriculture: a survey on current challenges and technological solutions. IEEE Access. 8: 38082–38097.

Sandal, Y.S., Pusane, A.E., Kurt, G.K., & Benedetto, F. (2020). Reputation based attacker identification policy for multi-access edge computing in Internet of Things. IEEE Trans. Veh. Technol. 69, 15346–15356.

Sengupta, A., Debnath, B., Das, A., & De, D. (2021). FarmFox: a quad-sensor based IoT box for precision agriculture. IEEE Consum. Electron. Mag. 10: 63–68.

Shafique, K., Khawaja, B.A., Sabir, F., & Qazi, S. (2020). Mustaqim, M. Internet of Things (IoT) for next-generation smart systems: a review of current challenges, future trends and prospects for emerging 5G-IoT scenarios. IEEE Access. 8: 23022–23040.

Shah, S.W.H., Mian, A.N., & Crowcroft, J. (2020). Statistical QoS guarantees for licensed-unlicensed spectrum interoperable D2D communication. IEEE Access. 8: 27277–27290.

Somani, G., Gaur, M.S., Sanghi, D., Conti, M., Rajarajan, M., & Buyya, R. (2017). Combating DDoS attacks in the cloud: requirements, trends, and future directions. IEEE Cloud Comput. 4: 22–32.

Song, Y., Yu, F.R., Zhou, L., Yang, X., & He, Z. (2021). Applications of the Internet of Things (IoT) in smart logistics: a comprehensive survey. IEEE Internet Things J. 8: 4250–4274.

Spachos, P., & Gregori, S. (2019). Integration of wireless sensor networks and smart UAVs for precision viticulture. IEEE Internet Comput. 23: 8–16.

Subahi, A.F., & Bouazza, K.E. (2020). An intelligent IoT-based system design for controlling and monitoring greenhouse temperature. IEEE Access. 8: 125488–125500.

Sundaram, J.P.S., Du, W., & Zhao, Z.A. (2020). Survey on LoRa networking: research problems, current solutions, and open issues. IEEE Commun. Surv. Tutor. 22: 371–388.

Tradigo, G., Vizza, P., Veltri, P., Lambardi, P., Caligiuri, F.M., Caligiuri, G., & Guzzi, P.H. (2019). An information system for the traceability of agricultural food production. Proc. IEEE Intel. Conf. Bioinformatics Biomedicine (BIBM), San Diego, CA, USA, Nov. 18–21, p. 2304–2309.

Tripathy, P.K., Tripathy, A.K., Agarwal, A., & Mohanty, S.P. (2021). MyGreen: an IoT-enabled smart greenhouse for sustainable agriculture. IEEE Consum. Electron. Mag. 10: 57–62.

Turgut, D., & Boloni, L. (2017). Value of information and cost of privacy in the Internet of Things. IEEE Commun. Mag. 55: 62–66.

Tyagi, S.K.S., Mukherjee, A., Pokhrel, S.R., & Hiran, K.K. (2020). An intelligent and optimal resource allocation approach in sensor networks for smart agri-IoT. IEEE Sens. J. 21: 17439–17446.

Udutalapally, V., Mohanty, S.P., Pallagani, V., & Khandelwal, V. (2020). sCrop: a novel device for sustainable automatic disease prediction, crop selection, and irrigation in internet-of-agro-things for smart agriculture. IEEE Sens. J. 21: 17525–17538.

Ullah, I., Fayaz, M., Naveed, N., & Kim, D. (2020). ANN based learning to Kalman filter algorithm for indoor environment prediction in smart greenhouse. IEEE Access. 8: 159371–159388.

Vadlamudi, S. (2015). Enabling trustworthiness in artificial intelligence: a detailed discussion. Engg. International, 3: 105-114. https://doi.org/10.18034/ei.v3i2.519

Vadlamudi, S. (2016). What impact does Internet of Things have on project management in project based firms?. Asian Business Review. 6: 179-186. https://doi.org/10.18034/abr.v6i3.520

Vadlamudi, S. (2017). Stock market prediction using machine learning: a systematic literature review. Amer. J. Trade and Policy, 4: 123-128. https://doi.org/10.18034/ajtp.v4i3.521

Wang, J., Hao, S., Wen, R., Zhang, B., Zhang, L., Hu, H., & Lu, R. (2021). IoT-praetor: undesired behaviors detection for IoT devices. IEEE Internet Things J. 8: 927–940.

Wang, L., Xu, L., Zheng, Z., Liu, S., Li, X., Cao, L., Li, J., & Sun, C. (2021). Smart contract-based agricultural food supply chain traceability. IEEE Access. 9: 9296–9307.

Wang, Y., Rajib, S.M.S.M., Collins, C., & Grieve, B. (2018). Low-cost turbidity sensor for low-power wireless monitoring of fresh-water courses. IEEE Sens. J. 18: 4689–4696.

Xing, C., & Li, F. (2020). Unlicensed spectrum-sharing mechanism based on Wi-Fi security requirements implemented using device to device communication technology. IEEE Access. 8: 135025–135036.

Yang, X., Shu, L., Chen, J., Ferrag, M.A., Wu, J., Nurellari, E., & Huang, K. (2021). A survey on smart agriculture: development modes, technologies, and security and privacy challenges. IEEE/CAA J. Autom. Sin. 8: 273–302.

Yang, Y., Ren, R., & Johnson, P.M. (2020). VetLink: a livestock disease-management system. IEEE Potentials. 39: 28–34.

Zhang, X., Cao, Z., & Dong, W. (2020). Overview of edge computing in the agricultural internet of things: key technologies, applications, challenges. IEEE Access. 8: 141748–141761.

Zhou, X., & Zhou, J. (2020). Data-driven driving state control for unmanned agricultural logistics vehicle. IEEE Access. 8: 65530–65543.

Chapter-9

Artificial Intelligence for Crop Breeding

Abstract

Artificial intelligence (AI) has emerged as a revolutionary field, providing a great opportunity to shape modern crop breeding, and is extensively used indoors for plant science. Advances in crop phenomics, enviromics, together with the other "omics" approaches are paving the way for elucidating the detailed complex biological mechanisms that motivate crop functions in response to environmental trepidations. These "omics" approaches have provided plant researchers with precise tools to evaluate the important agronomic traits for larger-sized germplasm at a reduced time interval in the early growth stages. However, the big data and the complex relationships within impede the understanding of the complex mechanisms behind genes driving the agronomic trait formations. AI brings huge computational power and many new tools and strategies for future breeding. It is important to understand how applications of AI technology can be utilized for current breeding practices, and assist in solving the problem in high-throughput phenotyping, and gene functional analysis. Advances in AI technologies have brought new opportunities for future breeding, and made envirotyping data widely utilized in breeding progeammes. In the current breeding methods, linking genotype to phenotype remains a massive challenge and impedes the optimal application of high-throughput field phenotyping, genomics, and enviromics. Therefore, the application of AI will be the preferred tool to increase the accuracy in high-throughput crop phenotyping, genotyping, and envirotyping data. Integration of AI with "omics" tools shall assist rapid gene identification and accelerate the process of crop-improvement programmes.

Keywords: *Artificial intelligence, AI technology, envirotyping, phenomics, smart breeding, gene function, biochemical phenotyping*

Introduction

Plant breeding practice was initiated by Humans soon after the initial domestication of cereal grains, from which they stored the seeds of the best plants, to grow the next generation of crops. Early plant-breeding technologies were the forerunners to such rudimentary selection strategies (Allard 2014). People from all across the world explored and cultivated nearly 7000 varieties of food plants during Breeding 1.0, which began 10–12 thousand years ago (Mayer et al. 2012). Breeding stage 2.0 started in the late nineteenth and early twentieth centuries when an inbreeding depression was discovered. During this time, many advances in plant breeding were made in the science of breeding itself, such as replicated field trials, controlled crossings, statistical analyses, formal experimental designs, hybrid breeding, pedigree-based estimates of breeding values, and precise yield measurement at scale (e.g., with multirow combines) (Moran and Smith 1918). Breeding 3.0 reached to about 30 years ago, when molecular markers and genomic data began to supplement phenotypic data (Meuwissen et al. 2001). Breeding 4.0 is rapidly approaching, with huge omics data and the rapid progress of informatic technologies (Wallace et al. 2018). Since the beginning of the plant breeding process, starting from crop domestication, new approaches mediated by different technological revolutions are being enriched in the science continuously to increase the pace, accuracy, and precision in plant breeding (Zargar et al. 2015). This resulted into the green revolution by developing semidwarf, nutrient-responsive, and hybrid cultivars (Bhat et al. 2021). Considering the population growth, decreasing arable land, and climate changes, it has become necessary to develop high-throughput approaches that can mediate to develop crop cultivars at a greater pace, with higher accuracy, and precision. Accordingly, the field of artificial intelligence (AI) has recently emerged, which has the potential to possess extraordinary power to assist in breeding climate-resilient smart crops. Climate-resilient crops are implied to maintain or increase crop yields and quality under various climatic and environmental changes, and possess the ability to resist multiple biotic and abiotic stresses. Accordingly, climate-resilient smart crops will allow crops to address the interlinked challenges of food security and climate change (Zohry and Ouda 2022).

The aim of artificial intelligence (AI) is to replicate some features of human intelligence using technology (Yang and Siau 2018). AI can be defined as a set of studies and techniques dealing with the computer science and mathematical aspects of statistical modelling, with significant economic and social implications, and the goal of developing technological systems capable of solving problems and performing tasks or duties normally performed by the human mind (Konar 2018). The increased interest in AI in breeding is due to the technological advances made during recent years, which led to the development of the ability to analyze large amounts of data in a short period of time to reveal unexpected linkages. From a breeding point of view, AI allows individuals to systematize information that is typically already available, although in a disaggregated form, and transform data into breeding decisions. AI technologies help to solve problems of high-throughput phenotyping and gene functional analysis in current breeding practices, as well as tackle the current massive data processing bottlenecks in both phenotyping and genotyping, and bring new avenues for future breeding to make envirotyping data extensively used in breeding.

AI for Crop Breeding

Artificial intelligence uses computers and technology to simulate the human mind's problem-solving and decision-making skills (Jarrahi 2018). AI, often known as machine intelligence, is an area of computer science that focuses on developing and managing technology, which can learn to make decisions and carry out activities independently without the need for human effort (Dwivedi et al. 2021). AI is a broad term that encompasses a wide range of technologies; it is a catch-all word for any software or hardware component that helps with machine learning, computer vision, natural language comprehension, and natural language processing (NLP) (Li et al. 2021). Traditional complementary metal-oxide-semiconductor (CMOS) hardware and the same fundamental computational processes that drive traditional software are used in today's AI (Kok et al. 2009). AI is the most rapidly emerging technology in computer science in today's digital world, and it creates intelligent computers that replicate the intellect of the human mind (Harfuche et al. 2019). For instance, the deep neural network (DNN), artificial neural network (ANN), random forest (RF), and support vector machine (SVM) are a few examples of machine-learning algorithms, as well as advanced hi-tech equipment such as the internet of things (IoT) (Baduge et al. 2022). AI is a fascinating hi-tech system that provides an endless opportunity as far as its agricultural applications are considered; hence, this opens up new frontiers for digital breeding (Monfesions-Lopez et al. 2018). Future AI generations are projected to inspire new sorts of brain-inspired circuits and architectures capable of making data-driven judgments faster and more precisely than humans can (Xu et al. 2021). Furthermore, artificial intelligence, big data, machine learning, and data analytics are all terms that appear often in current academic and corporate writings that deal with data (Duan et al. 2019).

Big data, machine learning, and AI are some of the terms used to characterize modern computer processes (Cravero and Sepulveda 2021). Big data is concerned with the use of huge data of diverse types and complex structures that cannot be handled well when analyzed through classical approaches (Supriya and Deepa 2020). In this context, the AI trains a computer to perform jobs that are beyond human efforts, especially by considering the time and labour involved, which are typically involved in decision-making in a variety of situations (Berente et al. 2021). Machine learning (ML) is a branch of AI in which computers discover relationships from massive training datasets. For environment and weather applications, a simple definition is: firstly, big data involves the collection of meteorological or Earth

System-related measurements, as well as high spatial and temporal resolution Earth System model (ESM) outputs for analysis; secondly, ML is the refining or discovery of new linkages between locations, times, and quantities in the datasets (e.g., where sea surface temperature features aid the weather prediction for months over land regions); thirdly, AI is a means of providing automatic warnings and guidance to society in the event of oncoming weather extremes, based on the links discovered by machine learning (Huntingford et al. 2019). The current ease of application of ML methods due to improved computing capabilities is aided in part by the unique usage of computer graphics processing units (GPUs), with GPU speed improving at a quicker rate than ordinary central processing units (Raschka et al. 2020). This is an innovative use of computer memory to make calculations both more efficient and considerably closer to the data storage location (Huntingford et al. 2019). The main emphasis of employing AI in breeding is that it complements the work of the breeder by guaranteeing continuous farm monitoring. Indeed, with the automation of farms and the generalization of data, breeders may dedicate more time to higher-value jobs by spending less time in their buildings. AI saves time in data identification and processing, which is of considerable benefit. Breeders and technical advisors acquire confidence and reactivity, allowing them to act when it is most appropriate (Talaviya et al. 2020).

AI technology has been used to accelerate the process of breeding new plant varieties, such as high-throughput genomics and phenomics to advanced breeding (Crossa et al. 2017, Harfouche et al. 2019, Esposito et al. 2020, Reinoso-Pelaez et al. 2020). Increasingly, ML methodology has been used in genomic prediction, genomic selection, and marker-assisted selection (Crossa et al. 2017, Reinoso-Pelaez et al. 2020). Many agricultural companies such as Monsanto and John Deere have already invested hundreds of millions of dollars to develop such technologies that can utilize extensive data on soil type, seed variety, and weather to help farmers reduce costs and enhance yields (Faulkner et al. 2014). Many of the same data sources, such as weather forecasts and Google Maps, are used to fuel both of their businesses. In addition, they may access farm equipment data that are wirelessly sent to the cloud (Stergiou and Psannis 2017). As part of a precision-farming experiment in Romania, companies like Nippon Electric Company, Limited (NEC; headquartered in Minato, Tokyo, Japan) and Dacom (headquarter in Santa Clara, USA) employed environmental sensors and huge data analytics tools to increase yields. The use of current technologies and information systems enhances the overall productivity of agriculture (El Bilali and Allahyari 2018). Due to the agricultural data sets' complexity, novel architecture, and frameworks, algorithms, as well as analytics face several obstacles in extracting the value and hidden information from this data (Sharma and Ramesh 2020). The recent research on AI tools, including ML, deep learning, and predictive analysis intended toward increasing planning, learning, reasoning, thinking, and action-taking abilities (Shaw et al. 2019). Plant Breeders are developing systems to aid in a better understanding of plant behaviour under a variety of climatic situations (Jeong et al. 2016). Summit, the world's most powerful supercomputer, was recently unveiled with the potential to hold 27,000 GPUs, paving the way for a bright future. AI has the potential to be a game-changer for bringing an agricultural revolution and global food security (Streich et al. 2020).

The application of Next-Gen AI in plant breeding requires intelligent and efficient mining of breeding datasets by employing relevant models and definitive (Harfouche et al. 2019). There have been constant efforts to innovate and improve the efficiency of AI to enable high-definition image recognition for analyzing complex data sets and therefore has become a prime target for accelerating the crop improvement process (Godwin et al. 2019, Harfouche et al. 2019). AI, such as neural networks (NN) and deep learning (DL), are currently being exploited to improve the efficiency and accuracy of multi-omics data (Parmley et al. 2015).

The mechanisms by which these two AI functions are often opaque involve multiple nonlinear hierarchical methods to build nodes for easy classification of datasets mimicking brain neurons (Parmley et al. 2015, Niazian and Niedbala 2020). Plant breeders are conceptualizing a Next-Gen AI that will analyze breeding values and provide a comprehensive analysis of complex traits under changing environmental conditions (Niazian and Niedbala 2020). AI is also expected to improved iteratively to improve data mining accuracy and efficiency to predict better the factors underlying disease resistance/agronomic traits, thereby accelerating breeding programs. Extensive hybridization and rigorous selection parameters have significantly altered the phenotypic plasticity of crop plants (Parmley et al. 2015). Phenotypic plasticity of economically important traits is also substantially reduced upon genotypic variation occurring among the genotypes as a direct consequence of their interaction with the environment (Niazian and Niedbala 2020). Current breeding programs aim extensively to improve the abiotic stress tolerance of crop plants by bridging the genotype-phenotype gap that has occurred due to alteration in the phenotype plasticity (Niazian and Niedbala 2020).

Integration of genotypic and environmental data with the observed phenotype, has been used to strengthen the agronomic abiotic stress breeding program to identify the best genotype with critical agronomic traits (van Dijk et al. 2021). As these are complex traits governed by several factors, sophisticated monitoring systems are required to record tiny changes/alterations occurring in the plants (Table 9.1). To resolve such issues, an AI-based physiological gravimetric system that can measure the slightest change occurring in plants concerning soil and atmosphere called the soil-plant-atmosphere continuum (SPAC) is used (Godwin et al. 2019, Harfouche et al. 2019). This system offers ease of phenotyping the slightest variations among the complex traits at different plant growth and development (Harfouche et al. 2019, van Dijk 2021). Constant and rigorous monitoring of these phenotypic data and their subsequent analysis by employing the Next-Gen AI approach can facilitate the identification of stress-responsive QTLs or QTLs related to important agronomic traits (Parmley et al. 2015, Niazian and Niedbala 2020). A field phenomics suite has also been devised to accelerate breeding programs by providing high-resolution images for easy discrimination of better-performing genotypes in large populations (Niazian and Niedbala 2020). The field phenomics suite incorporates a machine-learning approach to capture high-throughput phenotypic data relevant to breeding programs using unmanned aerial vehicles (UAV) and ground-based equipment. This UAV and ground-based equipment are fitted with high-resolution cameras and sensors to generate comprehensive data from thousands of field-grown plants (van Dijk et al. 2021). The data generated are then analyzed by the AI or specific software that enables breeders to identify superior genotypes displaying the best agronomic/disease-resistant traits (Figure 9.1). This advanced phenotypic tool can be combined with MAS and GWAS approaches to dissect plants at the molecular level to identify novel genes/QTLs (Esposito et al. 2020, van Dijk et al. 2021).

In *Glycine max* significant progress has been made in field phenomics implemented to study stress-responsive traits (Harfouche et al. 2019). Although, the barrier still exists to linking phenomic data generated with the help of AI to the genotype, leading to identifying genotype with higher genetic gain. Harnessing complex traits and their subsequent correlation with the environmental variables is of utmost requirement to remove the above barrier is a challenge to the breeders. Future research directed at the generation of next-gen AI is an essential prerequisite to bridging the phenotype-genotype gap to facilitate crop improvement programs. An overview of how AI and speed breeding can lead to improved cultivar development within a short period is depicted in Figure 9.1.

Table 9.1 Successful implementation of artificial intelligence/machine learning models in plant breeding studies.

Crops	Machine learning techniques	Algorithm used	Trait studied	Predictable function	References
Glycine max	Best linear unbiased prediction (BLUP), Neural networks (NNs), Kernel methods	Multilayer perceptron (MLP), support vector machine (SVM), ensemble-stacking (E-S) and random forest (RF), Stochastic gradient descent (SGD)	Pre-harvest, Yield performance	Selection of high yielding cultivars	Yoosefzoideh-Najafabadi et al. 2018, Xavier 2021
Glycine max	Convolutional Neural Networks (CNNs)	Batch Normalization (BN)	Seed per pod estimation	Prediction of seed characters under changing environment	Uzal et al. 2018
Phaseolus vulgaris	Artificial neural networks (ANNs)	Mean square deviation (MSD) and mean square of residue (MSR)	Average yield	High adaptability and phenotypic stability under stress conditions	Coriea et al. 2016
Zea mays, Triticum aestivum	Neural networks (NNs), Deep NNs, CNNs	Generalized matrix factorization (GMF). MLP, SVM	Yield performance, salt stress tolerance	Identification of best performing parental populations, enhanced genomic selection for stress resistance	Ravori et al. 2016, Altunfas et al. 2019, Montesions-Lopez et al. 2019, Khaki et al. 2020, Niedbata et al. 2019
Brassica rapa	Artificial neural networks (ANNs)	MLP	Yield performance	Prediction of seed setting	Niedbata et al. 2019
Abelmoschus esculentus	Deep neural networks (DNNs)	Image processing (IP)	Yield performance under salt stress	Tolerance to salt stress	Feng et al. 2020

Crops	Machine learning techniques	Algorithm used	Trait studied	Predictable function	References
Carum copticum, Trachyspermum ammi, Sprague	Artificial neural networks (ANNs)	Multiple regression analysis	Oil content, physical properties of Callus	Prediction of secondary metabolite production and somatic embryos	Niazian et al 2018a,b
Oryza sativa	Deep CNN	Video detection metrics	Pest and disease resistance	Tolerance to biotic stress	Li et al. 2020
Lycopersicm esculentum	Artificial neural networks (ANNs)	IP, SVM	Callus regeneration and late blight infection	Induction of callus and disease resistance	Niazian et al 2018a
Arabidopsis thaliana	Deep learning	SVM, Naïve Bayes	Stress tolerance	Prediction of miRNA expression for enhance stress tolerance	Vakilian 2020
Daucus carota	Random forest	-	Yield potential	Precision agriculture for yield enhancement	Wei et al. 2020
Solanum tuberosum	Artificial neural networks (ANNs)	IP	Agronomic traits	Identification of superior genotypes	Azizi et al. 2016
Carthamus tinctorius, Sesamum indicum	Artificial neural networks (ANNs)	Multiple regression analysis	Seed yield, oil content	Identification of superior genotypes	Abdipour et al. 2019
Pennisetum glaucum	Deep CNNs	IP, SVM	Disease identification	Identification of disease resistant genotypes	Coulibaly et al. 2019

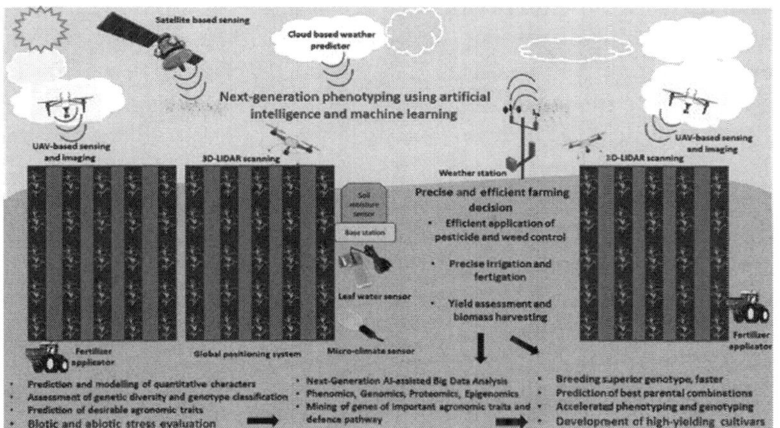

Figure 9.1. An overview of the potential application of artificial intelligence in augmenting plant breeding technology for easy, precise, and early prediction of genotypes/parental combinations for varietal development. (Reproduced from Rai 2022. Mol. Biol. Report. 49: 11385-11402. https://doi.org/10.1007/s11033-022-07769-4)

AI for Overcoming the Bottlenecks of Phenomics

Plant phenomics has advanced rapidly in recent years providing the scope for precision breeding. This progress can be ascribed to an increase in the invention and availability of new technologies that allow for high-throughput phenotyping of complex plant features. The features AI, viz., computer vision, ML, and deep learning have been effectively integrated into non-invasive imaging procedures. Using ML for robust picture analysis, this integration is steadily enhancing the efficiency of data gathering and analysis. Moreover, AI has aided the development of software and tools for data gathering and management in field phenotyping. These include open-source devices and platforms that allow for community-driven research and data sharing, providing the enormous amounts of data needed for reliable phenotyping research (Nabwire et al. 2021). AI is used in three critical components of phenomic data management (Nabwire et al. 2021):

- Algorithms and programs to convert sensory data into phenotypic information,
- Model development to understand genotype-phenotype relationships with environmental interactions, and
- Database management to allow information and resources to be shared

Experiments involving repeated trials in diverse conditions (considering the statistical need for an unbiased estimation) are required in order to screen plants for desirable features (such as grain size, abiotic stress tolerance, product quality, or yield potential). The measuring of individual plants in controlled conditions has been the subject of considerable phenotyping discussion, since plant developed under open-air circumstances is not accurately represented in controlled environments (Yang et al. 2020). Accordingly, a large gap has been seen practically regarding the performance of plants from lab to field (Bhat et al. 2021). The ongoing integration of AI into various technologies promises a development toward smarter, faster, and lower-cost solutions. Compared to other imaging techniques, the integration of AI into the data management pipeline of tomography and thermography is rather on a smaller scale in the area of phenotyping image data analysis. Deep learning has been successfully used in the analysis of composite materials (Bang et al. 2020). Even though field phenotyping is a practical need in crop breeding, application of the high-throughput phenotyping under field conditions lags behind the indoor phenotypic facilities currently available. Thus, to increase their accuracy, AI technologies require a significant amount of data from numerous sources. Accordingly, it is important to the customization of current technologies for field data collection, and the use of already available AI adaptable technology, such as smartphones, to boost the number and quality of data collection (Kumar et al. 2022). Smartphones have become common consumer items and the ease with which their sensors may be used suggests their great application in agriculture (Nabwire et al. 2021). The main purpose of using these approaches and technology is to offer the infrastructure for tracking how plants progress during the growing season and to make the data analysis, management, and use of results via AI methods easier (Song et al. 2021).

Recent reports have shown that the phenotyping of crops with the help of AI, can generate improvement in crop phenotyping and predictions (Sadeghi-Tehran et al. 2017, Brichet et al. 2017, Abdalla et al. 2019, Mahlein et al. 2019, Mochida et al. 2019, Selvaraj et al. 2020, Jung et al. 2021). For example, Selvaraj et al. (2020) reported that ML algorithms, viz., k-Nearest Neighbours (kNN), RF, and SVM revealed the best performance for root yield prediction in the cassava (*Manihot esculenta* Crantz), with the highest accuracy of R2 = 0.67, 0.66, and 0.64, respectively. AI-assisted high-throughput phenotyping systems have been successfully applied to wheat and maize to identify the plant growth stage (Sadeghi-Tehran et al. 2017), and plant image segmentation (Brichet et al. 2017); to oilseed crops for semantic segmentation of the crops and weeds (Abdalla et al. 2019); to the phenotyping of disease resistance of crops (Mahlein et al. 2019); and for improvement of plant productivity (Mochida et al. 2019, Jung et al. 2021).

Ai for Gene Function Analysis

The rapid development of high-throughput technologies in biological sciences has resulted in the generation of massive data in recent decades. Disciplines that attempt to collect and analyze enormous volumes of biological data are often referred to as "omics", which is used to indicate the total quantity of DNA contained in each cell of an organism (Jung et al. 2021). "Omics" data has become too large and complicated to be analyzed visually or by using statistical correlations. This has generated the application AI, which manages large amounts of data that are insurmountable for human minds. The application of AI has helped to extract information that goes beyond our current understanding of the system under investigation and, analyze it automatically to draw logical conclusions (Grapov et al. 2018).

Several ML tools and algorithms are available for different kinds of bioinformatics analysis, such as protein-coding gene identification, cis-regulatory element identification, gene expression, subcellular location, protein-protein interaction, gene ontology, metabolic pathways, phenotypes, and genomic prediction (Mahood et al. (2020). AI algorithms might potentially be used to address comparative genomic investigations or information transfer from a model plant to a crop of interest (Caudai et al. 2021). Two models namely, DeepBind (Alipanahi et al. 2018) and DeepSEA (Zhou and Troyanskaya 2015) have been created to predict and analyze genetic features (Esposito et al. 2020). Various expressions or sequencing data analysis can be generated with the goal of predicting gene functions or the differential effects of gene expression on a trait (Dassari et al. 2021).

The generation of huge amounts of genomic data, as a result of the fruitful breakthroughs of high-throughput sequencing technology, has created a huge problem for the storage and examination of the data (Esposito et al. 2020). The AI technology of bioinformatics enables the measurement of simultaneous expressions of many genes, or each and every gene that is included in the genome under a wide range of situations (Huang et al. 2009, Lemmon and Lemmon 2013). This has helped the breeders to combine the more relevant representation of their data and the ability to integrate them, which enables them to examine their genomic data, test and confirm their assumptions throughout the experimental cycle, and ultimately obtain useful results (Hu et al. 2018, Xia et al. 2019).

AI for Linking Crop Genome to Phenome

Modern breeding approaches are focused on linking the genotype with the crop phenotype accurately and precisely. However, linking the whole of the genome information to high-throughput phenotypes remains a massive challenge, and is impeding the optimal application of field phenotyping and omics (Harfouche et al. 2019). Germplasm collection and mapping populations can efficiently differentiate the phenomics and genomics data through AI. Crop diversity, single nucleotide polymorphisms (SNPs) detection and selection, quantitative trait loci (QTL) analysis, genome-wide association study (GWAS) analysis, and genomic selection and sequences generate a large amount of data. AI can evaluate and link the phenomics and genomics data from these big data to improve breeding approaches. AI-related to a computation and training model can predict the gene functional analysis and high-throughput crop phenotyping and also predict the performance of yield and traits of the crop (Batley and Edwards 2018, Mochida et al. 2019, Wang et al. 2020, Jung et al. 2021, Caudia et al. 2021). Hence, the integration of AI with phenomics and genomics tools can allow for rapid gene identification associated with the crop phenotypes that eventually accelerate crop improvement programs. How the application of AI technology can be linked to high-throughput genomics and phenomics to derive results in the production of better breeding strategies is summarized in Figure 9.2.

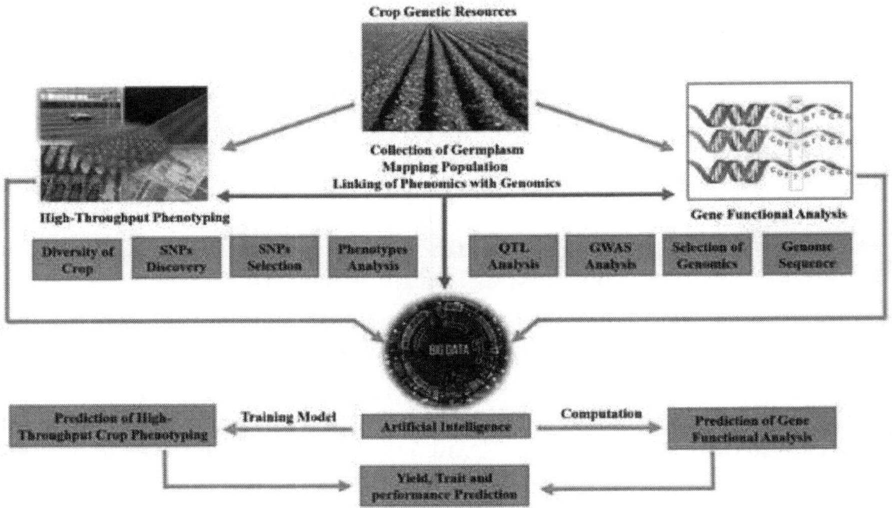

Figure 9.2. Artificial Intelligence used as a powerful tool for the prediction of high-throughput crop phenotyping and gene functional analysis in modern crop breeding. The high-throughput phenotypic and genotypic data were collected from large crop germplasm and breeding populations. The massive comprehensive database could integrate various resources with AI technology, such as phenotypic diversity of crops, SNPs polymorphisms, QTL analysis, GWAS analysis, genomics selection, and genome sequence. AI technologies are applied to predict the crop phenotype with whole genome prediction, the novel breeding strategies are produced through AI related to computation and training models. (Reproduced from Khan et al. Int. J. Mol. Sci. 2022. 23: 11156. https://doi.org/10.3390/ijms231911156).

Research on crop genomics is not only understanding the molecular mechanisms of phenotypes but also using technical data and bioinformatics techniques to analyse and understand the molecular mechanisms behind phenotypes (Altman et al. 2021). AI is a fascinating approach to bringing out these tasks inevitably (Williamson et al. 2022). AI approaches provide the platform to analyze huge, various, and useless datasets such as the generation of genome sequencing/photo imaging over conventional analytical strategies (Libbrecht and Nobel 2015, Harfouche et al. 2019). The AI approach has been explicitly employed in varied research fields of phenomics and genomics, such as: analysing genome assembly and genome-specific algorithms (Esposito et al. 2020), broad-range data analysis to mitigate multiplex biological complications in metabolomics, proteomics, genomics, transcriptomics, as well as systematic biology (Libbrecht and Nobel 2015, Xu and Jackson 2019), interpretation of gene expression cascades (Ni et al. 2016, Kwon et al. 2020), identification of significant SNPs in polyploid plants (Korani et al. 2019). high-throughput crop stress phenotyping (Zhao et al. 2016, Selvaraj et al. 2020).

AI has been employed for the development of models to modulate the flow of information from generic DNA to genetic-based phenotypes, to investigate the potential variants in natural populations (Mahood et al. 2020). More specifically, for breeders, AI will assist in further investigation of genetic loci to facilitate the agricultural output by triggering the genome algorithms and allowing high-throughput crop phenotyping in quantitative traits for open-field and controlled environments (Mahood et al. 2020, Razzaq et al. 2021). AI can be cohesively combined with bioinformatics and genome sequencing analysis to interpret various molecular repertories such as transcription factor binding sites (Long et al. 2020), long non-coding RNAs (lncRNAs) (Sun et al. 2015), microRNA (miRNAs), epistatic modifications, coding genes, targeted polyadenylation sites (Gao et al. 2018), as well as *cis*-regulatory elements (CREs) (Earnest and Kellis 2017, Mahood et al. 2020).

Various crop databases insert a huge amount of heterogeneous-related phenotypic and genotypic data (big data), for providing insight into potential resources to untangle novel trait-identified candidate genes (Tong and Nikoloski 2021). Further, AI provides a novel benchmark summary for analytical and computational methods for the integrated analysis of such enormous datasets based on the big-data spectrum (Mahood et al. 2020, Tong and Nikoloski 2021). Employing AI to conclude the interrelations between candidate genes and CREs is a novel approach for categorizing and identifying previously unknown genes for significant crop improvements (Li et al. 2018). It has been reported that AI strategies have more potential for interpretation of the crop yield, variation in climatic assessment, high-throughput crop stress phenotyping, climate temperature, ultraviolent (UV) radiation, wind, and hail (Crane-Droesch 2018, Esposito et al. 2020, Tong and Nikoloski 2021). The role of AI is becoming more and more important in obtaining, analyzing, integrating, and managing genomic and phenomic data to increase agricultural climate resilience (Marsh et al. 2021, Shen et al. 2022).

Next-generation sequencing (NGS)-based genotyping methods have helped to improve gene-mapping resolution, gene identification, and NGS-based genotyping for GWAS analysis, used in crop improvement (Razzaq et al. 2021). For example, in soybean, such studies have been widely used to identify genetic loci and candidate genes for seed weight (Jing et al. 2018), seed protein and oil contents (Li et al. 2019), pod dehiscence (Hu et al. 2019), nitrogen fixation (Torkamanch et al. 2020), soybean plant height and primary branches (Torkamanch et al. 2020), agronomic traits (Sonah et al. 2015), disease resistance (Zhao et al. 2015), and tocopherol concentration (Sui et al. 2020). Bulk segregant analysis (BSA) and its modified methodologies are being used in many crops for their improvement (Giovannoni et al. 1991, Michelmore et al. 1991, Yang et al 2013, Takagi et al. 2013, Majeed et al. 2021, Zhon et al. 2021). The NGS-based BSA has been used to identify candidate genes for various traits, such as the soybean mosaic virus (Yang et al. 2020), charcoal rot resistance (de Silva et al. 2020), flowering time (Watanabe et al. 2017), phytophthora resistance (Cheng et al. 2017), and powdery-mildew resistance (Jiang et al. 2019). The deep-learning algorithm for BSA (DeepBSA) has been developed for QTL mapping and functional gene cloning in maize (Li et al. 2021).

AI for Accessing Envirotyping Data for Crop Breeding

Climate change has a great impact on the environment and crop production for the present and future. The concept of envirotyping is proposed as a third "typing" technology, accompanying phenotyping and genotyping to decode environmental influences on crop breeding (Xu et al. 2016). Envirotyping plays a key role in crop modelling and the prediction of phenotypes through its efficient components, including the genotype-by-environment interaction (GEI), environmental signals, responsive genes, biotic and abiotic stresses, as well as integrative phenotyping (Marsh et al. 2021). Cortes et al. (2022) discuss the state of the emerging field of study known as "genome-environment associations", which combines ecological climatic data with evolutionary genomics (GEAs). They advocate for the community to begin collecting genomic estimated adaptive values (GEAVs) for genomic prediction (GP) and multi-dimensional ML models in order to take polygenic evolutionary adaptation into consideration. Xu et al. (2022) proposed an integrated genomicenviromics prediction breeding scheme using integrated multi-omics information, big data technology, and artificial intelligence.

Strategies need to be developed to maximize the limited resources and their utilization for crop breeding and land management due to climate change (Resende et al. 2021). Climate-smart crops and climate-smart soils have been adapted to the environment for more effective breeding programs, and breeders may use this knowledge to generate new smart crops for the new climate. Genomic technologies together with high-throughput phenotyping are providing researchers and farmers with the information required to guide and notify the breeding methods for climate-smart breeding (Marsh et al. 2021). AI plays a vital role in integrating and manipulating this fast-collecting abundance of data by conducting association studies to identify genomic targets which are related to adaptive climate-resilient traits (Shafiekhani et al. 2017). Cortes and Lopez-Hernandez (2021) put forward a roadmap to use ML, GP, and multi-trait gene editing approaches to capture novel abiotic stress tolerance variations from wild crop relatives to utilize these variants for bread drought-tolerant crops. Genomic and phenomic data will need to be integrated into comprehensive clade-specific databases and platforms, as well as accessible tools that breeders may utilize to inform the selection of climate-adapted characteristics, to effectively translate research into the field.

Khan et al (2022) proposed a breeding scheme (Figure 9.3) that integrates genotypic, phenotypic, and envirotypic information to improve efficacy. The phenotypic data of crop plants both for indoor and outdoor environments are collected by high-throughput robotic systems (Shafiekhani et al. 2017) and the phenotypic information from various environments will transfer to a high-throughput phenotyping (HTP) server via the Internet. The multiple datasets will take the genotypic, phenotypic, and envirotypic information together, and the G × E interaction (GEI) will quantify by multiple environments. AI technology, particularly ML and deep learning (DL), shall be used in cultivar selection for specific or multiple major environments. This approach is expected to help in making decisions about the selected cultivars, and whether it is suitable for cultivation in limited environments or all major environments (Khan et al. 2022).

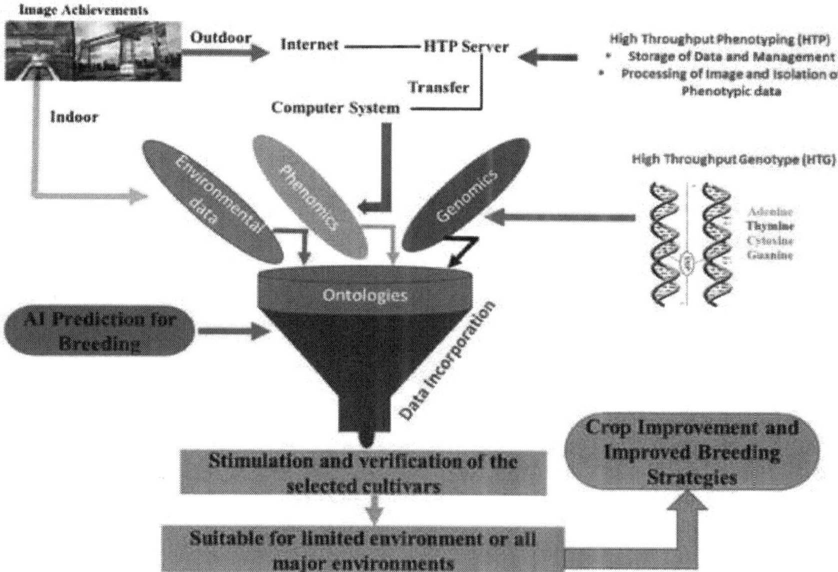

Figure 9.3. Integration and management of Genomics, Phenomics, and Enviromics data by artificial intelligence for crop-breeding improvement. The phenotypic data of crops are collected from both indoor and outdoor environments, the information of phenotypic, genotypic and environmental are combined together with AI technology. With mathematical modelling, logical deduction, and decision-making, the AI-assisted breeding system will simulate and verify the selected cultivars, whether it is suitable for cultivation in limited environments or all major environments. (Reproduced from Khan et al. Int. J. Mol. Sci. 2022. 23: 11156. https://doi.org/10.3390/ijms231911156).

AI for Studying Biochemical Phenotypes

Advances in the technological have made recording genotypic and phenotypic variation in plants more sophisticated and precise, leading to easy extraction of valuable information within complex datasets (Esposito et al. 2020, Jung et al. 2021, van Dijk et al. 2021). Accordingly, AI is now being utilized to analyze complex biochemical pathway data sets to help decipher the real-time changes occurring at the molecular level under abiotic stress conditions (Esposito et al. 2020). Several biochemical/metabolic changes are governed by discrete changes occurring at the genomics (gene expression), proteomics (protein distribution), metabolomics (expression of metabolites), and epigenomics (DNA/histone modification) level. Technological advancements have now made it possible to measure critical metabolic traits at the OMICS level (Parmley et al. 2015, Niazian and Niedbala 2020). The data generated by these instruments, such as Next Generation Sequencing (NGS), Chromatin Immunoprecipitation (ChIP), Matrix-assisted laser desorption and ionization-Time of Flight-Mass Spectrophotometry (MALDI-TOF-MS), etc. are so vast and complex that it would take a lot of time and effort to decipher and conclude the final results (van Dijk et al. 2021). Therefore, AI have now been utilize to analyze large/complex datasets due to a lack of technological knowledge and understanding to analyze complex data sets for accurate interpretation of given biochemical traits (van Dijk et al 2021).

Several reports have shown the potential application of AI to interpret biochemical data to enhance the understanding of plant stress biology. For example, the application of AI successfully predicted genomic crossovers occurring in the maternal and parental maize plants and helped predict probable genomic regions displaying high mutation rates (Demirci et al. 2021). Similarly, AI have been used to classify and characterize genomic regions by analyzing the DNA methylation pattern of maize plants grown under stress conditions into functional genes and pseudogenes (Sartor et al. 2019). Uygun et al. (2019) reported employment AI to predict gene promoters and cis-regulatory elements in Arabidopsis and maize plants by studying the expression pattern of essential genes. Several studies have also affirmed the usefulness of AI in learning plant metabolic regulatory networks by deciphering tissue-specific changes occurring in the biosynthetic genes, such as nitrogen use efficiency, starch biosynthesis, and other secondary metabolites in *Arabidopsis* and rice (Varala et al. 2018, Li et al. 2020). Meena et al. (2021) reported use of AI to stimulate the production of biofuels by enhancing biomass production using plant species and algal bloom, thereby emphasizing the vital contribution of AI for managing bioenergy production.

Application of AI in studying single-cell RNA sequencing, DNA methylation, and post-translational modifications, which can provide testable insight into a specific region of the genome or candidate gene governing secondary metabolite production under stress conditions have been reported (Sartor et al. 2019). Testing AI to predict complex genomic traits such as photosynthesis, hormonal changes, and yield have also been initiated (Rajasundaram and Selbig 2016). The application of AI in breeding programs could also help identify QTLs by analyzing a complex region of the genome associated with a specific trait via facilitating MAS and GWAS (Gupta et al. 2010). AI in breeding could provide an in-depth understanding of genetic architecture, revealing the position/localization of essential genes governing economically crucial traits. The exploitation of AI could also integrate genomics, transcriptomics, proteomics, and metabolomics data to analyze macroscopic biochemical features controlling plant growth and development in response to environmental stimuli.

Ai For Integrating Phenomics With Genomics For Smart Breeding

One of the significant limitations of the classical breeding approach is its inability to provide substantial insight into the genomic architecture of plant species due to the lack of correlation between genotype and phenotype (van Dijk et al. 2021). Next-Gen AI can significantly facilitate understanding of genomic architecture by linking phenome with genome by generating genomic selection models for the particular crop species. These genomic selection models are prediction-based models developed by estimating marker-trait association via genetic markers followed by extensive phenotyping of test populations (Niazian and Niedbala 2020). These models often reflect important genomic regions or loci present in each haplotype and regulate the trait of interest. Most current breeding programs focused on developing climate-smart crops exploit genomic selection models to analyze genetic components such as SNPs and InDels associated with specific characteristics (van Dijk et al. 2021). The models developed by analyzing SNP and InDels are exploited as potential tools for a breeding population that helps plant breeders predict their phenotype before reaching maturity (Parmley et al. 2015).

The data generated by these models are then analyzed by AI to predict heritable components accurately, decreasing the breeding cycle and increasing plant yield (Nabwrie et al. 2021). Most genomic selection models are based on the correlation between genotype and phenotype data generated by genetic/linkage disequilibrium mapping development, which is challenging and prone to error (Shen et al. 2020, Nabwire et al. 2021). Neural networks amalgamating various AI-based algorithms have been developed to improve the accuracy of data interpretation of genomic selection models (Parmley et al. 2015, Nizian and Niedbala 2020). These neural networks are computational models displaying output as neuron-like nodes linking and analyzing information by communicating with refined production (van Dijk et al. 2021).

The neural networks have emerged as a successor for genomic selection models. However, it has failed to analyze complex data sets and, unable to improve the accuracy, for most plant species (Uzal et al. 2018). Thus, it has been realized that fully automated AI is insufficient to analyze big data, and human inputs are critical to improving AI-based models' accuracy (Uzal et al. 2018, Shen et al. 2020). The human inputs are essential because, from the breeder's perspective, they can manipulate the complex sets of OMICS data as per the goal of a breeding program in a much better way than AI alone (Uzal et al. 2018, Montesions-Lopez et al. 2019). Accordingly, the breeders are developing and testing various AI-based algorithms capable of analyzing a large variety of data that can demarcate specific features as per the need of the experimental programmes. Several transfer learning approaches circumventing published data into machine learning format have shown promising results (Montesions-Lopez et al. 2019).

A new deep transfer learning approach called ARIGAN (*Arabidopsis rosette* image generator AN) has been successfully used to generate synthetic rose-shaped plants by integrating in-silico data with field-based data using generative adversarial networks (Valerio Giuffrida et al. 2017). ARIGAN was also used to analyze complex multi-omics data, which successfully rendered extensive gene expression data to provide a glimpse of transcriptional regulation in a predictive model (Valerio Giuffrida et al. 2017, Ubbens et al. 2018). However, there is an urgent need for developing the black-box nature of AI models and their potential application in plant breeding to develop climate-smart crops. It has become essential to create a more sophisticated Next-Gen AI-based system capable of screening a sizeable multi-omics data set that will open a new realm for plant breeders, to achieve desired goals in their breeding programmes.

Future Prospects

Many sectors including agriculture breeding, are benefiting from greater profitability and have high economic growth rates as a result of the introduction and successful deployment of AI technologies. AI will focus on developing novel, human-centered techniques, and assessing the application of robotic technology to a variety of industries and businesses. AI will also transform the way different companies expand and compete by representing new manufacturing concepts that will result in breeding profitability. To take full advantage of such prospects, most firms will need to be more involved in the creation of various AI methods. It will also concentrate on developing a variety of responsible AI machines with moral and ethical ideals, which will lead to positive outcomes as well as to enable individuals to perform tasks they are familiar with. The development of various AI systems will assist the global agricultural breeding sector in assuming the availability of symbolic structures, such as reasoning ability and knowledge existence.

In the near future, AI is likely to be used to address a variety of plant science genomic concerns. It is possible that one area of plant genomics that may be addressed is how to deal with several species (such as wheat, soybean, rice, maize, tomato, and oilseed rape) at the same time, as intercropping is becoming important for covering crops of mixed species. AI algorithms might potentially be used to address comparative genomics investigations or information transfer from a model plant to a crop of interest.

It is expected that the farmers and breeders will be able to feed the data into cloud-based AI applications via portable devices, drones, and agriculture-equipment platforms, thereby making AI applications more widely accessible. The phenomics and genomics data obtained with ML and DL are accurate, but not good enough to totally rely on the technology to speed up breeding, as they are time-consuming, and costly processes.

AI will efficiently revolutionize the "omics" approaches and breeding management with high-technology methods. AI will develop many genotypes and phenotypes that must be screened using model-based envirotyping for a wide range of adaptive genotypes and phenotypes. The segregating material has to be developed and advanced using AI-based management systems to shorten the breeding cycle and improve genetic gain.

Conclusions

Plant breeders have stumbled to develop and breed high-yielding cultivars that can withstand abiotic and biotic stresses. AI has emerged as a potential alternative for reducing time, space, and cost to develop, release, and commercialize new/ improved cultivars with improved accuracy and predictability. Plant growth and developmental conditions are the critical factors that govern plant performance under changing environmental conditions. Application of Next-Gen AI has opened a new realm for speed breeding and agriculture that will enable decision-making and handling of big OMICS data with great precision, which will help get novel insight into plant functions under climate extremes. Implementation of AI-based technologies requires extensive planning and a continuous supply of physical resources. Therefore, it has become necessary to develop public and private ventures to facilitate capacity building, technology transfer, and finance, coupled with AI-driven research to facilitate crop improvement programs. These public-private partnerships will also help create a framework for successfully implementing AI-augmented plant breeding research and innovation for the betterment of humans, animals, and the environment.

References

Abdalla, A., Cen, H., Wan, L., ... Rashid, R. (2019). Fine-tuning convolutional neural network with transfer learning for semantic segmentation of ground-level oilseed rape images in a field with high weed pressure. Comput. Electron. Agric. 167: 105091.

Abdipour, M., Younessi-Hmazekhanlu, M., & Ramazani, S.H.R. (2019). Artificial neural networks and multiple linear regression as potential methods for modeling seed yield of safflower (Carthamus tinctorius L.). Ind. Crops Prod. 127:185–194.

Alipanahi, B., Delong, A., Weirauch, M.T., Frey, & B.J. (2015). Predicting the sequence specificities of DNA-and RNA-binding proteins by deep learning. Nat. Biotechnol. 33: 831–838.

Allard, R. (2014). Plant breeding: encyclopedia Britannica. Chicago, IL, USA.

Altman, A., Fan, L., Foyer, C., Cowling, W., ... Mittler, R. (2021). Past and future milestones of plant breeding. Trends Plant Sci. 26: 530–538.

Altuntaş, Y., Cömert, Z., & Kocamaz, A.F. (2019). Identification of haploid and diploid maize seeds using convolutional neural networks and a transfer learning approach. Comput. Electron. Agric. 163:104874. https://doi.org/10.1016/j.compag.2019.104874

Azizi, A., Abbaspour-Gilandeh, Y., Nooshyar, M., & Afkari-Sayah, A. (2016). Identifying potato varieties using machine vision and artificial neural networks. Int. J. Food Prop. 19: 618–635. https://doi. org/10.1080/10942912.2015.1038834

Baduge, S.K., Thilakarathna, S., Perera, J.S., ... Arashpour, M. (2022). Artificial intelligence and smart vision for building and construction 4.0: Machine and deep learning methods and applications. Autom. Constr. 141: 104440.

Bang, H.T., Park, S., & Jeon, H. (2020). Defect identification in composite materials via thermography and deep learning techniques. Compos. Struct. 246: 112405.

Batley, J., & Edwards, D. (2016). The application of genomics and bioinformatics to accelerate crop improvement in a changing climate. Curr. Opin. Plant Biol. 30: 78–81.

Berente, N., Gu, B., Recker, J., & Santhanam, R. (2021). Managing artificial intelligence. MIS Q. 45: 1433–1450.

Bhat, J.A., Yu, D., Bohra, A., Ganie, S.A., & Varshney, R.K. (2021). Features and applications of haplotypes in crop breeding. Commun. Biol. 4: 121266.

Brichet, N., Fournier, C., Turc, O., Strauss, O., ... Artzet, S. (2017). A robot-assisted imaging pipeline for tracking the growths of maize ear and silks in a high-throughput phenotyping platform. Plant Methods. 13: 96.

Caudai, C., Galizia, A., Geraci, F., ... Le Pera, L. (2021). AI applications in functional genomics. Comput. Struct. Biotechnol. J. 19: 5762–5790.

Cheng, Y., Ma, Q., Ren, H., Xia, Q., ... Song, E. (2017). Fine mapping of a Phytophthora-resistance gene RpsWY in soybean (Glycine max L.) by high-throughput genome-wide sequencing. Theor. Appl. Genet.130: 1041–1051.

Corrêa, A.M., Teodoro, P.E., Gonçalves, M.C., ... Barroso, L.M.A. (2016). Artificial intelligence in the selection of common bean genotypes with high phenotypic stability. https://locus.ufv.br//handle/123456789/12869

Cortés, A.J., & López-Hernández, F. (2021). Harnessing crop wild diversity for climate change adaptation. Genes. 12: 783.

Cortés, A.J., López-Hernández, F., & Blair, M.W. (2022). Genome–environment associations, an innovative tool for studying heritable evolutionary adaptation in orphan crops and wild relatives. Front. Genet. 13: 910386.

Coulibaly, S., Kamsu-Foguem, B., Kamissoko, D., & Traore, D. (2019). Deep neural networks with transfer learning in millet crop images. Comput. Ind. 108:115–120. https://doi.org/10.1016/j. compind.2019.02.003

Crane-Droesch, A. (2018). Machine learning methods for crop yield prediction and climate change impact assessment in agriculture. Environ. Res. Lett. 13: 114003.

Cravero, A., & Sepúlveda, S. (2021). Use and adaptations of machine learning in big data: applications in real cases in agriculture. Electronics. 10: 552.

Crossa, J., Pérez-Rodríguez, P., Cuevas, J., ... Montesinos-López, O. (2017). Genomic selection in plant breeding: Methods, models, and perspectives. Trends Plant Sci. 22: 961–975.

da Silva, M.P., Zaccaron, A.Z., Bluhm, B.H., & Rupe, J.C. (2020). Bulked segregant analysis using next-generation sequencing for identification of genetic loci for charcoal rot resistance in soybean. Physiol. Mol. Plant Pathol. 109: 101440.

Dasari, C.M., Amilpur, S., & Bhukya, R. (2021). Exploring variable-length features (motifs) for predicting binding sites through interpretable deep neural networks. Eng. Appl. Artif. Intell. 106: 104485.

Demirci, M., Gozde, H., & Taplamacioglu, M.C. (2021). Comparative dissolved gas analysis with machine learning and traditional methods. 3rd International Congress on Human-Computer Interaction, Optimization and Robotic Applications (HORA) IEEE. P. 1–6. https://doi.org/10.1109/HORA52670.2021.9461371

Duan, Y., Edwards, J.S., & Dwivedi, Y.K. (2019). Artificial intelligence for decision making in the era of Big Data–evolution, challenges and research agenda. Int. J. Inf. Manag. 48: 63–71.

Dwivedi, Y.K., Hughes, L., Ismagilova, E., ... Aarts, G. (2021). Artificial Intelligence (AI): multidisciplinary perspectives on emerging challenges, opportunities, and agenda for research, practice and policy. Int. J. Inf. Manag. 57: 101994.

El Bilali, H., & Allahyari, M.S. (2018). Transition towards sustainability in agriculture and food systems: role of information and communication technologies. Int. Process. Agric. 5: 456–464.

Ernst, J., & Kellis, M. (2017). Chromatin-state discovery and genome annotation with ChromHMM. Nat. Protoc. 12: 2478–2492.

Esposito, S., Carputo D., Cardi T., & Tripodi P. (2020). Applications and trends of machine learning in genomics and phenomics for next-generation breeding. Plants 9: 34. https://doi.org/10.3390/ plants9010034

Esposito, S., Carputo, D., Cardi, T., & Tripodi, P. (2020). Applications and trends of machine learning in genomics and phenomics for next-generation breeding. Plants. 9: 34.

Faulkner, A., Cebul, K., & McHenry, G. (2014). Agriculture gets smart: the rise of data and robotics. Cleantech Agriculture Report; Cleantech Group: San Francisco, CA, USA.

Feng, X., Zhan, Y., Wang, Q., Yang, X., Yu, C., ... Wang, H. (2020). Hyperspectral imaging combined with machine learning as a tool to obtain high-throughput plant salt-stress phenotyping. Plant J. 101: 1448–1461. https://doi. org/10.1111/tpj.14597

Gao, X., Zhang, J., Wei, Z., & Hakonarson, H. (2018). DeepPolyA: A convolutional neural network approach for polyadenylation site prediction. IEEE Access. 6: 24340–24349.

Giovannoni, J.J., Wing, R.A., Ganal, M.W., & Tanksley, S.D. (1991). Isolation of molecular markers from specific chromosomal intervals using DNA pools from existing mapping populations. Nucleic Acids Res. 19: 6553–6558.

Godwin, I.D., Rutkoski, J., Varshney, R.K., & Hickey, L.T. (2019). Technological perspectives for plant breeding. Theor. Appl. Genet. 132: 555–557. https://doi.org/10.1007/s00122-019-03321-4

Grapov, D., Fahrmann, J., Wanichthanarak, K., & Khoomrung, S. (2018). Rise of deep learning for genomic, proteomic, and metabolomic data integration in precision medicine. OMICS. 22: 630–636.

Gupta, P.K., Kumar, J., Mir, R.R, & Kumar, A. (2010). Marker-assisted selection as a component of conventional plant breeding. Plant Breed Rev. 33: 145. https://doi.org/10. 1002/9780470535486.ch4

Harfouche, A.L,, Jacobson, D.A., Kainer, D., ... Romero, J.C. (2019). Accelerating climate resilient plant breeding by applying next-generation artificial intelligence. Trends Biotechnol. 37: 1217–1235. https://doi.org/10.1016/j.tibtech.2019.05.007

Hu, D., Kan, G., Hu, W., Li, Y., Hao, D., ... Li, X. (2019). Identification of loci and candidate genes responsible for pod dehiscence in soybean via genome-wide association analysis across multiple environments. Front. Plant Sci. 10: 811.

Hu, H., Scheben, A., & Edwards, D. (2018). Advances in integrating genomics and bioinformatics in the plant breeding pipeline. Agriculture. 8: 75.

Huang, D.W., Sherman, B.T., & Lempicki, R.A. (2009). Bioinformatics enrichment tools: Paths toward the comprehensive functional analysis of large gene lists. Nucleic Acids Res. 37: 1–13.

Huntingford, C., Jeffers, E.S., ... Bonsall, M.B. (2019). Machine learning and artificial intelligence to aid climate change research and preparedness. Environ. Res. Lett. 14: 124007.

Jarrahi, M.H. (2018). Artificial intelligence and the future of work: Human-AI symbiosis in organizational decision making. Bus. Horiz. 61: 577–586.

Jeong, J.H., Resop, J.P., Mueller, N.D., Fleisher, D.H., ... Yun, K. (2016). Random forests for global and regional crop yield predictions. PLoS ONE. 11: e0156571.

Jiang, B., Li, M., Cheng, Y., Cai, Z., ... Ma, Q. (2019). Genetic mapping of powdery mildew resistance genes in soybean by high-throughput genome-wide sequencing. Theor. Appl. Genet. 132: 1833–1845.

Jing, Y., Zhao, X., Wang, J., ... Teng, W. (2018). Identification of the genomic region underlying seed weight per plant in soybean (Glycine max L. Merr.) via high-throughput single-nucleotide polymorphisms and a genome-wide association study. Front. Plant Sci. 9: 1392.

Jung, J., Maeda, M., Chang, A., & Bhandari, M. (2021). The potential of remote sensing and artificial intelligence as tools to improve the resilience of agriculture production systems. Curr. Opin. Biotechnol. 70: 15–22.

Khaki, S, Khalilzadeh, Z., & Wang, L. (2020). Predicting yield performance of parents in plant breeding: A neural collaborative filtering approach. PLoS ONE. 15: e0233382. https://doi.org/10.1371/

Khan, M.H.U., Wang, S., Wang, J., Ahmar, S., ... Saeed, S. (2022). Applications of artificial intelligence in climate-resilient smart-crop breeding. Int. J. Mol. Sci. 23: 11156. https://doi.org/10.3390/ijms231911156

Kok, J.N., Boers, E.J., Kosters, W.A., Van der Putten, P., & Poel, M. (2009). Artificial intelligence: definition, trends, techniques, and cases. Artif. Intell. 1: 270–299.

Konar, A. (2018). Artificial intelligence and soft computing: behavioral and cognitive modeling of the human brain. CRC Press: Boca Raton, FL, USA.

Korani, W., Clevenger, J.P., Chu, Y., & Ozias-Akins, P. (2019). Machine learning as an effective method for identifying true single nucleotide polymorphisms in polyploid plants. Plant Genome. 12: 180023.

Kumar, R.S., Keerthana, G., Murali, L., ... Chidambaranathan, S. (2022). Enabling artificial intelligence and cyber security in smart manufacturing. In: Artificial intelligent techniques for wireless communication and networking. John Wiley & Sons, Ltd.: Hoboken, NJ, USA. p. 269–286.

Kwon, M.S., Lee, B.T., Lee, S.Y., & Kim, H.U. (2020). Modeling regulatory networks using machine learning for systems metabolic engineering. Curr. Opin. Biotechnol. 65: 163–170.

Lemmon, E.M., & Lemmon, A.R. (2013). High-throughput genomic data in systematics and phylogenetics. Annu. Rev. Ecol. Evol. Syst. 44: 99–121.

Li, D., Wang, R., Xie, C., Liu, L., Zhang, J., Li, R., Wang, F, Zhou, M., & Liu W. (2020). A recognition method for rice plant diseases and pests video detection based on deep convolutional neural network. Sensors. 20: 578. https://doi.org/10.3390/s20030578

Li, D., Zhao, X., Han, Y., Li, W., & Xie, F. (2019). Genome-wide association mapping for seed protein and oil contents using a large panel of soybean accessions. Genomics. 111: 90–95.

Li, Y., Shi, W., & Wasserman, W.W. (2018). Genome-wide prediction of cis-regulatory regions using supervised deep learning methods. BMC Bioinform. 19: 202.

Li, Y., Zhou, X., & Li, G. (2021). Bridging natural language and graphical user Interfaces. In: Artificial intelligence for human computer interaction: a modern approach. Springer: Berlin/Heidelberg, Germany, p. 463–493.

Li, Z., Chen, X., Shi, S., Zhang, H., ... Wang, X. (2022). DeepBSA: A deep-learning algorithm improves bulked segregant analysis for dissecting complex traits. Molecular Plant. 15: 1418–1427.

Libbrecht, M.W., & Noble, W.S. 2015. Machine learning applications in genetics and genomics. Nat. Rev. Genet. 16: 321–332.

Long, P., Zhang, L., Huang, B., Chen, Q., & Liu, H. (2020). Integrating genome sequence and structural data for statistical learning to predict transcription factor binding sites. Nucleic Acids Res. 48: 12604–12617.

Mahlein, A.K., Kuska, M.T., Thomas, S., ... Wahabzada, M. (2019). Quantitative and qualitative phenotyping of disease resistance of crops by hyperspectral sensors: Seamless interlocking of phytopathology, sensors, and machine learning is needed! Curr. Opin. Plant Biol. 50: 156–162.

Mahood, E.H., Kruse, L.H., & Moghe, G.D. (2020). Machine learning: A powerful tool for gene function prediction in plants. Appl. Plant Sci. 8: e11376.

Majeed, A., Johar, P., Raina, A., ... Salgotra, R.K. (2022). Harnessing the potential of bulk segregant analysis sequencing and its related approaches in crop breeding. Front. Genet. 13: 944501.

Marsh, J.I., Hu, H., Gill, M., Batley, J., & Edwards, D. (2021). Crop breeding for a changing climate: integrating phenomics and genomics with bioinformatics. Theor. Appl. Genet. 134: 1677–1690.

Meuwissen, T.H., Hayes, B.J., & Goddard, M.J.G. (2001). Prediction of total genetic value using genome-wide dense marker maps. Genetics. 157: 1819–1829.

Meyer, R.S., DuVal, A.E., & Jensen, H.R. (2012). Patterns and processes in crop domestication: a historical review and quantitative analysis of 203 global food crops. New Phytol.196: 29–48.

Michelmore, R.W., Paran, I., & Kesseli, R.V. (1991). Identification of markers linked to disease-resistance genes by bulked segregant analysis: a rapid method to detect markers in specific genomic regions by using segregating populations. Proc. Natl. Acad. Sci. USA. 88: 9828–9832.

Mochida, K., Koda, S., Inoue, K., ... Hirayama, T. (2019). Computer vision-based phenotyping for improvement of plant productivity: a machine learning perspective. GigaScience. 8: giy153.

Montesinos-López, O.A., Martín-Vallejo, J., Crossa, J., ... Gianola, D. (2019). A benchmarking between deep learning, support vector machine and Bayesian threshold best linear unbiased prediction for predicting ordinal traits in plant breeding. G3 Genes Genomes Genetics 9: 601–618. https://doi.org/10.1534/g3.118.200998

Montesinos-López, O.A., Montesinos-López, A., ... Crossa, J. (2018). Multi-trait, multi-environment deep learning modeling for genomic-enabled prediction of plant traits. G3 Genes Genomes Genet. 8: 3829–3840.

Moran, P., & Smith, C. (1918). The correlation between relatives on the supposition of mendelian inheritance. Trans. R. Soc. Edinb. 52: 399–438.

Nabwire, S., Suh, H.K., Kim, M.S., Baek, I., & Cho, B.K. (2021). Review: application of artificial intelligence in phenomics. Sensors. 21: 4363.

Ni, Y., Aghamirzaie, D., Elmarakeby, H., ... Collakova, E. (2016). A machine learning approach to predict gene regulatory networks in seed development in Arabidopsis. Front. Plant Sci. 7: 1936.

Niazian, M., & Niedbała, G. (2020). Machine learning for plant breeding and biotechnology. Agriculture. 10: 436. https·//doi. org/10.3390/ agriculture10100436

Niazian, M., Sadat-Noori, S.A., & Abdipour, M. (2018a). Artificial neural network and multiple regression analysis models to predict essential oil content of ajowan (Carum copticum L.). J. Appl. Res. Med. Aromat. Plants. 9:124–131. https:// doi.org/10.1016/j. jarmap.2018.04.001

Niazian, M., Sadat-Noori, S.A., Abdipour, M., Tohidfar, M., & Mortazavian, S.M.M. (2018b). Image processing and artificial neural network-based models to measure and predict physical properties of embryogenic callus and number of somatic embryos in ajowan (Trachyspermum ammi (L.) Sprague). Vitr. Cell Dev. Biol. Plant. 54: 54–68. https://doi.org/10.1007/s11627-017-9877-7

Niedbała, G., Piekutowska, M., Weres, J., ... Korzeniewicz, R. (2019). Application of artificial neural networks for yield modeling of winter rapeseed based on combined quantitative and qualitative data. Agronomy. 9: 781. https://doi. org/10.3390/agronomy9120781

Parmley, K.A., Higgins, R.H., Ganapathysubramanian, B., Sarkar, S., & Singh, A.K. (2019). Machine learning approach for prescriptive plant breeding. Sci. Rep. 9: 1–12. https://doi.org/10.1038/ s41598-019-53451-4

Rai, K.K. (2022). Integrating speed breeding with artificial intelligence for developing climate-smart crops. Molecular Biology Reports. 49:11385–11402.

Rajasundaram, D., & Selbig J. (2016). More effort—more results: recent advances in integrative 'omics' data analysis. Curr. Opin. Plant Biol. 30: 57–61. https:// doi.org/10.1016/ j.pbi.2015.12.010

Raschka, S., Patterson, J., & Nolet, C. (2020). Machine learning in python: main developments and technology trends in data science, machine learning, and artificial intelligence. Information. 11: 193.

Ravari, S.Z., Dehghani, H., & Naghavi, H. (2016). Assessment of salinity indices to identify Iranian wheat varieties using an artificial neural network. Ann. Appl. Biol. 168: 185–194. https://doi.org/10.1111/ aab.12254

Razzaq, A., Kaur, P., Akhter, N., Wani, S.H., & Saleem, F. (2021). Next-generation breeding strategies for climate-ready crops. Front. Plant Sci. 12: 620420.

Reinoso-Peláez, E.L., Gianola, D., & González-Recio, O. (2022). Genome-enabled prediction methods based on machine learning. In: Genomic prediction of complex traits; methods in molecular biology. Ahmadi, N., Bartholomé, J. (Eds). Humana: New York, NY, USA, Vol. 2467.

Resende, R.T., Piepho, H.P., Rosa, G.J.M., ... Silva-Junior, O.B. (2021). Enviromics in breeding: Applications and perspectives on envirotypic-assisted selection. Theor. Appl. Genet. 134: 95–112.

Sadeghi-Tehran, P., Sabermanesh, K., Virlet, N., & Hawkesford, M.J. (2017). Automated method to determine two critical growth stages of wheat: heading and flowering. Front. Plant Sci. 8: 252.

Sartor, R.C., Noshay, J., Springer, N.M, & Briggs, S.P. (2019). Identification of the expressome by machine learning on omics data. Proc. Natl. Acad. Sci. USA. 116:18119–18125.

Selvaraj, M.G., Valderrama, M., Guzman, D., ... Valencia, M. (2020). Machine learning for high-throughput field phenotyping and image processing provides insight into the association of above and below-ground traits in cassava (Manihot esculenta Crantz). Plant Methods. 16: 87.

Shafiekhani, A., Kadam, S., Fritschi, F.B., & DeSouza, G.N. (2017). Vinobot and vinoculer: Two robotic platforms for high-throughput field phenotyping. Sensors. 17: 214.

Sharma, R.P., & Ramesh, D. (2020). ML based sustainable precision agriculture: a future generation perspective. Sustain. Comput. Inform. 28: 100439.

Shaw, J., Rudzicz, F., Jamieson, T., & Goldfarb, A. (2019). Artificial intelligence and the implementation challenge. J. Med. Internet Res. 21: e13659.

Shen, Y., Zhou, G., Liang, C., & Tian, Z. (2022). Omics-based interdisciplinarity is accelerating plant breeding. Curr. Opin. Plant Biol. 66: 102167.

Sonah, H., O'Donoughue, L., Cober, E., Rajcan, I., & Belzile, F. (2015). Identification of loci governing eight agronomic traits using a GBS-GWAS approach and validation by QTL mapping in soya bean. Plant Biotechnol. J. 13: 211–221.

Song, P., Wang, J., Guo, X., Yang, W., & Zhao, C. (2021). High-throughput phenotyping: breaking through the bottleneck in future crop breeding. Crop J. 9: 633–645.

Stergiou, C., & Psannis, K.E. (2017). Recent advances delivered by mobile cloud computing and internet of things for big data applications: a survey. Int. J. Netw. Manag. 27: e1930.

Streich, J., Romero, J., Gazolla, JGFM., Kainer, D., ... Cliff, A. (2020). Can exascale computing and explainable artificial intelligence applied to plant biology deliver on the United Nations sustainable development goals? Curr. Opin. Biotechnol. 61: 217–225.

Sui, M., Jing, Y., Li, H., Zhan, Y., Luo, J., & Teng, W. (2020). Identification of loci and candidate genes analyses for tocopherol concentration of soybean seed. Front. Plant Sci. 11: 539460.

Sun, L., Liu, H., Zhang, L., & Meng, J. (2015). lncRScan-SVM: A tool for predicting long non-coding RNAs using support vector machine. PLoS ONE. 10: e0139654.

Supriya, M., & Deepa, A. (2020). Machine learning approach on healthcare big data: a review. Big Data Inf. Anal. 5: 58–75.

Takagi, H., Abe, A., Yoshida, K., ... Kosugi, S. (2013). QTL-seq: Rapid mapping of quantitative trait loci in rice by whole genome resequencing of DNA from two bulked populations. Plant J. 74: 174–183.

Talaviya, T., Shah, D., Patel, N., Yagnik, H., & Shah, M. (2020). Implementation of artificial intelligence in agriculture for optimisation of irrigation and application of pesticides and herbicides. Artif. Intell. Agric. 4: 58–73.

Tong, H., & Nikoloski, Z. (2021). Machine learning approaches for crop improvement: Leveraging phenotypic and genotypic big data. J. Plant Physiol. 257: 153354.

Torkamaneh, D., Chalifour, F.P., & Beauchamp, C.J. (2020). Genome-wide association analyses reveal the genetic basis of biomass accumulation under symbiotic nitrogen fixation in African soybean. Theor. Appl. Genet.133: 665–676.

Uzal, L.C., Grinblat, G.L., Namías, R., ... Larese, M.G. (2018). Seed-per-pod estimation for plant breeding using deep learning. Comput. Electron. Agric. 150:196–204. https://doi.org/10.1016/j.compag.2018.04.024

Vakilian, K.A. (2020). Machine learning improves our knowledge about miRNA functions towards plant abiotic stresses. Sci. Rep. 10: 1–10. https://doi.org/10.1038/s41598-020-59981-6

van Dijk, A.D.J., Kootstra, G., Kruijer, W., & de Ridder D. (2021). Machine learning in plant science and plant breeding. iScience. 24: 101890. https://doi.org/10.1016/j.isci.2020.101890

Wallace, J.G., Rodgers-Melnick, E., & Buckler, E.S. (2018). On the Road to Breeding 4.0: Unraveling the Good, the Bad, and the Boring of Crop Quantitative Genomics. Annu. Rev. Genet. 52: 421–444.

Wang, H., Cimen, E., Singh, N., & Buckler, E. (2020). Deep learning for plant genomics and crop improvement. Curr. Opin. Plant Biol. 54: 34–41.

Watanabe, S., Tsukamoto, C., Oshita, T., ... Yamada, T. (2017). Identification of quantitative trait loci for flowering time by a combination of restriction site–associated DNA sequencing and bulked segregant analysis in soybean. Breed. Sci. 24: 17013.

Wei, M.C.F., Maldaner L.F., Ottoni, P.M.N., & Molin J.P. (2020). Carrot yield mapping: A precision agriculture approach based on machine learning. AI. 1: 229–241. https://doi.org/10.3390/ ai1020015

Williamson, H.F., Brettschneider, J., Caccamo, M., & Davey, R.P. (2021). Data management challenges for artificial intelligence in plant and agricultural research. F1000Research. 10: 324.

Xavier, A. (2021). Technical nuances of machine learning: implementation and validation of supervised methods for genomic prediction in plant breeding. Crop Breed. Appl. Biotechnol. 21. https:// doi.org/10.1590/1984-70332021v21Sa15

Xia, E.H., Lil, F.D., Tong, W., Li, P.H., ... Wu, Q. (2019). Tea plant information archive: a comprehensive genomics and bioinformatics platform for tea plant. Plant Biotechnol. J. 17: 1938–1953.

Xu, C., & Jackson, S.A. (2019). Machine learning and complex biological data. Springer: Berlin/Heidelberg, Germany.

Xu, Y. (2016). Envirotyping for deciphering environmental impacts on crop plants. Theor. Appl. Genet. 129: 653–673.

Xu, Y., Liu, X., Cao, X., Huang, C., Liu, E., ... Qian, S. (2021). Artificial intelligence: a powerful paradigm for scientific research. Innovation. 2: 100179.

Xu, Y., Zhang, X., Li, H., Zheng, H., ... Zhang, J. (2022). Smart breeding driven by big data, artificial intelligence and integrated genomic-enviromic prediction. Mol. Plant 16: 225.

Yang, Q., Jin, H., Yu, X., Fu, X., Zhi, H., & Yuan, F. (2020). Rapid identification of soybean resistance genes to soybean mosaic virus by SLAF-seq bulked segregant analysis. Plant Mol. Biol. Rep. 38: 666–675.

Yang, W., Feng, H., Zhang, X., Zhang, J., ... Doonan, J.H. (2020). Crop phenomics and high-throughput phenotyping: past decades, current challenges, and future perspectives. Mol. Plant. 13: 187–214.

Yang, Y., & Siau, K.L. (2018). A qualitative research on marketing and sales in the artificial intelligence age. Proceedings of the Thirteenth Midwest Association for Information Systems Conference, Saint Louis, MO, USA, 17–18 May.

Yang, Z., Huang, D., Tang, W., ... Zheng, Y. (2013). Mapping of quantitative trait loci underlying cold tolerance in rice seedlings via high-throughput sequencing of pooled extremes. PLoS ONE. 8: e68433.

Yoosefzadeh-Najafabadi, M., Earl, H.J., Tulpan, D., Sulik, J., & Eskandari, M. (2021). Application of machine learning algorithms in plant breeding: predicting yield from hyperspectral reflectance in soybean. Front. Plant Sci. 11:2169. https://doi.org/10.3389/ fpls.2020.624273

Zargar, S.M., Raatz, B., Sonah, H., Bhat, J.A., ... Dar, Z.A. (2015). Recent advances in molecular marker techniques: Insight into QTL mapping, GWAS and genomic selection in plants. J. Crop Sci. Biotechnol.18: 293–308.

Zhao, J., Bodner, G., & Rewald, B. (2016). Phenotyping: using machine learning for improved pairwise genotype classification based on root traits. Front. Plant Sci. 7: 1864.

Zhao, X., Han, Y., Li, Y., Liu, D., & Sun, M. (2015). Loci and candidate gene identification for resistance to Sclerotinia sclerotiorum in soybean (Glycine max L. Merr.) via association and linkage maps. Plant J. 82: 245–255.

Zhou, H., Tang, K., Li, G., Liu, W., Yu, H., & Yuan, X. (2021). A robust and rapid candidate gene mapping pipeline based on M2 populations. Front. Plant Sci. 12: 681816.

Zhou, J., & Troyanskaya, O.G. (2015). Predicting effects of noncoding variants with deep learning–based sequence model. Nat. Methods. 12: 931–934.

Zohry, A.E.H., & Ouda, S. (2022). Climate-resilient crops. In: climate-smart agriculture: reducing food insecurity. Springer International Publishing: Cham, Germany, p. 115–135.

Chapter-10

Software Tools for Crop Breeding

Abstract

Plant breeders are usually faced with the problem of predicting the performance of new individuals with untested gene combinations. Therefore, it is important to follow an integrated breeding approach by combining molecular tools, molecular mapping, and MAS. It is also required to develop tools for modeling and simulation analysis by utilizing all pre-existing and newly generated data. Several software tools have been developed that integrate breeding simulations and phenotype prediction models using genomic information. Reliable phenotype prediction models for the simulation were constructed from actual genotype and phenotype data. Such a simulation-based genome-assisted approach to breeding will help optimize plant breeding in all important agricultural crops. Software tools have also been developed for designing target sites or evaluating the outcome of genome/gene editing systems. This chapter provides an overview of the key software support tools that will assist the plant breeders in decision-making during the process of conducting various breeding programs.

Keywords: *Software tools, modelling and simulation, predicting models, genomic information, decision making, breeding by design*

Introduction

Important steps involved in molecular breeding include: identification and selection of beneficial genetic variations, and utilization of such variations more effectively and efficiently for crop improvement. Marker-assisted selection (MAS) has been recognized as a reliable identification tool and thus has a lot of potential for its utilization in crop breeding. To achieve the desired goal, MAS has to be combined with other advanced techniques such as genome scans, advanced biometrical analysis, and quantitative genetics modeling that will require complex software. The effective molecular breeding procedure will depend on the following parameters (Singh et al. 2010):

Identification of new sources of variation and development of strong marker-trait associations,

- Management of large numbers of genotypes and manipulation of genotype, phenotype and pedigree data,
- Selection of desirable recombinants by combining genotypic and phenotypic information, and
- Development of breeding systems to minimize population size, number of generations and overall cost, but to maximize genetic gain for traditional and novel traits.

Supporting tools are required to manage and optimize various components of plant molecular breeding procedures. Many of these tools come in the form of software. Some of the important software having applications in molecular plant breeding can be obtained from the sites: http://bioreseach.ac.uk/browse/mesh/D012984.html, http://linkage.rockefeller.edu/soft/list.html.

Management and Evaluation of Germplasm

With the increase in the size of germplasm collections, genebank curators find it difficult to manage and provide the required information to breeders and researchers. MAGE (marker-assisted germplasm evaluation) has played an important role in the process of acquisition, maintenance, distribution, and use of germplasm in gene banks. For efficient management of the various activities of the gene bank through MAGE it is important to have the following resources in place (Xu 2014):

- Characterization of suitable genetic markers for large number alleles, polymorphic information content (PIC value), size and range of the alleles, signal strength, working conditions and information for multiplexing,
- Generate high-density molecular maps for selection of markers evenly spread over the entire genome or densely spread over the specific region of interest,
- Establishment of association of the marker and the trait of agronomical importance,
- Establishment of high-throughput genotyping system, and
- Establishment of an efficient data management and analysis system.

Several open-source and commercial software packages such as GeneFlow SAS, STATISTICA, NTSYS, JMP, STRUCTURE, PowerMaker etc. are available for evaluation and analysis of germplasm. STRUCTURE can be used to investigate population structure using multi-locus genotype data. Which in turn provide information about the presence of distinct populations, study hybrid zones, identify migrants and admixed individuals, and estimate the frequency of the alleles in specific populations. With PowerMaker it is possible to perform statistical analysis of marker data derived from germplasm accessions, and deliver a data-driven integrated analysis environment (IAE) for marker data. It also can handle a variety of data from SSRs, SNPs, and RFLPs. A host of other statistical analyses, having relevance to the management and utilization of germplasm, can be carried out through PowerMaker. GGT (Geographical Genotype) software allows for transforming molecular marker data into simple chromosomal drawings. Through POPDIST it is possible to calculate the number of different genetic identities, reconstruction of phylogeny measures, and distance measures. ADEGENET is used to handle molecular markers data for multivariate analysis.

Management of the Breeding Population

Computational tools have been progressively used to assist plant breeders in decision-making during various breeding programs. These include a choice of parental lines, types of crosses to be undertaken, nature of the breeding system to be followed, establishment and maintenance of heterozygotic groups, selection of lines for development of synthetic cultivars, prediction of the performance of progeny and hybrid, etc. (Damme et al. 2011, Lorenz 2013, Fu et al. 2017)

Establishing Heterotic Patterns

Development of a highly heterotic hybrid depends largely on having large genetic diversity in the pool of germplasm of potential parents. However, in many crops, it is not possible to predict the level of hybrid vigor, based on the analysis of the parental lines. For example, crosses between inbred lines derived from complementary heterotic groups are performed to generate commercial maize hybrids. Therefore, one of the primary strategies in maize hybrid breeding is to construct heterotic groups. But, which genotype from the heterotic group will provide the maximum heterosis is not possible to predict in many instances.

Availability of gene-based markers may provide an opportunity to establish parent-hybrid performance relationships at the molecular level. The establishment of genome-wide heterozygosity and specific allelic (linkats) combinations may be utilized to determine maximum heterosis and hybrid vigor in some crops. Determination of heterotic patterns is a continuous process and each cycle is composed of the following steps: identification of a broad heterotic group through cluster analysis, determination of heterotic pattern through combining ability and heterosis analysis, and updating and maintenance of heterotic groups (Kizilkaya et al. 2010).

Predicting Hybrid Performance

Breeding for hybrid cultivars involves two major steps: identification of parental lines and selection of the best combinations of these parental lines for hybrid development. These procedures demand a large amount of work for field evaluation, test crossing, and progeny tests. In the breeding process, testcrossing has to be carried out at many stages starting from the very first generation. Since hybrid performance is highly unpredictable in most crops, breeders have to spend a lot of time in testcrossing, as currently there is no alternative to evaluate the hybrid performance. Therefore, the development of alternative methods for predicting hybrid performance is one of the challenges for hybrid plant breeders.

Development of a reliable hybrid performance method without performing hundreds or thousands of single crossings has been the objective of many studies using marker data and a combination of phenotypic and marker data in rice and maize (Technow et al. 2012, Technow 2014, Technow 2019). Hybrid performance is believed to be governed by many genes. Therefore, genome-wide genotyping of the parental lines should provide an opportunity to establish parent-hybrid performance relationships at the molecular level. Analysis of genome-wide heterozygosity and allelic combinations may provide clues for breeding more heterotic and vigorous hybrid crop plants. Thus, the utilization of parental genotyping should substantially reduce the quantum of testcross-based phenotyping analysis (Huang et al. 2015, Kadam et al. 2016, Lin et al. 2016).

Animal breeders have successfully used Best Linear Unbiased Prediction (BULP) procedure for evaluating the genetic merits of dairy cattle. Traditionally, intrapopulation additive genetic models have been used with BLUB in animal breeding. In 1994, BLUP was used in maize breeding with interpopulation genetic models involving general and specific combining ability tests, and was found to be useful for predicting single-cross performance (Technow 2019). The predicted single-cross performance can be used to predict the performance of F_2 x tester combinations, double-crosses, and three-way crosses as well (Andorf et al. 2019).

In certain situations, tools are required for selective genotyping and pooled DNA analysis. The software package GenePool (http://genepool.tgen.org/) provides analytical tools which can be used for the detection of shifts in relative allele frequency between pooled genomic DNA of specific cases and their controls through SNP-based genotyping microarrays. The Pooled DNA Analyser (PDA) is another package that can be used for the analysis of pooled DNA data (http://www.ibms.sinica.edu.tw/~csjfann/ first%20flow/programlist.htm).

Construction of Genetic Maps Through Molecular Markers

The prerequisite steps for the application of MAS are: the construction of genetic maps through molecular markers and the utilization of these maps in marker-trait association analysis. Some of the tools used for the above-mentioned purpose are discussed in the following section.

Tools for Genetic Map Construction

Conventionally genetic maps are constructed using information obtained from the segregating populations of specific crosses. The MAPMAKER/EXP was the first and most widely used software for the construction of genetic maps, which was developed in 1987 by the Whitehead Institute. Almost all genetic maps constructed using first-generation molecular markers, RFLP, have utilized this software. In 1993, alternative software MAP MANAGER CLASSIC was developed which has an interactive program to map Mendelian loci using intercross with codominant markers, backcross, or recombinant inbred lines (RILs) (http://www.mapmanager.org/mapmgr.html).

For the construction of genetic maps from distorted segregation information obtained from backcrossing, double haploid, and RIL population, specialized software such as MAPDISTO (we/ftp:http://mapdisto.free.fr/) is used. This software can analyze marker data showing segregation distortion due to differential viability of gametes or zygotes, and compute and draw genetic maps using graphical interface.

It is also possible to combine maps or data from multiple populations, derived from different crosses, into a single map through the software JOINMAP. The mapping population may consist of BC_1, F_2, RIL, F_1 and F_2 derived double haploid (DH) and out-breeder full-sib family. JOINMAP can also integrate maps derived from several other functions such as automatic phase determination for out-breeder full-sib family, linkage group determination etc. The CMap is another software package with comparative functions, developed as a web-based tool. It provides information on the comparison of genetic and physical maps.

QTL Mapping Based on Genetic Linkage

The association between target traits and molecular markers is based on genetic linkage. There exist several software packages for studying the association between marker genotype and phenotypic traits. Some of the commonly used software is: QTL, MAPQTL, Cartographer, Qgene, and PLAQTL. While MCQTL performs QTL mapping in multi-multi-dimensional allelic situations, all the others are capable of handling bi-allelic populations. The MAPMAKER/QTL is the most frequently used software, which is based on maximum likelihood estimation of linkage between marker and phenotype using interval mapping. Another popular software MAPL can be used to get information on segregation ratio, linkage tests, value of recombination, marker groups, and marker orders by metric-dimensional scaling through analysis of variance (ANOVA) and interval mapping.

The QTL Cartographer (http://statgen.ncsu.edu/qtlcart/cartographrer.html) is most widely used software which performs several statistical methods using multiple markers simultaneously. The software package OneMap (http://www.ciagre.usp.br/~aafgarci/ OneMap/) can be used for the construction of linkage maps in out-crossing plant species.

To meet the requirements of special situations of QTL, several mapping software were developed. These include MCQTL, MapPop, and QTLNetwork. The MCQTL allows the analysis of populations derived from inbred lines and can link the families on the assumption that QTL locations are the same in all families. MapPop is used for selective mapping and bin mapping by selecting good samples from the mapping populations. QTLNetwork can be used for mapping and visualizing the genetic architecture of complex traits on populations derived from crossing inbred lines.

A web-based tool WEBQTL was developed for exploring the genetic modulation of several thousand phenotypes collected over 30 years period by several hundred investigators, for mice. Similar web-based tools should be developed for plants as well.

eQTL Mapping

The availability of whole genome sequences in many crop plants has opened additional scope for revealing links between phenotypes and genotypes (genes) through tools such as linkage analysis, positional cloning, and microarray. It has become essential to develop a convenient bioinformatics tool to display the relationships between eTraits, markers, and genes. It has also become essential to integrate current results with the information generated from earlier studies on the organism. To address these issues, Muller et al. (2006) developed the eQTL Explorer. This provides scope to store expression profiles, linkage data, and information derived from external sources from a relational database. This helps in the simultaneous visualization and interpretation of the combined data through a Java geographical interface. In 2007, a web-based eQTL Viewer was developed which plots the results of eQTL mapping. With these thousands of eTraits can be viewed in a single display, with readily identifiable *cis*- and *trans*- regulation patterns. It is also empowered to present annotations, and organize eTrait in biological groups such as biochemical pathways. All these features have made eQTL viewers amenable to understanding genome-wide transcriptional regulation patterns.

The web-based software PhenoGen can be used to identify candidate genes that control complex characters, based on co-occurrence of differentially expressed genes in microarray experiments and phenotypic QTL expression. The PGMapper can automatically match phenotypes to genes from a defined genome region or a group of given genes derived from known databases such as OMIM and PubMed.

Linked-Disequilibrium Based QTL Mapping

Linked-disequilibrium (LD) or association mapping has become increasingly popular in recent years. Unstructured populations having unrelated individuals or randomly selected individuals can be used for such mapping. Before initiating LD mapping, the genotyped units are subjected to statistical analysis to remove the population structure, as it can cause false positive associations due to virtual correlations rather than actual linkage. The software STRUCTURE can be used for this purpose.

Genome-Wide Association Mapping

To determine the linkage between genetic variations and agronomic traits, genome-wide association (GWA) studies are widely undertaken. A powerful GWA study shall include the measurement of hundreds of thousands of SNPs in several thousand individuals. Highly sophisticated analyzing tools are required for high-volume data (Spindel et al. 2016). GOLDSURFACE2 (GS2) can be used for the analysis and visualization of GWA studies. The program is developed in Java and can be used on all platforms. Other tools available for GWA studies are GENOMIZER (http://www.ikmb.uni-kiel.de/genomizer), PLANK (http://pngu.mgh.harvard.edu/ purcell/ plink), MAPBUILDER (http://bios.ugr.es/ MBapBuilder), CATS (Calculator for Association with Two Stage design, http://www.sph. umich.edu/csg/abecasis/ CaTS) (Massman et al. 2013, Perez and Compos 2014).

Integrated Haplotype and LD Analysis

Usually, it becomes difficult to handle large amounts of SNP data for analysis of haplotypes and their association with the traits of interest. To overcome such problems, HaploBuild (http://snp.bumc.u.edu/modules.php?name=HaploBuild) was developed which can be used for the construction and testing of haplotypes for SNPs found in close proximity but may not necessarily be contiguous. The software, HAPLOVIEW (http://www.broad.mit.edu/ personal/jcbarret/haploview) can simplify and expedite the process of haplotype analysis. HAPSTAT (http:// www.bios.unc.edu/~lin/hapstat/) can be used for statistical analysis of haplotype-disease association. Otter related software are: DPPH (Direct method for Preferred Phylogeny Haplotyping, http://www.c-sif.cs.ucdavis.edu/~gusfield/dpph.html), EHAP (detecting association between haplotypes and phenotypes, http://www. compgen.pitt.edu), HAPLOBLOCK (http://bioinfo.cs.technion.ac.il/haploblock), HAPLOT, HAPLOREC, and HAP (http://research.calit2.net/hap).

Support Tools for Marker-Assisted Selection

MAS being a major procedure of molecular breeding need various decision making tools at various stages. Stages at which these tools shall be useful include foreground and background selection, identification of recombinants having favorable allelic combinations etc. In large scale MAS huge data are generated, which need to be analyzed and integrated into other types of data, within a short time. Although the software has been developed to assist in decision-making in such situations, they do not fulfill the breeder's requirements, as it can resolve only some procedures of MAS. Thus, development of support tools for MAS is one of the challenges for the efficient and quick adoption MAS.

Support tools for MAS are required for the following purposes (Perez-de-Castro et al. 2012).

- To determine minimum sample size for foreground and background selection,
- To estimate genetic gains,
- To construct selection indices for multiple traits and whole genome selection,

- To estimate the content of the recipient genome of selected individuals at each generation of integration,
- To identify desirable plants based on genotype and phenotype,
- To estimate cost-benefit, and
- To make simulation studies.

PLABSIM a simulation tool used for MAS programs. This can be used to investigate the effect of varying population size, determine the position and density of markers, selection strategies to be adopted on the genetic composition of the breeding products and on number of marker data point required. PLABISM has the following features: simulate of any diploid genome with an arbitrary number of loci at arbitrary positions on arbitrary chromosome number, combine an arbitrary number of selection steps with a selection strategy, reproduce schemes for all common breeding methods, select genotypes at defined loci, calculate selection indices from allelic frequencies at several loci, and analyze simulated data for a broad range of genetic parameters including population size, position and density of markers, number of marker data points required and selection strategies for the genetic composition of the breeding product.

iMAS (www.generationcp.org) can assist in developing and applying MAS. iMAS integrates freely available software for the identification and application of trait-linked markers. This also helps the user to operate the software and interpret the results correctly. Other software applicable to MAS include: POPMIN a program that allows numerical optimization of population sizes in marker-assisted backcross breeding programs, BCSIM, a backcross simulation software for evaluation of marker-assisted backcross programs, the GGT (Graphical Genotypes software) which allows transforming molecular marker data to simple chromosomes (He et al. 2014).

Breeding by Design

The concept of 'breeding by design' offers the possibility to predict the outcome of a set of crosses on the basis of relevant information on molecular markers. The process involves three steps: mapping all the loci involved in determining all the relevant phenotypic traits, determination of the allelic variations at those loci, and making crosses as per the design. The first step can be completed either by using mapping populations segregating for the phenotypic trait of interest or candidate gene approach (in which information from model plant species is exploited) or linkage disequilibrium (LD) mapping. However, it has been found that application of GWA (Genome Wide Association) studies allow more efficient way to accomplish the first step. The second step cannot be completed on the basis of bi-parental populations, as only two alleles per locus can segregate. Therefore, the analysis should include plant materials representing the variability of the species (Mossaman et al. 2013, Rabier et al. 2016).

In 'breeding by design', after the loci of interest have been mapped, and contribution of each allelic variant determined, crosses can be designed by combining favorable alleles to obtain superior genotypes (Yamamoto et al. 2015). According to the breeding objective, this breeding strategy can be defined and executed. So far, this procedure has been used for seed length in soybean, heading date in rice, and improvement of quality in maize (Spindel et al. 2015, Fritsche-Neta et al. 2018a, Muleta et al. 2019).

Genome-Wide Selection

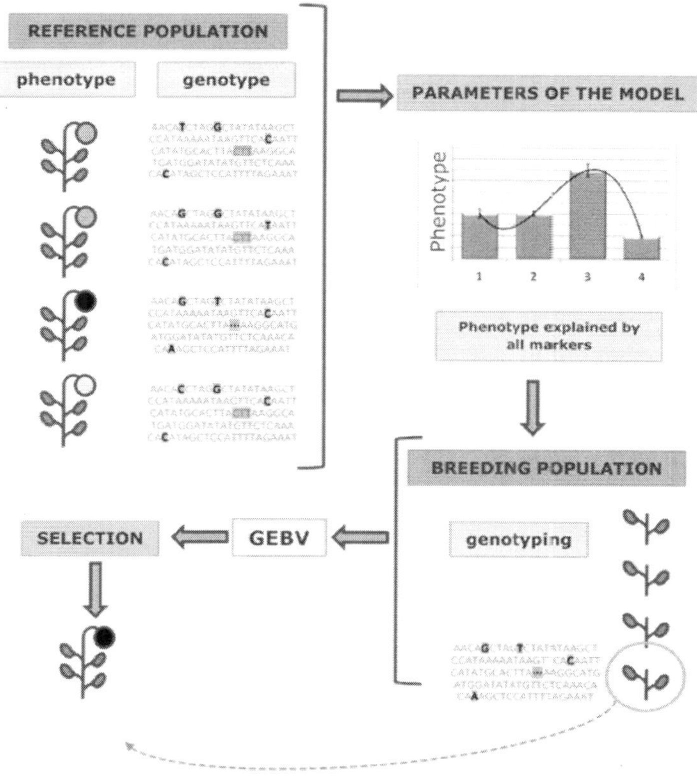

Figure 10.1. Genome selection scheme. The model predicts the phenotype of plants in a breeding population on the basis of the genotyping results: this is the genomic estimated breeding value (GEBV), used to select the desired phenotype. (Reproduced from: Perez-de-Castro et. al. Current Genomics, 2012).

MAS requires the identification of markers associated with the trait of interest. This is considered to be one of the weaknesses of MAS. The approach through this step can be eluded, is called genome-wide selection (Figure 10.1). In genome-wide selection, estimation of effects of all loci on phenotypes, haplotypes, and available markers are considered simultaneously. It is required to have both phenotypic and genotypic data for the reference population. The parameters for the simulation model can be determined based on available data set. After establishing the model, it becomes possible to determine the genomic value of each individual, representing the breeding populations. Availability of good number of molecular markers shall ensure accurate prediction (Schoop et al. 2015, Gorjanc et al. 2017).

It has been possible to compare the predictions of genotypes values obtained using genome-wide selection with true genotypic value due to availability of large phenotypic databases for different crops. In several studies conducted on *Arabidopsis*, maize and barley, it was observed that the results obtained through genome-wide study was more accurate compared to results obtained from previous studies conducted on the basis of selection of markers with effects on the phenotypes (Jan et al. 2016).

Models used in genome-wide selection can predict breeding values, and in some instances, can detect the regions associated to the trait. Usually, from the breeders' perspective, application of MAS with the available molecular markers is sufficient. However, with the availability of new high throughput–omics technologies, breeders have can adopt new strategies to search for candidate genes, based on microarray for differential expression of genes. Exploitation of these strategies could facilitate in candidate gene identification involved in the trait of interest and make MAS even more efficient (Massman et al. 2013, Desta and Ortiz 2014).

Simulation and Modeling

Plant breeding through simulation modeling has become necessary due to generation of vast and diverse information. Simulation modeling can assist in decision making at different stages of breeding processes. In this section present status on the application of simulation modeling in plant breeding has been discussed.

Importance of Simulation Modeling

Over time, the number of genes and QTLs for various traits identified has increased substantially. Therefore, there is a need for software tools to handle and analyze multitude of information for crop improvement. Computer simulation can help the breeders to obtain information on probable crossings with selected of genotypes, which will drastically reduce the time and resource requirements of field experiments. Usually, in quantitative genetics various assumptions (on: linkage, multiple alleles, epistasis, pleiotropy, experimental design) are made to test the validity of mathematical, and statistical theories. Computer simulation can relax the use of some of the assumptions and their effects on breeding programs (Sun et al. 2011).

The whole plant physiology modeling can be used to partition complex traits into simple components and to know how these components interact with each other and contribute towards overall expression of the trait in different environmental conditions. The areas in which crop modeling can help to improve the efficiency of plant breeding program are: characterize environment to define the target population for each component of the environment, assessing the value of specific putative traits in improved plant type, and enhancing integration of molecular genetic methodologies.

Simulation Through Genetic Models

Simulation, using simple genetic models has been used to study many special plant breeding issues. Genetic details identified for simulation to elucidate differences in growth and development among cultivars is as follows:

- ❑ Genetic model without any reference to species,
- ❑ Genetic model having specific reference to species but without reference to genotypes,
- ❑ Genetic model representing genetic differences by cultivar-specific parameters,
- ❑ Genetic model representing genetic differences by specific alleles, having linear effects of the genes on the parameters of the model,
- ❑ Genetic model representing genetic differences by genotypes with simulated gene action based on knowledge about gene regulation and effect of gene products, and
- ❑ Genetic model representing genetic differences by genotypes with simulated gene action at the level of interactions of regulators, products of the gene and other metabolites

The level i and ii are used only when genetic representation of species are required. Level iii and iv (GeneGro Version 1) are used for the current crop models. Level v is used on the basis of knowledge on gene action (represented in GeneGro Version 2). Level vi is used for unicellular organisms.

It is possible to create a QTL-based crop physiology model, by combining eco-physiological modeling with genetic mapping, which could be used for resolving genetic basis of complex environment dependent yield related traits. Using this tool it has been possible to predict specific leaf area in barley, leaf growth response to varying temperature and water deficit in maize, stay-green response to nitrogen in sorghum, and duration of pre-flowering in barley. By excluding environment and gene dependence parameters, it was possible to develop breeding strategies to enhance rate of yield over several cycles of selection. By combining CROPGRO-Soybean with a linear model, which can predict cultivar-specific parameters as a function of E-loci, it has been possible to predict cultivar performance and refine crop breeding systems.

It is possible to predict how different genotypes interact with environments to produce different phenotypes through the genotype-to-genotype (GP) model. Based on the information on genes, characteristics of germplasm, parents, breeding objectives, target environments etc., the breeding procedures and selection methods can be simulated and optimized, and predict the development of new cultivars.

The simulation platform QUGENE has been developed for quantitative analysis of genetic models, through its two-stage architecture. The first stage deals with define the genotype-environment (GE) system and generate the starting population of individuals. In the second stage, application modules are included, for investigation and analysis of the starting population of individuals within the GE system. Several breeding modules were developed using QUGENE software. These modules can simulate the performance of breeding lines in a given environment and also predict the effect of long-term selection over many breeding cycles and seasons, to optimize and improve the efficiency breeding methodology.

QULINE can be used for simulation for mass selection, pedigree selection, bulk selection, backcross breeding, top cross breeding, DH breeding and MAS. To simulation in QULINE, method of propagation applicable for the crop must be defined. The propagation methods with decreasing genetic diversity are: Non-self-random mating, Random mating, Double cross, Top cross, Backcross, Single cross, Self-pollinated, Double haploid (DH), and clonal (asexual) reproduction. For each cross, QULINE randomly determine the male and female parents, from a defined initial population or may select few preferred parents from a crossing block. Identification of the preferred parents from the crossing blocks can be made based on the defined selection criteria.

Application Of Genotyping-By-Sequencing In Plant Breeding

Genotyping-by-sequencing (GBS) is one of the most powerful platforms for undertaking plant breeding programs, from single gene markers to whole genome profiling. GBS is a faster and low-cost tool, which allows plant breeders to implement GWAS, genetic linkage analysis, genome diversity study, molecular marker discovery, and genome selection (GS). GBS is a robust method across a range of species, as prior knowledge about the genome of the species is not required, and SNP discovery and genotyping can be completed together (He et al. 2014).

Since GWAS require hundreds of thousands of markers to generate enough information and coverage, the emergence of NGS technologies has greatly improved such marker resolution. Progressively more and more crops (rice, maize, wheat, barley, potato, and cassava) are being optimized by GBS for the low-cost, efficient, and large scale of genome sequencing.

GBS has also been shown to be an efficient and valid tool for genomic diversity studies. Among other applications of GBS in plant breeding, identification of high-density SNP markers to construct genetic linkage maps has been found to be very useful. The GBS technology has been found to be useful for genetic analysis and marker development in maize, soybean, lupine, lettuce, and rapeseed. GBS can be effectively applied even in the absence of reference genome sequences or without having previous DNA polymorphism data.

Application of GS through GBS has become a major supplement to traditional breeding approaches, and is an important feature to convert genome-assisted breeding into commercial crops having large and complex genomes. Application GBS approach in barley and wheat has resulted development of high-density markers in species without having a reference genome.

Although GBS has been found to have novel qualities for enhancing the efficiency of plant breeding approaches, some potential drawbacks have been identified. First, it is difficult to align true alleles of each single locus in the case of large, complex and polyploidy genomes. However, among all the tools, GBS offers highest potential to resolve such issues. Second, in the case of mutation at the restriction site, the genomic DNA of that site cannot be PCR amplified. Consequently, the SNPs of this region will not be available. In such a scenario, a heterozygote may appear as a homozygote. Although, the drawback stated above is common to all the other methods applicable for such analysis (He et al. 2014).

Software Toolkit Crispr-Ge

CRISPR-GE is a set of powerful tools and following functions can be carried-out through this: (i) design target single guide RNAs (sgRNAs) (targetDesign), (ii) predict off-target sites (offTarget), (iii) design primers for construction of the sgRNA expression cassettes (primerDesign), (iv) amplify target site-containing genomic fragments, (v) determine mutant sequences from the sequencing chromatograms of genome PCR amplicons containing target site(s) (DSDecodeM), and (vi) download genomic sequences of certain regions from reference genomes (seqDownload). By using CRISPR/Cas9/Cpf1 vector system, it is possible to achieve complete solution for genome editing in plants through CRISPR-GE (Xie et al. 2017).

The tool targetDesign of the CRISPR-GE software assist in choosing appropriate target site(s) for the Cas9/Cpf1 nucleases. The associated program offTarget helps in the selection of specific target site(s) and prevents cleavage at non-target sites (sequences) of the genome. The targetDesign tool assist to find all the possible target sites in a given sequence and thereby predict the off-target sequences (sites). With the help of the offTarget algorithm it is possible to estimate scores for predicting all the off-target sites in the assigned genome (Barrangou and Horvath 2014).

There exist several other genome editing design tools such as E-CRISPR, CRISPR-P, and Breaking-Cas (Heigwar et al. 2014, Oliveros et al. 2016). However, CRISPR-GE has the following advantages over the other software: (i) through targetDesign and offTarget programs it is possible to design target sites and predict potential off-target sites, (ii) in case reference genome sequence or the genome sequence of any close related species is not available, it is possible to design target site(s) through targetDesign. However, in such cases prediction of potential off-target sites may not be possible in the target genome (Xie et al. 2017).

Conclusions

Plant breeders or agro-technicians are using plant breeding software for their **daily activities in breeding, testing,** and analytical. They use digital tools to collect data, validate the information, take decisions and monitor their activities. The plant breeding software is not only their daily work instrument, but also provides the dashboards to organize and monitor the breeding programs.

During breeding programs several operations have to be followed: crosses and hybridization, inbred line production, test crosses and trials, and analysis. The plant breeding software supports all such activities at every key stage. The plasticity of the software enables to configure it according to the specific requirements of the breeders. For example, when breeders choose parental lines for your next crossings, they would like to do it their way: selection of specific males and females parents which fulfill their criteria, provide attention to specific traits and combining abilities to make their selection etc. With plant breeding software, these activities can be done quickly, without hazardous data manipulation, on a powerful crossing matrix. Sometimes even new useful features are discovered through such systems.

To successfully implement genomics-assisted breeding (GAB) in crop improvement programs, it is important to have efficient and effective analytical and decision support tools (ADSTs), to evaluate and select plants for developing next-generation crops. Although phenotyping remains expensive and time-consuming procedure, availability of software for prediction of allelic effects on phenotypes opens new doors to enhance genetic gain across crop cycles, building on reliable phenotyping approaches, and building pedigree information.

References

Andorf, C., Beavis, W., Hufford, D., SMITH, S., Suza, W.P., Wang, K., ... Lubberstedt, T. (2019). Technological advances in maize breeding: past, present and future. Theoretical and Applied Genetics. Retrieved from: https://doi.org/10.1007/s00122-019-03306-3.

Barrangou, R., & Horvath, P. (2014). Functions and applications of RNA-guided CRISPER-Cas immune system. In R. A. Meyers (Ed.) Encyclopedia of molecular cell biology and molecular medicine: RNA biology (pp. 34-56). New York: Wiley-VCH Verlag GmbH & Co.

Damme, V.V., Gomez-Paniagua, H., & Vicente, M.C. (2011). The GCP molecular marker toolkit, an instrument for use in breeding food security crops. Molecular Breeding, 28, 597-610.

Desta, Z.A., & Ortiz, R. (2014). Genomic selection: genome-wide prediction in plant improvement. Trends in Plant Science, 19, 592-601.

Fritsche-Neto, R., Akdemir, D.,& Jannink, J.L. (2018). Correction to: accuracy of genomic selection to predict maize single-crosses obtained through different mating designs. Theoretical and Applied Genetics, 131, 1603-1612.

Fritsche-Neto, R., Akdemir, D.,& Jannink, J.L., (2018a). Accuracy of genomic selection to predict maize single-crosses obtained through different mating designs. Theoretical and Applied Genetics, 131, 1153–1162.

Fu, Y.B., Yang, M.H., Zeng., & Biligetu, B. (2017). Searching for an accurate marker-based prediction of an individual quantitative trait in molecular plant breeding. Frontiers in Plant Breeding, 8, 1-12.

Garber, M., Grabherr, M.G., Guttman, M., & Trapnell, C. (2011). Computational methods for transcriptome annotation quantification using RNA-Seq. Natural Methods, 8, 469-477.

Gorjanc, G., Battagin, M., Dumasy, J.F., Antolin, R., Gaynor, R.C., & Hickey,J.M. (2017). Prospects for cost-effective genomic selection via accurate within-family imputation. Crop Science, 57, 216–228.

He, J., Zhao, X., Laroche, A., Lu, Z.Y., Liu, H., & Li, Z. (2014). Genotyping by sequencing (GBS), an ultimate marker assisted selection (MAS) tool to accelerate plant breeding. Frontiers in Plant Science, 5, 1-9.

Heigwar, F., Kerr, G., & Boutros, M. (2014). E-CRISPER: fast CRISPER target site identification. Natural Methods, 11, 122-123.

Hickey, J.M., Dreisigacker, S., Crossa, J., Hearne, S., Babu, R., Prasanna, B.M., ... Gorjanc, G. (2014). Evaluation of genomic selection training population designs and genotyping strategies in plant breeding programs using simulation. Crop Science, 54, 1476-1488.

Huang, X., Yang, S., Gong, J., Zhao, Y., Feng, Q., Gong, H., ... Han, B. (2015). Genomic analysis of hybrid rice varieties reveals numerous superior alleles that contribute to heterosis. Nature Communications, 6, 6258. doi:10.1038/ncomms7258

Jan, H.U., Abbadi, A., Lucke, S., Richard, A.N., & Showdon, R.J. (2016). Genomic prediction of testcross performance in canola (Brassica napus). Public Library of Science (PLoS) ONE, 11, e0147769.

Kadam, D.C., Potts, S.M., Bohn, M.O., Lipka, A.E., & Lorenz, A.J. (2016). Genomic prediction of single crosses in the early stages of a maize hybrid breeding pipeline. G3: Genes, Genomes and Genetics, 6, 3442–3453.

Kizilkaya, K., Fernando, R.L., & Garrick, D.J. (2010). Genomic prediction of simulated multibreed and purebred performance using observed fifty thousand single nucleotide polymorphic genotype. Journal of Animation Science, 88, 544-551. Larièpe, A., Moreau, L., Laborde, J., Bauland, C., Mezmouk, S., Bauland, C., Charcos set, A. (2017). General and specific combining abilities in a maize (Zea mays L.) test-cross hybrid panel: relative importance of population structure and genetic divergence between parents. Theoretical and Applied Genetics, 130, 403–417.

Lin, G., Zhao, Y., Gowda, M., Friedrich, C., Longin, H., Reif, J.C., & Mette, M.F. (2016). Predicting hybrid performance for quality traits through genomic-assisted approaches in central European wheat. Public Library of Science (PLoS) ONE, 1, e0158635.

Liu, G., Zhao, Y., Gowda, M., Longin F.H., Reif, J.C., & Mette, M.F. (2016). Predicting hybrid performance for quality traits through genomic-assisted approaches in central European wheat. Public Library of Science (PLoS) ONE, 11, e0158635.

Longin, C. F. H., Mi, X., & Würschum, T. (2015). Genomic selection in wheat: optimum allocation of test resources and comparison of breeding strategies for line and hybrid breeding. Theoretical and Applied Genetics, 128, 1297–1306.

Lorenz, A.J. (2013). Resource allocation for maximizing prediction accuracy and genetic gain of genomic selection in plant breeding: a simulation experiment. G3: Genes, Genomes and Genetics, 3, 481–491.

Massman, J.M., Jung, H.J.G., & Bernardo, R. (2013). Genomewide selection versus marker-assisted recurrent selection to improve grain yield and stover quality traits for cellulosic ethanol in maize. Crop Science, 53, 58-66.

Massman, J.M., Gordillo, A., Lorenzana, R.E., & Bernardo R. (2013) Genomewide predictions from maize single-cross data. Theoretical and Applied Genetics, 126, 13–22.

Metje-Sprink, J., Menz, J., Modrzejewski, D., & Sprink, T. (2019). DNA-free genome editing: past, present and future. Frontiers in Plant Science, 9, 1-9.

Meuwissen, T., Hays, B., & Goddrd, M. (2001). Prediction of total genetic value using genome-wide dense marker maps. Genetics, 157, 1819-1829.

Muleta, K.T., Pressoir, G., & Morris, G.P. (2019). Optimizing genomic selection for a sorghum breeding program in Haiti: a simulation study. G3: Genes, Genomes and Genetics, 9, 391-401.

Muller, M., Goal, A., Thimma, M., Dickens, N.J., Hitman, T.J., & Mangion, J. (2006). eQTL explorer: integrated mining of combined genetic linkage and expression experiments. Bioinformatics, 2, 309-311.

Oliveros, J.C., Franch, M., Tabas-madrid, D., San-Leon D., Montoliu, L., Cubas, P., & Pazos, F. (2016). Breaking-Cas-interactive design of guide RNAs for CRISPER-Cas experiments for ENSEMBL genomes. Nucleic Acid Research. 44, W267-W271.

Pérez, P., & Campos, G. (2014). Genome-wide regression & prediction with the BGLR statistical package. Genetics, 198, 483–495.

Perez-de-Castro, A.M., Vilanova, S., Canizares, J., Pascual, J.M., Blanca, M.J., Prohens, J., & Pico, B. (2012). Application of genomic tools in plant breeding. Current Genomics, 13, 179-195.

Rabier, C.E., Barre, P., Asp, T., Charnet, G., & Mangin, B. (2016). On the accuracy of genome selection. Public Library of Science (PLoS) ONE, 11, e0156086.

Schoop, P., Riedelsheimer, C., Utz, H.F., Schon, C.C., & Melcchiner. A.E. (2015). Forecasting the accuracy of genome prediction with different selection targets in the training and prediction set as well as truncation selection. Theoretical and Applied Genetics, 128, 2189-2201.

Singh, B., Bohra, A., Mishra, S., Joshi, R., & Pandey, S. (2015). Embracing new-generation 'omics' tools to improve draught tolerance in cereal and food-legume crops. Biologia Plantarum, 59, 413-428.

Singh, R.K., Singh, R., Ye, G.U., & Selvi, A. (2010). Molecular plant breeding, methods and application. Houston, Studium Press LLC.

Spindel, J., Begum, H., Akdemir, D., Virk, P., Collard, B., Redona, E., ... McCouch, S.R. (2015). Genome selection and association mapping in rice (Oryza sativa): effect of trait genetic architecture, training population composition, marker number and statistical model on accuracy of rice genome selection in elite, tropical rice breeding lines. Public Library of Science (PLoS) Genet, 11, e1004982.

Spindel, J.E., Begum, H., Akdemir, D., Coolard, B., Redona, E., Jannink J.L., & McCouch, S. (2016). Genome-wide prediction models that incorporate de novo GWAS are a powerful new tool for tropical rice improvement. Heridity, 116, 395-408.

Sun, X., Peng, T., & Mumm, R.H. (2011). The role and basics of computer simulation in support of critical decisions in plant breeding. Molecular Breeding. 28, 421-436.

Technow, F. (2019). Use of F2 bulks in training sets for genomic prediction of combining ability and hybrid performance, G3: Genes, Genomes, Genetics, 9, 1557-1569.

Technow, F., Riedelsheimer, C., Schrag, T.A., & Melchinger, A.E. (2012). Genomic prediction of hybrid performance in maize with models incorporating dominance and population specific marker effects. Theoretical and Applied Genetics, 125, 1181–1194.

Technow, F., Schrag, T.A.,Schipprack, W., Bauer, E., Simianer, H., & Melchinger, A.E. (2014). Genome properties and prospects of genomic prediction of hybrid performance in a breeding program of maize. Genetics 197, 1343–1355.

Thavamanikumar, S., Dolferus, R., & Thumma, B.R. (2015). Comparison of genomic selection models to predict flowering time and spike grain number in two hexaploid wheat double haploid popunations. G3, Gene, Genomes and Genetics, 5, 1991-1998. doi: 10.1534/g3.115.019745

Varshney, R.K., Singh, V.K., Hickey, J.M., Xun, X., Marshall, D.F., Wang, J., ... Ribaut, J.M. (2016). Analytical and decision support tools for genomics-assisted breeding. Trends in Plant Sciences, 21, 354-360.

Varshney, R.K., Terauchi, R., & McCouch, S.R. (2014). Harvesting the promising fruits of genomics: applying genome sequencing technologies to crop breeding. Public Library of Science (PLoS) ONE, 12, e1001883.

Wang, M., Mao, Y., Lu, Y., Tao, X., & Zhu, J.K. (2017). Multiplex gene editing in rice using the CRISPER-Cpf1 system. Molecular Plant, 10, 1217-1222.

Windhausen, V.S., Atlin, G.N., Hicky, J.M., Crossa, J., Jannink, J.L., Sorrells, M.E.,... Melchinger, A.E. (2012). Effectiveness of genome prediction of maize hybrid performance in different breeding populations and environments. G3: Gene, Genomes and Genetics, 2, 1427-1436. doi:10.1534/g3.112.003699.

Xie, X., Ma, X., Zhu, Q., Zeng, D., Li, G., & Liu, Y.G. (2017). CRISPER-GE: a convenient software toolkit for CRISPER-based genome editing. Molecular Plant, 10, 1246-1249.

Xu, Y. (2014). Molecular plant breeding. New York: CAB International.

Yamamoto, E., Matsunaga, H., Onogi, A., Kajiya-Kanegae, H., Minamikawa, M., Suzuki, A., ... Fukuoka, H. (2015) A simulation-based breeding design that uses whole-genome prediction in tomato. Scientific Reporter, 6, 1-11.

Index